Digital Logic Design

In memory of Brian Holdsworth, 1925–2001

Clive Woods and the staff at Newnes would like to dedicate this book to the memory of Brian Holdsworth, who wrote the first edition of *Digital Logic Design* for Butterworth Scientific in 1982 while he was a Senior Lecturer at Chelsea College, University of London. Brian's academic career spanned lecturing posts in the UK, the University of Science & Technology in Kumasi, Ghana, Ahmadu Bello University, Zaria, Northern Nigeria and the University of Zambia, in Lusaka. Brian was closely involved in the preparation of this fourth edition and died shortly before its publication. We trust that this book will bring Brian's talent for explaining complex subjects in straightforward terms to yet a new generation of students.

Digital Logic Design

Fourth edition

B. HOLDSWORTH, BSc (Eng), MSc, FIEE

and

R.C. WOODS, MA, DPhil

621.395

Q.

188125

Newnes

OXFORD   AMSTERDAM   BOSTON   LONDON   NEW YORK   PARIS
SAN DIEGO   SAN FRANCISCO   SINGAPORE   SYDNEY   TOKYO

Newnes is an imprint of Elsevier
Linacre House, Jordan Hill, Oxford OX2 8DP, UK
30 Corporate Drive, Suite 400, Burlington, MA 01803, USA

First edition 1982
Second edition 1987
Third edition 1993
Fourth edition 2002
Reprinted 2006

Notice
No responsibility is assumed by the publisher for any injury and/or damage to persons
or property as a matter of products liability, negligence or otherwise, or from any use
or operation of any methods, products, instructions or ideas contained in the material
herein. Because of rapid advances in the medical sciences, in particular, independent
verification of diagnoses and drug dosages should be made

British Library Cataloguing in Publication Data
A catalogue record for this book is available from the British Library

Library of Congress Cataloging-in-Publication Data
A catalog record for this book is available from the Library of Congress

ISBN–13: 978-0-7506-4582-9
ISBN–10: 0-7506-4582-2

For information on all Newnes publications
visit our website at www.newnespress.com

Printed and bound in *China*

06 07 08 09 10   10 9 8 7 6 5 4 3 2

Working together to grow
libraries in developing countries

www.elsevier.com | www.bookaid.org | www.sabre.org

ELSEVIER   BOOK AID International   Sabre Foundation

Contents

Preface to the fourth edition

In this newly revised edition of *Digital Logic Design*, we have taken the opportunity to undertake extensive revisions of much material contained in the third edition, whilst retaining its comprehensive coverage of the subject. Like the previous editions, the current edition is intended to cover all the material that is needed in a typical undergraduate or Master's course on Digital Logic Systems, and also to act as a reference text for graduates working in this field. To this end, we have retained all elementary material assuming little or no background, but the advanced chapters have accordingly been revised to take account of recent trends in hardware availability. A number of additional problems have been set at the end of some of the chapters, sometimes without answers, in order to allow the reader to exercise his/her design capabilities without the luxury of being able to refer to worked solutions.

The chapter on instrumentation and interfacing is almost entirely new, and the chapters on programmable logic devices, and on fault diagnosis and testing, have been considerably enlarged as a result, on the one hand, of significant advances in the technology and the range of devices now available to the designer, and on the other hand to emphasise that logical fault-finding methods, far from being esoteric, impossible to apply in practice, trivial, or demeaning for a professional engineer to use, are actually worthy of serious study and application.

> Material enclosed in boxes in this manner is usually not needed later in this text, and is not as important as the main narrative, or sometimes summarises work in the main text. This material may be rather more demanding than the main text, or be unusual or obscure in some other manner; generally speaking, proofs of results in these sections and subsections are not given in detail, and are left as more of a challenge for the interested reader to work out in full. The first-time reader, or a reader not aiming for complete coverage of all the material in this text, may safely ignore these sections and subsections.

Throughout the main part of this edition, we have used the 'old' IEEE logic symbols rather than the 'new' BS3939 symbols; this is a result of a perceived shift in attitudes in the engineering profession, and the IEEE symbols are now recommended alongside the BS symbols. Modern CAD systems are capable of printing the 'old' symbols with ease, eliminating the major initial advantage of the 'new' symbols when first introduced. However, as an understanding of the 'new' symbols is also a useful accomplishment, a summary of the 'new' system is included as an Appendix.

Acknowledgments

Figures based on or adapted from figures and text owned by Xilinx, Inc., courtesy of Xilinx, Inc. © Xilinx, Inc. 1999 All rights reserved.

We are also grateful to Texas Instruments plc for allowing us to use their diagrams.

1 Number systems and codes

1.1 Introduction

A digital logic system may well have a numerical computation capability as well as its inherent logical capability and consequently it must be able to implement the four basic arithmetic processes of addition, subtraction, multiplication and division. Human beings normally perform arithmetic operations using the decimal number system, but, by comparison, a digital machine is inherently binary in nature and its numerical calculations are executed using a binary number system.

Since the decimal system has ten digits, a ten-state device is required to represent the decimal digits, one state being allocated to each of the decimal digits. Ten-state devices are not readily available in the electrical world, however two-state devices such as a transistor operating in a switching mode are, and it is for this reason that the binary number system is of great importance to the digital engineer. In addition to the binary system, a number of other systems such as the hexadecimal system are used in conjunction with programmable logic devices, consequently the digital engineer must be familiar with a variety of different number systems.

It is also true that arithmetic processes executed by a digital machine are not necessarily identical to the pencil and paper methods which are normally employed by humans. For example the process of subtraction is carried out as an addition and this involves the use of complement arithmetic.

Again, a frequent requirement is that the output of a digital machine should be a decimal display, for obvious reasons. Since the machine normally computes in pure binary, a way has to be found to represent decimal numbers in terms of binary digits and this requires a binary coded decimal system. Methods have to be devised so that any numerical computations carried out in pure binary can be converted into binary coded decimal so that at the interface with the outside world a decimal display or readout is available.

Coding of information is a basic consideration in the use of a digital system. Codes are required for decimal numbers, the letters of the alphabet and a variety of other well used symbols such as =, ?, etc. We previously referred to binary coded decimal as a coded representation for decimal numbers. This is an example of a weighted code of which there are a number of examples. In addition to weighted codes there are a variety of other codes available, for example the XS3 code, and the choice of a suitable code is not arbitrary. Its properties have to be considered before selection for use. In practice the most widely used code is the 8-4-2-1 weighted code which is referred to as *naturally binary coded decimal.*

The aim of this chapter is to describe the various number systems in common usage and to develop methods for implementing the four fundamental arithmetic operations on a machine. Additionally, a brief survey of some of the more common codes will be presented.

1.2 Number systems

The number system most familiar to man is the decimal system. A decimal number such as $(473.85)_{10}$ may be expressed in the following form:

$$(N)_{10} = 4 \times 10^2 + 7 \times 10^1 + 3 \times 10^0 + 8 \times 10^{-1} + 5 \times 10^{-2}$$

The number $(N)_{10}$ consists of a series of decimal digits multiplied by the number $(10)_{10}$ raised to some power that depends upon the position of the decimal digit in the number. The number $(10)_{10}$ is termed the *base* or *radix* of the number system and is equal to the number of distinguishable digits in the system. For example, in the decimal system there are ten digits, 0 to 9 inclusive. However, the binary number system has a base of 2 and has only two digits 0 and 1.

The decimal magnitude $(N)_{10}$ of a number in any system can be expressed by the equation:

$$(N)_{10} = a_{n-1}b^{n-1} + a_{n-2}b^{n-2} + \ldots a_0 b^0 + a_{-1}b^{-1} \ldots + a_{-m}b^{-m}$$

where n is the number of integral digits and m the number of fractional digits. The base of the system is b and a is a digit in the number system whose base is b. Using this equation the binary number $(101.11)_2$ is evaluated as follows:

$$(N)_{10} = 1 \times 2^2 + 0 \times 2^1 + 1 \times 2^0 + 1 \times 2^{-1} + 1 \times 2^{-2}$$

$$= 4.0 + 0.0 + 1.0 + 0.5 + 0.25$$

$$= (5.75)_{10}$$

Two other number systems of some importance are the *octal*, or base 8 system, and the *hexadecimal*, or base 16 system. The octal system has eight digits, 0 to 7 inclusive. A typical octal number is $(27.2)_8$ and its decimal value is given by

$$(N)_{10} = 2 \times 8^1 + 7 \times 8^0 + 2 \times 8^{-1}$$

$$= 16.0 + 7.0 + 0.25$$

$$= (23.25)_{10}$$

In the hexadecimal system there are 16 digits and since there are only ten digits available some additional ones have to be invented. The additional six digits are by convention represented by the first six letters of the alphabet, A to F inclusive, whose corresponding decimal values are

$$(A)_{16} = (10)_{10} \qquad (B)_{16} = (11)_{10} \qquad (C)_{16} = (12)_{10}$$
$$(D)_{16} = (13)_{10} \qquad (E)_{16} = (14)_{10} \qquad (F)_{16} = (15)_{10}$$

A typical hexadecimal number $(A2.C)_{16}$ has a decimal value which is given by:

$$(N)_{10} = A \times 16^1 + 2 \times 16^0 + C \times 16^{-1}$$

$$= 160.0 + 2.0 + 0.75$$

$$= (162.75)_{10}$$

1.3 Conversion between number systems

A number in any base can be divided into two parts, (a) the integral part to the left of the *radix point*, and (b) the fractional part to the right of the radix point. The process of conversion to another base is different for the two parts of the number.

The decimal value of the integral part $(N_I)_{10}$ of a base b number is given by:

$$(N_I)_{10} = a_{n-1}b^{n-1} + a_{n-2}b^{n-2} + \ldots a_1 b^1 + a_0 b^0$$

Dividing both sides of the equation by the base b gives:

$$\frac{[N_i]_{10}}{b} = a_{n-1}b^{n-2} + a_{n-2}b^{n-3} + a_{n-3}b^{n-3} + \ldots a_1 b^0 + \frac{a_0}{b}$$

The result of dividing by the base is to leave the least significant digit of the number a_0 as the remainder after the first division. Subsequent repeated divisions will produce remainders of $a_1, a_2 \ldots a_{n-1}$. As an example of the process of repeated division by the required base the decimal number $(100)_{10}$ is converted below to its binary, octal and hexadecimal equivalents:

```
2 | 100  0  ↑        8 | 100  4  ↑        16 | 100  4  ↑
2 |  50  0           8 |  12  4            16 |   6  6
2 |  25  1           8 |   1  1                    0
2 |  12  0           8 |   0
2 |   6  0
2 |   3  1
2 |   1  1
      0
```

$$(100)_{10} = (1100100)_2 \qquad = (144)_8 \qquad = (64)_{16}$$

The decimal value of the fractional part $(N_F)_{10}$ of a base b number is given by:

$$(N_F)_{10} = a_{-1}b^{-1} + a_{-2}b^{-2} + \ldots a_{-m}b^{-m}$$

and if both sides are multiplied by the base, then

$$b(N_F)_{10} = a_{-1} + a_{-2}b^{-1} + \ldots a_{-m}b^{-(m-1)}$$

and, the first multiplication reveals the coefficient a_{-1}. Subsequent multiplications will reveal the coefficients a_{-2}, a_{-3}, $\ldots a_{-m}$. As an example of this process $(0.265)_{10}$ is converted to its corresponding binary, octal and hexadecimal forms below:

$.265 \times 2$	$.265 \times 8$	$.265 \times 16$
0.530×2	2.120×8	4.240×16
1.060×2	0.960×8	3.840×16
0.120×2	7.680×8	$D.440 \times 16$
0.240×2	5.440×8	7.040×16
$\downarrow$ 0.480	$\downarrow$ 3.520	$\downarrow$ 0.640
$(0.265)_{10}=(0.0100)_2$	$(0.20753)_8$	$(0.43D70)_{16}$

and the number $(0.265)_{10}$ is expressed to five binary, octal and hexadecimal places respectively.

Octal	Binary
0	000
1	001
2	010
3	011
4	100
5	101
6	110
7	111

(a)

HD	Binary	HD	Binary
0	000	8	1000
1	001	9	1001
2	010	A	1010
3	011	B	1011
4	100	C	1100
5	101	D	1101
6	110	E	1110
7	111	F	1111

(b)

Figure 1.1 *(a) Octal/binary conversion table (b) Hexadecimal/binary conversion table*

Besides these conversions from decimal to binary, octal and hexadecimal, it is also possible to convert from both octal and hexadecimal to binary and vice versa.

The octal digits from 0 to 7 inclusive can each be represented by three binary digits as shown in Figure 1.1(a). To find the octal representation of a string of binary digits it is divided into groups of three, beginning from the least significant digit. The octal equivalent for each group of three digits is then written down with the aid of the conversion table as shown below:

$$\begin{array}{llll} (110 & 001 & 011 & 100)_2 \\ \downarrow & \downarrow & \downarrow & \downarrow \\ =(6 & 1 & 3 & 4)_8 \end{array}$$

If the binary number has a fractional part then, to find the octal fractional part, divide the binary fractional number into groups of three beginning at the binary point and moving to the right. The corresponding octal equivalents for each group of three are then found in the conversion table. For example:

$$\begin{array}{lllll} (100 & 001 & 010 & 100 & .010) \\ \downarrow & \downarrow & \downarrow & \downarrow & \downarrow \\ =(4 & 1 & 2 & 4 & .2) \end{array}$$

Octal numbers can also be converted to binary by replacing each octal digit with the corresponding three binary digits from the conversion table. For example:

$$\begin{array}{llll} (4 & 3 & 2 & .7)_8 \\ \downarrow & \downarrow & \downarrow & \downarrow \\ =(100 & 011 & 010 & .111)_2 \end{array}$$

Similarly, each of the sixteen hexadecimal digits can be represented by four binary digits as shown in Figure 1.1(b). To convert a binary number into hexadecimal, divide the integral digits into groups of four, beginning at the binary point and moving left; and divide the fractional digits into groups of four beginning at the binary point and moving right. Each group of four binary digits is then replaced by its hexadecimal equivalent from the conversion table as illustrated in the following example:

$$(1011 \quad 1010 \quad 0011 \quad .0010)_2$$
$$= (B \qquad A \qquad 3 \qquad .2)_{16}$$

For the reverse conversion, each hexadecimal digit can be replaced by the appropriate four binary digits from the conversion table. For example:

$$(4 \qquad F \quad C \qquad 2)_{16}$$
$$= (0100 \quad 1111 \quad 1100 \quad 0010)_2$$

1.4 Binary addition and subtraction

Addition

The rules for the addition of two single-bit numbers are defined by the table shown in Figure 1.2 and the addition of two positive 4-bit numbers using these rules is demonstrated in the following example:

Augend	Addend	Sum	Carry
0	0	0	0
0	1	1	0
1	0	1	0
1	1	0	1

Figure 1.2 *Rules for the addition of two binary digits*

	2^3	2^2	2^1	2^0	
Augend	1	0	1	1	11
Addend	0	1	1	1	+7
Sum	1	0	0	1	0	18
Carries	1	1	1	1		

where the weighting of the individual digits is shown above each pair of digits. It will be observed that the carry ripples through the addition from the 2^0 column to the 2^3 column. *Carry ripple* is a significant problem that has to be taken into account in the design of addition circuits.

When two *n*-bit numbers, whose most significant digits are 1, are added together they will generate an $(n + 1)$-bit sum. The additional bit generated is termed *arithmetic overflow*. In pencil and paper calculations the extra digit does not create a problem. However in a digital machine, prior to the addition, the *augend* and *addend* may be stored in separate registers and after it has been performed the sum may well be returned to one of these two registers. In this case an extra bit must be provided in the register containing the sum to store the overflow.

Subtraction

The rules for subtraction are summarised in Figure 1.3 and the subtraction of two 4-bit positive numbers, where the *subtrahend* is less than the *minuend*, is illustrated in the following example:

	2^3	2^2	2^1	2^0	
Minuend	1	1	0	0	12
Subtrahend	0	0	1	1	−3
Difference	1	0	0	1	+9
Borrows		1	1		

Minuend	Subtrahend	Difference	Borrow
0	0	0	0
0	1	1	1
1	0	1	0
1	1	0	0

Figure 1.3   *Rules for the subtraction of two binary digits*

If the subtrahend is greater than the minuend an *arithmetic underflow* occurs which results in a borrow-out at the most significant bit and the difference is negative in this case. The borrow-out can be used in a digital machine to set a 1-bit register and in doing so will indicate that the difference is negative. Arithmetic underflow is illustrated by the following example:

	2^3	2^2	2^1	2^0	
Minuend	0	0	1	1	3
Subtrahend	1	1	0	0	−12
Difference	0	1	1	1	−9
Borrows	1	1			

Subtraction is commonly used in a digital machine that performs numerical computations, for comparing the magnitude of two binary numbers. If arithmetic underflow occurs the borrow out indicates that the subtrahend is greater than the minuend. Otherwise the two numbers are either equal or the minuend is greater than the subtrahend.

1.5 Signed arithmetic

The previous section has dealt with positive numbers and only in the case where the subtrahend is greater than the minuend is the answer negative. It is important that there should be a distinction made between positive and negative numbers in a machine. A *sign digit* can be used to provide this distinction. A negative number is identified by a 1 that appears in the most significant bit (MSB) position whilst a positive number is identified by a 0 in that position, so that:

$$(-23)_{10} = (1, 0010111)_2 \qquad (+23)_{10} = (0, 0010111)_2$$
$$(-0)_{10} = (1, 0000000)_2 \qquad (+0)_{10} = (0, 0000000)_2$$

This is termed *signed magnitude* representation. The range of numbers available with an 8-bit signed integer is from -127 to $+127$ with two possible representations for zero.

Because the design of a logic circuit capable of numerical computation in signed magnitude representation is somewhat complex it is rarely used. In practice, numerical computation in a machine is performed using *complement arithmetic*.

1.6 Complement arithmetic

This is a powerful yet simple technique which minimises the hardware implementation of signed arithmetic operations in a digital machine. In practice, when using complement arithmetic, the process of subtraction becomes one of addition.

In any number system two complements are available. In the binary system they are (a) the *2's complement* or *radix complement*, and (b) *1's complement* or *diminished radix complement*. For the decimal system they are: (a) the *10's complement* or *radix complement* and (b) the *9's complement* or *diminished radix complement*. It is worth noting that the use of the 1's complement in the binary system raises certain hardware implementation difficulties so that signed arithmetic processes are invariably performed using 2's complement notation.

1.7 Complement representation for binary numbers

The 2's complement of a binary number X is defined by the equation

$$[X]_2 = 2^n - X$$

where $[X]_2$ is the 2's complement representation and n is the number of binary digits contained in X. For $X = 1010$ and $n = 4$ the 2's complement is given by:

$$[X]_2 = 2^4 - 1010$$
$$= 10000 - 1010$$
$$= 0110$$

Two other methods are available for determining the 2's complement of X. In the first method, all the digits are inverted and a 1 is added in the least significant place. For the second method, the lowest order 1 in X is sensed, and all succeeding higher digits are inverted. Examples of these two methods follow:

Method 1		Method 2
$X = 1010$		$X = 1010$
0101	Invert	↗ ↑
		Invert Sense
1	Add	
$[X]_2 = 0110$		$[X]_2 = 0110$

In the 2's complement representation a number is positive if its MSB is 0. Alternatively, it is negative if its MSB is 1. Examples of two 8-bit numbers in 2's complement form are given below:

$$+(19)_{10} = 0{,}0010011 \qquad\qquad -(19)_{10} = 1{,}1101101$$

S.D. Magnitude X   S.D. 2's comp $[X]_2$

An n-bit 2's complement number can be changed into an m-bit one, where $m > n$, by adding copies of the sign digit to the left of the MSB of the n-bit number. This process called *sign extension* is illustrated in the following examples:

$$n = 4 \qquad\qquad m = 8$$
$$+7 = 0{,}111 \qquad\quad = 0{,}0000111$$
$$-3 = 1{,}101 \qquad\quad = 1{,}1111101$$

The table shown in Figure 1.4 gives some of the 8-bit numbers available in 2's complement form with their corresponding decimal values. The range of these numbers is from -128 to $+127$ and it will be noticed that it is not symmetrical since there is no 2's complement number corresponding to -128. It will also be observed that zero in this system is regarded as positive since its sign bit is 0.

The diminished radix complement, as in all number systems, is one less than the radix complement. In the binary system the 1's complement $[X]_1$ is one less than the 2's complement and is found by inverting all the digits in the binary number, as shown in the following example:

$$X = 1010$$

$$[X]_1 = 0101 \quad \text{Invert}$$

Weighting	-2^7	2^6	2^5	2^4	2^3	2^2	2^1	2^0	Decimal value
Value	-128	64	32	16	8	4	2	1	
	0	1	1	1	1	1	1	1	+127
	0	1	1	1	1	1	1	0	+126
					.				
					.				
					.				
	0	1	0	0	0	0	0	0	+64
					.				
					.				
	0	0	0	0	0	0	0	1	+1
	0	0	0	0	0	0	0	0	+0
	1	1	1	1	1	1	1	1	−1
	1	1	1	1	1	1	1	0	−2
					.				
					.				
	1	0	0	0	0	0	0	0	−128

Figure 1.4   *Tabular representation of 8-bit 2's complement numbers*

Weighting	2^7	2^6	2^5	2^4	2^3	2^2	2^1	2^0	Dec. value
	0	1	1	1	1	1	1	1	+127
	0	1	1	1	1	1	1	0	+126
					.				
					.				
					.				
	0	1	0	0	0	0	0	0	+64
					.				
					.				
	0	0	0	0	0	0	0	1	+1
	0	0	0	0	0	0	0	0	+0
	1	1	1	1	1	1	1	1	−0
	1	1	1	1	1	1	1	0	−1
	1	1	1	1	1	1	0	1	−2
					.				
					.				
	1	0	0	0	0	0	0	0	−127

Figure 1.5 *Tabular representation of 8-bit 1's complement numbers*

A sign digit is added in the most significant place to distinguish between positive and negative numbers, 0 for a positive number and 1 for a negative number. The complement is only taken in the case of negative numbers. Examples of 8-bit numbers in the 1's complement representation follow:

$$+25 = 0,0011001 \qquad -72 = 1,0110111$$

Although complementation is easily achieved in hardware, the system has the disadvantage that there are both positive and negative representations of zero and in the cases of some numerical computations an *end-about carry* is generated which has to be added in at the least significant place. For these reasons the 2's complement representation is generally preferred for numerical computations in a digital machine.

The table in Figure 1.5 gives a list of some of the 8-bit numbers available in 1's complement form with their corresponding decimal values. The range of values is from -127 to $+127$.

1.8 The validity of 1's and 2's complement arithmetic

By definition $[X]_2 = 2^n - X$ and the subtraction $Y - X$ where Y and X are both binary integers may be written as the addition of Y and the 2's complement of X.

Hence $\qquad Y - X = Y + [X]_2 - 2^n$

where n is the number of binary digits contained in X.

The 1's complement $[X]_1$ is always one less than the 2's complement so that $[X]_2 = [X]_1 + 1$.

To establish that $[X]_1$ is the logical inversion of X it is only necessary to show that

$$X - X = 0 = X + \bar{X} + 1 - 2^n$$

or

$$X + \bar{X} + 1 = 2^n$$

Assuming $X = 1010$ and $\bar{X} = 0101$ then the sum of X, $\bar{X}$ and 1 is

$$
\begin{array}{ll}
1010 & X \\
0101 & \bar{X} \\
\underline{1} & \\
\underline{1}0000 &
\end{array}
$$

and the underlined digit in this sum has the significance of 2^n and it has been shown that $X + \bar{X} + 1 = 2^n$ as required.

1.9 Offset binary representation

This representation is useful in some applications, for example analogue-to-digital conversion and floating point arithmetic. Here the natural binary code is offset by shifting its origin to the most negative number in the range so that $(0)_{10}$ occurs near the mid-point of the range. For positive numbers the sign bit is 1 and for negative numbers it is 0. Hence:

$$(+6)_{10} = (1,110)_2 \qquad (-6)_{10} = (0,010)_2$$

A tabulation for excess binary in the range $(-8)_{10}$ to $(+7)_{10}$ is given in Figure 1.6.

In the four representations described, with the exception of offset binary, positive numbers remain unchanged when signed.

Decimal number	Offset binary	Decimal number	Offset binary
+7	1,111	−1	0,111
+6	1,110	−2	0,110
+5	1,101	−3	0,101
+4	1,100	−4	0,100
+3	1,011	−5	0,011
+2	1,010	−6	0,010
+1	1,001	−7	0,001
0	1,000	−8	0,000

Figure 1.6 *Tabular representation of 4-bit offset binary numbers*

1.10 Addition and subtraction of 2's complement numbers

Addition and subtraction in the 2's complement system are both carried out as additions. Subtrahends are regarded as negative numbers and are converted to their 2's complement form. They are then added to the positive minuend. When adding two negative numbers they are both converted to their 2's complement form before addition takes place. Six possible cases are considered for the addition and subtraction of two 8-bit numbers where the MSB represents the sign digit and is given a negative weighting of 2^7.

Case 1 Addition of two 8-bit numbers both of which are positive and whose sum is $\leq +127$

	-2^7	2^6	2^5	2^4	2^3	2^2	2^1	2^0	
	0,	0	1	0	1	1	0	0	+44
+	0,	0	1	1	0	1	0	0	+52
	0,	1	1	0	0	0	0	0	+96

Correct positive answer

Case 2 Addition of two 8-bit numbers both of which are positive and whose sum is
> +127.

-2^7	2^6	2^5	2^4	2^3	2^2	2^1	2^0	
0,	1	1	0	0	0	0	1	+97
+ 0,	0	1	1	0	0	0	0	+48
1,	0	0	1	0	0	0	1	145

This gives a negative answer which is clearly wrong, since both numbers are positive.
The incorrect answer is obtained because the sum, 145, cannot be represented by seven
binary digits and arithmetic overflow has occurred from the magnitude section into the
position occupied by the sign digit.

Case 3 Subtraction of two 8-bit numbers when the subtrahend is $\leq$ the minuend.
Subtrahend in 2's complement form. Difference found by addition.

	-2^7	2^6	2^5	2^4	2^3	2^2	2^1	2^0	
	0,	1	1	0	0	0	1	1	+99
+	1,	1	0	1	1	1	1	1	−33
Discard (1)	0,	1	0	0	0	0	1	0	66

Correct positive answer, but there is a carry out from the sign bit which has to
be discarded. If the working registers happen to be 8-bits wide the carry out is automatically lost.

It will be observed that the numerical value of the subtrahend (−33) can be obtained
directly from its 2's complement representation by including the negative weighting of
the sign digit in the numerical evaluation.

Case 4 Subtraction of two 8-bit numbers with subtrahend > minuend. Subtrahend
in 2's complement form. Difference found by addition.

-2^7	2^6	2^5	2^4	2^3	2^2	2^1	2^0	
0,	0	1	0	0	0	0	1	+33
+ 1,	0	0	1	1	1	0	1	−99
1,	0	1	1	1	1	1	0	−66

Answer is negative and is in 2's complement form. True magnitude is found by taking
the 2's complement of the sum as shown below.

-2^7	2^6	2^5	2^4	2^3	2^2	2^1	2^0	
1,	0	1	1	1	1	1	0	
0,	1	0	0	0	0	0	1	Invert
							1	Add
0,	1	0	0	0	0	1	0	= 66

Case 5 Addition of two negative numbers where the sum $\geq$ −127. Both numbers are
expressed in 2's complement form.

	-2^7	2^6	2^5	2^4	2^3	2^2	2^1	2^0	
	1,	1	1	0	0	0	1	1	−29
+	1,	1	1	0	0	0	0	0	−32
Discard (1)	1,	1	0	0	0	0	1	1	−61

The answer is negative. A carry is generated out of the sign bit position which has to be discarded. As in the previous case the magnitude is found by taking the 2's complement of the sum.

Case 6   Addition of two negative numbers where the sum is < -127. Both numbers are expressed in 2's complement form.

$$
\begin{array}{cccccccccr}
-2^7 & 2^6 & 2^5 & 2^4 & 2^3 & 2^2 & 2^1 & 2^0 & \\
1, & 1 & 0 & 1 & 0 & 1 & 1 & 1 & -41 \\
+\quad 1, & 0 & 1 & 0 & 0 & 0 & 0 & 1 & -95 \\
\hline
\text{Discard (1)}\quad 0, & 1 & 1 & 1 & 1 & 0 & 0 & 0 & -136
\end{array}
$$

The answer is positive which is clearly incorrect. The correct answer -136 cannot be represented by seven binary digits. Figure 1.4 shows that the maximum negative number that can be represented by eight binary digits is -128.

1.11 Graphical interpretation of 2's complement representation

The 2's complement number system can be represented by sixteen equally spaced points on the periphery of a circle, as shown in Figure 1.7(a). It will be observed that a decimal discontinuity occurs in the 2's complement scale between the points marked 0111 and 1000 where the corresponding decimal numbers are $+7$ and -8. For the addition of two numbers whose sum is ≤ 7, such as $(2 + 3)$, the point 0010 corresponding to 2 is first fixed on the 2's complement scale and a clockwise rotation of three points round the periphery of the circle is made to the point marked 0101 corresponding to the correct sum of 5.

It is clear that if the sum is $> +7$, for example $(2 + 9)$, a clockwise rotation through nine points on the periphery of the circle starting at 0010 crosses the decimal discontinuity into the negative region of the scale and an incorrect answer is obtained.

For subtraction, where the subtrahend $<$ the minuend, for example $(5 - 3)$, the point 0101 is fixed on the 2's complement scale and an anticlockwise rotation of three points gives the correct difference of 0010 corresponding to decimal 2 as illustrated in Figure 1.7(b). An alternative way of obtaining the same result is to make a clockwise rotation of $(2^n - X)$ points from the fixed position 0101 where $X = 3$. The final position reached will be 0010 on the 2's complement scale corresponding to $+2$. It will be recalled that $(2^n - X)$ has previously been defined as the 2's complement of X and it

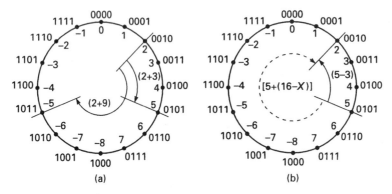

Figure 1.7   *Graphical interpretation of 2's complement arithmetic (a) Addition (b) Subtraction*

follows that a correct answer is obtained by adding the 2's complement of the subtrahend to the minuend.

1.12 Addition and subtraction of 1's complement numbers

A difference occurs in the addition and subtraction of 1's complement numbers when compared with 2's complement arithmetic in two cases only, namely *Case 3* and *Case 5*.

Case 3 Subtraction of two 8-bit numbers with the subtrahend < the minuend. Subtrahend in 1's complement form. Difference found by addition.

	-2^7	2^6	2^5	2^4	2^3	2^2	2^1	2^0	
	0,	1	1	0	0	0	1	1	+99
+	1,	1	0	1	1	1	1	0	−33
1	0,	1	0	0	0	0	0	1	
EAC								1	
	0,	1	0	0	0	0	1	0	+66

An *end-about carry* (EAC) added in at the least significant place gives the correct answer.

Case 5 Addition of two negative numbers whose sum ≥ -127. Both numbers expressed in 1's complement form.

	-2^7	2^6	2^5	2^4	2^3	2^2	2^1	2^0	
	1,	1	1	0	0	0	1	0	−29
+	1,	1	0	1	1	1	1	1	−32
1	1,	1	0	0	0	0	0	1	
EAC								1	
	1,	1	0	0	0	0	1	0	−61

An end-about carry is again generated and the magnitude is found by taking the 1's complement of the sum 1,1000010.

Multiplicand	Multiplier	Product
0	0	0
0	1	0
1	0	0
1	1	1

Figure 1.8 *Rules for binary multiplication*

1.13 Multiplication of unsigned binary numbers

The rules for binary multiplication are given in tabular form in Figure 1.8 and an example of the pencil and paper method of multiplication follows:

Multiplicand	1011	11
Multiplier	1101	13

$$
\begin{array}{l}
1011 \\
0000 \\
1011 \\
1011
\end{array}
\left.\vphantom{\begin{array}{l}1011\\0000\\1011\\1011\end{array}}\right\} \text{Partial Products}
$$

Product	10001111	143

It will be observed in this example that if two 4-bit unsigned numbers are multiplied together an 8-bit answer is generated. If an m-bit unsigned number and an n-bit unsigned number are multiplied together it is a general rule that the product will contain a maximum of $(m + n)$ *bits*.

A set of rules for the process of multiplication can be stated as follows:

1. If the least significant bit (LSB) of the multiplier is 1 write down the multiplicand and shift one place left.
2. If the LSB of the multiplier is 0 write down a number of 0s equal to the number of bits in the multiplicand and shift one place left.
3. For each bit of the multiplier repeat either (1) or (2).
4. Add all the partial products to form the final product.

Such a set of rules is called an *algorithm* which the digital designer can, if required, implement in hardware.

In practice, the hardware implementation of the multiplication of unsigned numbers differs from the pencil and paper method in one important aspect. The partial products are accumulated as they are generated rather than all being added together at the end. An example of the shift and add technique is given below:

Multiplicand (MD)	1011	11
Multiplier (MR)	1101	13
1st Partial Product (PP_1)	0000	
MR bit M = 1, add MD	1011	
PP_2	1011	
Shift PP_2 one place right	01011	
M = 0, MD not added. PP_3	01011	
Shift PP_3 one place right	001011	
M = 1. Add MD	1011	
PP_4	110111	
Shift PP_4 one place right	0110111	
M = 1. Add MD	1011	
	10001111	143

1.14 Multiplication of signed binary numbers

Multiplication in a computer must be implemented with signed arithmetic. Providing the multiplicand and the multiplier are both positive, the shift and add process is valid.

However, assuming that the multiplier or the multiplicand, or both, are negative, 2's complement arithmetic must be employed. The introduction of the sign digits and the use of the 2's complement form for negative numbers introduces a number of complications. Correction factors are required for certain cases and the required correction methods lead to complicated logic correction circuits. An alternative and more elegant method is due to A D Booth. With this scheme the procedure is the same regardless of signs. The method is beyond the scope of this introductory treatment of number systems and the reader is recommended to consult Lewin (see bibliography).

1.15 Binary division

The division process to be described here is based on a well known technique used in digital machines for comparing the magnitudes of two numbers relative to one another. The technique consists of subtracting the two numbers to be compared and if the minuend > the subtrahend, a carry is generated and the sign of the result is positive. Alternatively, if the minuend < the subtrahend, no carry is generated and the sign of the result is negative. Complement arithmetic is used so that the subtraction operation becomes an addition. This is illustrated in the following two examples which cover the two conditions described previously.

	Case 1		*Case 2*	
	Minuend > Subtrahend		Minuend < Subtrahend	
+ 6	0,110	+ 4	0,100
− 4	1,100	− 6	1,010
+ 2	10,010	− 2	1,110

carry generated — positive result
no carry generated — negative result

The rules for division of two single bit numbers are summarised in the table shown in Figure 1.9.

The division process can be regarded as one of repeated subtraction of the divisor X from the dividend Y. The number of times the divisor can be subtracted is the quotient Q and the residue after the last subtraction is the remainder R where $R < X$. The division equation may be written as:

$$Y = QX + R$$

where

$$R < X.$$

When the divisor is to be subtracted from the dividend or a partial remainder, there are only two possibilities. Either it will subtract and a positive result is obtained or it will

Dividend	Divisor	Quotient	Remainder
0	0	Indeterminate	Indeterminate
0	1	0	1
1	0	∞	Indeterminate
1	1	1	0

Figure 1.9   *Rules for binary division*

not subtract and a negative result is obtained. This leads to the *restoring division process* illustrated in the following example:

Divisor $= +6 = 0,110$. 2's complement of divisor $= 1,010$. Dividend $= +39 = 0,100111$

Carry out C_o

	0,100111	
	1,010	Subtract divisor
0	1,110111	$-$ve answer, $C_o = 0$, $Q = 0$
	0,100111	Restore Dividend
	1,00111x	Shift left
	1,010	Subtract divisor
1	0,01111x	$+$ve answer, $C_o = 1$, $Q = 1$
	0,1111xx	Shift left
	1,010	Subtract divisor
1	0,0011xx	$+$ve answer, $C_o = 1$, $Q = 1$
	0,011xxx	Shift left
	1,010	Subtract divisor
0	1,101xxx	$-$ve answer, $C_o = 0$, $Q = 0$
	0,011xxx	2's complement of $-$ve remainder

$Q = 0,0110$ $R = 0,011$

The algorithm used to perform the division process can be summarised as follows:

1. Align the most significant bits of the divisor and dividend.
2. Add the 2's complement of the divisor to the dividend.
3. If the most significant digit is 1 and $C_o = 0$ the answer is negative. Restore the dividend, shift it left and record the quotient bit $Q = C_o = 0$.
4. If the most significant digit is 0 and $C_o = 1$, the answer is positive, the subtraction is valid. Shift the dividend left and record the quotient bit $Q = C_o = 1$.
5. Repeat (2), (3), and (4) until the least significant digits of the dividend and divisor are aligned.

1.16 Floating point arithmetic

There are two possible methods that can be used for representing binary numbers in a computer. They are the *fixed point* and *floating point* systems. In practice, in a fixed point system, binary numbers are expressed as fractions with the radix point positioned

immediately right of the sign digit. For example, in a machine using 8-bit registers
1110.0 would be represented as

$$0.1110000 = 1110 \times 2^{-4}$$

by moving the radix point four places to the left.

Unfortunately there are problems associated with fixed point arithmetic. It has been
shown in the section on 2's complement arithmetic that if the sum of two 8-bit numbers
is > 127 or <-127 an additional bit is generated and an incorrect answer is obtained.
Assuming 8-bit registers are being used in the machine, the range of the registers has
been exceeded. The same problem exists for the multiplication and division operations.
If two 8-bit numbers are multiplied, one by the other, then in many cases a double-
length product will be formed and this would require a 16-bit register. Similarly, for the
division operations, a fractional quotient can only be formed if the divisor is greater
than the dividend.

To overcome the range problems experienced with fixed point representation
a floating point system can be used. Numbers in this system are expressed in the
following form:

$$n = m \times 2^e$$

where m, the mantissa, is the fractional representation of n and e is the exponent.

When performing a computation, a normalised form of the mantissa is used.
Normalisation is achieved by adjusting the exponent so that the mantissa has a 1 in
its most significant digit position. When this condition is satisfied:

$$0.5 \le |m| < 1$$

The exponent part of the number may be represented by a number which is the
sum of the exponent and a constant bias. The principle of a biased exponent is
perhaps more easily understood using the decimal system. Consider the following
two decimal numbers:

$$+1492.9187 = +.14929187 \times 10^{+4}$$
$$-.00034123 = -.34123000 \times 10^{-3}$$

which have been normalised. An alternative way of expressing these numbers would be

$$+.14929187 \times 10^{+4} = +.14929187e + 4$$
$$-.34123000 \times 10^{-3} = -.34123000e - 3$$

Assuming that the bias constant to be added to the exponent is 16 and that the
exponent part of the numbers is positioned to the left of the fractional part, the two
numbers would have the following form:

$$+.14929187 \times 10^{+4} = +20, 14929187$$
$$-.34123000 \times 10^{-3} = -13, 34123000$$

The addition of the constant 16 to the exponent expresses in two decimal digits any
exponent between 10^{+15} and 10^{-16} and consequently increases the range of numbers
the machine can handle.

1.17 Binary codes for decimal digits

Frequently there is a need for a decimal output even though digital machines operate in pure binary. As a result at the interface between a digital device and the outside world facilities must be provided to convert pure binary to a decimal representation. In practice, for example, calculators have been designed to work entirely in a decimal mode. In such cases decimal digits are represented by a string of binary digits referred to as a *code*. Four bits are required to represent the ten decimal digits, and since there are 2^4 combinations of four binary digits, six combinations are not used and the code is said to contain *redundancy*.

The four binary digits can be allocated to ten decimal digits in a purely arbitrary manner and it is possible to generate 2.9×10^{10} four-bit codes, only a few of which have any practical application. The most common group of codes for representing decimal numbers are *weighted* and there are 17 of these codes. For this group of codes the sum of the weights must be $\geq 9 \leq 15$ and examples of four of them are given in the tabulation shown in Figure 1.10.

Of this group the most commonly used weighted code is *naturally binary coded decimal* (NBCD) which uses the first ten combinations of the 4-bit binary count from 0000 to 1001 inclusive. The code weighting for NBCD is 8, 4, 2, 1 and this can be used to find the corresponding decimal value of a given code. For example:

$$1001 = 8 \times 1 + 4 \times 0 + 2 \times 0 + 1 \times 1 = (9)_{10}$$

Weighted codes having some negative weights are also available. Such a code is the 8, 4, -2, -1 which, like the 2, 4, 2, 1 code, has the useful property of *self-complementation*. By complementing each of the bits of a given codeword, a new codeword is formed which represents the 9's complement of the decimal digit represented by the original codeword. For example, in the 8, 4, -2, -1 code 0110 represents $(2)_{10}$ and, after self-complementation, 1001 represents $(7)_{10}$ which is the 9's complement of $(7)_{10}$. Another example of a self-complementing code is the XS3 code. This is not a weighted code but contains combinations of natural binary in the range $(3)_{10}$ to $(12)_{10}$. The decimal value allocated to each binary code is defined to be 3 less than its actual value. For example, $(1)_{10}$ is represented by 0100.

Decimal digit	NBCD 8,4,2,1	BCD 7,4,2,1	BCD 2,4,2,1	BCD 8,4,−2,−1	Excess 3 XS3
0	0000	0000	0000	0000	0011
1	0001	0001	0001	0111	0100
2	0010	0010	0010	0110	0101
3	0011	0011	0011	0101	0110
4	0100	0100	0100	0100	0111
5	0101	0101	1011	1011	1000
6	0110	0110	1100	1010	1001
7	0111	1000	1101	1001	1010
8	1000	1001	1110	1000	1011
9	1001	1010	1111	1111	1100

Figure 1.10   *Binary codes for the decimal digits*

Decimal digit	2-out-of 5	Biquinary 5043210
0	00011	0100001
1	00101	0100010
2	00110	0100100
3	01001	0101000
4	01010	0110000
5	01100	1000001
6	10001	1000010
7	10010	1000100
8	10100	1001000
9	11000	1010000

Figure 1.11 *Codes for the decimal digits using more than four bits*

There are some codes that use more than 4 bits to represent a decimal digit. Two examples of these are the *2-out-of-5* code and the *biquinary code* both of which are tabulated in Figure 1.11. It will be observed that each codeword in the 2-out-of-5 tabulation contains two 1's and a single error that complements one of the bits will generate an invalid code. The biquinary code is a weighted code where seven binary digits represent each of the decimal digits. The two most significant bits in each codeword, 01 and 10 indicate whether the digit represented is in the range $(0)_{10}$ to $(4)_{10}$ or $(5)_{10}$ to $(9)_{10}$ respectively. Each code combination contains only two 1's and the complementation of a single bit in a codeword will generate an invalid code.

1.18 *n*-cubes and distance

An *n*-bit string of binary digits can be visualised as being positioned at the vertex of what is termed an *n-cube*. Examples of 1, 2, and 3-cubes are illustrated in Figure 1.12. It will be observed from these diagrams that there is a single bit difference between the binary strings positioned at adjacent vertices. As the length of the string increases the number of vertices, 2^n, also increases and it becomes increasingly difficult to construct an *n*-cube for values of $n > 4$.

The *distance* between any two vertices on an *n*-cube is defined as the number of bit positions in which the two binary strings differ. Alternatively this is called the *Hamming distance*. A pair of adjacent vertices labelled 000 and 100 are a distance of 1 apart while the two binary strings 00101010 and 10111010 are a distance 2 apart. A more formal approach to the concept of distance follows:

The *modulo-2* sum of two binary digits is given in the four following equations:

$$0 \oplus 0 = 0 \qquad 1 \oplus 0 = 1$$
$$0 \oplus 1 = 1 \qquad 1 \oplus 1 = 0$$

The mod-2 sum is zero if the 2 bits are identical and is 1 if the 2 bits are different.

The *weight* of a codeword *g* is defined as the number of 1's contained in the word. For a combination of all the 0's $g(0) = 0$, and for the corresponding combinations of all the 1's $g(1) = k$, where *k* is the number of bits in codeword *g*.

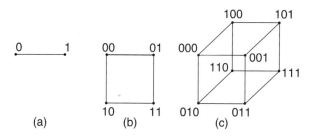

Figure 1.12 *n-Cubes for n = 1, 2, 3 (a) 1-cube (b) 2-cube (c) 3-cube*

The distance between two codewords W_1 and W_2 is defined as the number of disagreements between them so that

$$d(W_1, W_2) = \sum W_{1i} \oplus W_{2i}$$
$$= g(W_1 \oplus W_2)$$

where W_{1i} and W_{2i} are the ith bits of the two codewords.

For the distance between W_1 and W_0 where W_0 is the code combination consisting of all the 0's

$$d(W_1, W_0) = g(W_1)$$

and for the distance between W_1 and W_u where W_u is the code combination consisting of all the 1's

$$d(W_1, W_u) = k - g(W_1).$$

The minimum distance of a code d_{min} is the minimum value of $d(W_i, W_j)$ and for a complete code $d_{min} = 1$ when W_i and W_j are adjacent codewords.

1.19 Error detection and correction

An error occurs in a digital system because of the corruption of the data by some external influence such as noise. To improve the reliability of the system, methods are used to indicate the occurrence of an error and in some systems arrangements are made for both the detection and correction of errors. A single-bit error occurs when a 0 is converted to a 1 or vice versa. Multiple errors may also occur, but it is normally assumed that these are less likely to occur than single-bit errors.

The practical way of reducing error probability in a digital system is to introduce a controlled amount of redundancy. The 2-out-of-5 code is a typical example of such a code. In all, there are 2^5 combinations of five bits of which only ten are used, the remaining twenty two combinations being redundant. The ten combinations used, are the only combinations which contain two 1's and are tabulated in Figure 1.11. Any odd number of errors in a specified codeword will result in the received word having an odd number of 1's. Double or quadruple errors will also be detected unless a 1 is compensated by an error in a 0, thus ensuring the received codeword still contains two 1's.

The concept of distance is crucial in the design and understanding of error detecting codes. All single-bit errors will be detected if there is a minimum distance of 2 between all possible pairs of codewords. A minimum distance of 2 can be achieved by adding an extra bit to the transmitted word. This additional bit is called a *parity bit*. Two different parity systems are currently in use. In an *even parity* system the parity bit to be added to the codeword is chosen so that the number of 1's in the modified word is even, whilst in the *odd parity* system the added bit is chosen so that the number of 1's in the modified word is odd.

A 3-bit code is tabulated in Figure 1.13 alongside the modified codewords to which the even and odd parity bits have been added. It will be observed from this tabulation that a minimum distance of 2 is maintained between all adjacent pairs of modified codewords.

Original Code	Modified Code	
	Even parity	Odd parity
000	0000	0001
001	0011	0010
010	0101	0100
011	0110	0111
100	1001	1000
101	1010	1011
110	1100	1101
111	1111	1110

Figure 1.13 *Modification of 3-bit code by even and odd parity bits*

A 2-dimensional form of parity checking that will detect and correct single-bit errors is available for dealing with an array of words, for example, stored in memory. The technique is termed *iterative parity checking*. An array formed from 4-bit words is shown in Figure 1.14(a). Parity bits providing even parity are attached to each row and column. After attachment of the parity bits, the array shown in Figure 1.14(b) is obtained.

A single-bit error in this array can be both detected and corrected. The method allows the position in the array where the error has occurred to be identified and correction can then take place. For example, a single-bit

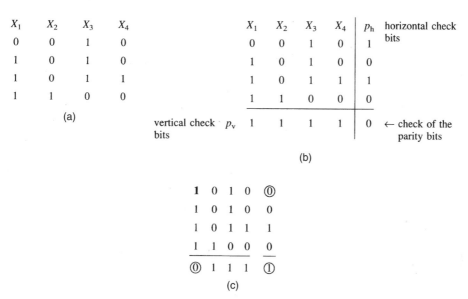

Figure 1.14 *(a) 2-dimensional code array (b) with horizontal and vertical check bits and (c) detection of the position of a single bit error*

error occurring in the most significant bit of the first 4-bit word in the array is identified as an error by using boldface, then re-computing the parity checks gives the array shown in Figure 1.14(c).

The error detection and correction procedure for the array consists, first, of checking the row parities and that reveals that there is an error in the top row of the array. At this point in the procedure it is not possible to determine which bit in the row is in error. However if a bit-by-bit XOR is taken of all the words in the array excepting the row in error but including the column check row the column in error is identified and the error corrected.

1.20 The Hamming code

This code provides a minimum distance of three between codewords, a necessary condition that must be provided in order to achieve single-bit error detection and correction. For any value of r, where r represents the number of check bits, $2^r - 1$ codeword bits can be formed consisting of r check bits and k message bits where $k = 2^r - 1 - r$. For $r = 3$, $k \leq 4$ so that four message bits are the maximum number that can be checked for $r = 3$.

The bit positions in the codeword are numbered from 1 to $2^r - 1$ and any position in the codeword whose number is a power of 2 contains a parity bit. For a 7-bit codeword the parity bits occupy positions 1, 2 and 4 so that the format of the transmitted codeword is:

bit position   7, 6, 5, 4, 3, 2, 1

$$C = k_4 \, k_3 \, k_2 \, r_3 \, k_1 \, r_2 \, r_1$$

The bit positions occupied by the parity bits 4, 2 and 1 when converted to binary are 100, 010 and 001. Each of these conversions contains a single 1 and are grouped with message bits $k_4 k_3 k_2 k_1$ whose numbers contain a 1 in the same bit position. For example, r_1 in bit position 001 is grouped with message bits that occupy the bit positions 011(3), 101(5) and 111(7). It is then arranged that for a given combination of message bits the parity bit is allocated so that even parity is achieved.

The value of parity bit r_1 is given by XORing the message bits in bit positions 7, 5 and 3. Hence:

$$r_1 = k_4 \oplus k_2 \oplus k_1$$

The value of parity bit r_2 is obtained by XORing the message bits in positions 7, 6 and 3 so that

$$r_2 = k_4 \oplus k_3 \oplus k_1$$

Finally r_3 is obtained by XORing the message bits in bit positions 7, 6 and 5

$$r_3 = k_2 \oplus k_3 \oplus k_4$$

Consider, as an example of the use of the Hamming code, the message bits $k_4 k_3 k_2 k_1 = 1101$. For this message the parity check bits to be transmitted with the message bits are:

$$r_{1s} = 1 \oplus 0 \oplus 1 = 0$$
$$r_{2s} = 1 \oplus 1 \oplus 1 = 1$$
$$r_{3s} = 1 \oplus 1 \oplus 0 = 0$$

and the transmitted codeword is

$$k_4 k_3 k_2 r_3 k_1 r_2 r_1 = 1100110$$

Assuming that the message bit k_1 is in error when the codeword is received, then at the receiving end

$$k_4 k_3 k_2 r_3 k_1 r_2 r_1 = 1100010$$

Re-computing the parity at the receiving end gives

$r_{1r} = 1 \oplus 0 \oplus 0 = 1$

$r_{2r} = 1 \oplus 1 \oplus 0 = 0$

$r_{3r} = 1 \oplus 1 \oplus 0 = 0$

The position of the error can be obtained by XORing the transmitted and received parity bits:

$$\left. \begin{array}{l} r_{1s} \oplus r_{1r} = 0 \oplus 1 = 1 \\ r_{2s} \oplus r_{2r} = 1 \oplus 0 = 1 \\ r_{3r} \oplus r_{3r} = 0 \oplus 0 = 0 \end{array} \right\} \text{syndrome indicating error}$$

The syndrome $S = 011$ and indicates there is an error in the third bit position.

1.21 Gray code

A Gray code is one in which only one digit changes as a transition is made from one code combination to the next in the sequence. In terms of distance, a Gray code is a unit distance code. One particular form of Gray code is called *reflected binary*, which can be constructed using the following technique. The two binary digits 0 and 1 are reflected about a horizontal line and the digits above the line are prefixed by 0 and below the line by 1 as shown below:

```
0   0
0   1
    ‾‾‾
1   1
1   0
```

It will be observed that in this tabulation of four 2-bit codes adjacent combinations differ in one digit place only. This process can now be extended by reflecting the four 2-bit combinations placed below the combination 10 and then proceeding as described previously. The eight 3-bit combinations generated are tabulated in Figure 1.15 alongside the eight 3-bit combinations of the binary number system. An alternative method of translating from the binary number system to the Gray code tabulated in Figure 1.15 is to use the expression:

$$g_i = b_i \oplus b_{i+1}$$

where the ith Gray code digit is found by taking the mod-2 sum of the ith and $(i + 1)$th digit of the binary number. Thus the Gray code corresponding to 110 in binary is generated as follows:

Reflected binary	Natural binary
000	000
001	001
011	010
010	011
1‾1‾0	100
111	101
101	110
100	111

Figure 1.15 *Tabulation of 3-bit natural binary and 3-bit reflected binary*

$g_0 = b_0 \oplus b_1 = 0 \oplus 1 = 1$

$g_1 = b_1 \oplus b_2 = 1 \oplus 1 = 0$

$g_2 = b_2 \oplus b_3 = b_2 = 1$

b_3 assumed to be 0

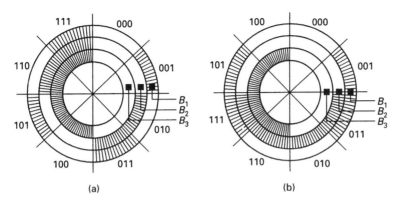

Figure 1.16   *(a) Natural binary coded disc (b) Gray coded disc*

When a transition is being made from 001 to 010 in natural binary, two digits should change simultaneously. If the two changes do not coincide, then transient states of 011 or 000 may occur. This generation of transient states is of significance in the design of angular digital encoders which are used to measure the angular position of a rotating shaft.

The encoder disc shown in Figure 1.16(a) has an arrangement of metallic areas placed on a non-conducting base. All the metallic areas are electrically interconnected and are supplied through a fixed brush in contact with a continuous metallic ring which rotates with the shaft. Three other brushes are positioned radially in fixed positions relative to the axis of the rotating shaft. As the disc rotates, the brushes are connected to the supply voltage whenever they are in contact with the metallic part of the disc.

Mechanical assembly is not perfect, and the two right-hand brushes, because of this imperfection, may simultaneously be in contact with metallic regions as the transition from 001 to 010 is made, thus generating the transient output 011. The solution to this problem is to employ Gray code encoding on the disc, as illustrated in Figure 1.16(b), so that at any boundary on the disc, contact with one brush only is changing.

A single disc with 10 tracks employing a 10-bit Gray code will give a resolution of 1 in 1024. One disadvantage of this type of mechanical encoding is associated with brush wear and mechanical vibration which can break contact between the brush and rotating disc.

1.22 The ASCII code

Codes are not only used to represent numerical data but can also represent non-numeric data such as the alphabet. The most common alphanumeric code in present use is the *American Standard Code for Information Interchange (ASCII)*. This code has been adopted by the computing fraternity as the basis for a standard alphanumeric code. It is a 7-bit code which provides 128 different characters, including upper-case alphabet, lower-case alphabet, decimal digits, punctuation symbols, and control characters. An eighth parity bit can also be used with the code to provide protection against errors. Information other than data is carried by the control characters. For example STX (start of text) and ETX (end of text) are used to define the limits of a block of data and EOT defines the end of transmission. The code is tabulated in Figure 1.17(a) where the more compact form of the hexadecimal code has been used. Control characters are listed in Figure 1.17(b).

Hex	ASCII	Hex	ASCII	Hex	ASCII	Hex	ASCII	Hex	ASCII	Hex	ASCII	Hex	ASCII	Hex	ASCII
00	NUL	10	DLE	20	SP	30	0	40	@	50	P	60	`	70	p
01	SOH	11	DC_1	21	!	31	1	41	A	51	Q	61	a	71	q
02	STX	12	DC_2	22	"	32	2	42	B	52	R	62	b	72	r
03	ETX	13	DC_3	23	£(#)	33	3	43	C	53	S	63	c	73	s
04	EOT	14	DC_4	24	$	34	4	44	D	54	T	64	d	74	t
05	ENQ	15	NAK	25	%	35	5	45	E	55	U	65	e	75	u
06	ACK	16	SYN	26	&	36	6	46	F	56	V	66	f	76	v
07	BEL	17	ETB	27	'	37	7	47	G	57	W	67	g	77	w
08	BS	18	CAN	28	(	38	8	48	H	58	X	68	h	78	x
09	HT	19	EM	29	)	39	9	49	I	59	Y	69	i	79	y
0A	LF	1A	SUB	2A	*	3A	:	4A	J	5A	Z	6A	j	7A	z
0B	VT	1B	ESC	2B	+	3B	;	4B	K	5B	[	6B	k	7B	{
0C	FF	1C	FS	2C	,	3C	<	4C	L	5C	\	6C	l	7C	|
0D	CR	1D	GS	2D	-	3D	=	4D	M	5D	]	6D	m	7D	}
0E	SO	1E	RS	2E	.	3E	>	4E	N	5E	^	6E	n	7E	~
0F	SI	1F	US	2F	/	3F	?	4F	O	5F	_	6F	o	7F	DEL

(a)

Code HD	Symbol	Function
00	NUL	All the 0's
01	SOH	Indicates start of header field
02	STX	Indicates start of text
03	ETX	Indicates end of text
04	EOT	Termination of transmission
05	ENQ	Enquire if terminal is on
06	ACK	Informs Tx of receipt of error free data
10	DLE	Data link escape
15	NACK	Informs Tx of receipt of data containing errors
16	SYN	Establishes bit and character synchronism
17	ETB	Indicates the end of block of data

(b)

Figure 1.17   *(a) The ASCII code (b) ASCII characters for control of communication*

Hex	00				01				10				11			
Bits 1	00	01	10	11	00	01	10	11	00	01	10	11	00	01	10	11
4567 ↓	0	1	2	3	4	5	6	7	8	9	A	B	C	D	E	F
0000 0	NUL	DLE			SP	&	-									0
0001 1	SOH	SBA					/		a	j			A	J		1
0010 2	STX	EUA		SYN					b	k	s		B	K	S	2
0011 3	ETX	IC							c	l	t		C	L	T	3
0100 4									d	m	u		D	M	U	4
0101 5	PT	NL							e	n	v		E	N	V	5
0110 6			ETB						f	o	w		F	O	W	6
0111 7			ESC	EOT					g	p	x		G	P	X	7
1000 8									h	q	y		H	Q	Y	8
1001 9		EM							i	r	z		I	R	Z	9
1010 A					©	!	≈	:								
1011 B						$	.	#								
1100 C		DUP		RA	<		%	@								
1101 D		SF	ENQ	NAK	(	)	-	.								
1110 E		FM			+	;	>	=								
1111 F		ITB		SUB	|	?	"									

Figure 1.18   *The EBCDIC code*

One of the most important alternatives to ASCII coding is the *Extended Binary Coded Decimal Interchange Code* (*EBCDIC*) which was used by IBM and ICL mainframes amongst others. As in the case of ASCII the decimal digits 0 to 9, the lower- and upper-case alphabet, special symbols and control codes are all assigned unique binary values although in the case of EBCDIC these values are all 8 bits in length. Consequently by comparison with ASCII the code contains a considerable amount of redundancy. The code, tabulated in Figure 1.18, is based upon an NBCD coding where the four least significant bits (nybble) do not take on a value greater than $(9)_{10}$. Because of this feature the coding of the alphabet is not contiguous.

Problems

1.1 Convert the following binary numbers to base 10:

    (a) 10101101   (b) 110110.1   (c) 1.00101

1.2 Convert the following octal numbers to base 10:

    (a) 273   (b) 1021   (c) 16.432

1.3 Convert the following hexadecimal numbers to base 10:

    (a) 145   (b) A2C1   (c) 1A.B2

1.4 Convert the following decimal numbers to base 2:

    (a) 122   (b) 98   (c) 48.45

1.5 Convert the following decimal numbers to octal:

    (a) 522   (b) 1119   (c) 129.25

1.6 Convert the following decimal numbers to hexadecimal:

    (a) 1145   (b) 2421   (c) 192.86

1.7 The following arithmetic operations are correct for at least one number system. Determine possible radices for the given operations.

    (a) $3142 + 2413 = 5555$ (b) $\dfrac{51}{3} = 13$

    (c) $23 + 44 + 14 + 32 = 223$ (d) $\sqrt{51} = 6$

1.8 Determine the base b in each of the following cases:

    (a) $(361)_{10} = (551)_b$ (b) $(859)_{10} = (5B7)_b$ (c) $(982)_{10} = (1726)_b$

1.9 Perform the following binary arithmetic operations showing all carries and borrows

    (a) $101011 + 10111$

    (b) $1101 + 1110 + 1001$

    (c) $11101 - 10110$

    (d) $1100.010 - 1000.111$

1.10 Write the 8-bit signed magnitude, 2's complement and 1's complement form of the following decimal numbers:

    (a) $+119$ (b) -77 (c) -3

1.11 Perform the following arithmetic operations using 2's complement arithmetic and assuming a word length of 8 bits:

79	64	87
-42	$+37$	-99

1.12 Form the radix complement and diminished radix complement for each of the following numbers:

(a) $(01011)_2$ (b) $(5291)_{10}$ (c) $(4723)_8$

(d) $(ABC1)_{16}$

1.13 Perform the multiplication of the following unsigned binary numbers:

(a) 1110×1101 (b) 10101×1110 (c) 11001×10101

1.14 Perform the following multiplications using signed binary numbers:

(a) $+7 \times -9$ (b) $+12 \times +9$ (c) -13×-8

1.15 Perform the following divisions using the restoring division process:

(a) $101010 \div 0101$ (b) $1100110 \div 1001$

1.16 Perform the following arithmetic operations using the NBCD code to represent the decimal numbers:

(a) $79 + 101$ (b) $87 + 179$ (c) $98 - 43$

1.17 Perform the following arithmetic operations using signed NBCD arithmetic:

(a) $85 + 67$ (b) $43 - 92$

1.18 Write out the following decimal weighted codes:

(a) 7, 4, 2, 1   (c) 2, 4, 2, 1

(d) 5, 2, 1, 1   (d) 8, 4, -2, -1

2 Boolean algebra

2.1 Introduction

In a digital system the electrical signals that are used have two voltage levels which may, for example, be 5 and 0 volts. The electrical devices used in these systems can generally exist indefinitely in one of these two possible voltage states, providing the power supply is maintained. For example, a bipolar transistor that is non-conducting in a 5 volt system will have approximately 5 volts between collector and emitter. However, when the transistor is turned on and is conducting, it can be arranged, with a suitable choice of load, that the voltage between collector and emitter is approximately zero. The two voltage levels employed in a digital circuit can be arbitrarily assigned values of 0 and 1. The two states defined in this way can have logical significance in that they can indicate the presence of a particular condition or, alternatively, its absence.

An algebra developed in the nineteenth century by George Boole (1815–1864), an English mathematician, is well suited for representing the situation above. This branch of mathematics, called Boolean algebra, is a discrete algebra in which the variables can have one of two values, either 0 or 1. Associated with the algebra is a number of theorems which allow the manipulation and simplification of Boolean equations.

Shannon, who was the first to develop information theory, became aware that Boolean algebra was useful in the design of switching networks. Initially, the algebra was used in the design of relay networks. More recently switching circuits were implemented using discrete components but rapid technological advances have seen the introduction of MSI, LSI and VLSI devices and because of the sophisticated and versatile nature of these components there have been significant changes in the design techniques used by engineers. In spite of these changes it is still essential for engineers to have a good working knowledge of traditional switching theory.

2.2 Boolean algebra

Any mathematical system has a minimal set of basic definitions which are assumed to be true and from which all information about the system may be determined. In the case of Boolean Algebra the three basic definitions are:

NOT. The NOT of a variable is 1 if, and only if, the variable itself is 0 and vice versa. NOT A is written as $\bar{A}$. Thus if $A = 0$, then $\bar{A} = 1$ and if $A = 1$ then $\bar{A} = 0$. Since A has only two possible values it follows that $\bar{\bar{A}} = A$.

To refer to both A and $\bar{A}$ which define the opposite values of the same variable, the term *literal* is used, where a literal is defined as a variable with or without a complement bar.

AND. The AND of two variables is 1 if, and only if, *both* the variables are 1. AND is written as A AND B or as $A \cdot B$ or alternatively as AB. Thus $AB = 1$ only when $A = B = 1$.

OR. The OR of two variables is 1 if *either* (or *both*) of the variables is 1. OR is written as A OR B or as $A + B$. Thus $A + B = 1$ if $A = 1$ or $B = 1$ or $A = B = 1$.

In addition to the above basic operations, one other function, the Exclusive-OR, is required for arithmetic-related operations.

XOR. The Exclusive-OR of two variables is 1 if *either* of them but *not both* is 1. The XOR operation is written as A XOR B or as $A \oplus B$. Thus $A \oplus B = 1$ if $A = 1$ and $B = 0$ or if $A = 0$ and $B = 1$.

2.3 Derived Boolean operations

The following Boolean operations are derived from the three basic operations by complementing or inverting those operations:

NAND. NOT − AND = NOT of (A AND B) or $\overline{AB}$

NOR. NOT − OR = NOT of (A OR B) or $\overline{A + B}$

XNOR. Exclusive NOR = NOT of (A XOR B) or $\overline{A \oplus B}$, which is sometimes referred to as the coincidence function and is written A $\odot$ B.

2.4 Boolean functions

A Boolean function consists of a number of Boolean variables joined by the Boolean connectives AND and OR. For example

$$f(A, B, C, D) = AB\bar{C} + CD + \bar{B}$$

or

$$g(A, B, C, D) = (A + B + C)(\bar{C} + \bar{D})(A + B)$$

The *dual* of a function is obtained by changing the AND operations to OR operations and vice versa, and simultaneously changing any 1's to 0's and vice versa. Thus the dual of the function $f = (AB\bar{C} + CD + \bar{B})$ is given by

$$f_d(A, B, C, D) = (A + B + \bar{C})(C + D)\bar{B}$$

Two functions are *equivalent* providing they have the same value (1 or 0) for each of the possible combinations of the variables.

Two functions are *complementary* if one function equals 1 when the other function equals 0 and vice versa. The complement of a function can be found by complementing each literal in the dual of that function. Thus the complement of $f(A, B, C, D) = AB\bar{C} + CD + \bar{B}$ is

$$\bar{f}(A, B, C, D) = (\bar{A} + \bar{B} + C)(\bar{C} + \bar{D})B$$

In evaluating Boolean equations AND operations are performed before OR operations unless the OR operation is enclosed within brackets.

2.5 Truth tables

A *truth table* provides the basic method of describing a Boolean function. It contains a row for every combination of the variables and prescribes the value of the function (0 or 1) for each of these combinations. For the 3-variable function $f(A, B, C)$ whose truth table appears in Figure 2.1, there are 2^3 combinations and the value of the function for each of these combinations is listed in the right hand column. The Boolean function described by the truth table is provided by the logical sum of those combinations for which the function has a value of $f = 1$. Hence

$$f(A, B, C) = \bar{A}\bar{B}C + \bar{A}B\bar{C} + \bar{A}BC + A\bar{B}C$$

Minterms	A	B	C	f
m_0	0	0	0	0
m_1	0	0	1	1
m_2	0	1	0	1
m_3	0	1	1	1
m_4	1	0	0	0
m_5	1	0	1	1
m_6	1	1	0	0
m_7	1	1	1	0

Figure 2.1   *Truth table for $f = \sum m_1, m_2, m_3, m_5$*

Each combination of the variables is called a *minterm*. For example, $m_1 = \bar{A}\bar{B}C$ and the function tabulated in Figure 2.1 can be described as a sum of minterms so that

$$f = \sum m_1 m_2 m_3 m_5$$

Alternatively, a minterm can be identified by its subscript and the function can be defined by the following equation

$$f = \sum 1, 2, 3, 5$$

A Boolean function expressed as a sum of minterms is termed the *canonical sum-of-products* form of the function.

The inverse function $\bar{f}$ is obtained by taking the logical sum of those combinations for which $f = 0$. From Figure 2.1

$$\bar{f} = \bar{A}\bar{B}\bar{C} + A\bar{B}\bar{C} + AB\bar{C} + ABC$$

and by inversion using the principle of duality

$$f = (A + B + C)(\bar{A} + B + C)(\bar{A} + \bar{B} + C)(\bar{A} + \bar{B} + \bar{C})$$

Each term in this equation is called a *maxterm* and the Boolean function is expressed as a product of maxterms. The resulting expression is called the *canonical product-of-sums* form of the function and it may be written as

$$f = \prod M_0 M_4 M_6 M_7$$

and in terms of the maxterm subscripts the function may be written

$$f = \prod 0, 4, 6, 7$$

The truth table representation of a Boolean function has strict limitations. The number of rows in the table for an *n*-variable function is 2^n and if $n \geq 5$ the construction of the table is tedious, time consuming and prone to error. For this reason this method of representation is of little practical use to the circuit designer.

2.6 The logic of a switch

Consider the circuit shown in Figure 2.2(a) consisting of a voltage source having an internal resistance R_i connected in series with a switch X and a resistance R.

With the switch closed

$$I = \frac{V}{R_i + R}$$

$$V_o = IR = \frac{VR}{R_i + R}$$

Assuming $R_i \ll R$ then $V_o \simeq V$ and the voltage across the switch is zero. If the switch is open $I = 0$ and the source voltage appears across the switch terminals so that $V_o = 0$. The two states of the output voltage can be defined in terms of a Boolean variable f. When $V_o = 0, f = 0$ and when $V_o \simeq V, f = 1$. These two states correspond to the two possible states of the switch, open and closed, which can also be described in terms of a Boolean variable A. When X is open $A = 0$ and when X is closed $A = 1$. These results are tabulated in the truth table shown in Figure 2.2(b) and from an inspection of the table it is clear that $f = A$.

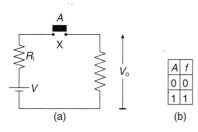

A	f
0	0
1	1

(a) (b)

Figure 2.2 *(a) An on/off switch circuit (b) the truth table for the switch circuit*

2.7 The switch implementation of the AND function

Two switches X and Y are connected in series with a voltage source V and resistance R. The two possible states of the switches X and Y are defined by the Boolean variables A and B. When the two switches are open, $A = B = 0$, and when they are closed $A = B = 1$. The output voltage can also be expressed in terms of a Boolean variable f whose value depends upon the absence or presence of the voltage V_o across the resistance R.

There are four possible combinations of the variables A and B, and these are tabulated in the truth table shown in Figure 2.3(a). For example, if X is open and Y is open, then $A = B = 0$ and $V_o = 0$, hence $f = 0$. If, however, X and Y are both closed, then $A = B = 1$, $V_o \simeq V$, and it follows that $f = 1$.

The truth table shown in Figure 2.3(b) is that of the AND function, sometimes referred to as the *Boolean multiplication function*. The rules of Boolean multiplication are identical to those of binary multiplication and they are summarised in Figure 2.3(c).

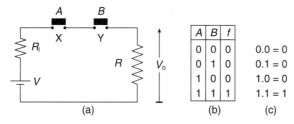

A	B	f	
0	0	0	0.0 = 0
0	1	0	0.1 = 0
1	0	0	1.0 = 0
1	1	1	1.1 = 1

(a) (b) (c)

Figure 2.3 *(a) Switch implementation of the AND function (b) Truth table for the AND function (c) Rules of binary and Boolean multiplication*

Figure 2.4 *Conventional symbol for the AND gate*

In practice, the AND function is implemented by a high speed electronic gate which is capable of operating in a few nanoseconds. One of the generally accepted symbols for the gate is shown in Figure 2.4. The output of the gate is 1 only if both inputs A and B are 1. For any other combination of the input variables the output is 0.

2.8 The switch implementation of the OR function

A parallel connection of two switches X and Y with a voltage source V and a resistor R is shown in Figure 2.5(a). The two possible states of the switches are defined by the Boolean variables A and B. When both switches are open $A = B = 0$ and when they are both closed $A = B = 1$. The corresponding states of the output voltage, 0 and V_0, are expressed in terms of the Boolean variable f whose value depends upon the absence or presence of the output voltage. If both switches are open $A = B = 0$ and it is then clear that the current in the circuit $I = 0$ and the output voltage $V_0 = 0$. For this condition $f = 0$. If switch X is closed and switch Y is open then $A = 1$ and $B = 0$, a conducting path is available through closed switch X, so $V_0 \simeq V$ and for this condition $f = 1$.

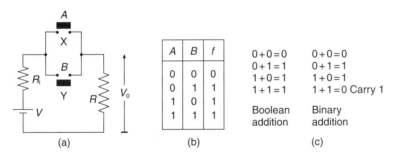

Figure 2.5 *(a) Switch implementation of the OR function (b) Truth table for the OR function (c) rules of binary and Boolean addition*

The truth table of the OR function, sometimes referred to as the *Boolean addition function*, is shown in Figure 2.5(b). There are four possible combinations of A and B which correspond to the four possible open and closed conditions of the two switches X and Y. Examination of the truth table shows that the Boolean variable representing the output voltage $f = 1$ if X is closed or Y is closed or if both X and Y are closed simultaneously. For the Exclusive-OR function if $A = B = 1$, $f = 0$, so strictly speaking, the OR function should be referred to as the Inclusive-OR function since $f = 1$ when $A = B = 1$.

The rules of Boolean addition are tabulated alongside the rules of binary addition in Figure 2.5(c) and it will be noted that they differ in one respect. For binary addition $1 + 1 = (10)_2$ where 0 is the sum digit and 1 is the carry to the next stage of the addition, whereas in the case of Boolean addition $1 + 1 = 1$.

Figure 2.6 *Conventional symbol for the OR gate*

The OR function is normally implemented by an electronic gate which can be represented by the symbol shown in Figure 2.6. The output of this 2-input gate will be 1 if input $A = 1$ or input $B = 1$ or if both inputs A and B are simultaneously 1. For the remaining combination $A = B = 0$ the output will be 0.

2.9 The gating function of the AND and OR gates

Control of the transmission of a string of digital data can be achieved by either of the two basic gates when they are operated in a gating mode. The B input in Figure 2.7 can be used as a control or mode input, while the data stream is applied at input A. If $B = 1$ the gate is open and the data stream is transmitted, as shown in Figure 2.7(a). If $B = 0$ the gate is closed and data transmission is inhibited, as shown in Figure 2.7(b). The 2-input OR gate shown in Figure 2.8 will perform the same gating function if the B input is used as a control input.

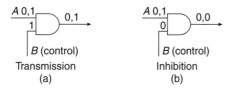

Transmission   Inhibition
(a)   (b)

Figure 2.7   *The gating function of an AND gate*

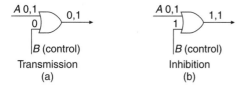

Transmission   Inhibition
(a)   (b)

Figure 2.8   *The gating function of an OR gate*

2.10 The inversion function

Consider the switching arrangement shown in Figure 2.9(a) connected in series with the voltage source V and resistance R. The switch has a pair of ganged contacts, one of which is normally open, whilst the other is normally closed. For the upper contact $A = 0$ when it is open and for the lower closed contact $\bar{A} = 1$. The variable $\bar{A}$ is the *inverse* of A and the inversion operation is defined by the bar over the Boolean variable A. The output voltage V_o, as in the case of the OR and AND functions, is represented by the Boolean variable f.

The truth table for the circuit is shown in Figure 2.9(b) and it is clear that the output voltage $V_o \simeq V$ irrespective of the state of the switch. Hence the equation of this circuit may be written as:

$$A + \bar{A} = 1$$

which is an algebraic statement of the *complementation theorem*. The dual of this equation is obtained by replacing the $+$ with a $\cdot$ and by changing 1 to 0. Hence:

$$A \cdot \bar{A} = 0$$

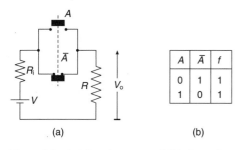

A	$\bar{A}$	f
0	1	1
1	0	1

(a)   (b)

Figure 2.9   *(a) Switch circuit and (b) the truth table for $f = A + \bar{A}$*

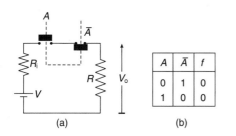

A	$\bar{A}$	f
0	1	0
1	0	0

(a)   (b)

Figure 2.10   *(a) Switch circuit and (b) the truth table for $f = A \cdot \bar{A}$*

The switch contacts for this circuit and the corresponding truth table are shown in Figure 2.10.

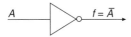

Figure 2.11 *Conventional representation of an inverter*

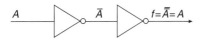

Figure 2.12 *The theorem of double inversion*

In practice, the inversion operation is implemented by an electronic gate which is represented symbolically in Figure 2.11. The inversion circle at the output of the gate is widely used in logic circuits to indicate the inversion of a Boolean variable. When two inverters are connected in series, as shown in Figure 2.12, a double inversion takes place and the output of the second inverter is the same as the input A to the first inverter.

2.11 Gate or switch implementation of a Boolean function

The implementation of the Boolean function $f = A(B + \bar{C}) + \bar{B}D$ using either switches or, alternatively, gates is illustrated in Figure 2.13. In the switch contact, circuit a $+$ in the equation is interpreted as a pair of parallel branches whilst a $\cdot$ is interpreted as a series connection. Normally closed switch contacts are identified by a bar over the switch variable.

2.12 The Boolean theorems

In the field of Logic Design it is the function of the designer to develop a Boolean expression which describes the required circuit performance. Algebraic manipulation of this expression with the aid of the Boolean Theorems can produce a simpler implementation of the circuit. There are a number of these theorems which can be used for simplifying a Boolean expression, and some of those in general use can be verified directly by using the method of *perfect induction*. The method involves inserting the two possible values 0 and 1 into a statement of the selected theorem in order to confirm the validity of the theorem in both cases. For example, the Idempotency theorem states that

$A + A = A$

For $A = 1$, $1 + 1 = 1$, and for $A = 0$, $0 + 0 = 0$ and these results verify that the theorem is true. Boolean theorems exist in pairs and the second form of Idempotency is obtained by taking the dual of $A + A = A$ and may be written as

$A \cdot A = A$

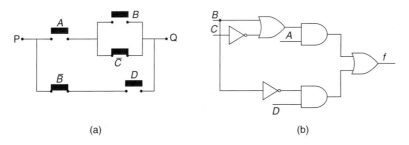

(a) (b)

Figure 2.13 $f = A(B + \bar{C}) + \bar{B}D$ *implemented (a) with switches and (b) with gates*

Theorem	Function	Dual
Idempotency	$A+A=A$	$A \cdot A = A$
Union & Intersection	$\left\{\begin{array}{l} A+0=A \\ A+1=1 \end{array}\right.$	$\begin{array}{l} A \cdot 1 = A \\ A \cdot 0 = 0 \end{array}$
Complementation	$A+\bar{A}=1$	$A \cdot \bar{A} = 0$
Inversion	$(\bar{\bar{A}})=A$	

Figure 2.14 *The single variable theorems*

There are a number of Boolean theorems, including Idempotency, which involve a single variable. These are listed in Figure 2.14, and it is left to the reader to verify them using the method of perfect induction.

Commutation, Association and *Distribution* are three of a group of theorems involved with more than one variable. Each of these theorems occurs as a pair, the original accompanied by its dual. The first two of these theorems are identical to the laws of commutation and association for addition and multiplication of integers. Commutation states

$$A + B = B + A \quad \text{and} \quad AB = BA$$

which follows directly from the truth tables that define the AND and OR functions. Association states that

$$A + (B + C) = (A + B) + C \quad \text{and} \quad A(BC) = (AB)C$$

and clearly the parentheses are unnecessary since the order in which two or more variables are ANDed or ORed is irrelevant.

Factorisation of Boolean functions can be achieved by the application of the distribution theorem whose two forms are

$$A + BC = (A + B)(A + C) \quad \text{and} \quad A(B + C) = AB + AC$$

A	B	C	BC	A+BC	A+B	A+C	(A+B)(A+C)
0	0	0	0	0	0	0	0
0	0	1	0	0	0	1	0
0	1	0	0	0	1	0	0
0	1	1	1	1	1	1	1
1	0	0	0	1	1	1	1
1	0	1	0	1	1	1	1
1	1	0	0	1	1	1	1
1	1	1	1	1	1	1	1

Figure 2.15 *Proof of the Distribution theorem by truth table*

The second form of distribution is identical to the same process in ordinary algebra, and a proof of the first form of the theorem by truth table is shown in Figure 2.15.

The Absorption or Redundancy theorem is a further example of a Boolean theorem which involves more than a single variable. The theorem states:

$$A + AB = A$$

and it may be proved by the application of a number of the single variable theorems as follows:

$$
\begin{array}{ll}
A + AB = A \cdot 1 + AB & \text{Intersection} \\
= A(B + \bar{B}) + AB & \text{Complementation} \\
= AB + A\bar{B} + AB & \text{Distribution} \\
= AB + A\bar{B} & \text{Idempotency} \\
= A(B + \bar{B}) & \text{Distribution} \\
= A & \text{Complementation}
\end{array}
$$

The equation $f = A \cdot 1 + AB$ is expressed in the sum-of-products (S-of-P) form, where each of the terms in the equation are called *product terms*. It follows that in any Boolean equation that is expressed in S-of-P form, a product term that contains all the factors of another product term is redundant. For example, in the equation $f = AD + ABD + ACD$

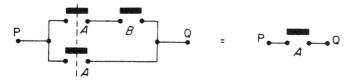

Figure 2.16   *Switch contact circuits illustrating the absorption theorem* $A = A + AB$

the product terms ABD and ACD can be eliminated because each contains all the factors present in the product term AD, and the expression can be reduced to $f = AD$.

The switch contact circuit for the equation $f = A + AB$ is illustrated in Figure 2.16. This diagram clearly shows that if switch contact A is made, then a connection exists between points P and Q irrespective of whether switch contact B is open or closed and consequently switch contact B is redundant.

Yet another example of a Boolean theorem which involves more than one variable is the *Consensus theorem*. Consider the function $f = AC + B\bar{C}$ which contains the variable C in one of the terms and its complement $\bar{C}$ in the other. An optional product or *consensus term* can be formed by taking the product of the remaining two variables, in this case A and B. Furthermore, the optional product can be added to the original function f to give a new function $f' = AC + B\bar{C} + AB$ whose value is identical to the original function for all values of the variables A, B and C. This can be proved algebraically as follows:

$$f' = AC + B\bar{C} + AB = AC + B\bar{C} + AB(C + \bar{C})$$
$$= AC + B\bar{C} + ABC + AB\bar{C}$$
$$= AC(1 + B) + B\bar{C}(1 + A)$$

but $(1 + A) = (1 + B) = 1$, hence $f' = f = AC + B\bar{C}$.

The consensus term can be defined as one whose presence in a Boolean function does not alter the value of the function, and in the example given the optional product AB is redundant. However, consensus terms have their uses, since when they are formed and introduced into a Boolean equation they may eliminate other terms and simplify the original equation. For example:

$$f = C + AB\bar{C}$$

The consensus term AB is added to the original equation to give

$$f = C + AB\bar{C} + AB$$
$$f = C + AB \qquad \text{Absorption theorem}$$

It should be noted that after the elimination of the original term $AB\bar{C}$, AB is no longer optional, it is now essential.

The technique of forming consensus terms and adding them to a function without altering its value is a useful one. It will be seen in a later chapter that the technique can be used to eliminate static hazards in combinational circuits.

A	B	$\bar{A}$	$\bar{B}$	$\overline{AB}$	$\bar{A}+\bar{B}$
0	0	1	1	1	1
0	1	1	0	1	1
1	0	0	1	1	1
1	1	0	0	0	0

Figure 2.17 *Proof of De Morgan's theorem for two variables*

The last theorem involving more than one variable is De Morgan's theorem, named after a mathematician Augustus De Morgan, a contemporary of Boole. The truth table in Figure 2.17 demonstrates the validity of one form of this theorem

$$\overline{AB} = \bar{A} + \bar{B}$$

whilst the second form of the theorem is obtained by applying the principle of duality to the above equation so that

$$\overline{A + B} = \bar{A}\bar{B}$$

These theorems express the complements of the AND and OR functions in terms of the complements of the variables *A* and *B*, and are probably the most useful of all the Boolean theorems.

De Morgan's theorem can be applied to any number of variables, and the truth table of Figure 2.17 only verifies the law for two variables. For an arbitrary number of variables the theorem requires a proof by *finite induction*. First, the theorem is proved for $n = 2$ using the method of perfect induction. An assumption is then made that the theorem is true for $n = h$, and if this is true it can be shown that it is also true for $n = h + 1$.

A summary of the Boolean theorems involved with more than a single variable is tabulated in Figure 2.18.

	Function	Dual
Commutation	$A + B = B + A$	$AB + BA$
Association	$A + (B + C) = (A + B) + C$	$A(BC) = (AB)C$
Distribution	$A + BC = (A + B)(A + C)$	$A(B + C) = AB + AC$
Absorption	$A + AB = A$	$A(A + B) = A$
De Morgan	$\overline{A+B} = \bar{A} \cdot \bar{B}$	$\overline{AB} = \bar{A} + \bar{B}$
Consensus	$AC + B\bar{C} =$	$(A + C)(B + \bar{C}) =$
	$AB + AC + B\bar{C}$	$(A + B)(A + C)(B + \bar{C})$

Figure 2.18 *Boolean theorems involving more than one variable*

The generalised statement of De Morgan's theorem along with Shannon's Expansion theorem are tabulated in Figure 2.19. The proof of Shannon's theorem by induction for the variable A_1 requires the substitution of the two possible values of A_1, 0 and 1, into either of the two forms of the theorem appearing in the tabulation. For $A_1 = 1$ in the first form of the theorem

$$f(1, A_2, A_3, \ldots A_n) = 1 \cdot f(1, A_2, A_3, \ldots A_n) + 0 \cdot f(0, A_2, A_3, \ldots A_n)$$
$$= f(1, A_2, A_3, \ldots A_n)$$

and for $A_1 = 0$

$$f(0, A_2, A_3, \ldots A_n) = 0 \cdot f(1, A_2, A_3, \ldots A_n) + 1 \cdot f(0, A_2, A_3, \ldots A_n)$$
$$= f(0, A_2, A_3, \ldots A_n)$$

De Morgan's theorem	$\overline{A_1 + A_2 + A_3 \dots A_n} = \overline{A_1} \cdot \overline{A_2} \cdot \overline{A_3} \dots \overline{A_n}$
	$\overline{A_1 \cdot A_2 \cdot A_3 \cdots A_n} = \overline{A_1} + \overline{A_2} + \overline{A_3} \cdots \overline{A_n}$
Expansion theorem	$f(A_1, A_2, A_3 \cdots A_n) = A_1 f(1, A_2, A_3, \cdots A_n) + \overline{A_1} f(0, A_2, A_3, \cdots A_n)$
	$f(A_1, A_2, A_3 \cdots A_n) = [A_1 + f(0, A_2, A_3, \cdots A_n)][\overline{A_1} + f(1, A_2, A_3, \cdots A_n)]$

Figure 2.19   *Boolean theorems involving n variables*

Similarly, the theorem can be verified for any of the $n - 1$ remaining variables using this method.

2.13 Complete sets

Any Boolean function can be implemented using only AND and INVERT gates since the OR function can be generated by a combination of these two gates, as shown in Figure 2.20(a). It follows that these two gates can implement any arbitrary Boolean function and they are said to form a *complete set*. Similarly, the OR and INVERT gates also form a complete set since the AND function can be implemented by a combination of these two gates, as shown in Figure 2.20(b).

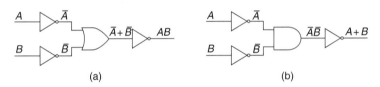

(a)   (b)

Figure 2.20   *Complete sets (a) OR/INVERT (b) AND/INVERT*

The derived gates NAND and NOR are in themselves a complete set since, for example, a series combination of two NAND gates will generate the AND function [Figure 2.21(a)]. In this connection the second NAND gate has all its inputs commoned and acts as an inverter. Similarly, the OR function can be generated by two NOR gates in series (Figure 2.21(b)), where the second NOR gate is implementing the inversion function. It follows that any arbitrary Boolean function can be implemented by either of these gates.

NAND   Inverter   NOR   Inverter

Figure 2.21   *Complete sets formed by (a) series combination of two NAND gates and (b) series combination of two NOR gates*

2.14 The exclusive-OR (XOR) function

The XOR or Mod-2 addition operation is defined by the equation

$$A \oplus B = \overline{A}B + A\overline{B}$$

An alternative way of expressing this relationship is

$$A \oplus B = (A + B)(\bar{A} + \bar{B})$$

The laws of Association, Commutation and Distribution are also valid for the XOR operation. They are

$$A \oplus (B \oplus C) = (A \oplus B) \oplus C \quad \text{Association}$$
$$A \oplus B = B \oplus A \qquad\qquad \text{Commutation}$$
$$A(B \oplus C) = AB \oplus AC \qquad \text{Distribution}$$

If Boolean algebraic equations are written in terms of the XOR function, the following identities may prove useful:

$$A \oplus A = 0 \quad A \oplus \bar{A} = 1$$
$$A \oplus 1 = \bar{A} \quad A \oplus 0 = A$$
$$A + B = A \oplus B \oplus AB$$
$$A + B = A \oplus B \text{ if } AB = 0$$

If $A \oplus B = C$ then $A = B \oplus C$, $B = A \oplus C$ and $A \oplus B \oplus C = 0$
$A_1 \oplus A_2 \oplus \ldots \oplus A_n = 0$ for an even number of variables of value 1, and 1 for an odd number of variables of value 1.

2.15 The Reed–Muller equation

A canonical equation can also be defined in terms of the AND and XOR functions, and for a single variable may be written as

$$f(A) = f_0 \bar{A} + f_1 A$$

and since A and $\bar{A}$ are mutually exclusive, the expression may be written as

$$f(A) = f_0 \bar{A} \oplus f_1 A$$
$$= f_0(1 \oplus A) \oplus f_1 A$$
$$= f_0 \oplus (f_0 \oplus f_1)A$$
$$= c_0 \oplus c_1 A$$

For two Boolean variables, the canonical sum-of-products equation may be written

$$f(A_2, A_1) = f_0 \bar{A}_2 \bar{A}_1 + f_1 \bar{A}_2 A_1 + f_2 A_2 \bar{A}_1 + f_3 A_2 A_1$$

where f_0, f_1, f_2 and f_3 take on the value of 0 or 1 depending upon whether their associated minterms are present. For example if $f_0 = 1$ and $f_1 = f_2 = f_3 = 0$ then

$$f(A_2, A_1) = \bar{A}_2 \bar{A}_1$$

Since $m_j m_k = 0$ for $j \neq k$ the logical addition symbol $+$ can be replaced by $\oplus$ and the canonical sum-of-products may now be written as

$$f(A_2, A_1) = f_0 \bar{A}_2 \bar{A}_1 \oplus f_1 \bar{A}_2 A_1 \oplus f_2 A_2 \bar{A}_1 \oplus f_3 A_2 A_1$$

The inverted variables A_2 and A_1 can be replaced by $(A_2 \oplus 1)$ and $(A_1 \oplus 1)$ respectively. Hence

$$f(A_2, A_1) = f_0(A_2 \oplus 1)(A_1 \oplus 1) \oplus f_1(A_2 \oplus 1)A_1 \oplus f_2 A_2(A_1 \oplus 1) \oplus f_3 A_2 A_1$$

Multiplying out and collecting terms produces the following equation

$$f(A_2A_1) = f_0 \oplus (f_0 \oplus f_2)A_2 \oplus (f_0 \oplus f_1)A_1 \oplus (f_0 \oplus f_1 \oplus f_2 \oplus f_3)A_2A_1$$

This is the canonical XOR sum for two Boolean variables and is called the *Reed–Muller* canonical equation. If $c_0 = f_0$, $c_1 = f_0 \oplus f_2$, $c_2 = f_0 \oplus f_1$, and $c_3 = f_0 \oplus f_1 \oplus f_2 \oplus f_3$, the equation may be written

$$f(A_2, A_1) = c_0 \oplus c_1 A_2 \oplus c_2 A_1 \oplus c_3 A_2 A_1$$

and it will be observed that the equation contains only the AND and XOR functions which, like the AND and INVERT functions, form a complete set.

2.16 Set theory and the Venn diagram

Set theory is concerned with the combination of sets and the theorems associated with the theory are identical to the theorems of Boolean algebra. In spite of their identical structures the algebra of sets looks somewhat different since the connectives used, $\cup$ and $\cap$, replace + and · in Boolean algebra.

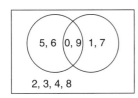

Figure 2.22   *Visual representation of Set theory*

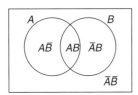

Figure 2.23   *Venn diagram illustrating the four minterms of two Boolean variables*

Consider the two sets of decimal digits $A = \{0, 5, 6, 9\}$ and $B = \{0, 1, 7, 9\}$. The *union* of these two sets, written $A \cup B$, is defined as the set that contains all the digits in A or B or both. Hence $A \cup B = \{0, 1, 5, 6, 7, 9\}$ and it is clear that union is analogous to the OR function. *Intersection* of the two sets, written $A \cap B$ is defined as that set which contains all those digits that are common to the two sets A and B. Hence $A \cap B = \{0, 9\}$ and this function is analogous to the AND function. If it happens that there are no common digits in these two sets, $A \cap B$ is referred to as the *null* set which is represented symbolically by ϕ. Finally, the set which contains all the decimal digits is referred to as the *universal* set and can be represented diagramatically by a rectangle.

These results can be represented on the Venn diagram shown in Figure 2.22 where the rectangle represents the universal set containing all the decimal digits, while the intersecting circles represent the sets A and B. Venn diagrams are also able to demonstrate graphically the meaning of Boolean functions. For example, all the minterms of two Boolean variables are displayed in Figure 2.23 whilst the three Venn diagrams in Figure 2.24 illustrate with shading some typical Boolean expressions. In Chapter 3 it will be shown how the

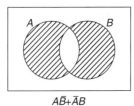

$$A\bar{B} + \bar{A}B$$

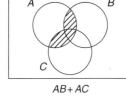

$$AB + AC$$

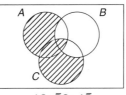

$$AC + \bar{B}C + A\bar{B}$$

Figure 2.24   *Venn diagrams for three Boolean expressions*

structure of the Venn diagram can be modified to form a Karnaugh map which is widely used for the simplification of Boolean functions.

Problems

2.1 Using the theorems of Boolean algebra simplify the following expressions:

$f_1(A, B, C, D) = B + BCD + \bar{B}CD + AB + \bar{A}B + \bar{B}C$

$f_2(A, B, C, D) = (AB + C + D)(\bar{C} + D)(\bar{C} + D)(\bar{C} + D + \bar{E})$

$f_3(A, B, C) = B\bar{C}(C + A\bar{C}) + (\bar{A} + \bar{C})(\bar{A}B + \bar{A}C)$

2.2 Simplify each of the following expressions using the method of optional products:

$f_1(A, B, C, D) = A\bar{C} + B\bar{C}D + A\bar{B}C + ACD$

$f_2(A, B, C, D) = B + \bar{A}\bar{B} + ACD + A\bar{C}$

$f_3(A, B, C, D) = B + \bar{A}B\bar{D} + AB\bar{C} + A\bar{B}D + A\bar{C}D$

2.3 Prove that $(A + B)(\bar{A} + C) = AC + \bar{A}B$ without using perfect induction.

2.4 Construct a truth table for the following functions and from the truth table obtain an expression for the inverse functions:

$f_1(A, B, C) = A + B\bar{C}$

$f_2(A, B, C) = AC + BC + AB$

$f_3(A, B, C) = (A + \bar{B})(\bar{A} + \bar{B} + C)$

$f_4(A, B, C, D) = BD + \bar{A}C + \bar{B}\bar{D}$

2.5 Find the inverse of the following expressions and do not simplify your result

$f_1(A, B, C) = A + \bar{B}C$

$f_2(A, B, C, D) = A(B + C) + B\bar{D}(\bar{A} + C)$

$f_3(A, B, C, D, E) = [AB + C(\bar{A} + DE)][\bar{B} + AC(\bar{E} + \bar{B}\bar{D})]$

2.6 Expand and simplify the following expressions using De Morgan's theorem.

$f_1(A, B, C) = (\overline{A + B})(\overline{ABC})(\overline{\bar{A}C})$

$f_2(A, B, C) = \overline{(AB + \bar{B}C) + (B\bar{C} + \bar{A}B)}$

$f_3(A, B, C) = \overline{(AB + \bar{B}C)(AC + \bar{A}\bar{C})}$

2.7 Prove the following identities

(1) $\bar{A}B + A\bar{B} = (\bar{A} + \bar{B})(A + B)$

(2) $(AB + C)B = AB\bar{C} + \bar{A}BC + ABC$

(3) $BC + AD = (B + A)(B + A)(B + D)(A + C)(C + D)$

2.8 For the following two 4-variable functions

$f_1 = \overline{\overline{A + B} + \overline{C + D}}$

$f_2 = A + C + BD$

how many of the input minterms are included in each of these functions and how many are not? What are the minterm expressions for the two functions? Simplify both functions using the theorems of Boolean algebra.

2.9 Given the timing diagram shown in Figure P2.9 find the displayed function expressed as a sum of minterms and also find the function as a product of maxterms. Simplify the minterm expression, using the Boolean theorems, and find the inverse of the simplified expression.

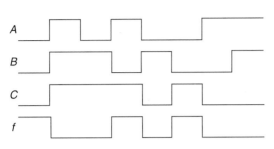

Figure P2.9

2.10 Draw (i) the switch contact circuits and (ii) the AND/OR implementations for the following Boolean functions.

$$f_1(A, B, C) = \bar{A} + B(\bar{C} + \bar{D})$$
$$f_2(A, B, C) = (\bar{A} + B)(\bar{B} + C) + (AB + C)$$
$$f_3(A, B, C, D) = (A + B + C)(\bar{A} + D) + B\bar{C} + A(B + D)(\bar{C} + D)$$

2.11 The main stairway in a block of flats has three switches for controlling the lights. Switch A is located at the top of the stairs, switch B is located halfway up the stairs and switch C is positioned at the bottom of the stairs. Design a logic network to control the lights on the staircase.

2.12 Sketch the following functions on a Venn diagram:

$$f_1(A, B) = A\bar{B} + \bar{A}B$$
$$f_2(A, B, C) = A\bar{B}\bar{C} + \bar{A}B\bar{C} + \bar{A}BC + AB\bar{C} + ABC + \bar{A}\bar{B}C$$
$$f_3(A, B, C) = \bar{A}\bar{C} + AC + B\bar{C} + \bar{A}B$$
$$f_4(A, B, C) = AB + \bar{A}\bar{C}$$

2.13 Prove

$$\bar{A} \oplus B = A \oplus \bar{B}$$
$$A \oplus B = \bar{A} \oplus \bar{B}$$
$$A\bar{B} + \bar{A}B = \overline{AB + \bar{A}\bar{B}}$$

2.14 A lift door control is to operate in the following manner. When the lift stops at a floor the door will open and a signal is generated that remains on until all the passengers are on or off the lift. An additional signal is also generated to ensure that the doors do not close on a passenger in the doorway. Doors will close if a call button has been pressed on another floor or if a lift passenger has pressed a button for another floor. Set up a truth table for the design of the lift control and derive the corresponding switching equation.

3 Karnaugh maps and function simplification

3.1 Introduction

One of the objectives of the digital designer when using discrete gates is to keep the number of gates to a minimum when implementing a Boolean function. The smaller the number of gates used, the lower the cost of the circuit. Simplification could be achieved by a purely algebraic process, but this can be tedious, and the designer is not always sure that the simplest solution has been produced at the end of the process.

A much easier method of simplification is to plot the function on a *Karnaugh map* (or 'K-map') and with the help of a number of simple rules to reduce the Boolean function to its minimal form. This particular method is very straightforward up to and including six variables. Above six variables it is better to use a tabulation method such as that due to Quine and McCluskey which, after programming, can be run on a computer.

3.2 Minterms and maxterms

As explained in section 2.5, a *minterm* (sometimes called a 'product term' or 'P-term') of n variables is the logical AND of all n variables where any of the n variables may be represented by the variable itself or its complement. In the case of two variables A and B there are four possible combinations of the variables, and these are tabulated in Figure 3.1. Corresponding to these four combinations of the variables there are four possible minterms which can be obtained as follows. In the first row of the table $A = 0$ and $B = 0$, hence $\bar{A}\bar{B} = 1$. The minterm is formed using the values of the variables which make the value of the minterm equal to 1, hence $m_0 = \bar{A}\bar{B}$. The other three minterms are obtained in the same way.

As also explained in section 2.5, a *maxterm* (sometimes called a 'sum term' or 'S-term') of n variables is the logical OR of all n variables where any one of the variables may be represented by its true or complemented form. The maxterms are formed using the values of the variables which make the value of the maxterm equal to 0.

A	B	Minterms	Maxterms
0	0	$m_0 = \bar{A}\bar{B}$	$M_0 = A + B$
0	1	$m_1 = \bar{A}B$	$M_1 = A + \bar{B}$
1	0	$m_2 = A\bar{B}$	$M_2 = \bar{A} + B$
1	1	$m_3 = AB$	$M_3 = \bar{A} + \bar{B}$

Figure 3.1 *The minterms and maxterms of two variables*

Now, for $A = 0$ and $B = 0$ we have that

$$m_0 = \bar{A}\bar{B} = 1, \text{ and } \bar{m}_0 = \overline{\bar{A}\bar{B}} = 0,$$

giving

$$\bar{m}_0 = M_0 = A + B,$$

i.e. the maxterm is the logical complement of its corresponding minterm. The other three maxterms can be obtained by the same method.

For three variables A, B, and C there are eight possible combinations of the variables and consequently there are eight minterms and eight maxterms. If there are n variables there are 2^n possible combinations of those variables and this leads to 2^n minterms and 2^n maxterms. It is clear that the number of minterms and maxterms rises exponentially with n.

One important property of minterms is that the logical OR of all 2^n minterms is equal to logical 1, i.e.

$$\sum_{i=0}^{2^n-1} m_i = 1$$

The dual of this equation is

$$\prod_{i=0}^{2^n-1} M_i = 0$$

where $\prod$ signifies the Boolean product (AND), so that the logical product of all the maxterms is equal to logical zero. For example, in the case of two variables the logical sum (OR) of all the minterms is given by the expression

$$\begin{aligned} \text{Sum} &= \bar{A}\bar{B} + \bar{A}B + A\bar{B} + AB \\ &= \bar{A}(\bar{B} + B) + A(\bar{B} + B) \\ &= \bar{A} + A \\ &= 1 \end{aligned}$$

Taking the dual of the expression for the sum gives

$$(\bar{A} + \bar{B})(\bar{A} + B)(A + \bar{B})(A + B) = 0$$

and this represents the logical product of all the maxterms of two variables.

3.3 Canonical forms

Also mentioned in section 2.5 is the concept of the *canonical form*, a term used to describe a Boolean function that is written either as a sum of minterms, or as a product of maxterms. For example, using three variables A, B, and C, the equation

$$f(A, B, C) = A(B \oplus C) + \bar{A}\bar{B}C$$

is not written explicitly as a sum of minterms (or a product of maxterms) and so is *not* in canonical form. Simple Boolean algebraic manipulation produces the same function in canonical form written as the logical sum of three minterms:

$$f(A, B, C) = AB\bar{C} + A\bar{B}C + \bar{A}\bar{B}C$$

while the following equation is written as the product of three maxterms and so is also in canonical form:

$$f(A, B, C) = (\bar{A} + \bar{B} + \bar{C})(A + B + \bar{C})(A + \bar{B} + C)$$

3.4 Boolean functions of two variables

There are a specific number of Boolean functions of two variables. Each Boolean function in its canonical form will consist of a certain number of minterms; for example, $f(A, B) = \bar{A}B + A\bar{B}$ is a Boolean function of two variables and contains two of the four available minterms. The total number of Boolean functions of two variables can be obtained in the following manner.

Figure 3.2 shows a table in which the presence of a minterm in a two-variable function is indicated by a 1, and its absence by a 0. For example, if the minterm $\bar{A}\bar{B}$ is included in the expression, its presence will be represented by a 1 in the position of that minterm in the table. If not included, its absence will be indicated by a 0. In the case where all four minterms are absent, this will be indicated by a column of four 0s, as shown in the table, and it follows that the corresponding Boolean function will be $f_0 = 0$.

Minterms	f_0	f_1	f_2	f_3	f_4	f_5	f_6	f_7	f_8	f_9	f_{10}	f_{11}	f_{12}	f_{13}	f_{14}	f_{15}
$m_0 = \bar{A}\bar{B}$	0	0	0	0	0	0	0	0	1	1	1	1	1	1	1	1
$m_1 = \bar{A}B$	0	0	0	0	1	1	1	1	0	0	0	0	1	1	1	1
$m_2 = A\bar{B}$	0	0	1	1	0	0	1	1	0	0	1	1	0	0	1	1
$m_3 = AB$	0	1	0	1	0	1	0	1	0	1	0	1	0	1	0	1

Figure 3.2   *Table for determining all the Boolean functions of two variables*

$f_0 = 0$	False
$f_1 = AB$	AND
$f_2 = A\bar{B}$	AND (not *B*)
$f_3 = A$	Identity
$f_4 = \bar{A}B$	AND (not *A*)
$f_5 = B$	Identity
$f_6 = \bar{A}B + A\bar{B}$	Exclusive OR
$f_7 = A + B$	OR
$f_8 = \bar{A}\bar{B} = \overline{A + B}$	NOR
$f_9 = \bar{A}\bar{B} + AB$	Equality
$f_{10} = \bar{B}$	NOT
$f_{11} = A + \bar{B}$	OR (not *B*)
$f_{12} = \bar{A}$	NOT
$f_{13} = \bar{A} + B$	OR (not *A*)
$f_{14} = \bar{A} + \bar{B} = \overline{AB}$	NAND
$f_{15} = 1$	True

Figure 3.3   *The 16 Boolean functions of two variables*

There are two ways in which the entry in the first row can be allocated: it can be either 0 or 1. There are also two ways in which the entry in the second row can be allocated. When combined with the first row allocation this leads to four ways in which the first two rows can be allocated with 0s and 1s. For four rows, it follows that there are $2^4 = 16$ ways in which the 0s and 1s can be allocated. These allocations are shown in Figure 3.2 and the 16 Boolean functions of two variables can be written down immediately from this table and are tabulated in Figure 3.3.

As the number of variables increases, the number of Boolean functions that can be formed increases rapidly. For three Boolean variables there are $2^8 = 256$ possible Boolean functions, for four variables there are $2^{16} = 65\,536$ possible Boolean functions and for n variables there are $2^{(2^n)}$ possible Boolean functions.

3.5 The Karnaugh map

For two variables there are four minterms and these can be conveniently placed on a 'map' as shown in Figure 3.4. The map consists of a square divided into four cells, one for each of the minterms. The possible values of the

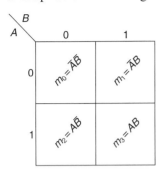

Figure 3.4   *The map for two Boolean variables*

variable A are written down the left hand side of the map, labelling the corresponding rows of the map, while the possible values of the variable B are written along the top of the map, labelling the corresponding columns of the map. Hence, the top left-hand cell represents the minterm where $A = 0$ and $B = 0$, i.e. the minterm $\bar{A}\bar{B}$. The bottom right-hand cell represents the minterm AB where $A = 1$ and $B = 1$. This kind of map is called a *Karnaugh map* or K-map.

Karnaugh maps can be labelled and marked in a variety of ways. For example, each cell can be numbered with the decimal subscript of the minterm that occupies the cell. In this case, the bottom right-hand cell would be numbered with 3, as shown in Figure 3.5(a). The cell numbering shown in Figure 3.5(a) assumes that A is the most significant bit in the binary to decimal conversion, and B the least significant bit. Since A has weighting 2 and B has weighting 1, this is sometimes indicated in abbreviated form as $A, B \equiv 2, 1$ (which is *not* a conventional equation, but merely indicates the respective weights of A and B). Alternatively, the cells can be marked with the binary representation of their corresponding subscript, as shown in Figure 3.5(b). A further possibility for the axis labels is to use $A, \bar{A}, B, \bar{B}$ instead of 0 and 1, as shown in Figure 3.5(c).

For three variables, the map contains eight cells, one for each of the possible minterms as shown in Figure 3.6(a), drawn for the weighting $A, B, C \equiv 4, 2, 1$. The variable A is

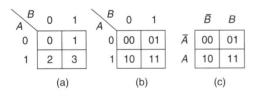

(a)   (b)   (c)

Figure 3.5   *Alternative methods for marking a Karnaugh map*

allocated to the two rows of the map, while the variables B and C are allocated to the four columns. There are four combinations of these two variables, and each combination is allocated to a column of the map.

The columns and rows are allocated in the way shown so that two adjacent columns are always associated with the true value of a variable or, alternatively, its complement. An examination of Figure 3.6(a) shows that the first two columns are associated with $\bar{B}$, the second and third columns are associated with C, and the third and fourth columns are associated with B. The reason for allocating the variables to the columns in this way will be clearer when the procedure for minimisation of a Boolean function is discussed later in this chapter. Note, however, that the column labels along the top of the K-map are the same as the Gray code order for two binary variables (see section 1.21). The reason for this is that the underlying principle of the K-map is that in moving from one cell to an adjacent cell either vertically or horizontally, the value of *one* (and *only one*) Boolean variable may change, and of course similarly Gray codes must change by just one digit only at each step. An alternative method of labelling the axes of a 3-variable K-map is shown in Figure 3.6(b), which

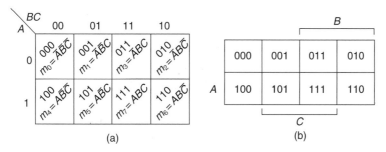

Figure 3.6   *Karnaugh maps for three variables*

makes clear that two adjacent columns are always associated with either the true value or the complement of a variable.

The 4-variable K-map is shown in two forms, differing only in the axis labelling method, in Figure 3.7. Since there are 16 minterms for four variables, the map contains 16 cells and each cell has been marked with the decimal subscript of its respective minterm, using the weighting $A, B, C, D \equiv 8, 4, 2, 1$. Note that in Figure 3.7(a), *both* axes are labelled in Gray code order.

In the case of five variables, it is convenient to use two 16-cell maps rather than one 32-cell map, as shown in Figure 3.8(a). The right-hand map is allocated to the true value of E, while the left-hand map is associated with the complement of variable E.

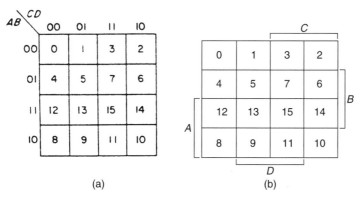

Figure 3.7   *Karnaugh maps for four variables*

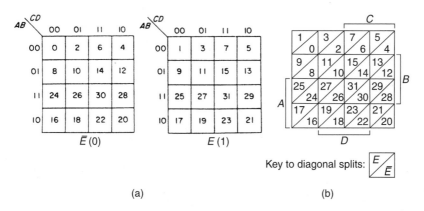

Figure 3.8   *Karnaugh maps for five variables, using the weighting $A, B, C, D, E \equiv 16, 8, 4, 2, 1$*

An alternative is to start with a single 4-variable K-map, and to subdivide each original square cell diagonally, as shown in Figure 3.8(b) to produce a single 32-cell map, so that the cells now become triangles; E is associated with the upper-left triangles, and $\bar{E}$ with the lower-right triangles.

For six variables, there are 64 minterms, and so 64 cells are required; the possibilities are to use four 16-cell maps, or two 32-cell maps, or a single 64-cell map produced by taking a 32-cell map and subdividing each original square cell diagonally again to produce four triangular cells in the space of each original square cell, as shown in Figure 3.9. In each case, all possible combinations of E and F are accommodated uniquely.

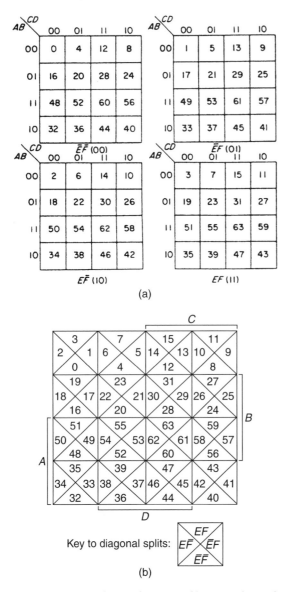

Figure 3.9 *Karnaugh maps for six variables, using the weighting $A, B, C, D, E, F \equiv 32, 16, 8, 4, 2, 1$*

3.6 Plotting Boolean functions on a Karnaugh map

For two variables, the K-map consists of four cells, one for each of the minterms. The function $f = \bar{A}B$ is shown plotted in Figure 3.10(a). It occupies the top right-hand cell of the map, this being indicated by marking the cell with a 1.

Consider now the function $f = A$ which does not depend upon the second variable, B; it is, of course, not in canonical form, as the minterms must involve both variables or their complements. Using the complementation theorem this may be expanded to give the canonical form

Figure 3.10 *(a)* $f = \bar{A}B$ *(b)* $f = AB + A\bar{B} = A$

$$f = A(B + \bar{B}) = AB + A\bar{B}$$

This function is plotted in Figure 3.10(b) and occupies the two cells on the bottom row of the map. Hence, this single-variable function occupies two *adjacent* cells when plotted on a 2-variable K-map.

Since for three variables there are eight minterms, a 3-variable function $f(A, B, C)$ requires an eight-cell K-map, as shown in Figure 3.11(a). The function $f = ABC + \bar{A}\bar{B}\bar{C}$ is shown plotted on this map. The marked cells in this case are not adjacent and this is an indication that the two terms which make up this function cannot be combined to form a simpler function.

The 2-variable term BC plotted on a 3-variable map occupies two adjacent cells, as shown in Figure 3.11(b). This is because BC is the logical sum of the two minterms ABC and $\bar{A}BC$, since

$$f = \bar{A}BC + ABC$$
$$= (\bar{A} + A)BC$$
$$= BC.$$

A one-variable term plotted on a 3-variable map occupies four adjacent cells. For example, the term $f = \bar{C}$ is shown plotted on Figure 3.11(c). An inspection of this map seems to indicate that the four cells are not adjacent. However, it is a fundamental

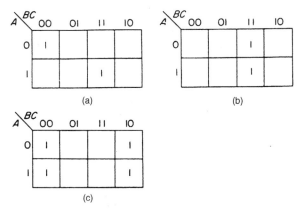

(a)

(b)

(c)

Figure 3.11 *(a)* $f = \bar{A}\bar{B}\bar{C} + ABC$ *(b)* $f = BC$ *(c)* $f = \bar{C}$

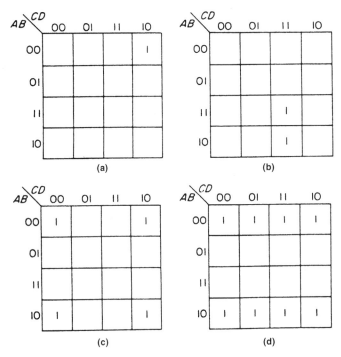

Figure 3.12 $(a) f = \bar{A}\bar{B}\bar{C}D$ $(b) f = ACD$ $(c) f = \bar{B}\bar{D}$ $(d) f = \bar{B}$

principle applying to K-maps for three and more variables that the map may '*wrap around*' in such a way that the right and left ends may be rolled over to form a cylinder with a vertical axis. Alternatively, it may be imagined that the map has been drawn upon a cylindrical tin. If this is done, it is clear that the left-hand and right-hand columns are now adjacent. Alternatively, it will be observed that in the top left-hand cell, the binary representation (000) of the minterm $\bar{A}\bar{B}\bar{C}$ differs by one digit only from 010, the binary representation of the minterm $(\bar{A}B\bar{C})$ in the top right-hand cell. It is a general rule that two minterms that differ in only one variable correspond to adjacent cells on the K-map, as also must binary representations that differ in only one digit. Clearly this also applies to the bottom left-hand and bottom right-hand cells.

As in the cases of the 2-variable and 3-variable maps, on a 4-variable map a minterm occupies one cell, as shown in Figure 3.12(a). Similarly, product terms of three, two and one variables, when plotted on a four-variable map, will occupy two, four and eight adjacent cells, respectively, as shown in Figures 3.12(b, c, d). Inspection of Figure 3.12(d) shows that the top and bottom rows of the K-map may be regarded as adjacent, and, as in the case of the 3-variable map, the first and last columns of the map may also be regarded as adjacent.

3.7 Maxterms on the Karnaugh map

As we have seen, minterms occupy just one cell on any K-map. For example, the minterm $A\bar{B}C$ of three variables is plotted on a 3-variable K-map in Figure 3.13(a).

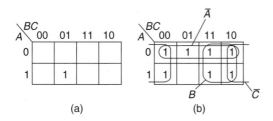

Figure 3.13 *K-map plot of (a) minterm $A\bar{B}C$ (b) maxterm $\bar{A}+B+\bar{C}$*

The name 'minterm' derives from the fact that it is represented by the smallest possible distinguishable area on the map.

A maxterm, such as $\bar{A}+B+\bar{C}$, from section 3.2 is the complement of the corresponding minterm $A\bar{B}C$. Plotting a maxterm on a Karnaugh map requires further consideration. It has been seen earlier that each individual term $(\bar{A},B,\text{and}\bar{C})$ corresponds to four adjacent cells on the map. As explained in section 2.16, the Boolean OR function corresponds closely to the 'set union' operation ($\cup$) performed on areas indicating sets in a Venn diagram, and so the required map area for the maxterm is the *combined* area formed by the *union* of the three areas, one for each individual term. The three individual areas are shown in Figure 3.13(b), and the combined area is that area whose cells are filled with 1. It is clear from Figure 3.13(b) that the resultant combined area for this maxterm is indeed the logical complement of the plot of the corresponding minterm, as shown in Figure 3.13(a). Wherever one of these maps has 1 marked in a certain cell, the other has 0, and wherever it has 0, the other has 1. The name 'maxterm' is obviously derived from the fact that the maxterm occupies all but one cell on any size of K-map; it represents the 'maximum distinguishable area' on the map.

3.8 Simplification of Boolean functions

The process of simplifying a Boolean function with the aid of a K-map is simply a process of finding adjacencies on the function plot. This is best explained with the aid of a very simple example. Suppose that it is required to simplify the Boolean function $f = \bar{A}\bar{B} + A\bar{B} + AB$. Using Boolean algebra alone, it can be readily found that

$$f = \bar{B}(\bar{A} + A) + AB = AB + \bar{B}$$

and at first sight it may be difficult to see any further simplification. However, suppose that f is plotted on a 2-variable K-map, as in Figure 3.14. The function f is specified in canonical sum-of-minterms form, so all that is necessary is to place 1 in each cell corresponding to the minterms in the given expression.

The next stage of the simplification process is to group together adjacent cells containing 1s. (In this context, note carefully that 'adjacent' means 'horizontally or vertically', *not* 'diagonally'.) Therefore, the bottom two cells, corresponding to A alone, may be grouped together. Similarly, the two left-hand cells, corresponding to $\bar{B}$ alone, may also be grouped together, as indicated in Figure 3.14.

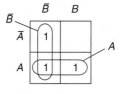

Figure 3.14 *Karnaugh map for the function $f = \bar{A}\bar{B} + A\bar{B} + AB$*

The final stage is to write down the final simplified expression for the function obtained from the groupings thus identified. In this case, therefore, $f = A + \bar{B}$. This is certainly simpler than the previous 'simplest' expression $f = AB + \bar{B}$ obtained using Boolean algebra, but it still may not be obvious that these two expressions are actually representations of the same function. This can be shown immediately by expanding the expression obtained from the K-map as

$$f = A + \bar{B}$$
$$= A(\bar{B} + B) + \bar{B}(\bar{A} + A)$$
$$= A\bar{B} + AB + \bar{B}\bar{A} + \bar{B}A$$
$$= \bar{A}\bar{B} + A\bar{B} + AB$$

and so the original expression for f has been proven equal to the simpler expression obtained from the K-map.

Of course, this is a trivial example, and serves only to illustrate the procedure and its validity. Its power may be demonstrated by examining a more complex example, such as

$$f = \sum 0, 1, 2, 3, 4, 6, 7, 8, 12, 13.$$

Here, the function f has been specified using the 'numerical minterm' canonical form introduced in section 2.5, where f is specified as a sum of the minterms indicated in decimal form. The first stage in the minimisation is to plot f on a Karnaugh map. In doing so, it is necessary to specify the relative weightings of the map variables, and here the weighting $A, B, C, D \equiv 8, 4, 2, 1$ is used.

The K-map form with numerical labels (e.g., Figure 3.7(a)) and the K-map form with direct symbolic axis labels (e.g., Figure 3.7(b)) are alternatives that correspond to the numerical and algebraic methods of expressing Boolean functions. Of course, the final simplified function is always independent of the mechanics used in finding it. Here, as the function is specified in numerical form, the numerical labelling of the map is used. Each numerical minterm corresponds to one cell on the map. The plot of the function is shown in Figure 3.15, in which it is clear that there are four separate encircled adjacencies or 'groups'; three of these are 4-cell adjacencies and one is a two-cell adjacency. The four cells on the top row of the map can be represented by the term $\bar{A}\bar{B}$, the four cells in the first column of the map by $\bar{C}\bar{D}$, the top right-hand four cells by $\bar{A}C$, and the 2-cell adjacency by $AB\bar{C}$. Hence the simplified function may be written

$$f = \bar{A}\bar{B} + \bar{C}\bar{D} + \bar{A}C + AB\bar{C}$$

The enclosed adjacencies are termed the *prime implicants* of the function.

If a prime implicant is needed to ensure full coverage of the plotted function it is termed an *essential prime implicant*. In the preceding example $AB\bar{C}$ is essential, since it is the only prime implicant selected that covers the cell containing the minterm $AB\bar{C}D$ (1101). The other three prime implicants are also essential since they too are covering cells not covered by any other prime implicant. For example, $\bar{C}\bar{D}$ is the only prime implicant covering cell 0100.

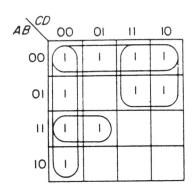

Figure 3.15 $f = \bar{A}\bar{B} + \bar{C}\bar{D} + \bar{A}C + AB\bar{C}$

It is now clear why the four variables were allocated to the columns and rows in the manner shown. The allocation used ensures that cells associated with the variables $\bar{C}, C, \bar{D}$ and D always lie in two adjacent columns while the cells associated with the variables $\bar{A}, A, \bar{B}$ and B always lie in two adjacent rows. If the allocation had been made in strict numerical order, i.e. 00, 01, 10, 11, then the cells associated with D, for example, would not have been in adjacent columns and simplification would no longer have been a process of looking for adjacencies.

Simplification of 5-variable functions is a little more complicated. As an example, consider the function

$$f = \sum 0, 1, 2, 3, 4, 5, 10, 11, 13, 14, 15, 16, 20, 21, 24, 25, 26, 29, 30, 31$$

which is shown plotted in Figure 3.16. The simplification procedure can be carried out as follows:

1. First find the simplified functions for each of the two maps $f_{\bar{E}}$ and f_E in the way previously described:

   $$f_{\bar{E}} = \bar{B}\bar{D}\bar{E} + BD\bar{E} + A\bar{C}\bar{D}\bar{E} + \bar{A}\bar{B}\bar{C}\bar{E}$$

   and

   $$f_E = C\bar{D}E + BCE + AB\bar{D}E + \bar{A}\bar{C}DE + \bar{A}\bar{B}\bar{C}E$$

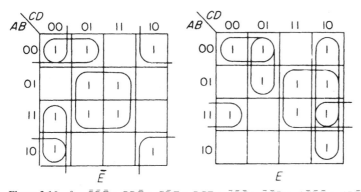

Figure 3.16 $f = \bar{B}\bar{D}\bar{E} + BD\bar{E} + C\bar{D}E + BCE + \bar{A}\bar{B}\bar{C} + \bar{A}\bar{C}D + A\bar{C}\bar{D}\bar{E} + AB\bar{D}E$

2. The second step is to look for possible combinations between prime implicants identified on the two maps that will result in an overall simplification of the logical sum of the two functions, $f_{\bar{E}} + f_E$. For example, $\bar{A}\bar{B}\bar{C}\bar{E}$ is a prime implicant of $f_{\bar{E}}$ and $\bar{A}\bar{B}\bar{C}E$ is a prime implicant of f_E. These two will combine to form one 3-variable term $\bar{A}\bar{B}\bar{C}$. It is also possible to add a *non-essential prime implicant* to the equation for $f_{\bar{E}}$, namely $\bar{A}\bar{C}D\bar{E}$. The cells corresponding to $\bar{A}\bar{C}D\bar{E}$ on the $f_{\bar{E}}$ map have already been covered by prime implicants $\bar{A}\bar{B}\bar{C}\bar{E}$ and $BD\bar{E}$. The non-essential prime implicant $\bar{A}\bar{C}D\bar{E}$ will combine with the essential prime implicant $\bar{A}\bar{C}DE$ on the f_E map to form one 3-variable term $\bar{A}\bar{C}D$. Hence, the equations for $f_{\bar{E}}$ and f_E may be written as follows:

$$f_{\bar{E}} = \bar{B}\bar{D}\bar{E} + BD\bar{E} + A\bar{C}\bar{D}\bar{E} + \underset{\downarrow}{\bar{A}\bar{B}\bar{C}\bar{E}} + \underset{\downarrow}{\bar{A}\bar{C}D\bar{E}}$$
$$f_E = C\bar{D}E + BCE + AB\bar{D}E + \underset{\downarrow}{\bar{A}\bar{B}\bar{C}E} + \underset{\downarrow}{\bar{A}\bar{C}DE}$$

$$\text{forms} \qquad \text{forms}$$
$$\bar{A}\bar{B}\bar{C} \qquad \bar{A}\bar{C}D$$

and the simplified function is

$$f = \bar{B}\bar{D}\bar{E} + BD\bar{E} + C\bar{D}E + BCE + \bar{A}\bar{B}C + \bar{A}CD + A\bar{C}\bar{D}\bar{E} + AB\bar{D}E$$

3.9 The inverse function

In some cases it is more economical to implement the inverse of the function rather than implement the given function. For example, suppose the given function is

$$f(A, B, C, D) = \sum 2, 6, 7, 8, 12, 13$$

This is plotted in Figure 3.17(a) and the simplified function obtained from this map is

$$f = AB\bar{C} + \bar{A}BC + \bar{A}C\bar{D} + A\bar{C}\bar{D}$$

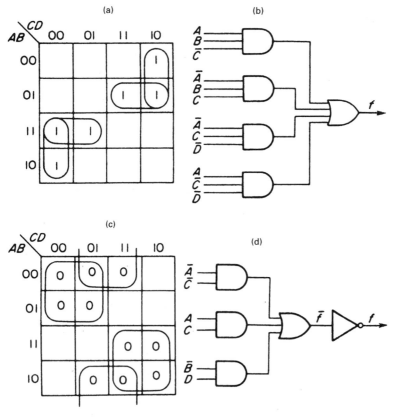

Figure 3.17   *(a) Plot of $f = AB\bar{C} + \bar{A}BC + \bar{A}C\bar{D} + A\bar{C}\bar{D}$ and (b) its implementation (c) Plot of the inverse function $\bar{f} = \bar{A}\bar{C} + AC + \bar{B}D$ and (d) the implementation of f from the inverse function*

To implement this function, four 3-input AND gates and one 4-input OR gate are required as shown in Figure 3.17(b). Besides these gates, the inverses of A, C, and D must be produced, needing three logic inverters, and there are 17 logic signal interconnections in the circuit.

The inverse function is represented by 0s plotted in the unmarked cells of Figure 3.17(a). These cells represent those combinations of the variables for which $f = 0$. For clarity, a separate map in Figure 3.17(c) shows the plot of the inverse function. From this map the simplified form of the inverse function is obtained and is given by the equation

$$\bar{f} = \bar{A}\bar{C} + AC + \bar{B}D.$$

This implementation of f is shown in Figure 3.17(d). In order to generate f, the inverse function $\bar{f}$ is inverted using an inverter. This implementation requires three 2-input AND gates, one 3-input OR gate, and four inverters including the three needed to produce $\bar{A}, \bar{B}$ and $\bar{C}$. The number of logic signal interconnections is 11. Because of the particular form of $\bar{f}$ in this case, this is a much simpler circuit than the original circuit generating f directly in Figure 3.17(b), although generation of the complement of a function may not always be advantageous; selection of the best implementation is discussed further in section 3.24.

3.10 'Don't care' terms

In some logic problems certain combinations of the variables may never occur. For example, the NBCD code tabulated in Figure 3.18(a) is frequently used to represent the decimal digits. This 4-bit code has 16 possible combinations, only ten of which are used. The remaining six combinations, namely 1010, 1011, 1100, 1101, 1110 and 1111, cannot occur in practice unless fault conditions exist, and as a consequence can be used for simplification purposes. Such terms are usually referred to as *'don't care'* or *'can't happen'* terms.

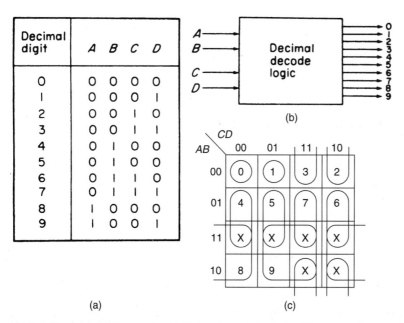

Figure 3.18 *(a) The NBCD code (b) Block schematic for the NBCD to decimal converter (c) Karnaugh map for determining the decimal decode logic*

If the NBCD code is to be converted to give a decimal output, as indicated by the block diagram in Figure 3.18(b), decode logic has to be used. Ten individual decode circuits will be required, one for each decimal digit.

The K-map used for determining the individual functions is shown in Figure 3.18(c). On this map the 'can't happen' or 'don't care' terms are marked X which indicates that the entry in the cell can be 0 or 1, whichever suits the designer. The remaining 10 cells are marked with the decimal digit corresponding to the cell code. For example, the cell defined by $ABCD = 0000$ is marked with the decimal digit 0. This cell cannot be combined with any of the cells marked X, so that the output indicating the decimal digit 0 is given by the minterm $\bar{A}\bar{B}\bar{C}\bar{D}$. Similarly, the cell corresponding to decimal digit 1 cannot be combined with any adjacent cell marked X, so that the output indicating the decimal digit 1 is given by the minterm $\bar{A}\bar{B}\bar{C}D$. However, the cell corresponding to decimal digit 2 is adjacent to one of the 'don't care' cells marked X as shown on the K-map, so that the output indicating the decimal digit 2 is given by the expression $\bar{B}C\bar{D}$ which is simpler than a minterm.

The equations for the remaining decimal digits can be found in the same way, and are given by the following Boolean expressions.

$3 = \bar{B}CD$ $\quad\quad$ $4 = B\bar{C}\bar{D}$ $\quad\quad$ $5 = B\bar{C}D$ $\quad\quad$ $6 = BC\bar{D}$

$7 = BCD$ $\quad\quad$ $8 = A\bar{D}$ $\quad\quad$ $9 = AD$

With the exception of the results for decimal digits 0 and 1, all of these expressions are simpler than the corresponding minterms. A significant simplification has been achieved by exploiting the 'don't care' states in this K-map. This would be reflected in a corresponding simplification of the logic circuits used for performing this decoding.

This particular example is somewhat unusual in the sense that all of the 'don't care' states have been included in the final answer. However, since the whole point of 'don't care' states is that it is irrelevant whether or not they are included in the final simplified expressions, it is quite possible that in other problems some 'don't cares' will not be included. The logic designer has complete freedom to choose whether or not any particular 'don't care' is included, according to the best way of simplifying the final result, as long as it is *absolutely certain* that these states will *never occur*.

Summary of rules for simplifying functions using Karnaugh maps

1. **All 1s** and **no 0s** must be included in groups of cells (unless the inverse function is being implemented, in which case **all 0s** and **no 1s** must be included).
2. Group adjacent (horizontal or vertical) cells only.
3. The allowable group sizes are **1**, **2**, **4**, or **8** (or higher powers of 2) only.
4. To obtain the simplest form, use the **largest size** groups possible. Use the **fewest** groups possible.
5. Use **overlaps** freely to achieve the goals of point 4 above.
6. The 3-variable map can **'wrap around'** horizontally, and the 4-variable map can **'wrap around'** both vertically and horizontally.
7. Include **'don't cares'** within groups as needed to achieve the goals of point 4 above. 'Don't cares' should *not* be included if by so doing the groups are not made larger or fewer.

3.11 Simplification of products of maxterms

As explained in section 2.5, Boolean expressions can be expressed as products of maxterms, sometimes referred to as the 'product of sums', 'P-of-S' or 'POS' form. Except in the simplest cases, these types of expressions are not easy to plot directly on a K-map. However, the inverse function will be directly expressed as a sum of minterms ('sum of products', 'S-of-P' or 'SOP' form) which can then be plotted immediately. The complement of this map (i.e., the cells corresponding to 0s) then represents the complement of the inverse function which is, of course, the original function. For example, suppose

$$f = \bar{A} + \bar{B}$$

and by De Morgan's theorem

$$\bar{f} = AB.$$

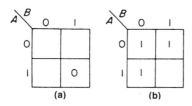

Figure 3.19 (a) $\bar{f} = AB$ (b) $f = \bar{A} + \bar{B}$

This function can be plotted directly on a K-map but in this case the cell containing the term AB is marked with a 0 as shown in Figure 3.19(a). In this simple case the cells representing the original function $f = \bar{A} + \bar{B}$ are marked with 1s and have been plotted directly on the map shown in Figure 3.19(b). It will be observed that the marked cells in this diagram are the unmarked cells in Figure 3.19(a).

If a Boolean function is expressed as a sum of minterms, perhaps with a number of 'don't care' minterms, and the simplest equivalent product of sums is required, then the way to proceed is to plot as 0s the minterms *missing* from the sum of minterms. For example, the zeros of the function $f = \sum 0, 2, 4, 6, 7, 8, 11, 14, 15$, together with 'don't care' minterms 1, 9, 10, 13, are plotted in Figure 3.20; the zeros correspond to the three missing minterms 3, 5, and 12.

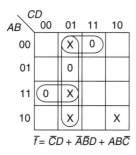

$\bar{f} = \bar{C}D + \bar{A}\bar{B}D + AB\bar{C}$

Figure 3.20 *Plot of the inverse function* $\bar{f} = \bar{C}D + \bar{A}\bar{B}D + AB\bar{C}$, *using the weighting* $A, B, C, D \equiv 8, 4, 2, 1$

The inverse function is shown plotted in Figure 3.20. Simplifying, using the techniques previously described, gives the minimal inverse function:

$$\bar{f} = \bar{C}D + \bar{A}\bar{B}D + AB\bar{C}.$$

Note that in this case, the 'don't care' minterm $(10)_{10}$ is *not* included in the minimised groupings, whereas the other three 'don't cares' *are* included. Taking the complement of this expression (see section 2.4) gives the required simplest (minimal) product of sums:

$$f = (C + \bar{D})(A + B + \bar{D})(\bar{A} + \bar{B} + C).$$

3.12 The Quine–McCluskey tabular simplification method

When a function of more than six variables has to be simplified, mapping techniques become increasingly difficult to employ and alternative methods have to be considered. A commonly used alternative is the Quine–McCluskey tabular method. This technique is tedious, time consuming, and subject to error when performed by hand. However, these difficulties can be overcome by writing a program which allows the simplification process to be run on a computer. The method is based on the complementation theorem which can be applied to the simplification process systematically. This theorem is illustrated by the simple case

$$f = ABC + AB\bar{C}$$

which can be expressed as

$$f = AB(C + \bar{C})$$

and hence the function can be simplified immediately as

$$f = AB.$$

When the two terms ABC and $AB\bar{C}$ are plotted on a K-map (Figure 3.21(a)) it will be observed that since they are occupying adjacent cells they are combinable and will form one 2-variable term.

An alternative way of identifying Boolean terms that will combine is to examine their binary equivalents. For example, the binary equivalents of the two terms in the given equation are $ABC = 111$ and $AB\bar{C} = 110$, and it will be noted that they differ in one digit place only. It is a general rule that if the binary equivalents of two Boolean terms differ in one digit place only, they are combinable. An examination of the K-map in Figure 3.21(a) confirms the rule, since it can be seen that any pair of adjacent cells on this map have a single digit difference between their corresponding binary representations.

If the given equation had been

$$f = \bar{A}\bar{B}C + A\bar{B}\bar{C}$$

then the binary equivalents of the two minterms are $\bar{A}\bar{B}C = 001$ and $A\bar{B}\bar{C} = 100$. Since the binary equivalents of the two terms differ in two digit places they are not

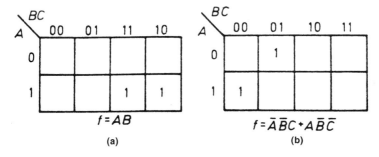

Figure 3.21 *(a) Plot of $f = AB\bar{C} + ABC$, two combinable terms (b) Plot of two non-combinable terms*

combinable. Furthermore, when plotted on the K-map (see Figure 3.21(b)) it will be observed that they do not occupy adjacent cells.

The first step in the Quine–McCluskey method is to tabulate the function to be simplified in sectionalised form such that section 1 lists the single minterm, if present in the function, containing no 1s. Then, section 2 lists any minterms containing one 1, and so on, until section n lists the minterms containing $(n-1)$ 1s. As an example, the 4-variable function

$$f(A, B, C, D) = \sum 0, 1, 2, 5, 6, 7, 9, 10, 11, 14$$

is shown tabulated in Figure 3.22(a) with corresponding binary and decimal equivalents of the minterms in adjacent columns.

The next step is to form all possible combinations between the terms in sections 1 and 2. For example, the term 0000 combines with the term 0001 to form 000–,

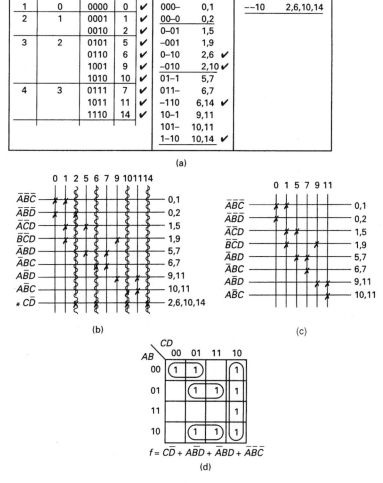

Figure 3.22 *Simplification using Quine–McCluskey tabulation method (a) Tabulation (b) Prime implicant table (c) Reduced prime implicant table (d) Plot of function*

a dash being placed in the position where the combination has occurred. This term appears at the top of the second column of the tabulation and the decimal equivalents 0 and 1 of the combining terms are placed by it. At the same time, terms 0 and 1 are checked off in the first column of the tabulation. When all combinations between sections 1 and 2 have been generated, then all possible combinations between sections 2 and 3 are formed. These combinations are tabulated in the next section of the second column.

When the second column is complete, a third column is formed by combining terms in adjacent sections of the second column. A combination of two terms is only possible if the dash in both terms occupies the same position and only one bit differs. Terms in the second column used to form terms in the third column are checked off and the decimal equivalents of the combining terms are placed at the side of the generated term.

After the tabulation is completed, all those terms that are not checked off are prime implicants of the function. The Boolean form of the prime implicants can be obtained from their binary representations. For example, the first term in the second column of the tabulation is $000- = \bar{A}\bar{B}\bar{C}$, the dash indicating that variable D is missing from this prime implicant. The decimal numbers to the right of the prime implicant indicate the cells it covers on the K-map.

The extracted prime implicants are now used to form the *prime implicant table* shown in Figure 3.22(b). In this table each column represents a minterm of the function and the column is headed by its decimal equivalent. Additionally, a row is placed in the table for each of the prime implicants with their Boolean form appearing at the left-hand end of the rows and the cells that they cover at the right-hand end of the rows. Crosses are entered where a cell column and a prime implicant row intersect provided the cell allocated to the column is covered by the prime implicant allocated to the row.

If a column has only one X in it then the prime implicant corresponding to that X is 'essential'. In Figure 3.22(b) the column headed '14' contains only one X which appears at the intersection with the row allocated to prime implicant $C\bar{D}$ and it follows that $C\bar{D}$ is essential. To indicate this, it is marked with an * in the table. There are four Xs in the $C\bar{D}$ row and the columns associated with them may now be removed from the table since the cells allocated to these columns are covered by this prime implicant.

The table is now redrawn in Figure 3.22(c) with columns 2, 6, 10 and 14 removed, as well as the row for the essential prime implicant $C\bar{D}$. If $A\bar{B}D$ is selected as one of the required prime implicants then cells 9 and 11 are covered, and rows $A\bar{B}D$ and $A\bar{B}C$ can be removed from the table as well as the columns headed 9 and 11. To cover cells 5 and 7 prime implicant $\bar{A}BD$ is selected. Rows $\bar{A}BC$ and $\bar{A}BD$ and also the columns headed 5 and 7 can now be removed from the table. The remaining two cells, 0 and 1, are covered by prime implicant $\bar{A}\bar{B}\bar{C}$ thus eliminating the prime implicants $\bar{A}\bar{B}\bar{D}$, $\bar{A}\bar{C}D$ and $\bar{B}\bar{C}D$ from the solution.

The selected prime implicants are $C\bar{D}, A\bar{B}D, \bar{A}BD$ and $\bar{A}\bar{B}\bar{C}$ and the simplified function is

$$f = C\bar{D} + A\bar{B}D + \bar{A}BD + \bar{A}\bar{B}\bar{C}.$$

The solution is shown plotted in Figure 3.22(d).

3.13 Properties of prime implicant tables

There are two features of prime implicant tables that can be utilised during the function simplification process:

1. *Dominating rows*: an example of a dominating row is shown in Figure 3.23(a). Row P contains all of the minterms contained in row Q, and so row P is said to *dominate* row Q. (The columns associated with row Q are a *subset* of the columns associated with row P.) If row P were selected then the minterms associated with prime implicant Q would be covered, so that therefore row Q can be removed from the table.
2. *Dominating columns*: an example of a dominating column is illustrated in Figure 3.23(b). Minterm S is covered by all of the prime implicants which cover minterm R, so that coverage of cell S is guaranteed by selecting a row that covers minterm R. (The rows covering column R are a *subset* of the rows covering column S.) Therefore, the dominating column, S, may be removed from the table.

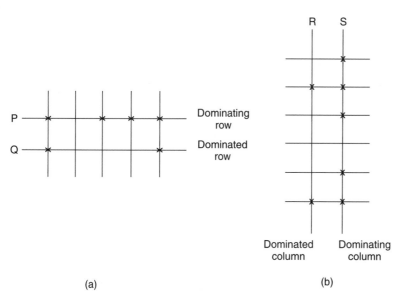

Figure 3.23   *(a) An example of a dominating row (b) An example of a dominating column*

3.14 Cyclic prime implicant tables

A prime implicant table is said to be *cyclic* if

1. It does not have any essential prime implicants, which implies that there are two Xs in every column.
2. There are no dominance relations among the rows and columns.

A typical example of a function which generates a cyclic prime implicant table is shown in Figure 3.24(a). The equation of the function is

$$f = \sum 0, 1, 3, 4, 7, 12, 14, 15.$$

All possible prime implicants of this function are given below:

$$a = \bar{A}\bar{B}\bar{C}(0,1) \qquad e = ABC(14,15)$$
$$b = \bar{A}\bar{B}D(1,3) \qquad g = AB\bar{D}(12,14)$$
$$c = \bar{A}CD(3,7) \qquad h = B\bar{C}\bar{D}(4,12)$$
$$d = BCD(7,15) \qquad i = \bar{A}\bar{C}\bar{D}(0,4)$$

The prime implicants table for the function is given in Figure 3.24(b) and it will be observed that all columns in this table contain two Xs. For this example the prime

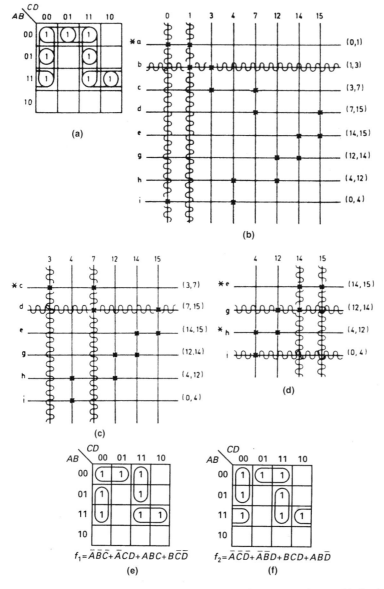

$$f_1 = \bar{A}\bar{B}\bar{C} + \bar{A}CD + ABC + B\bar{C}\bar{D}$$
(e)

$$f_2 = \bar{A}\bar{C}\bar{D} + \bar{A}\bar{B}D + BCD + AB\bar{D}$$
(f)

Figure 3.24 *(a) Plot of f = Σ 0, 1, 3, 4, 7, 12, 14, 15 (b) Prime implicant table for function f (c) Reduced prime implicant table (d) Further reduced prime implicant table (e) Plot of the minimal function f_1 (f) Plot of the alternative minimal function f_2*

implicant associated with row *a* is selected as one of the required prime implicants. Since it covers cells 0 and 1, the corresponding two columns headed 0 and 1 can be deleted from the table. After their removal, row *c* dominates row *b*, so that row *b* may be also deleted. The reduced prime implicants table after these deletions is given in Figure 3.24(c).

Column 3 in the reduced table contains only one X, hence the prime implicant associated with row *c* will form one of the terms of the simplified function and columns 3 and 7 can now be removed from the reduced table. A further reduced prime implicant table is shown in Figure 3.24(d).

Column 15 in this table contains only one X, consequently the prime implicant associated with row *e* will be a term in the simplified function and columns 14 and 15 may now be removed from this table. After removal, row *h* dominates both rows *g* and *i* and they can be deleted too, leaving the prime implicant associated with row *h* as the last term required for a minimal sum which is given by

$$f_1 = a + c + e + h$$
$$= \bar{A}\bar{B}\bar{C} + \bar{A}CD + ABC + B\bar{C}\bar{D}$$

The simplified function is shown plotted in Figure 3.24(e).

If the prime implicant associated with row *i* had been selected initially then the following alternative minimal function would have been obtained:

$$f_2 = \bar{A}\bar{C}\bar{D} + \bar{A}\bar{B}D + BCD + AB\bar{D}$$

and is shown plotted in Figure 3.24(f).

3.15 Semi-cyclic prime implicant tables

A semi-cyclic prime implicant table differs from a cyclic prime implicant table in one respect only. In the cyclic table, the number of cells covered by each prime implicant is identical. For the semi-cyclic table, the prime implicants do not necessarily cover the same number of cells.

An example of the K-map plot of a 5-variable function which generates a semi-cyclic table is shown in Figure 3.25(a), and the corresponding prime implicant table appears in Figure 3.25(b). Examination of the prime implicant table reveals that rows *a, b, c* and *d* each contain four Xs, which means that the corresponding prime implicants consist of three Boolean variables. The remaining rows in the table all contain two Xs and the corresponding prime implicants consist of four Boolean variables. Since each column contains two Xs, a prime implicant has to be selected to start the simplification process. The correct procedure is to select a row containing four Xs. Such a selection may lead to a reduced chip count and certainly would reduce the number of inter-connections. Examination of the prime implicant table shows that if prime implicants *c* and *d* are selected then in order to ensure that all of the cells 0, 2, 8, 9, 10, 11, 16, 17, 18 and 19 are covered, it is only necessary to select one of the prime implicants *a* and *b*. When the columns for these cells are removed from the table simplification continues by making one of the following two selections:

1. Select *h*, then *i* can be removed from the table and the solution is

$$f = (a \text{ or } b) + c + d + (e \text{ or } g) + h + (j \text{ or } k)$$

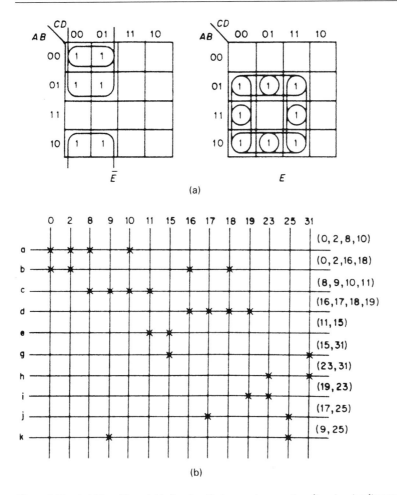

Figure 3.25   *(a) Plot of 5-variable function that generates a semi-cyclic prime implicant table (b) Semi-cyclic prime implicant table*

2. Select *g*, then row *e* can be removed from the table and the solution is

$$f = (a \text{ or } b) + c + d + g + h + (j \text{ or } k)$$

3.16 Quine–McCluskey simplification of functions containing 'don't care' terms

When the initial tabulation is drawn up, the 'don't care' or 'can't happen' terms should be included, since such terms may be covered by the prime implicants of the function. However, when the prime implicant table is constructed, columns do not have to be included for the 'can't happen' terms. These terms do not necessarily have to be covered, although they may be, for the reason given above. All other terms require columns in the prime implicant table since, of necessity, they must be covered. An example of handling 'don't care' terms is given below in the next section, section 3.17.

3.17 Decimal approach to Quine–McCluskey simplification of Boolean functions

The decimal approach to the Quine–McCluskey simplification of Boolean functions provides a simpler tabulation; also, since decimal representation of Boolean terms is employed, errors are less likely to occur, and when they do they are easier to spot. However, the rules for the implementation of this approach are somewhat different. For example, if the decimal difference between a pair of numbers in adjacent sections of the tabulation is a power of 2 and the one in the upper section is less than the one in the lower section, then the terms will combine. An example of the application of this rule follows:

No. of 1s	Decimal	Binary	
0	0	0000	Upper section
1	1	0001	Lower section
	Difference $= 1 = 2^0$	000–	

This alternative approach will be demonstrated by simplifying the function

$$f(A, B, C, D) = \sum 0, 3, 5, 6, 7, 8, 12, 15$$

together with 'don't care' (or 'can't happen') minterms 2, 9, 11, 13, and using the weighting $A, B, C, D \equiv 8, 4, 2, 1$. The first step is to tabulate in decimal form all the minterms specified, including the 'don't cares', sectionalised according to the number of 1s in each minterm, as shown in Figure 3.26(a).

In the second column of the table all those terms that differ by a power of 2 are combined and tabulated, providing the number in the upper section of the first column tabulation is less than the number in the adjacent lower section. The numbers in parentheses by the side of each term in this column represent the power of 2 by which the two combining terms differ. As the terms in the first column are used to form a combination in the second column, they are checked off.

The numbers in parentheses in the second column also indicate which digit has disappeared in the process of combination. To obtain the combined term, the decimal values of the weightings for each binary variable must be used. For example, the first entry in the second column of the tabulation is 0, 2(2), where the (2) indicates that the Boolean variable that is weighted 2 has been eliminated from these two minterms in this combination. The weighting key $A, B, C, D \equiv 8, 4, 2, 1$ shows that it is variable C that has been eliminated. The two original terms in the combination are $0 \equiv \bar{A}\bar{B}\bar{C}\bar{D}$ and $2 \equiv \bar{A}\bar{B}C\bar{D}$, so that after the removal of C the result is $\bar{A}\bar{B}\bar{D}$.

When preparing the third column in the table the conditions for combination are:

1. The number in parentheses for the term in the upper section shall be the same as for the term in the adjacent lower section. For example, the two terms 2, 3(1) (Figure 3.26(a), column 2, section 2) and 6, 7(1) (Figure 3.26(a), column 2, section 3) are candidates for combination as the (1) in both terms indicates that the same digit is missing in each.
2. The difference between the first two digits and the second two digits in the two terms to be combined shall be the same power of 2. For example, for the terms 2, 3(1) and 6, 7(1), the difference between each pair of digits is $4 = 2^2$.

Number of   decimal
   1 s

Number of 1 s	decimal			

```
Number of   decimal
   1 s
                                0,2 (2)
    0        0  √              0,8 (8)              2,3,6,7     (1,4)
    1        2  √              2,3 (1)   √          2,6,3,7     (4,1)
             8  √              2,6 (4)   √          8,9,12,13  (1,4)
    2        3  √              8,9 (1)   √          8,12,9,13  (4,1)
             5  √              8,12 (4)  √          3,7,11,15  (4,8)
             6  √              3,7 (4)   √          3,11,7,15  (8,4)
             9  √              3,11 (8)  √          5,7,13,15  (2,8)
            12  √              5,7 (2)   √          5,13,7,15  (8,2)
    3        7  √              5,13 (8)  √          9,11,13,15 (2,4)
            11  √              6,7 (1)   √          9,13,11,15 (4,2)
            13  √              9,11 (2)  √
    4       15  √              9,13 (4)  √
                              12,13 (1)  √
                               7,15 (8)  √
                              11,15 (4)  √
                              13,15 (2)  √
```

(a)

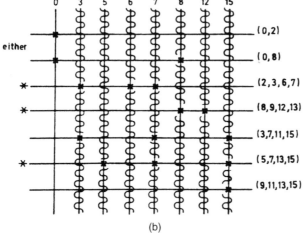

(b)

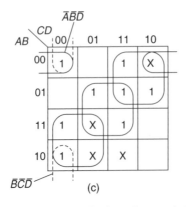

(c)

Figure 3.26 *Simplification using the decimal approach (a) Tabulation (b) Prime implicant table (c) Plot of function $f = \bar{A}C + BD + A\bar{C} + \bar{A}B\bar{D}$ (or $\bar{B}\bar{C}\bar{D}$)*

The second number included in parentheses in the third column indicates the position of the second variable that has been removed by the combination. In the example given, the combined term is 2, 3, 6, 7(1, 4).

The Boolean expression corresponding to minterm 2 is $\bar{A}\bar{B}C\bar{D}$ and the eliminated digits (1, 4), together with the weighting key $A, B, C, D \equiv 8, 4, 2, 1$, indicate that B and D are to be eliminated, and hence the reduced term formed by the combination is $\bar{A}C$.

Terms in the second column are checked off as they are used to form further reduced terms in the third column. When the third column is complete, no further tabulation in a fourth column is possible since there are no terms in the third columns in adjacent sections that have the same numbers in parentheses.

The unchecked terms represent the prime implicants of the function. The prime implicant table shown in Figure 3.26(b) is now constructed and it will be noted that there are no columns for 'can't happen' terms. This table is reduced in the manner previously described and provides two equally simple solutions. The essential prime implicants are

$$2, 3, 6, 7(1, 4) = \bar{A}C$$
$$8, 9, 12, 13(1, 4) = A\bar{C}$$
$$5, 7, 13, 15(2, 8) = BD$$

and either $0, 2(2) = \bar{A}\bar{B}\bar{D}$ or $0, 8(8) = \bar{B}\bar{C}\bar{D}$.

There are two equally simple solutions and they are

$$f_1 = \bar{A}C + A\bar{C} + BD + \bar{A}\bar{B}\bar{D}$$
$$f_2 = \bar{A}C + A\bar{C} + BD + \bar{B}\bar{C}\bar{D}$$

A K-map of the function is plotted in Figure 3.26(c) and the map simplification confirms the two solutions given above.

3.18 Multiple output circuits

Suppose that it is required to produce a circuit with two outputs, one equal to the function $f_1 = \bar{A}\bar{B} + \bar{B}C$ and the other equal to the function $f_2 = AC + BC$. The K-maps for these two functions are shown in Figures 3.27(a) and (b) respectively, and as no further simplification is possible the simplest circuits are shown in Figures 3.27(c) and (d) respectively. Assuming such an independent implementation, a total of six gates, two inverters, and 14 interconnections are required.

However, independent implementation has overlooked the possibility of finding a joint optimal implementation requiring a smaller number of gates and interconnections.

Examination of the two K-maps shows that the minterm $A\bar{B}C$ is common to both maps and is therefore common to both functions. Bearing this commonality in mind, the two functions can be rewritten as

$$f_1 = \bar{A}\bar{B} + A\bar{B}C$$

and

$$f_2 = BC + A\bar{B}C$$

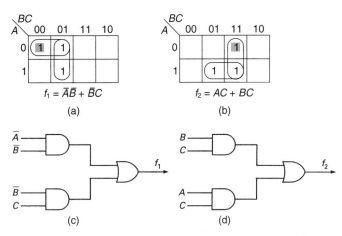

Figure 3.27   *(a) and (b) Karnaugh maps for the functions $f_1 = \bar{A}\bar{B} + \bar{B}C$ and $f_2 = AC + BC$ respectively; (c) and (d) independent implementations of f_1 and f_2 respectively*

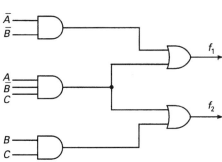

Figure 3.28   *Multiple output circuit showing the optimum implementation of the two functions f_1 and f_2*

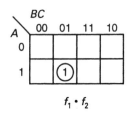

$f_1 \cdot f_2$

Figure 3.29   *Karnaugh map for the product function $f_1 \cdot f_2$*

These modified functions can now be implemented using a common gate to produce the minterm $A\bar{B}C$, as shown in Figure 3.28. This optimum implementation requires only five gates, two inverters, and 13 interconnections.

In general, to minimise several *multiple output functions* simultaneously in this way, the K-maps for *all possible Boolean output products* (ANDs of outputs) must be plotted and examined in addition to the K-maps for each individual function. In the example considered above, this means that the K-maps for the two functions in Figures 3.27(a) and (b) must be supplemented by the K-map of the product function $f_1 \cdot f_2$ as shown in Figure 3.29.

The K-map of the product function results in the multiple output prime implicant $A\bar{B}C$ that is common to both functions, and since this cell has only to be covered once it does not have to be covered again when considering the maps of the individual functions. In fact, if the multiple output prime implicant is definitely to be covered, the corresponding cells can be treated as 'don't cares' on the individual function maps. On the individual function maps (Figures 3.27(a) and (b)) the cells that are shaded are both covered by just one of the prime implicants of the single-output functions. It follows that an essential prime implicant for a particular single-output function must contain a minimum of one cell distinguished in this way.

To minimise multiple output functions with three output lines, f_1, f_2 and f_3, then there are *four* possible product terms that must be considered, i.e. $f_1 \cdot f_2$, $f_1 \cdot f_3$,

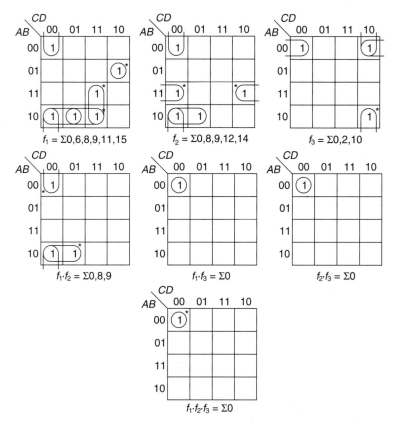

Figure 3.30   *Function and product maps for determining multiple output circuit*

and $f_2 \cdot f_3$, and the three-term product $f_1 \cdot f_2 \cdot f_3$, as well as the three individual function K-maps. A typical problem is illustrated in Figure 3.30, where the prime implicants are shown as loops on each K-map.

The prime implicant of the $f_1 \cdot f_2 \cdot f_3$ map is $\bar{A}\bar{B}\bar{C}\bar{D}$ and this term must appear in each of the three function equations. The selection of this prime implicant implies that it has been covered on all seven maps. Since this is the only term entered on the $f_2 \cdot f_3$ and $f_1 \cdot f_3$ maps no further consideration of these two maps is required. However, on the $f_1 \cdot f_2$ map, cells 8 and 9 remain to be covered so that the term $A\bar{B}\bar{C}$ corresponding to these two cells must appear in the function equations for f_1 and f_2.

The same procedure for selecting prime implicants can be adopted for the three function maps, and the tabulation below gives the prime implicants that have been selected in this particular solution of the problem:

f_1, f_2, f_3 $\bar{A}\bar{B}\bar{C}\bar{D}$

f_1, f_2 $A\bar{B}\bar{C}$

f_3 $\bar{B}C\bar{D}$

f_2 $AB\bar{D}$

f_1 $ACD, \bar{A}BC\bar{D}$

The selected prime implicants are marked with asterisks in Figure 3.30, and combining terms from the above tabulation leads to the following three function equations:

$$f_1 = \bar{A}\bar{B}\bar{C}\bar{D} + A\bar{B}\bar{C} + ACD + \bar{A}BC\bar{D}$$
$$f_2 = \bar{A}\bar{B}\bar{C}\bar{D} + A\bar{B}\bar{C} + AB\bar{D}$$
$$f_3 = \bar{A}\bar{B}\bar{C}\bar{D} + \bar{B}C\bar{D}$$

The implementation of the three functions appears in Figure 3.31. Nine gates and 29 inputs are required for this implementation. If the functions had been individually implemented, 12 gates and 37 inputs would have been required.

It will be seen later, in Chapter 11, that in practice the use of programmable devices such as logic arrays offers an efficient approach to implementing multiple output circuits, but the same design principles are still valid because the most efficient use of a programmable device is obtained by collectively optimising the multiple outputs as explained in this section.

Number of K-maps needed to find minimal implementation of n output functions

To design a minimal implementation for n output functions, clearly a K-map is needed for all the possible product terms involving any *pair* of functions, and any *three* functions, and so on up to and including the final K-map for *all* the functions ANDed together, as well as for all the individual functions concerned. Imagine that each of these K-maps is labelled with an n-bit binary integer, the bits numbered from 1 to n inclusive, corresponding to functions f_1 to f_n respectively. This label is determined by writing a binary 1 in all bits corresponding to the output function(s) contained in the product or function plotted in that K-map, and binary 0 is written in all the other bits. For example, in the case where there are five output functions, the K-map for f_3 is labelled 00100, the map for $f_2 \cdot f_4$ is labelled 01010, and the map for $f_1 \cdot f_2 \cdot f_3 \cdot f_4 \cdot f_5$ is labelled 11111. There is, of course, no K-map labelled 00000. The total number of K-maps is therefore the total number of distinct arrangements of 0 and 1 in n places, excluding the case of all 0s, and therefore corresponding to all possible integers from $(1)_2$ up to and including the largest value label which corresponds to the K-map for the product of all the functions. Adding the number $(1)_2$ to the largest label gives the binary integer represented by 1 followed by n 0s. This binary integer has a value 2^n, so that the number of distinct n-bit labels, including the disallowed value of all 0s, is therefore 2^n. Hence the *total* number of *allowed* K-maps is $2^n - 1$. It follows that the number of *extra* K-maps needed to consider the product terms is given by $2^n - n - 1$.

For four output functions, 15 K-maps in total must be considered and for five output functions, 31 K-maps in total must be considered. Clearly this is a very unwieldy process and for greater than four output functions the minimisation procedure described in this section using K-maps is impractical. In this case, a tabulation method based upon the Quine–McCluskey method should be used.

3.19 Tabular methods for multiple output functions

As a simple example of using tabular methods for large numbers of simultaneous output functions, the same multiple-output problem solved above in section 3.18 using map techniques will be reworked using decimal representation for the minterms.

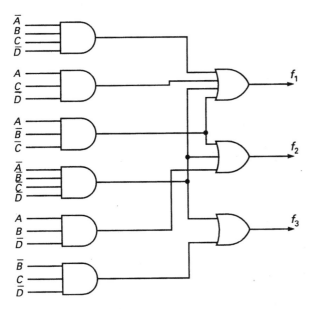

Figure 3.31 *Multiple output implementation for functions defined in Figure 3.30*

As in Figure 3.30, the three required output functions are:

$$f_1 = \sum 0, 6, 8, 9, 11, 15$$
$$f_2 = \sum 0, 8, 9, 12, 14$$
$$f_3 = \sum 0, 2, 10$$

For each decimal minterm, there is a binary *tag* that identifies the output functions which include that minterm. For a total of *n* output functions, the tag is an *n*-bit binary integer where the bits are numbered from 1 to *n* inclusive, corresponding to functions f_1 to f_n respectively. The tag is determined by writing a binary 1 in all bits corresponding to the output function(s) containing that minterm, and writing binary 0 in all the other bits. For example, in the present problem, decimal minterm 14 is included only in function f_2, so its tag is 010.

The method proceeds as shown in Figure 3.32. The first column in Figure 3.32(a) contains all the minterms in the three output functions expressed in decimal form and sectionalised according to the number of 1s in each term. Beside each minterm are the tag columns for f_1, f_2, and f_3.

In the second column all those terms in adjacent sections differing by a power of 2 are combined, provided the number in the upper section is less than the number in the adjacent lower section and that they have 1 entries in corresponding positions in the tag columns. The terms in the first column used to form a combination tabulated in the second column can then be checked off, providing the combination formed in the second column has 1s in the same position in the tag columns as the terms from which it is formed. The number in parentheses by the side of each term in the second column is the power of 2 by which the two combining terms differ. Additionally, this number also indicates the digit that disappeared from the binary representation of the minterms when combination took place.

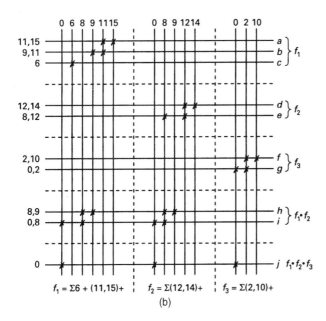

(a)

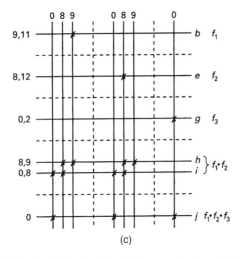

Figure 3.32 *Simplification of multiple output functions (a) Tabulation (b) Prime implicant table (c) Reduced prime implicant table*

To form a third column, the number in parentheses for a term in the upper section must be identical to that of a term in an adjacent lower section. Examination of the second column in Figure 3.32(a) shows that this condition is never satisfied and a third column cannot be formed. The unchecked terms, marked with an asterisk, are the prime implicants of the three output functions.

The prime implicant table shown in Figure 3.32(b) has been divided into three vertical sections, one for each of the output functions. The individual sections contain a vertical line for each of the minterms associated with that function. A row is provided for each of the prime implicants and they have also been sectionalised so that the first horizontal section consists of the prime implicants which are associated with f_1 only, and lower down the table sections have been allocated to those prime implicants which are associated with more than one of the output functions. Xs have been inserted in the table in accordance with the rules described in section 3.12 earlier in this chapter.

Prime implicants a and c can be removed from the table since they are essential for the output function f_1 but for neither of the other two functions. Since cells 6, 11 and 15 are covered by these prime implicants the columns headed by these numbers in the first vertical section of the table can also be removed. Prime implicant d is essential for f_2 only, and can be removed from the table. This leads to the removal of the columns headed 12 and 14 in the second vertical section of the table. Similarly, prime implicant f is essential for f_3 only, and can be removed from the table, allowing the columns headed 2 and 10 in the third vertical section to be removed.

A reduced prime implicant table is shown in Figure 3.32(c). Prime implicant h is essential to cover cell 9 in f_2. h is common to both f_1 and f_2 and will appear in those two output functions. Only cell 0 remains to be covered, and since the corresponding minterm is common to all three functions, the prime implicant j will appear in each of the output equations.

Hence in terms of the prime implicants:

$$f_1 = a + c + h + j$$
$$f_2 = d + h + j$$
$$f_3 = f + j$$

These equations, when written in terms of the Boolean variables A, B, C and D, can be shown to be identical to those obtained in the previous section. Implementation of these functions is illustrated in Figure 3.31.

3.20 Reduced dimension maps

K-maps are only useful up to and including functions with six variables, but in the case of a function having a larger number of variables and providing the function does not contain too many terms it can be useful to plot it on a *reduced dimension map* (RDM). Such a map is one in which the individual cells can now contain variables, so that a map for n variables can be used to represent functions having $n + 1$ or even $n + 2$ variables.

Consider the four-variable map shown in Figure 3.33(a). The two-cell loops on this map occupy those cells where the combinations of the variables A, B, and C are constant. For example, in the top left-hand corner of the map the two cells looped correspond to $A = 0, B = 0$, and $C = 0$. In effect, this K-map consists of eight sub-maps, each sub-map being used to plot the single variable D. However, the

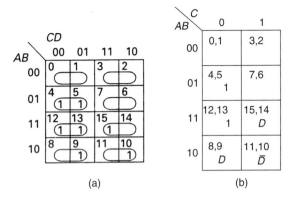

Figure 3.33 *(a) 4-variable map showing 2-cell loopings (b) Reduced dimension map for function plotted in (a)*

same information may also be displayed in a map of eight cells by indicating the value of the function in each sub-map by a *map-entered variable* (MEV) written in one cell. So, to reduce the dimensionality of the map by the single variable D, all the MEVs must be one of the four possibilities D, $\bar{D}$, 0, or 1 as determined by the following conversion table applied to each sub-map in turn and using the sub-map's values of A, B, and C.

Value of function for...		
$D=0$	$D=1$	MEV
0	0	0
0	1	D
1	0	$\bar{D}$
1	1	1

The resultant map is known as a *reduced dimension map* or RDM, and for the function shown in Figure 3.33(a) the RDM is shown in Figure 3.33(b). Each cell of the RDM corresponds to the appropriate two adjacent looped cells in the full map shown in Figure 3.33(a). By using this nomenclature, the same function has now been plotted on a 3-variable K-map whereas before, it was plotted on a 4-variable K-map.

It is also possible to start with a map that already contains MEVs and reduce its dimensions further. For example, the RDM shown in Figure 3.34(a) contains the MEV E and it is required to reduce the map dimension from four to three. Looping cells, four separate terms can be identified, and they are:

Term $p = (\bar{A}\bar{B}C)\bar{D}E$

Term $q = (\bar{A}B\bar{C})\bar{D}E + (\bar{A}B\bar{C})D$

$\qquad = (\bar{A}B\bar{C})(D + E)$

Term $r = (AB\bar{C})\bar{D}E$

Term $s = (A\bar{B}\bar{C})\bar{D} + (A\bar{B}\bar{C})D\bar{E}$

$\qquad = (A\bar{B}\bar{C})(\bar{D} + \bar{E})$

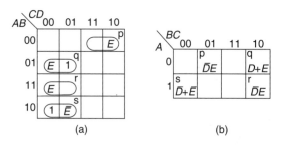

(a)   (b)

Figure 3.34 *(a) Five-variable function plotted on a 3-variable map (b) Further dimension reduction of function plotted in (a)*

This 5-variable function plotted on a 3-variable RDM is shown in Figure 3.34(b). In this case the axes of the K-map have been labelled in the same way as previous 3-variable K-maps (e.g., Figure 3.6) which disguises the one-to-one correspondence between cells in the RDM and pairs of cells in the previous 4-variable map.

3.21 Plotting RDMs from truth tables

A 3-variable function is defined by the truth table shown in Figure 3.35. It is required to plot this function on a 2-variable map with the third variable C being designated as the MEV. The table is first divided horizontally and vertically by the dotted lines to give the four possible combinations of the variables A and B. Entries are then made in the appropriate cell of the RDM for each of these four combinations. For example, with $A = 0$ and $B = 1$, the table shows that $f = 1$ when $C = 1$, and that $f = 0$, when $C = 0$, so C is entered in the appropriate cell of the RDM. The other entries are obtained in the same way. For both minterms $AB\bar{C}$ and ABC, f is listed as a 'don't care' and so this is the entry on the RDM. Using the same method, a more complex example (that of plotting a 4-variable function on a 2-variable RDM) is shown in Figure 3.36. Firstly, each cell on the K-map is identified with a 4-line section of the truth table, where variables A and B are constant. Next, the entry in the RDM is calculated as shown. 'Don't cares' are indicated by writing X after the variables giving the 'don't care' state. For the cell identified by $A = 0$, $B = 1$ (the top right-hand cell), the original function is indicated by the entry $\bar{C}\bar{D} + C\bar{D}$X. A simple 2-variable K-map of variables C and D shows that this may be simplified to $\bar{D}$ although this loses some information (i.e., the 'don't care' minterm $\bar{A}BC\bar{D}$) about the original function f.

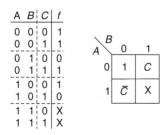

A	B	C	f
0	0	0	1
0	0	1	1
0	1	0	0
0	1	1	1
1	0	0	1
1	0	1	0
1	1	0	X
1	1	1	X

Figure 3.35 *Development of an RDM from a truth table*

A	B	C	D	f
0	0	0	0	0
0	0	0	1	1
0	0	1	0	0
0	0	1	1	1
0	1	0	0	1
0	1	0	1	0
0	1	1	0	X
0	1	1	1	0
1	0	0	0	1
1	0	0	1	0
1	0	1	0	1
1	0	1	1	1
1	1	0	0	0
1	1	0	1	0
1	1	1	0	0
1	1	1	1	X

Figure 3.36 *Plotting a 4-variable function on a 2-variable RDM*

Therefore, in this case, either entry can be used depending on the degree of sophistication required.

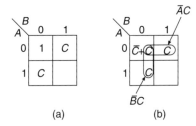

(a)   (b)

Figure 3.37   *Reading an RDM*
(a) The original plot (b) Looping the
Cs to form the terms $\bar{A}C$ and $\bar{B}C$

3.22 Reading RDM functions

A three-variable function has been plotted on a 2-variable RDM in Figure 3.37(a) with C being the MEV. The entry 1 on the map can be replaced by $C + \bar{C}$ and adjacent identical MEVs can be looped as shown in Figure 3.37(b). The loops thus formed represent the terms $\bar{A}C$ and $\bar{B}C$, so that the function can be written as

$$f = \bar{A}\bar{B}(\bar{C} + C) + \bar{A}BC + A\bar{B}C$$
$$= \bar{A}\bar{B}(\bar{C} + C + C + C) + \bar{A}BC + A\bar{B}C$$
$$= \bar{A}\bar{B}(\bar{C} + C) + \bar{A}C(\bar{B} + B) + \bar{B}C(\bar{A} + A)$$
$$= \bar{A}\bar{B} + \bar{A}C + \bar{B}C$$

and the final result can also be written down directly from Figures 3.37(a) and (b) by inspection.

3.23 Looping rules for RDMs

To develop the principles presented in elementary form in the previous section (section 3.22), this section lists the rules for reading and simplifying a function presented on an RDM. Figure 3.38 shows some typical examples.

1. Cells containing identical entries may be looped together according to the usual K-map rules (see section 3.10) and the entry (common to all the looped cells)

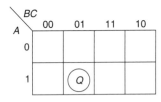

(a) Encircled term read as $A\bar{B}CQ$

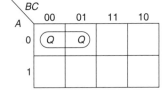

(b) Encircled term read as $\bar{A}\bar{B}Q$

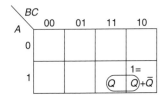

(c) Encircled term read as ABQ

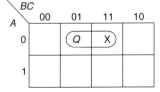

(d) Encircled term read as $\bar{A}CQ$

Figure 3.38   *Looping rules for RDMs*

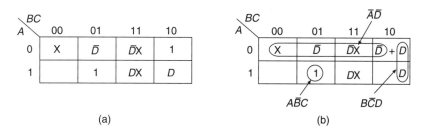

Figure 3.39 *(a) Typical 3-variable RDM for a 4-variable function (b) RDM after minimisation*

is ANDed with the usual result from the looping. Thus, in Figures 3.38 (a) and (b) the contributions to the function are $A\bar{B}CQ$ and $\bar{A}\bar{B}Q$ respectively.

2. A cell entry of Q adjacent to a cell containing 1 may be looped together to give Q ANDed with the OR of the two cells, together with the usual result from the cell containing 1. This is the situation treated in section 3.22, and further illustrated in Figure 3.38(c) where the contribution to the function is $Q(AB\bar{C} + AB\bar{C}) + AB\bar{C} = ABQ + AB\bar{C}$. Similar results apply to larger groups of cells with identical entries, if they are adjacent to cells containing 1 that may be used to obtain an allowable group size.

3. A cell entry of Q adjacent to a cell containing a 'don't care' or X may be looped together to give Q ANDed with the OR of the two cells. This is illustrated in Figure 3.38(d) where the contribution to the function is $\bar{A}CQ$. Larger groups of cells with either entries of Q or entries of X give the usual looping result ANDed with Q.

To illustrate these rules further, Figure 3.39 shows an RDM for a 4-variable function. Term $\bar{A}\bar{D}$ is obtained by looping the entire top row; term $B\bar{C}D$ is obtained by looping the two right-hand cells, and the term $AB\bar{C}$ is not needed as essentially it has been included in the two previous loops in the form $AB\bar{C}(\bar{D} + D)$. Finally, the term $A\bar{B}C$ cannot be combined with any other cells in a way that simplifies the result. Therefore, the final result is $f = \bar{A}\bar{D} + B\bar{C}D + A\bar{B}C$.

3.24 Criteria for minimisation

During the infancy of Digital Logic design, Boolean functions were typically implemented by using individual logic gates, perhaps made using discrete components. The use of K-maps and Quine–McCluskey minimisation techniques were of direct importance in developing all but the most trivial of designs, to economise on the number of components used. As the technology developed from those early days, firstly integrated circuit gates of various types were produced, and then successively larger scale integration has been used in developing integrated circuits containing greater numbers of logic gates, and capable of progressively more sophisticated logic functions. 'SSI' (small scale integration) chips contained just a handful of individual gates on one chip; 'MSI' (medium scale integration) chips contained a number of more complex functions, such as flip-flops; 'LSI' (large scale integration) and 'ELSI' (extra-large scale integration) chips contained the equivalent of several thousand conventional gates, typically arranged to function as a specialised logic unit, such as a basic calculator, and 'VLSI' (very large scale integration) chips contain much greater numbers of logic gates, as exemplified by current microprocessor chips and similar components containing the equivalent of millions of logic gates.

Amongst VLSI chips are the so-called *programmable logic devices* or PLDs which can be used to implement custom Boolean logic functions, and for all but the smallest logic designs these chips are currently the method of choice for implementing a new logic design. These chips are of enormous importance and are covered in detail in Chapter 11. However, a logic designer still needs to be familiar with the basics of function minimisation in order to understand the fundamental processes involved in programming a PLD, and also if the designer becomes involved in the gate-level design of a new custom VLSI chip. In addition, for smaller designs which, in principle, can use any type of logic gate and that are impractical or uneconomic to implement using PLDs, optimisation using Karnaugh maps or Quine–McCluskey minimisation in order to obtain the simplest possible circuit is one of several principles that might be used in the design. However, other design criteria include:

1. minimal cost – it may be cheaper to use certain components as opposed to others,
2. minimised number of gates – one interpretation of the 'simplest' solution,
3. smallest size – depending upon whether chips are available that can directly implement parts of the functions required,
4. minimised chip count – depending upon which chips are available and how their internal gate structure can be used,
5. minimised number of chip-to-chip interconnections, which are a source of unreliability,
6. the use of only one type of gate (e.g. only NAND gates) in order to reduce the number of standard parts that must be stored in case of malfunction (see also Chapter 4), and
7. minimal propagation delay (see chapter 4) – in cases where the very fastest circuit operation is required.

Which of these design principles is used in practice depends largely upon the function of the circuit being designed, its intended use, and its intended market. In some cases, it may be necessary to experiment with several equivalent circuit designs in order to choose the one that best meets the chosen criteria.

Problems

3.1 Expand the following Boolean functions into their canonical form:

(a) $f_1(A, B, C) = \bar{A}B + C$

(b) $f_2(A, B, C) = AB + \bar{A}C + A\bar{B}C$

(c) $f_3 = B + CD + AB\bar{D} + \bar{A}\bar{B}C\bar{D}$

3.2 Simplify the following three-variable Boolean functions algebraically:

(a) $f_1 = \sum 1, 2, 5, 6$

(b) $f_2 = \sum 0, 1, 2, 3, 7$

(c) $f_1 = \sum 3, 5, 6, 7$

3.3 (a) Express the three-variable function $f = \sum 0$, 1 as a product of maxterms.

(b) Express the three-variable function $f = \prod 0$, 1, 2, 5, 6, 7 as a sum of minterms.

(c) Determine the inverse function of $f = \sum 3$, 5, 6, 7 and express it as a product of maxterms.

3.4 Find the minimised sum-of-products expression equal to:

(a) $f_1(A, B, C) = \sum 0$, 1, 3, 4, 6, 7
(b) $f_2(A, B, C, D) = \sum 0$, 1, 2, 3, 7, 8, 9, 11, 12, 15
(c) $f_3(A, B, C, D) = \prod 0$, 4, 5, 6, 7, 8, 9, 10
(d) $f_4(A, B, C, D, E) = \sum 0$, 1, 3, 5, 6, 7, 8, 9, 10, 15, 16, 20, 21,
$\qquad\qquad$ 22, 23, 24, 25, 28, 29, 30, 31.

3.5 Minimise the following functions using the 'don't care' terms for simplification wherever possible:

(a) $f(A, B, C) = \sum 3$, 5 with 'don't care' terms 0, 7
(b) $f(A, B, C, D) = \sum 1$, 2, 3, 5, 6, 7, 10, 11 with 'don't care' terms 9, 12, 15
(c) $f(A, B, C, D) = \prod 0$, 4, 7, 11, 14 but terms 6, 8, 9, 13 are 'don't cares'
(d) $f(A, B, C, D, E) = \sum 4$, 5, 6, 7, 12, 14, 16, 20, 21, 24, 26, 27, 31 with
$\qquad\qquad$ 'don't care' terms 0, 11, 19, 22, 30

3.6 Find the minimised product-of-sums expression equal to:

(a) $f(A, B, C) = \sum 0$, 1, 2, 5, 7
(b) $f(A, B, C, D) = \sum 0$, 1, 9, 10, 11
(c) $f(A, B, C, D, E) = \sum 1$, 2, 5, 6, 10, 11, 14, 15, 16, 17, 20, 21
(d) $f(A, B, C, D) = \sum 5$, 7, 9, 10, 11 with 'don't care' terms 2, 13, 15

3.7 Find the minimised sum-of-products expression for the logical product $F = F_1 F_2$ of the following pairs of functions:

(a) $F_1(A, B, C, D) = \sum 1$, 3, 5, 7
$\quad F_2(A, B, C, D) = \sum 2$, 3, 6, 7
(b) $F_1(A, B, C, D) = \sum 1$, 3, 5, 6, 8, 10, 11, 12, 13
$\quad F_2(A, B, C, D) = \sum 0$, 3, 5, 8, 9, 11, 13, 15
(c) $F_1(A, B, C) = \prod 0$, 3, 6, 7
$\quad F_2(A, B, C) = \prod 1$, 3, 7

3.8 The XS3 code is used to represent the ten decimal digits. Develop the decode logic for converting from XS3 to decimal.

3.9 Minimise the following functions using the Quine–McCluskey tabular method:

(a) $f(A, B, C, D) = \sum 0$, 1, 3, 6, 9, 10, 11, 12, 14, 15
(b) $f(A, B, C, D, E) = \sum 0$, 1, 5, 8, 11, 12, 14, 16, 20, 21, 25, 27, 28, 30, 31
$\qquad\qquad$ with 'don't care' terms 2, 7, 13, 22, 23
(c) $f(A, B, C, D) = \prod 0$, 3, 4, 5, 11, 12, 13, 15 but terms 2, 6, 8 'can't happen'.

3.10 Minimise the following functions using the Quine–McCluskey decimal tabulation method:

(a) $f(A, B, C, D) = \sum 2, 3, 4, 7, 8, 11, 13, 14$ with 'can't happen' terms 1, 5, 10

(b) $f(A, B, C, D, E) = \sum 0, 1, 2, 3, 5, 11, 12, 13, 17, 19, 20, 22, 23, 25, 27, 28,$ 29, 31 with 'can't happen' terms 7, 15, 21

(c) $f(A, B, C, D) = \prod 2, 4, 6, 7, 8, 11, 12, 13, 15$

3.11 Plot the following 4-variable function on a 3-variable RDM.

$f(A, B, C, D) = \sum 1, 3, 7, 8, 10, 11, 13, 14$

3.12 Plot the following 4-variable function on a 2-variable RDM.

$f(A, B, C, D) = \sum 0, 2, 5, 7, 9, 10$

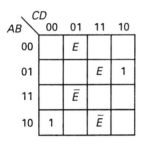

Figure P3.13

3.13 Reduce the 4-variable RDM shown in Figure P3.13 to a three-variable RDM.

3.14 Construct a truth table for the function $f(A, B, C, D) = \sum 0, 1, 5, 6, 11, 12, 14, 15$ with 'don't care' terms 3, 7, 9 and develop a 2-variable RDM with the aid of the truth table.

3.15 Determine the Boolean function represented by the following RDM maps.

(a)

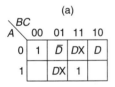

(b)

(c)

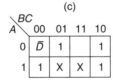

Figure P3.15

4 Combinational logic design principles

4.1 Introduction

The gates dealt with in the two preceding chapters have been the AND, OR and NOT gates. These gates are the easiest to handle using the formal methods of Karnaugh maps and Quine–McCluskey minimisation, but in practice, logic circuits are often actually implemented using NAND and NOR gates. Historically this was because these gates were the easiest to fabricate using readily available logic technologies, and in the case of certain technologies currently at the research stage these types of limitations are still present. Although AND and OR gates are also available using most types of SSI technologies, there is a smaller selection of them, they may be more expensive, and they may have slightly poorer performance (e.g. longer *propagation delay*, the short delay time introduced by a logic gate). Simple combinations of gates are also available in the mature SSI technologies, such as the AND-OR-INVERT (AOI) function and the expandable AND-OR (AO) function. Other derived logic functions are also commonly available now, such as Exclusive-OR (XOR) gates and Exclusive-NOR (XNOR) gates. The purpose of this chapter is to describe design techniques used for implementing actual logic designs, building upon the theoretical approaches of Chapter 3.

4.2 The NAND function

The NAND function is defined by the equation

$$f = \overline{(A \cdot B)}$$

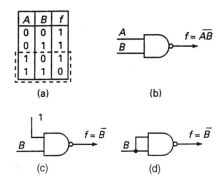

(a)

(b)

(c)

(d)

Figure 4.1 *(a) Truth table for the NAND function (b) Conventional circuit symbol for a NAND gate (c) and (d) The NAND gate used as an inverter*

and the truth table for the function is given in Figure 4.1(a). This table shows that the output of the gate is 1 if either or both inputs are 0 and that the output is 0 only if both inputs are 1. Using de Morgan's theorem (section 2.12) it is clear that an alternative equation for the NAND function is

$$f = \overline{(A \cdot B)} \equiv \bar{A} + \bar{B}.$$

The usual symbol for a two-input NAND gate is shown in Figure 4.1(b).

If the A input of the gate is permanently connected to logic 1 level, then clearly the output is given by

$$f = \overline{(A \cdot B)} = \overline{(1 \cdot B)} = \bar{B}$$

so that the NAND gate is now acting as an inverter. This can also be observed directly from the truth table. The only relevant rows in the truth table are those enclosed by the dotted lines in Figure 4.1(a). An examination of these rows shows that if $B = 0$ then $f = 1$, and if $B = 1$ then $f = 0$. Another way of achieving logic inversion using a NAND gate is by connecting both inputs to the same logic level, whence if $A = B$ then

$$f = \overline{(A \cdot B)} = \overline{(B \cdot B)} = \bar{B}.$$

These two connections are illustrated in Figures 4.1(c) and (d). There are two important provisos that must be emphasised at this point:

1. All inputs to real logic gates must be connected to a well-defined logic level, either 0 or 1, at all times. If a logic gate input is left unconnected, the gate will either operate erratically or may even be destroyed through excessive power dissipation caused by transient input voltage levels outside the gate's design limits. Any unused inputs to a NAND gate should be connected to logic 1 level, and this is often achieved by connecting all the appropriate inputs to the positive supply rail through resistors whose exact value is unimportant but is usually around 10kΩ. In principle, gate inputs may be connected directly to the supply rail without using a resistor, but this is not usually recommended as the resistor affords a measure of protection to the delicate gate input against large voltage surges and spikes on the power supply rails. Alternatively, an unused input may be connected to one of the used inputs, but with some risk of reduced performance, as explained in point 2 below.
2. In some logic families, there is a small speed penalty, usually measured in nanoseconds, but significant in certain circumstances if logic inputs are commoned (i.e., using the inversion circuit of Figure 4.1(d)). The technical reason for this is the increased effective capacitance of two gate inputs, compared with that of a single gate input, that must be driven by the preceding gate, and which therefore takes slightly longer to charge to the correct logic level. It is therefore preferable to use only the inversion circuit of Figure 4.1(c) if operating speed is likely to be an important issue in the design.

4.3 NAND logic implementation of AND and OR functions

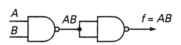

Figure 4.2 *NAND logic implementation of the AND function*

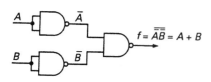

Figure 4.3 *NAND logic implementation of the OR function*

Implementation of the AND function using NAND gates alone is achieved by connecting two NAND gates in cascade, as shown in Figure 4.2. The second NAND gate acts as an inverter, and converts the circuit's function from NAND to AND.

If $f = \overline{(\bar{A} \cdot \bar{B})} \equiv A + B$, then the OR function can be implemented by performing the NAND operation on the inverted variables. The same deduction can also be made from a Karnaugh map of the OR function. The implementation of the OR function using NAND gates is illustrated in Figure 4.3.

4.4 NAND logic implementation of sums-of-products

Figure 4.4(a) shows a straightforward implementation of the function $f = AB + CD$ using AND/OR logic. The diagram shows that there are two levels of logic in this circuit, the first level consisting of the OR gate and the second of the two AND gates; therefore, this function and similar functions are referred to as *two-level-sum-of-products* expressions.

This circuit can be translated into a NAND circuit by using the transformations developed in section 4.3 above. The translation is shown in Figure 4.4(b), where the first block enclosed by dotted lines represents the two AND gates, and the second block constitutes the OR gates. It can be seen that in both branches of the circuit there are two single input NAND gates in cascade and these will simply produce a double inversion of the signals AB and CD. As a consequence, the four gates shown crossed through are redundant, and the circuit reduces to that shown in Figure 4.4(c). This diagram shows that there is a one-to-one translation from the AND/OR configuration to the corresponding NAND configuration.

An even more complicated function such as

$$f = (A + \bar{B}D)C + (C + \bar{D})(A + C)B$$

can be regarded as a two-level sum-of-products since it can be expressed in the following form:

$$f = PQ + RST$$

where

$$P = A + \bar{B}D, \quad Q = C$$

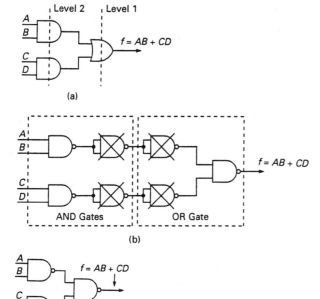

(a)

(b)

(c)

Figure 4.4   *(a) The function $f = AB + CD$ implemented with AND/OR logic (b) Transformation of the AND/OR circuit to a NAND circuit (c) The simplest NAND implementation of $f = AB + CD$*

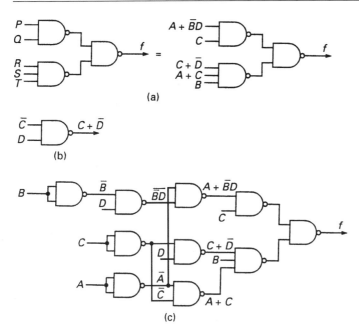

Figure 4.5   *(a) Basic circuits for the implementation of* $f = (A + \bar{B}D)C + (C + \bar{D})(A + C)B$ *(b) The NAND logic implementation of* $C + \bar{D}$ *(c) The NAND logic implementation of* $f = (A + \bar{B}D)C + (C + \bar{D})(A + C)B$

and

$$R = C + \bar{D}, \quad S = A + C, \quad T = B.$$

Hence, the implementation must be of the form shown in Figure 4.5(a).

In order to generate a term such as $R = C + \bar{D}$ using a NAND gate, the required expression is rewritten (using De Morgan's theorem) as $R = \overline{(C + \bar{D})} = \overline{(\bar{C} \cdot D)}$. That is, to the inputs of a NAND gate are connected the *inverses* of the variables that must be summed by the NAND gate, as shown in Figure 4.5(b). The complete circuit for the implementation of the given function is shown in Figure 4.5(c).

The technique described above for the implementation of a Boolean function using NAND gates alone does not necessarily lead to the minimal NAND implementation. However, sometimes by using a factorisation process it is a simple matter to produce a NAND implementation which leads to a circuit that requires a smaller number of gates. For example, consider the function

$$f = A\bar{C} + A\bar{B} + \bar{C}D.$$

Direct implementation of this function as a two-level sum-of-products circuit leads to the circuit shown in Figure 4.6(a) which requires one 3-input NAND gate, three two-input NAND gates and two NAND gates connected as inverters.

However, the expression can also be written in the form

$$f = A(\bar{B} + \bar{C}) + \bar{C}D$$

and this function can be implemented using five NAND gates, as shown in Figure 4.6(b). Alternatively, the expression may be written in the form

$$f = \bar{C}(A + D) + A\bar{B}$$

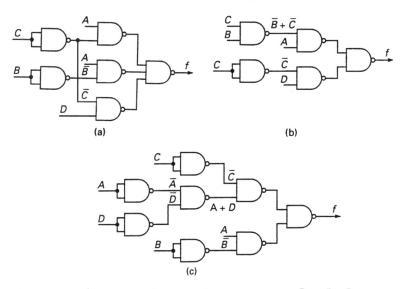

Figure 4.6   *(a), (b) and (c) Three ways of implementing $f = A\bar{C} + A\bar{B} + \bar{C}D$*

and this can be implemented in the form shown in Figure 4.6(c), which requires eight NAND gates.

In this example the implementation shown in Figure 4.6(b) uses the smallest number of NAND gates; it requires three levels of logic, as does the circuit in Figure 4.6(a). Factorisation of a Boolean function will lead to an increase in the number of logic levels required, and consequently this will increase the propagation delay through the circuit. The shortest delay time is always obtained with the two-level sum-of-products implementation. If complemented variables are available then the circuit of Figure 4.6(a) would provide minimum propagation delay since only two levels of logic would be required.

4.5 The NOR function

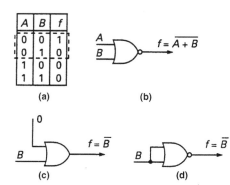

Figure 4.7   *(a) Truth table for the NOR function (b) Conventional circuit symbol for a NOR gate (c) and (d) The NOR gate used as an inverter*

The NOR function is defined by the equation

$$f = \overline{A + B}$$

which, by using De Morgan's theorem, can be alternatively expressed as

$$f = \bar{A} \cdot \bar{B}.$$

The truth table is shown in Figure 4.7(a) and the conventional symbol used to represent the gate is shown in Figure 4.7(b). An examination of the truth table shows that if any one, or both, of the inputs are 1 the gate output is 0, while the output is only 1 provided both inputs are 0.

If the input A of the gate is permanently connected to logic 0 level then clearly the output is given by

$$f = \overline{(A + B)} = \overline{(0 + B)} = \bar{B}$$

so that the NOR gate is now acting as an inverter. This can also be observed directly from the truth table. The only relevant rows in the truth table are those enclosed by the dotted lines in Figure 4.7(a). An examination of these rows shows that if $B = 0$ then $f = 1$, and if $B = 1$ then $f = 0$. Another way of achieving logic inversion using a NOR gate is by connecting both inputs to the same logic level, whence if $A = B$ then

$$f = \overline{(A + B)} = \overline{(B + B)} = \bar{B}$$

These two connections are illustrated in Figures 4.7(c) and (d).

As noted above in section 4.2, unused inputs to any gate must never be left 'floating' or unconnected. Unused NOR gate inputs are often connected to logic 0 level, usually achieved in practice simply by connecting the input directly to the ground (0V) of the digital logic circuit. Alternatively, unused inputs in a NOR gate can be connected to one of the used inputs, but again with the risk of reduced performance, as explained above in section 4.2.

4.6 NOR logic implementation of AND and OR functions

The implementation of the OR function using NOR gates is achieved by connecting two NOR gates in cascade, as shown in Figure 4.8(a). In this arrangement the first NOR gate performs the NOR operation on the two input variables A and B while the second gate acts as an inverter, as described in the previous section.

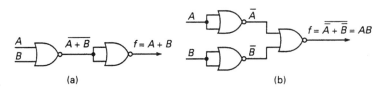

Figure 4.8   *Implementation of (a) the OR function and (b) the AND function using NOR gates*

The circuit for generating the AND function can be developed as follows. Since NOR gates are being used for the implementation of the function, the output gate will be a NOR gate whose output is $f = AB$, as shown in Figure 4.8(b). In order to obtain this output, the inputs to the gate should be $\bar{A}$ and $\bar{B}$, since (by De Morgan's theorem) $\overline{(\bar{A} + \bar{B})} = \bar{\bar{A}} \cdot \bar{\bar{B}} = AB$. Therefore, the output NOR gate is preceded by two further NOR gates, both used as inverters, one for each variable.

4.7 NOR logic implementation of products-of-sums

A function such as $f = (A + B)(C + D)$ is called a *two-level product-of-sums* expression. A possible implementation of this function using OR/AND logic is

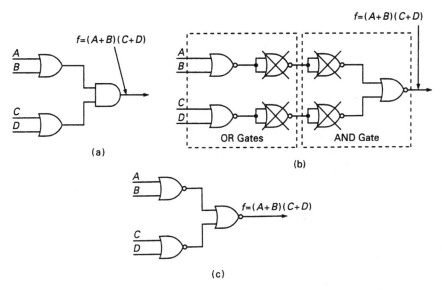

Figure 4.9 *(a) $f = (A + B)(C + D)$ implemented with OR/AND logic (b) Transformation of the OR/AND circuit to a NOR circuit (c) The NOR implementation of $f = (A + B)(C + D)$*

shown in Figure 4.9(a). This circuit can be converted to a circuit using NOR gates only by using the transformations developed in the previous section, as shown in Figure 4.9(b). An examination of this circuit shows that two pairs of the NOR gates in this implementation are redundant since they are merely producing a pair of double inversions, and therefore they have been crossed through in Figure 4.9(b).

The simplest form of the circuit using NOR gates is shown in Figure 4.9(c), and it can be seen that there is a one-to-one transformation from the OR/AND circuit to the corresponding NOR circuit.

4.8 NOR logic implementation of sums-of-products

It frequently happens that a Boolean function is expressed as a sum of product terms (sometimes, but not necessarily, minterms) and if this function is to be implemented using NOR gates then it must first be converted to the product-of-sums form. For example, suppose that it is required to implement, using only NOR gates, the function

$$f = \sum 0, 1, 3, 4, 5, 8, 12, 13, 15.$$

The absent minterms in this summation represent the inverse function, $\bar{f}$, and are plotted as 0s on the K-map shown in Figure 4.10(a). Simplifying,

$$\bar{f} = C\bar{D} + \bar{A}BC + A\bar{B}D.$$

Hence, by De Morgan's theorem,

$$f = (\bar{C} + D)(A + \bar{B} + \bar{C})(\bar{A} + B + \bar{D}).$$

This is the minimal product-of-sums form of the original Boolean function and it is shown implemented using NOR gates in Figure 4.10(b).

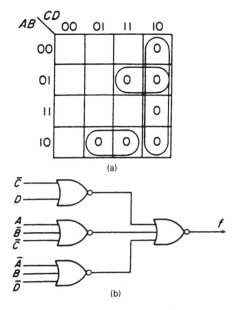

(a)

(b)

Figure 4.10   *NOR implementation of a sum-of-products expression (a) Plot of the inverse function (b) Implementation of minimised product-of-sums expression*

4.9 Boolean algebraic analysis of NAND and NOR networks

Analysis of NAND and NOR networks is often much more time-consuming than the analysis of comparable AND/OR networks because of the inversions that take place at the outputs of each NAND or NOR gate. For example, consider the NAND network implemented in Figure 4.5(c) and redrawn for convenience without the inverters in Figure 4.11. The outputs of the various NAND gates in the network are labelled p, q, r, s, t, u and f.

There are two main approaches to analysing this circuit using Boolean Algebra. In the method shown below, complementation bars (generated by the inversion at the output of each gate) are removed using De Morgan's theorem, as the analysis proceeds. Firstly, expressions are derived for the intermediate circuit outputs:

$$p = \overline{\bar{B}D} = B + \bar{D}$$

$$q = \overline{\bar{A}p} = \overline{\bar{A}(B + \bar{D})} = A + \overline{(B + \bar{D})} = A + \bar{B}D$$

$$r = \overline{\bar{C}q} = \overline{\bar{C}(A + \bar{B}D)} = \bar{C} + \overline{(A + \bar{B}D)} = \bar{C} + \bar{A} \cdot \overline{(\bar{B}D)} = \bar{C} + \bar{A}(B + \bar{D})$$

$$s = \overline{\bar{C}D} = C + \bar{D}$$

$$u = \overline{A\bar{C}} = A + C$$

$$t = \overline{sBu} = \overline{(C + \bar{D})B(A + C)} = \overline{(C + \bar{D})} + \bar{B} + \overline{(A + C)} = \bar{C}D + \bar{B} + \bar{A}\bar{C}.$$

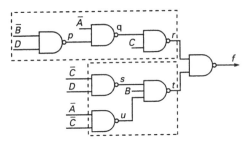

Figure 4.11   *Circuit to illustrate the analysis of a NAND gate network*

Then, the expression for the circuit output can be constructed:

$$f = \overline{rt}$$

$$= \overline{[\bar{C} + \bar{A}(B + \bar{D})] \cdot [\bar{C}D + \bar{B} + \bar{A}\bar{C}]}$$

$$= \overline{[\bar{C} + \bar{A}(B + \bar{D})]} + \overline{[\bar{C}D + \bar{B} + \bar{A}\bar{C}]}$$

$$= C(A + \bar{B}D) + (C + \bar{D})B(A + C).$$

In an alternative approach, the complementation bars may be retained until the final expression for the output has been obtained. By repeated application of De Morgan's theorem, the expression for the output can then be reduced to that obtained using the first method. The steps in this method will not be shown in detail as the final result must be the same as that deduced above, but if this method is used particular care is needed to ensure that the complementation bars are applied to the correct parts of their corresponding expressions.

For both of these methods it will be seen that a considerable amount of algebraic manipulation is needed; however, the same basic approaches are applicable to NOR networks as well as to NAND networks.

4.10 Symbolic circuit analysis for NAND and NOR networks

This alternative method of analysing a logic network depends upon the fact that the NAND function can be implemented by an AND gate in cascade with a NOT gate (see Figure 4.12(a)) or, alternatively, it can be implemented by an OR gate whose inputs are inverted, as shown in Figure 4.12(b).

The lower section of Figure 4.11, enclosed by the dotted lines (see Figure 4.13(a)), can now be represented by the network shown in Figure 4.13(b). Double inversions appear on two of the input lines to the OR gate and can be eliminated, while the inversion on the B input line can be represented by an inversion circle (or 'inversion bubble') at the OR gate input. This modified form of the circuit is shown in Figure 4.13(c) and the output t of this section of the original network can immediately be written down as $t = \bar{C}D + \bar{B} + \bar{A}\bar{C}$, which agrees with the expression obtained for t using Boolean algebra in the preceding section.

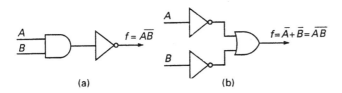

Figure 4.12   *Alternative implementations of a NAND gate*

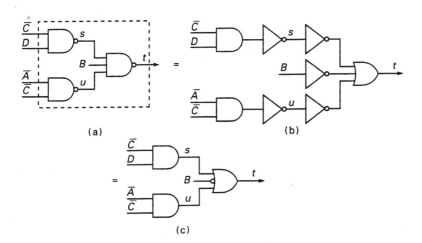

Figure 4.13   *Transformation from NAND/NAND to AND/OR configuration*

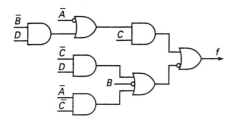

Figure 4.14   *The NAND network of Figure 4.11 transformed into a more readily analysable network*

Using the same transformations, the upper part of the network shown in Figure 4.11 can also be modified and the whole network can be redrawn in a form which is easier to analyse, as shown in Figure 4.14.

Similar transformations are available for NOR gates, as shown in Figures 4.15(a) and (b), and the method of analysis for a NOR network is then analogous to that used in the NAND network.

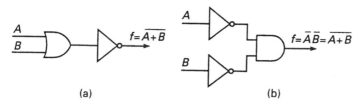

(a)   (b)

Figure 4.15   *Alternative implementations of a NOR gate*

4.11 Alternative function representations

The alternative representations for the NAND and NOR functions developed in section 4.10 can be shown in a more compact form using inversion circles. For example, using De Morgan's theorem, the NAND function may be expressed as

$$f = \overline{(A \cdot B)} = \bar{A} + \bar{B},$$

and an alternative representation for the NAND function consists of an OR gate with inversion circles at its inputs, as shown in Figure 4.16(a). The NOR function, again using De Morgan's theorem, is given by

$$f = \overline{A + B} = \bar{A} \cdot \bar{B},$$

and an alternative representation of this function requires an AND gate with inversion circles at its inputs (see Figure 4.16(b)).

For the AND and OR functions, alternative representations are obtained by inverting the defining equations for the NAND and NOR functions. Hence

$$AB = \overline{\overline{AB}} = \overline{\bar{A} + \bar{B}}$$

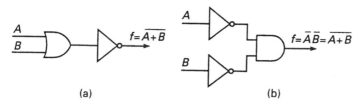

(a)   (b)

(c)   (d)

Figure 4.16   *Alternative representations for (a) the NAND function, (b) the NOR function, (c) the AND function, (d) the OR function*

and

$$A + B = \overline{\overline{A + B}} = \overline{\overline{A} \cdot \overline{B}}$$

For the AND function, the alternative representation consists of an OR gate with inversion circles at each of its inputs and also at its output (see Figure 4.16(c)), while the OR function requires an AND gate with inversion circles at each of its inputs and also at its output (see Figure 4.16(d)).

One way to remember De Morgan's theorem is that in an AND, NAND, OR, or NOR combination of Boolean variables or inverses, an inversion bar across *all* the variables may be split or joined at will, provided the operator combining them is changed simultaneously (i.e. '+' is changed to '·', or '·' is changed to '+'). This rule corresponds precisely with using alternative representations based upon De Morgan's theorem in circuit diagrams. An AND gate symbol may be swapped for an OR gate symbol, and vice-versa, provided that simultaneously the inversion circles are swapped either from the output to all the inputs, or from all the inputs to the single output. This procedure will often make the circuit diagram easier to understand but will not affect the Boolean operation of the circuit.

4.12 Gate signal conventions

In a practical gate network it is always assumed that the AND and OR functions are implemented within the confines of their *distinctive shape* symbols. The *absence* of inversion circles drawn at the input(s) to a gate circuit symbol indicates that the corresponding gate input is *active high*, i.e. the usual case, where logic 0 and 1 are passed unchanged to the gate itself. However, the *presence* of inversion circle(s) drawn at the input(s) to a gate circuit symbol indicates that the corresponding gate input is *active low*, and logic levels are *inverted* before being presented to the basic gate symbol. (In practice, the inversion(s) and the gate function are often undertaken by an integrated circuit whose operations cannot be physically separated in this manner, but the method of analysis and the nomenclature are still often used.) Similarly, at a gate output, an inversion circle indicates an *active low* output, i.e. when the gate is activated (the basic gate alone has output 1) the final output is 0. Finally, the *absence* of the inversion circle indicates an active high output, i.e. an output of 1 when the gate is activated. Sometimes, the use of active low logic lines is described in terms of the *negative logic convention*, where logic 1 is represented by 0V and logic 0 is represented by (for example) +5 V, but this approach will not be developed further here. The usual voltage representations (logic $0 \equiv 0V$, logic $1 \equiv +5V$), used throughout this book, are referred to as the *positive logic convention*.

4.13 Gate expansion

Suppose that a NAND gate is required with more inputs than are available from one device. The number of inputs can be increased by the use of AND circuits synthesised from NAND gates, as shown in Figure 4.17(a) or, if they are available, AND gates can be used to achieve the same effect. However, an alternative method of obtaining the logical AND of many signals is shown in Chapter 13. A similar

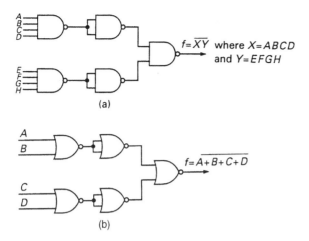

Figure 4.17 *(a) Expansion of NAND gate inputs by means of AND circuits synthesised from NAND gates (b) Expansion of NOR gate inputs by means of OR circuits synthesised from NOR gates*

technique may be used to increase the number of inputs to a NOR gate, as illustrated in Figure 4.17(b).

The extent to which the number of inputs to a gate may be expanded depends upon the 'fan-in', defined as the number of inputs available on each gate, and the type of gates available in a particular logic family. For example, in the type 74TTL families, NAND gates with up to eight inputs are available; hence, using a two-level expansion, a NAND equivalent with up to 64 inputs could be obtained using 17 NAND gates (or eight AND gates plus one NAND gate).

4.14 Miscellaneous gate networks

A gate network performing the AND-OR-NOT (or AND-OR-INVERT) operation is illustrated in Figure 4.18(a), and this network is sometimes available as a circuit

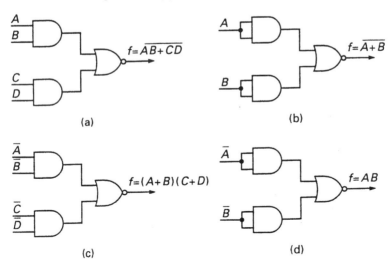

Figure 4.18 *(a) The AND-OR-NOT module (b) connected as a NOR gate (c) generating a 2-level product-of-sums and (d) connected as an AND gate*

element in its own right (for example, as the type number 74XX51 in the 74TTL logic families where XX indicates the particular type of technology used, such as Low Power Schottky (LS or ALS)). It forms the complement of a two-level sum-of-products, as shown in Figure 4.18(a). The network can also be used as a NOR gate either by commoning the inputs to each AND gate, as shown in Figure 4.18(b) or, better, by connecting one of each AND gate's inputs to logic 1 level. Alternatively, if inverted variables are connected to the AND gate inputs, then the network will form a two-level product-of-sums, as shown in Figure 4.18(c) since, using De Morgan's theorem:

$$f = \overline{(\bar{A} \cdot \bar{B}) + (\bar{C} \cdot \bar{D})} = \overline{(\bar{A} \cdot \bar{B})} \cdot \overline{(\bar{C} \cdot \bar{D})} = (A + B) \cdot (C + D).$$

If the network is used as a NOR gate but inverted variables are connected to the input of each AND gate, the network generates the AND function (see Figure 4.18(d)), since

$$f = \overline{\bar{A} + \bar{B}} = AB.$$

The AND-OR configuration, without a final inversion, is also sometimes available as a unit (for example, type number 74XX52 in the 74TTL logic families). This network is illustrated in Figure 4.19(a), and its basic use is to form a two-level sum-of-products.

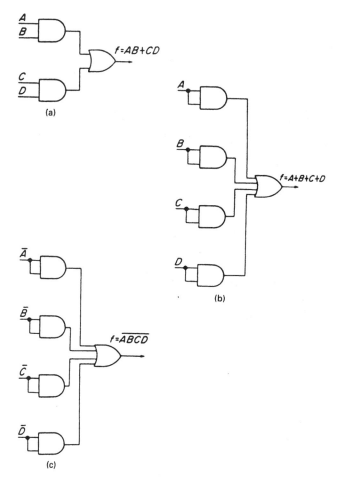

(a)

(b)

(c)

Figure 4.19 *(a) The AND-OR network (b) The AND-OR network used as an OR gate (c) The AND-OR network used as a NAND gate*

The network generates the OR function either by commoning the inputs to each AND gate (Figure 4.19(b)) or, better, by connecting one of each AND gate's inputs to logic 1 level. For single inverted inputs this connection generates the NAND function as shown in Figure 4.19(c).

Some of the networks described are capable of expansion. Expander chips are available which generate the AND function for a specified number of input variables. For example, type 74XX61 in the 74TTL family consists of three AND gates, each of which generates a function $f = ABC$. This output can then be used as an additional input to the OR gate in an expandable AND-OR network or, alternatively, as an additional input to the NOR gate in an expandable AND-OR-NOT network such as the 74XX53.

Other gates available include gates provided with a *strobe* input which can be regarded as an input that either *enables* or *disables* the gate. For example, the 74XX25 consists of twin four-input NOR gates with a strobe input. The output of each gate on this chip is given by $f = \overline{G(A + B + C + D)}$, where G is the strobe input. If $G = 0$ then the gate is disabled (giving an output $f = 1$) and conversely if $G = 1$ then the gate is enabled to give $f = \overline{(A + B + C + D)}$.

4.15 Exclusive-OR and exclusive-NOR

The Exclusive-OR (XOR) function was defined in section 2.14 by the Boolean equation

$$f = \bar{A}B + A\bar{B} = A \oplus B$$

where the symbol $\oplus$ is used to indicate the XOR operation. The truth table for this operation is given in Figure 4.20(a) and the conventional symbol for the practical logic gate that implements the XOR operation is shown in Figure 4.20(b).

The XOR operation is identical to the conventional Boolean OR operation using variables A and B except that it *excludes* the case $f = 1$ when $A = B = 1$, hence the name Exclusive-OR. When the XOR operation is performed on all possible combinations of two binary digits (see Figure 4.20(c)) the *modulo-2* sum is obtained, where the modulo-2 sum is defined as the conventional numerical sum of the two digits but ignoring the carry-out bit. For example, the modulo-2 sum of $1 + 1 = 0$.

Since the XOR function generates the modulo-2 sum of two binary digits it is apparent that it has a direct application in the design of arithmetic circuits. It also has applications in fault-detection systems (see section 13.13) and in error detection and correction circuits found in data-transmission systems. Here, the modulo-2 sum of a number of binary digits is obtained and generates the parity function which is commonly used as an error-control function.

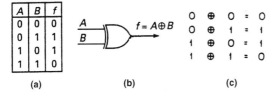

(a)   (b)   (c)

Figure 4.20   *(a) Truth table for XOR function (b) Conventional symbol for XOR gate (c) The modulo-2 sum of two binary digits*

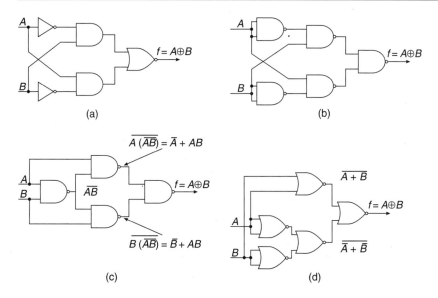

Figure 4.21 *Basic gate implementations of the XOR function: (a) using AND/OR gates (b) using NAND gates only (c) minimal NAND implementation and (d) NOR implementation*

There are a number of ways of implementing the XOR function and these include implementation with AND, OR and NOT gates, as illustrated in Figure 4.21(a). The minimised implementation is obtained by algebraic manipulation of the XOR function $f = \bar{A}B + A\bar{B}$. Adding $A\bar{A}$ and $B\bar{B}$ to the right-hand side of the equation gives

$$f = A \oplus B$$
$$= \bar{A}B + B\bar{B} + A\bar{B} + A\bar{A}$$
$$= B(\bar{A} + \bar{B}) + A(\bar{A} + \bar{B}).$$

This can be implemented with just four NAND gates (see Figure 4.21(c)). From Figure 4.21(c), it is clear that the output of the circuit is given by:

$$f = \overline{(\bar{A} + AB)(\bar{B} + AB)}$$
$$. = \overline{\bar{A}\bar{B} + AB}$$
$$= \overline{A \oplus \bar{B}}$$
$$= A \oplus B.$$

Alternatively, rewriting $B(\bar{A} + \bar{B}) + A(\bar{A} + \bar{B}) = (A + B)(\bar{A} + \bar{B})$ gives a two-level product-of-sums that can be implemented using five NOR gates (see Figure 4.21(d)). The output of this circuit is

$$f = \overline{\overline{(\bar{A} + \bar{B})} + \overline{(A + B)}}$$
$$= (\bar{A} + \bar{B})(A + B)$$
$$= \bar{A}B + A\bar{B}$$
$$= A \oplus B.$$

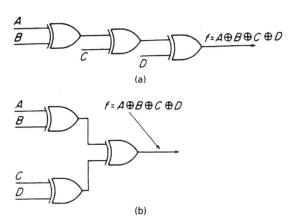

(a)

(b)

Figure 4.22   *(a) Serial cascade of Exclusive-OR gates (not recommended) (b) Parallel cascade of Exclusive-OR gates (preferred)*

Although the XOR operation can be implemented in a number of ways by a combination of discrete gates as shown in Figure 4.21, it is available directly on SSI chips in the 74TTL family; for example, the 74XX86 chip provides four two-input XOR gates. In many cases, the easiest way to handle the XOR operation in Boolean expressions is to substitute the defining Boolean equation $(A \oplus B = \bar{A}B + A\bar{B})$ and then to use the usual rules of Boolean Algebra, but section 2.14 includes some additional useful results that may shorten such analysis.

If the XOR is indeed of a greater number of variables than can be accommodated by one XOR gate, it is necessary to cascade XOR gates. Two possible approaches are illustrated in Figure 4.22. In the first method (Figure 4.22(a)) the XOR gates are connected serially to produce the XOR of four variables, and in the second method two XOR gates operate 'in tandem' to feed the third. Both methods use exactly the same number of gates but the method of Figure 4.22(a) requires three levels of logic, whereas the method of Figure 4.22(b) requires only two levels of logic. If the time delay introduced by each gate is important then the total propagation time delay through the second configuration (Figure 4.22(b)) will clearly be less than the time delay for the first circuit (Figure 4.22(a)), and so generally the parallel configuration of Figure 4.22(b) is preferred. If the XOR of eight variables is required, then for the serial method seven levels of logic are needed, whereas the parallel connection requires only three levels.

The XOR gate can also be used as a *controlled inverter*. This is illustrated in Figure 4.23 where one input to the 2-input gate is for binary data, while the second input is supplied with a control signal M which may be either 0 or 1. If $M = 0$ the gate transmits the input data unchanged to its output, but for $M = 1$ the gate inverts the input data. This particular connection can be used in conjunction with an adder circuit, and the combination of adder and controlled inverter can then be used as an adder/subtractor (see later). The controlled inverter also has applications in data processing circuits where it is required to complement data under external control.

Figure 4.23   *The XOR gate as a controlled inverter (a) Transmission (b) Inversion*

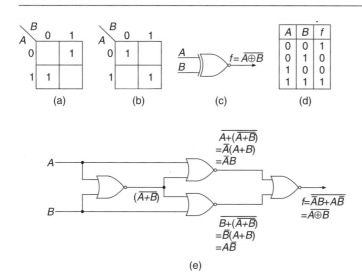

Figure 4.24 *K-maps for (a) the Exclusive-OR function, and (b) the Exclusive-NOR function (c) Circuit symbol of the Exclusive-NOR function (d) Truth table for the coincidence (XNOR) operation (e) Minimal NOR gate implementation of the XNOR operation*

K-maps for the XOR function and its complement are shown in Figures 4.24(a) and (b). Selecting those combinations of the variables which make the value of the complement function equal to 1 leads to the Boolean equation

$$f = \overline{A \oplus B} = \bar{A}\bar{B} + AB = A \oplus \bar{B} = \bar{A} \oplus B$$

where $\overline{A \oplus B}$ indicates the EXCLUSIVE-NOR (XNOR) function, sometimes written as $A \odot B$. The XNOR function has the value $f = 1$ when $A = B = 0$ or $A = B = 1$ (i.e., when $A = B$), and hence is sometimes alternatively termed the 'coincidence' function. The conventional circuit symbol for the XNOR gate is simply an XOR gate followed by an inversion circle, as shown in Figure 4.24(c), and the truth table for the function is shown in Figure 4.24(d).

The XNOR operation is, like the XOR operation, also Commutative and Associative. This is clear from its close relation to the XOR operation. In addition, since it is clear that

$\overline{(A \oplus B)} \oplus C$

$= (A \oplus \bar{B}) \oplus \bar{C}$ (using the Boolean relation for the XNOR operation),

$= A \oplus (\bar{B} \oplus \bar{C})$ (using the Associative property of the XOR operation),

$= A \oplus \overline{(\bar{B} \oplus C)}$ (using the Boolean relation for the XNOR operation),

$= A \oplus (\bar{\bar{B}} \oplus C)$ (using the Boolean relation for the XNOR operation),

$= A \oplus (B \oplus C),$

it has therefore been proved that the XNOR of three variables is equal to the XOR of the same three variables:

$$\overline{(A \oplus B)} \oplus C = A \oplus B \oplus C.$$

XNOR gates are available in the mature logic technologies but may also be implemented using other gates. Apart from simply inverting the output of an XOR gate, the XNOR may be implemented using only four NOR gates, as shown in Figure 4.24(e). The XNOR operation gives an output of logic 1 whenever the two input binary digits are equal, and consequently it has an application in those circuits that are designed to compare the magnitudes of two equal length strings of binary digits. This theme will be developed further in Chapter 5.

4.16 Noise margins

In Chapter 2 it was established that logic 0 and logic 1 can be represented by two voltages, usually 0 V and 5 V. The required power supply voltage (V_{CC}) is tightly specified for the common 74TTL transistor–transistor logic family of discrete gates, the original versions of which used bipolar transistor technology. However, in practice manufacturers design their logic gates to accept and operate correctly with logic voltage values considerably different from these ideal values.

Acceptable values for the low level (logic 0) and for the high level (logic 1) are as defined in Figure 4.25 for the 74LS series of logic gates. The symbol $V_{OL(max)}$ denotes the maximum output voltage that any gate will produce when it is in logic state 0, and $V_{OH(min)}$ is the minimum gate output voltage that any gate will produce when it is in logic state 1. Both $V_{OL(max)}$ and $V_{OH(min)}$ are defined for worst-case loading conditions. The symbol $V_{IL(max)}$ denotes the maximum gate input voltage guaranteed to be recognised as logic 0, and $V_{IH(min)}$ is the minimum gate input voltage guaranteed to be recognised as logic 1.

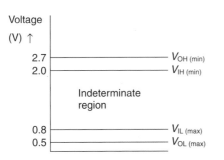

Figure 4.25 *74LS(TTL) series logic levels*

The region between these two voltage levels (either at input or output) is *indeterminate*. If a voltage in this range is presented to a logic gate input then its operation is not guaranteed to be sensible as neither a correct logic 0 nor a correct logic 1 is being applied. In practice it often happens that the gate output will oscillate at high frequency in this case, or perhaps will stay at a constant value that is itself within the indeterminate region, thus presenting further problems to the next gate in the logic system.

Since the voltages specifying logic state 1 are minimum values, in practice any voltage between the specified minimum and the supply voltage V_{CC} may be produced and will be recognised correctly as denoting logic state 1. Also, since the voltages specifying logic state 0 are maximum values, in practice any voltage between the specified maximum and 0 V may be produced and will be recognised correctly as denoting logic state 0.

Suppose a 74LS series gate gives a logic 0 output of +0.5 V (*just* within the specification); then, a corrupting noise voltage of more than +0.3 V superimposed on this value will result in the input to the next gate being in the indeterminate region where it will not be recognised as logic 0. Therefore, the logic 0 *noise margin* (or *noise immunity*) is defined as

$$N_L = V_{IL(max)} - V_{OL(max)}.$$

Similarly, the logic 1 noise margin is defined as

$$N_H = V_{OH(min)} - V_{IH(min)}.$$

Many digital logic devices internally use field-effect transistors which are made using a metal-oxide-semiconductor (MOS) structure. The most important MOS logic technology employs complementary metal oxide semiconductor (CMOS) transistors. In practice, the noise margins for the CMOS family can be much greater than those for the 74TTL family. The noise margins for typical members of the two families are tabulated for comparison:

Parameter	74LS (TTL)	CMOS (4000 series)
$V_{OH(min)}$	2.7 V	4.95 V
$V_{IH(min)}$	2.0 V	3.50 V
$V_{IL(max)}$	0.8 V	0.05 V
$V_{OL(max)}$	0.5 V	1.45 V
N_H	0.7 V	1.25 V
N_L	0.3 V	1.20 V

4.17 Propagation time

Suppose that a rectangular voltage pulse is applied to the input of a logic inverter, as shown in Figure 4.26. For any practical logic gate there will be a time delay or *propagation time* between the change in the input voltage to the corresponding change

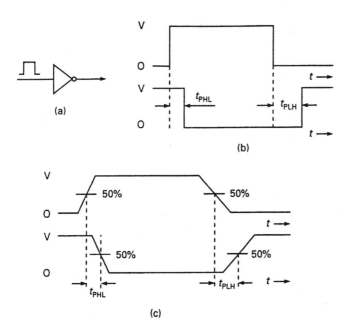

Figure 4.26 *(a) Rectangular voltage pulse applied at the input of an inverter (b) Idealised timing diagrams (c) Practical timing diagrams*

in the output voltage, and this delay is denoted by t_{PHL} when the output voltage changes from a high to a low level. When the output voltage changes from a low to a high level, the propagation delay time is denoted by t_{PLH}. These two propagation delays may, in principle, have different values. Although described here only in terms of a simple logic inverter, all logic components show propagation time effects to varying degrees, and in the case of complex components there may be differing values of t_{PHL} and t_{PLH} according to which inputs and which outputs are being considered.

The timing diagrams of Figure 4.26(b) are somewhat idealised since they imply that all the voltage transitions take place instantaneously. In practice, the input and output voltages will not change instantaneously, and the propagation times t_{PHL} and t_{PLH} are therefore usually defined as the time delays between the voltages halfway between the steady voltage levels achieved, sometimes called the '50% points', as shown in Figure 4.26(c).

The propagation delays specified by manufacturers usually fall into three categories: minimum, typical and maximum. This is because there is a manufacturing spread for these parameters. In effect, the manufacturer is stating that the maximum delay will never be exceeded, and the wise logic designer will ensure that the design operates correctly if the gates used only meet the maximum quoted values (i.e. 'worst case design').

For the 74TTL logic family, typical values of propagation delay lie in the range 2 to 33 ns depending upon the particular type of technology being employed (the most common of which are currently LS or ALS, and high-speed CMOS gates (HCT) that are designed to be compatible and interchangeable with TTL gates). Reduction in the propagation delay using bipolar technology can be achieved by employing emitter coupled logic (ECL) where propagation delays as low as 1 ns can be achieved. However, CMOS circuits are widely used in a great number of present system designs. They have the advantages of cheapness, low power consumption per gate and considerably higher packing densities (the number of gates manufactured per chip).

Electronic engineers are also interested in the *rise* and *fall times* of the voltage waveforms. The *rise time* is defined as the time taken for the voltage to change from 10% to 90% of its final value, while the *fall time* is defined as the time taken to change from 90% to 10% of its initial value. This parameter is also frequently referred to as the *transition time*.

4.18 Speed-power products

The propagation time per gate (i.e. the delay time introduced into the signal path by using a logic gate) multiplied by the electrical power dissipated in each gate (fed from the power supply) is approximately equal to the energy stored within the gate as a result of maintaining either a 0 or 1 logic level at the output. This is not an absolute or accurate measure, for the obvious reasons that the power dissipation depends to a greater or lesser extent upon the logic state at the gate output, and also because some power is lost in the gate circuitry not directly associated with the bit storage within the gate. However, this product is a useful 'figure of merit' for a family of logic gates.

The goal of many technologists designing logic gate families is to make this speed-power product or 'energy per gate' as small as possible in order to produce large logic systems with minimised power consumption. The speed-power product also indicates how successful the particular technology has been in reducing the stored energy in each gate. Assuming that using a certain technology the speed-power product has been reduced as far as is commercially practical, then fast operation of a logic circuit (i.e. small gate delay) requires a correspondingly large power dissipation per gate, and if the number of logic gates is increased as well then the total power consumption on the chip may increase so drastically that special cooling measures must be taken, including the use of heat-sinking and forced cooling. One very visible manifestation of the problems of using large numbers of logic gates is the fact that certain versions of the Intel 'Pentium' processor now require a small electrically driven fan mounted on top of the actual processor chip. Larger computing systems sometimes require the use of liquid coolant.

Experimental, and usually very simple, logic systems in research laboratories are currently able to establish logic levels based upon the storage of individual electrons, which at current gate size limits corresponds to a speed-power product of around 2×10^{-21} J. Using current technology, this represents the ultimate limitation on the energy stored per gate. Some values of the speed-power product for some representative commercial logic families are tabulated below. In this table, the standard, S, LS and ALS types refer to the original 74(TTL) series of conventional silicon ICs. The HCT logic family uses silicon CMOS technology but is compatible with TTL gates as its logic voltage levels are similar. The 4000 series is an older CMOS logic family; and GaAs logic devices are at the experimental, research, or low-volume development stages at the time of writing. The values given are typical only, as the precise values frequently depend upon the output state of the gate concerned, the logic function in question, and also the power supply voltages (e.g., in the case of CMOS gates).

Gate type	Typical gate delay/ns	Typical power per gate/μW	Typical maximum clock speed/MHz	Typical speed-power product/10^{-15} J
Standard	10	10000	35	100000
S	3	19000	125	57000
LS	5	2000	60	10000
ALS	5	1300	60	6500
HCT	7	2.5	50	17.5
4000	60	6	5	360
ECL	2	25000	200	50000
GaAs	0.08	1000	4000	80

4.19 Fan-out

The number of gate inputs that can be connected to a single driving gate output without overloading the driving gate is termed the *fan-out*. The limitation is usually that of the available current drive from the gate output compared to the current required to drive the gate inputs. Information on current capabilities and requirements, supplied on the manufacturers' data sheets, allows the designer to calculate fan-out values.

For example, in the 74LS(TTL) series, the maximum low state current required by a gate input, $I_{\text{IL(max)}}$, is 0.4 mA whilst a 74LS series gate output is capable of sinking a current of at least $I_{\text{OL(min)}} = 8$ mA. Hence, the ratio of currents in the low state is given by

$$\frac{I_{\text{OL(min)}}}{I_{\text{IL(max)}}} = \frac{8 \, \text{mA}}{0.4 \, \text{mA}} = 20$$

In the logic high state, the maximum current required at the input of a 74LS series gate is $I_{\text{IH(max)}} = 20 \, \mu\text{A}$ while a 74LS series gate output is capable of sourcing a current of at least $I_{\text{OH(min)}} = 400 \, \mu\text{A}$. Hence

$$\frac{I_{\text{OH(min)}}}{I_{\text{IH(max)}}} = \frac{400 \, \mu\text{A}}{20 \, \mu\text{A}} = 20$$

The *fan-out* is defined as the *worst-case* (i.e., the least value) obtained from these calculations, so that in this case the *fan-out* is 20; up to 20 gate inputs can be connected to one gate output in the 74LS series. Similar calculations can be made for other variants in the type 74 families. By coincidence, in the 74LS series the possible fan-out in both states is identical; in some logic families the calculations analogous to those above yield different results for the two states, in which case the lower value must be quoted as the fan-out. In certain CMOS technologies at low frequencies the gate input current is essentially zero since the gates have an extremely high input impedance, and so the fan-out is infinite (as many gate inputs as desired may be connected to one output). However, at high frequencies, the gate input capacitance becomes an important consideration (current must be supplied to charge the effective input capacitance of the gates sufficiently quickly) and the fan-out is reduced to a finite value.

Sometimes the term 'fan-out' is used in a more informal sense to indicate the number of gate inputs actually connected to a given output. For example, if a single gate output is connected to three gate inputs then the fan-out may be said to be three, irrespective of the maximum number of inputs that *could* be connected to one output.

Problems

4.1 Implement the following functions using only NAND gates:

(a) $f_1 = A\bar{B} + (\bar{B} + \bar{C})\bar{A}$

(b) $f_2 = (AB + C)(B + \bar{D}) + A(\bar{B} + C)(D + \bar{E})$.

4.2 Minimise the following functions and implement the minimised function using only NAND gates:

(a) $f(A, B, C) = \sum 0, 1, 2, 3, 4, 5, 6$

(b) $f(A, B, C, D) = \sum 0, 2, 8, 9, 10, 12, 13, 14$

(c) $f(A, B, C, D, E) = \sum 8, 9, 10, 11, 15, 16, 17, 18, 19, 20, 21, 22, 23, 24, 25, 26, 27, 31$

4.3 Implement the following Boolean functions using only NOR gates:

(a) $f_1 = A(\bar{A} + \bar{B})(B + \bar{C}D)$

(b) $f_2 = A(B + C + DE)(\bar{B} + CD + \bar{A}E)$

4.4 Find the minimum product-of-sums form for each of the following functions and implement the functions using NOR gates only:

(a) $f(A, B, C) = \sum 0, 2, 4, 6, 7$

(b) $f(A, B, C, D) = \sum 0, 1, 2, 3, 4, 9, 10, 13, 14$

(c) $f(A, B, C, D, E) = \sum 0, 1, 2, 3, 4, 6, 10, 11, 12, 13, 14, 15, 16, 29, 31$

4.5 Using a simple factoring technique, implement each of the following functions in as many ways as possible using only NAND gates:

(a) $f_1 = B\bar{C}D + \bar{B}CD + A$

(b) $f_2 = AC + BC + \bar{A}\bar{D}$

(c) $f_3 = A\bar{B}D + A\bar{B}\bar{C} + \bar{C}D$

4.6 Implement the following functions using only NAND gates having a maximum fan-in of three:

(a) $f_1 = AB\bar{C} + AD + B\bar{C}\bar{D} + AC\bar{D}$

(b) $f_2 = AB + \bar{A}D + BD + \bar{C}D + AC$

(c) $f_3 = ABCD + \bar{A}B\bar{C}D + AB\bar{D} + CD$

4.7 Analyse the circuits shown in Figures P4.7(a), (b) and (c), to produce Boolean algebraic expressions for the circuit outputs:

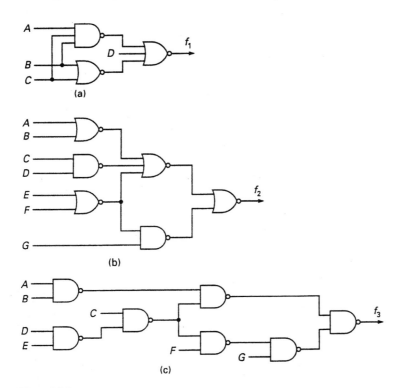

Figure P4.7

4.8   Analyse the circuits shown in Figures P4.7(a), (b) and (c) by removing as many inversion circles as possible.

4.9   Implement the following functions using only NOR gates having a maximum fan-in of three:

(a) $f_1 = (\bar{A} + B)(C + D)(B + \bar{C})(A + D)(\bar{A} + C)$

(b) $f_2 = (A\bar{C} + BC)(\bar{A} + C)$

(c) $f_3 = AB + B\bar{C}D + A\bar{B}\bar{D}$

4.10 Express the following equations in their minimal sum-of-products form:

(a) $f_1 = A(A \oplus B \oplus C)$

(b) $f_2 = A \cdot [\overline{(A \oplus B) \oplus C}]$

(c) $f_3 = A + (A \oplus B \oplus C)$

(d) $f_4 = A + \overline{(A \oplus B) \oplus C}$

4.11 Prove the following identities:

(a) $\overline{(A \oplus B \oplus C)} = A \oplus \overline{(B \oplus C)}$

(b) $(A \oplus B \oplus AB)(A \oplus C \oplus AC) = A + BC$

5 Combinational logic design with MSI circuits

5.1 Introduction

Since the introduction of MSI and LSI circuits, the traditional methods of logic design have largely been superseded. Traditionally, the design engineer has developed a Boolean equation as the solution to a particular problem. This function has then been minimised and implemented using SSI circuits.

In practice, many combinational circuits may have a large number of inputs and outputs, and consequently the use of truth tables in the design of such circuits is impractical. Furthermore, it is not economical to provide sufficient pins on an IC package to allow access to each of the gates that can be provided on a single chip. Many functions such as counting, adding and parity checking are common in a large number of designs, and a useful library of digital circuits for implementing these functions has been developed. As fabrication techniques improved it became possible to implement these functions on a single chip.

The development of MSI circuits has led to the technique of splitting a complex design into a number of sub-systems. This leaves the designer the task of inter-connecting available MSI functions in a manner which satisfies the initial design specification.

5.2 Multiplexers and data selection

A multiplexer (MUX) selects 1-out-of-n lines where n is usually 2, 4, 8 or 16. A block diagram of a multiplexer having four input data lines d_0, d_1, d_2 and d_3 and complementary outputs f and $\bar{f}$ is shown in Figure 5.1(a). The device has two *control* or *selection* lines A and B and an enable line E. Gate implementation of a 4-to-1 multiplexer is shown in Figure 5.1(b). In essence, the circuit is an AOI module having complementary outputs. The *characteristic equation* of the multiplexer is

$$f = \bar{A}\bar{B}d_0 + \bar{A}Bd_1 + A\bar{B}d_2 + ABd_3$$

Individual data lines are selected by the application of the appropriate binary signal to control lines A and B. When $\bar{A}\bar{B} = 1$ the output of the MUX is d_0, and when $\bar{A}B = 1$ the output is d_1, etc. When the input enable is $E = 1$ the four AND gates are enabled. With $E = 0$ multiplexer operation is inhibited.

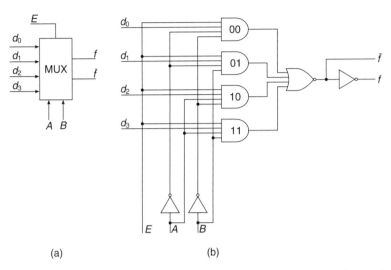

(a)   (b)

Figure 5.1   *(a) Block diagram of a 4-input multiplexer and (b) its gate implementation*

5.3 Available MSI multiplexers

The sizes of multiplexer available in the TTL family are limited by pin availability on standard MSI chips. With 16-pin chips, multiplexers having 2, 4, and 8 data lines are available and with 24-pin chips it is possible to provide a multiplexer with 16 data lines.

The 74157, a 16-pin chip, provides quadruple 2-to-1 multiplexers, where each multiplexer consists of a 2-wide, 3-input AO gate, with one input for data, one for selection, and the third which is the strobe or enable line. A logic diagram, the truth table and a traditional block diagram are shown in Figure 5.2. The numbers in parentheses on the input and output lines are the pin numbers.

The 16-pin 74353 is a dual 4-to-1 data selector/multiplexer. Each multiplexer consists of a 4-input 4-wide AOI gate with tri-state control (see Chapter 10) on the NOR output gates. The output control lines also act as separate enable inputs for the two devices, both of which are controlled by the common select lines. A logic diagram for the 74353, its function table and block diagram are shown in Figure 5.3.

At the upper end of the scale, the 74251 is an 8-to-1 multiplexer having complementary tri-state outputs. It consists of a 4-input 8-wide AOI gate with an enable strobe. There is also the 24-pin 74150, a 16-to-1 multiplexer/data selector which consists of a 5-input 16-wide AOI gate. The number of pins available on this chip limits the device to a single-output line.

5.4 Interconnecting multiplexers

Data within a digital system is normally processed in parallel form in order to increase the speed of operation. If the output of the system has to be transmitted over a relatively long distance then a parallel-to-serial conversion will take place so that the data can be transmitted serially over a single transmission line. This eliminates the requirement of individual transmission lines, one for each bit. The arrangement of Figure 5.4 can be regarded as an example of parallel-to-serial conversion. An 8-bit

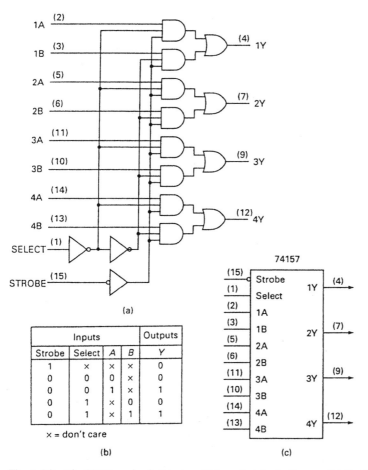

Figure 5.2 *The 74157 quadruple 2-to-1 multiplexer (a) Logic diagram (b) Truth table (c) Traditional logic symbol*

word is presented in parallel at the data inputs of the two multiplexers and is clocked from the the output in serial form. The MUX not enabled will give an output of logic 0. In the absence of the binary counter this arrangement can be used for the selection of 1-out-of-8 data lines. Selection of the required data line is made by the selection inputs *A, B* and *E*.

The principle of data selection can be extended to allow the selection of 1-out-of-64 lines. This can be achieved using nine 8-to-1 multiplexers (see Figure 5.5) arranged in two levels of multiplexing.

If $ABC = 001$ Multiplexer M1 is enabled and input D1 is selected on multiplexer M8
If $DEF = 111$ Data $X15$ is selected and is output on line $X01$
Then $ABCDEF = 001111$ selects $X15$ and outputs it at Z on M8.

5.5 The multiplexer as a Boolean function generator

For a 4-to-1 MUX the characteristic equation is

$$f = \bar{A}\bar{B}d_0 + \bar{A}Bd_1 + A\bar{B}d_2 + ABd_3$$

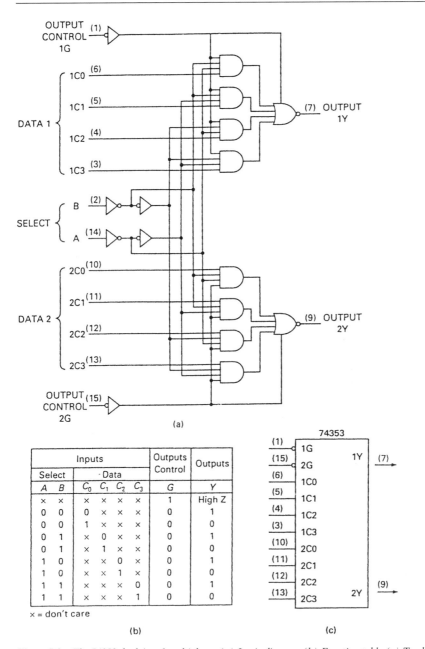

Figure 5.3 *The 74353 dual 4-to-1 multiplexer (a) Logic diagram (b) Function table (c) Traditional logic diagram*

where *A* and *B* are Boolean variables, applied at the select inputs, which can be factored out of any Boolean function of *n* variables, as shown below. The remaining *n*−2 variables, referred to as the *residue variables*, can be formed into *residue functions* which can then be applied at the data inputs. In practice, the residue functions can be implemented by discrete logic gates or, alternatively, by other multiplexers. If, for example, a 3-variable function $f(A, B, C)$ is to be generated and the variables *A* and *B* are applied at the select inputs, the residue functions expressed in terms of the variable *C* can be applied, one at each of the data inputs. The four available residue functions are $C, \bar{C}, 1$ and 0. In all,

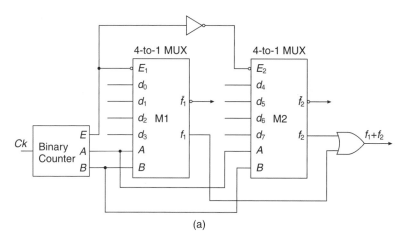

(a)

Inputs					Outputs
E	A	B	E_1	E_2	f_1+f_2
0	0	0	1	0	d_0
0	0	1	1	0	d_1
0	1	0	1	0	d_2
0	1	1	1	0	d_3
1	0	0	0	1	d_4
1	0	1	0	1	d_5
1	1	0	0	1	d_6
1	1	1	0	1	d_7

(b)

Figure 5.4 *(a) Combination of two MUXs providing a 1-out-of-8 Data*

there are $4^4 = 256$ possible combinations of the four residue functions and a multiplexer with four data inputs can generate any of the 256 possible Boolean functions of 3 variables.

For a 4-to-1 line MUX there are three possible choices for the Boolean variables to be applied at the selection inputs. They are AB, AC and BC. These combinations can be associated with individual data lines, as shown in Figure 5.6. Assuming that A and B are chosen as selection inputs, then for the condition $AB = 00$ the top two left-hand cells on the K-map in Figure 5.6(a) are associated with the data line d_0. Similarly, for the condition $AB = 01$ the top two right-hand cells on the K-map are specified and are associated with data line d_1. In effect, the 3-variable map has been divided into four 1-variable maps each of which are associated with one of the four data input lines and also with the residue variable C. Association of selection inputs AC and BC is shown in Figures 5.6(b) and (c).

As an example, the 3-variable function

$$f(A, B, C) = \sum 0, 1, 3, 4, 7$$

will be implemented using a 4-to-1 MUX. The function is first plotted on the K-map in Figure 5.7, and an arbitrary choice of selection variables is made, in this case A and B. Simplification takes place on each of the 1-variable maps and the resulting residue functions are:

$$d_0 = 1 \quad d_1 = C \quad d_2 = \bar{C} \quad d_3 = C$$

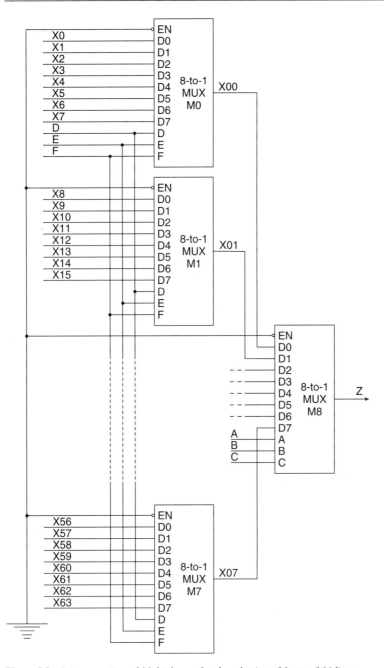

Figure 5.5 *Interconnection of Multiplexers for the selection of 1-out-of-64 lines*

If the complement of the variable C is available, implementation of this function can be achieved with a single MUX such as the 74353 dual 4-to-1 MUX. If C is not available an inverter is required, as shown in Figure 5.7.

If the choice of selection variables had been A and C then the inputs to the data lines would be $d_0 = \bar{B}, d_1 = 1, d_2 = \bar{B}$ and $d_3 = B$. For selection variables B and C the input to the data lines would be $d_0 = 1, d_1 = \bar{A}, d_2 = 0$ and $d_3 = 1$.

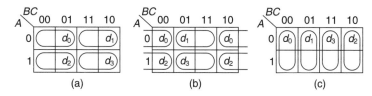

Figure 5.6 *Association of data lines with control signals for a four-input multiplexer. The control variables are A and B in (a), A and C in (b), and B and C in (c)*

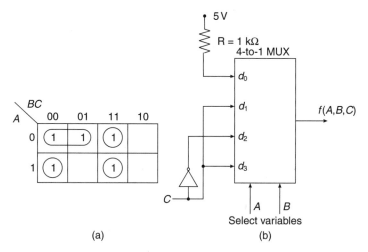

Figure 5.7 *(a) K-map plot of $f = \sum 0, 1, 3, 4, 7$ (b) implementation of the function by 4-to-1 MUX*

Alternatively, the residue functions can be found directly from the truth table of the Boolean function to be implemented. For the function

$$f(A, B, C, D) = \sum 0, 1, 3, 4, 5, 9, 10, 11, 14, 15$$

the truth table is listed in Figure 5.8. The selection variables A, B and C are sectionalised and isolated from the residue variable D. For the combination of these variables $ABC = 000$, $f = 1$ for both $D = 0$ and $D = 1$ and the residue function is $d_0 = 1$. The method used here is analogous to plotting an RDM map from a truth table (see Chapter 3). Similarly, for $ABC = 100$, $f = 1$ for $D = 1$ and residue function $d_4 = D$. Implementation of the function using an 8-to-1 multiplexer is shown in Figure 5.8. Other implementations can be found using an alternative choice of the selection variables.

5.6 Multi-level multiplexing

The implementation of Boolean functions may be achieved more economically and with fewer interconnections by using more than one level of multiplexing. Using the method described in the previous example for the function

$$f(A, B, C, D) = \sum 0, 1, 2, 5, 7, 9, 15, \quad \text{can't happen terms } 4, 11, 13$$

the residue functions are found to be

$$d_0 = d_2 = 1, \quad d_5 = d_6 = 0, \quad d_3 = d_4 = d_7 = D, \quad d_1 = \bar{D}$$

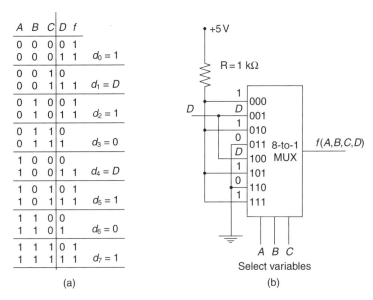

A	B	C	D	f	
0	0	0	0	1	
0	0	0	1	1	$d_0 = 1$
0	0	1	0		
0	0	1	1	1	$d_1 = D$
0	1	0	0	1	
0	1	0	1	1	$d_2 = 1$
0	1	1	0		
0	1	1	1		$d_3 = 0$
1	0	0	0		
1	0	0	1	1	$d_4 = D$
1	0	1	0	1	
1	0	1	1	1	$d_5 = 1$
1	1	0	0		
1	1	0	1		$d_6 = 0$
1	1	1	0	1	
1	1	1	1	1	$d_7 = 1$

(a)   (b)

Figure 5.8   *(a) Truth table for $f = \sum 0, 1, 3, 4, 5, 9, 10, 11, 14, 15$ and (b) function implementation*

assuming A, B and C have been chosen as the selection variables. The implementation of the function using an 8-to-1 MUX is shown in Figure 5.9. A typical 8-to-1 MUX that could be used in this design is the 74251. This is a 16 pin device having 8 data inputs, 3 selection inputs, true and complemented outputs and a strobe line not shown in Figure 5.9 which would be held at logic 0 level.

To implement the same function using two levels of multiplexing, a 4-to-1 multiplexer M5 is used to generate the function output. The function and the 'can't happen' terms are listed in the left-hand column of Figure 5.9(c). The selection variables for the output MUX are C and D and the variables A and B form the residue functions required at the four inputs.

The first level of multiplexing will consist of four 2-to-1 multiplexers, each of them having B as the select variable. Their inputs can be determined by examining the listings in each of the four right-hand columns in Figure 5.9(c). For the column headed $\bar{C}\bar{D}$ there are two terms, $\bar{A}\bar{B}$ and the 'can't happen' term $\bar{A}B$. When the selection variable $B = 0$, the required input is $\bar{A}$ and when the selection variable $B = 1$, the required input is 0 since $\bar{A}B$ is a 'can't happen' term. The first level inputs obtained from the three remaining columns of Figure 5.9(c) are marked on the function implementation diagram shown in Figure 5.9(d). It is immediately apparent from an inspection of this diagram that multiplexers M2, M3 and M4 are redundant. The number of interconnections required is ten and an inverter for the variable A may be required. Two multiplexer packages are needed, neither of them being fully utilised; this may or may not be a disadvantage as far as space requirements are concerned.

It is also possible to implement the function using three levels of multiplexing. For this arrangement the conventional architecture requires four 2-to-1 multiplexers at the first level, two at the second level and one at the output level although some of these multiplexers may be found to be redundant prior to implementation. The technique used to find the MUX inputs at the first level is identical to that used for two level

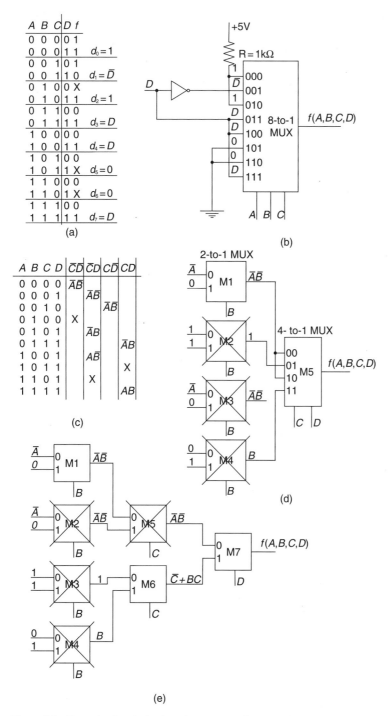

Figure 5.9 *Example of multi-level implementation of a 4 variable function (a) and (b) single-level (c) and (d) 2-level and (e) 3-level*

multiplexing. It consists of first finding the residue functions at the inputs to the output and second level multiplexers and finally determining the input residue functions at the first level from the second level listings. Implementation of the function using three levels of multiplexing is shown in Figure 5.9(e). The residue functions at each level of multiplexing are marked on the diagram and it will be observed that M2, M3, M4 and M5 are redundant. The implementation requires a single quad 2-to-1 MUX and an inverter if the complement of variable A is not available. The number of pin connections used for this implementation is ten, but only one multiplexer package is needed. In general the possibility of redundancy in a multi-variable function is highest when the smallest multiplexer elements are used.

5.7 Demultiplexers

As the name infers, a demultiplexer performs the opposite function to that of a multiplexer. A single data line can be connected to any one of the output lines provided by the choice of an appropriate select signal. If there are s select inputs then the number of output lines to which the data can be routed is $n = 2^s$. The structure of a demultiplexer is identical to that of a decoder. A basic 2-to-4 line decoder and its associated enable line is shown in Figure 5.10. If the enable line is now used as a data input the data can be routed to any one of the outputs. If, for example, $A = B = 0$ and the data input $= 1$ then output $Y0 = 1$ and the remaining three outputs are 0. Any decoder having an enable line can function as a demultiplexer and for this reason they are listed as decoder/demultiplexers in manufacturers' catalogues.

A typical example of a 3-to-8 line decoder/demultiplexer, the 74138, is shown in Figure 5.11. The 74138 with three enable inputs has a flexible enabling system. If the package is to be used as a demultiplexer then input lines G2A and G2B can be grounded and G1 can be used as the input line for data. When the input $G1 = 1$ the eight output NAND gates are enabled. If the select signal is now $ABC = 000$, the output $Y0 = 0$ while all other results remain at 1.

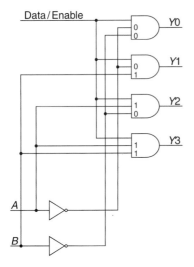

Figure 5.10 *Basic demultiplexer*

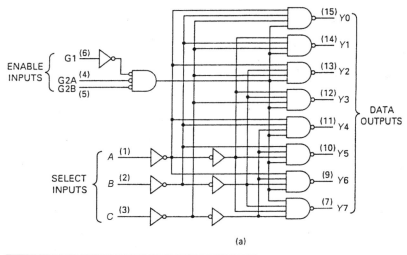

(a)

Inputs						Outputs							
Enable			Select										
G1	G2A	G2B	C	B	A	Y0	Y1	Y2	Y3	Y4	Y5	Y6	Y7
×	1	1	×	×	×	1	1	1	1	1	1	1	1
0	×	×	×	×	×	1	1	1	1	1	1	1	1
1	0	0	0	0	0	0	1	1	1	1	1	1	1
1	0	0	0	0	1	1	0	1	1	1	1	1	1
1	0	0	0	1	0	1	1	0	1	1	1	1	1
1	0	0	0	1	1	1	1	1	0	1	1	1	1
1	0	0	1	0	0	1	1	1	1	0	1	1	1
1	0	0	1	0	1	1	1	1	1	1	0	1	1
1	0	0	1	1	0	1	1	1	1	1	1	0	1
1	0	0	1	1	1	1	1	1	1	1	1	1	0

x = don't care

(b)

Figure 5.11   *The 74138 3-to-8 line decoder/demultiplexer*

5.8 Multiplexer/demultiplexer data transmission system

A simple data transmission system can be implemented using a multiplexer and a demultiplexer in conjunction with an interconnecting single line link. Such a system used over a relatively short distance such as 500 metres can result in a significant reduction in the number of lines required to transmit the data. A block diagram of the system is shown in Figure 5.12 where the 74251 8-to-1 multiplexer is linked to the 74138 3-to-8 line decoder, operating as a demultiplexer, by a single cable. The data presented in parallel at the MUX inputs is converted into a serial format for transmission, while at the receiving end the demultiplexer routes the serial data, in the correct sequence, to one of the eight output lines. The transmitted data is said to have been *time division multiplexed* since the eight input bits are appearing on the interconnecting link at different times.

For satisfactory communication, the select signals at the two ends of the link must be identical at any given instant. The common select signals generated by a Mod-8 counter at the transmitting end also have to be transmitted to the receiving end. The use of TDM has reduced the total number of lines required for the interconnection from eight to four.

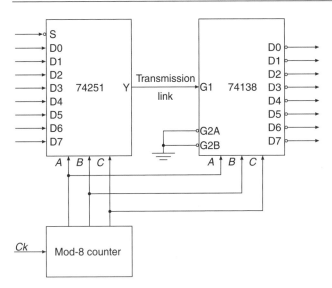

Figure 5.12   *A TDM transmission link*

A typical example where a short range transmission link might be employed is a security system where access to a building is monitored at a number of sensitive points. Signals from the outstations are time division multiplexed and transmitted to the demultiplexer at a central security office where a visual display will reveal any breach of security at the points of access.

5.9 Decoders

The basic function of an MSI decoder having n inputs is to select 1-out-of-2^n output lines. For example, if there are three inputs, the decoder will select 1-out-of-8 lines. In this case there is one output line for every input combination and the device is called a *complete decoder*. The selected output is identified either by a 1, when all other outputs are 0, or by a 0 when all other outputs are 1. In the first case the output is said to be active high while in the second case it is said to be active low.

The structure of a 2-to-4 line decoder is illustrated in Figure 5.13. It consists of an array of four NAND gates, one of which is selected for each combination of the input signals A and B. When $AB = 00$ the gate marked 00 is selected, and provided the chip has been enabled, the output of the gate marked 00 will be 0 while the outputs of the other three gates in the array are 1. A function table and a block diagram are also shown in Figure 5.13. Symbol X/Y on the block diagram indicates that the device converts from code X to code Y. Inputs A and B are allocated weights such that the weighting of B is 2^1 while that of A is 2^0. Hence the range of the sum of the weights is 0 to 3 and this is indicated on the block diagram by the notation $\frac{0}{3}$.

It will be seen that the logic diagram of the basic decoder is identical to that of the basic demultiplexer (see Figure 5.10) provided the data line is used to enable the decoder. The decoder may also be regarded as a minterm generator. Each output generates one minterm. An alternative way of looking at the decoder circuit is to regard A and B as address signals. Each combination of A and B defines a unique address which can access a location having that address. An example of this application occurs

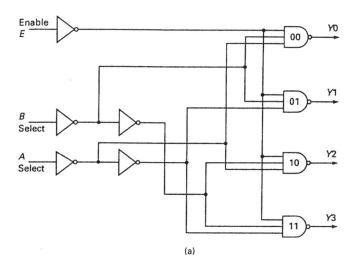

(a)

Inputs			Outputs			
E	B	A	Y0	Y1	Y2	Y3
1	x	x	1	1	1	1
0	0	0	0	1	1	1
0	0	1	1	0	1	1
0	1	0	1	1	0	1
0	1	1	1	1	1	0

x = don't care

(b)

(c)

Figure 5.13 *(a) Basic structure of a 2-to-4 line decoder (b) Function table (c) Block diagram*

in computer memories where an address decoder is used to access data stored in an address location identified by an address signal.

The basic decoder has only one level of logic. One n-input NAND or AND gate is required for each of the 2^n output lines. As n becomes large, the fan-in of the gates used also becomes large. For example, a 1Kbyte memory has 1024 memory locations, and using the simple structure in Figure 5.13 the address decoder would require 1024 ten-input AND or, alternatively, NAND gates. To alleviate the fan-in problem *tree* and *coincident* architectures are employed (see section 5.10).

A typical example of a decoder available in the TTL family is the 74138 3-to-8 line decoder. The gate level circuit, along with the function table, is shown in Figure 5.11. More flexible enabling arrangements are provided on this chip in that there are three independent enable pins. The Boolean function for enabling the chip is

$$En = G1 \cdot \overline{G2A} \cdot \overline{G2B} = \overline{\overline{G1} + G2A + G2B}$$

To enable the gates in the array $G1 = 1$ and $G2A = G2B = 0$. The selected output then depends upon the input combination of A, B and C. For $C = 0$ and $B = A = 1$ output $Y3 = 0$ and all the other outputs are 1.

Another commonly used decoding module in the TTL family is the 74154 4-to-16 line decoder. The logic circuit of the decoder and its function table are shown in Figure 5.14. A decoder can be used for converting any 4-bit code which is used to represent the decimal digits to give a decimal output. The Gray code tabulated in Figure 5.15 can be converted to decimal by selecting the appropriate outputs.

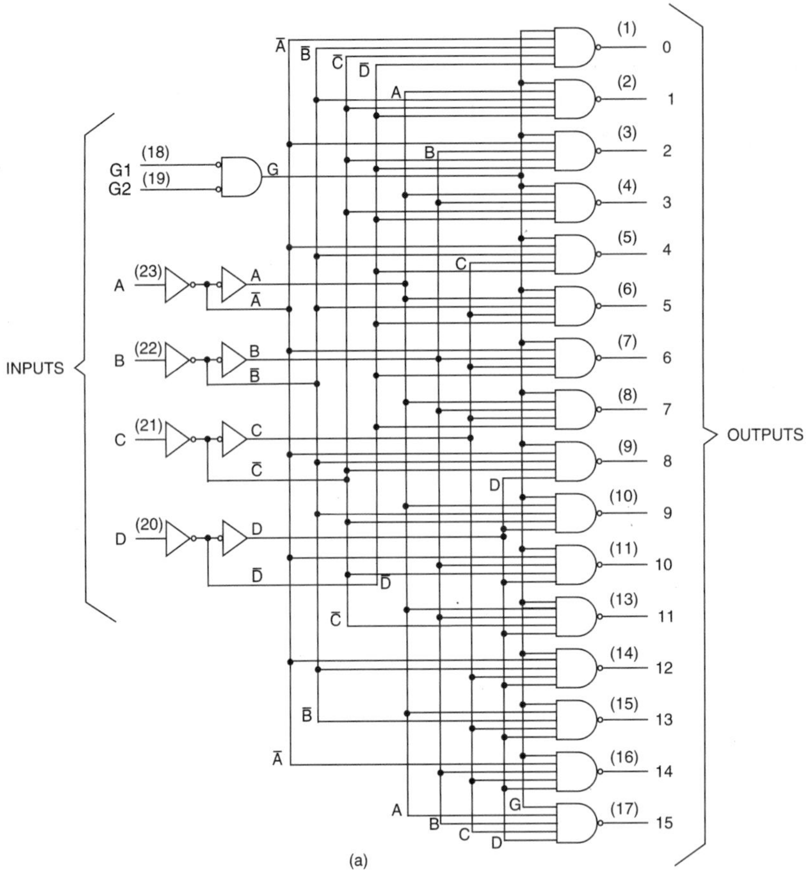

(a)

INPUTS						OUTPUTS															
G1	G2	D	C	B	A	0	1	2	3	4	5	6	7	8	9	10	11	12	13	14	15
L	L	L	L	L	L	L	H	H	H	H	H	H	H	H	H	H	H	H	H	H	H
L	L	L	L	L	H	H	L	H	H	H	H	H	H	H	H	H	H	H	H	H	H
L	L	L	L	H	L	H	H	L	H	H	H	H	H	H	H	H	H	H	H	H	H
L	L	L	L	H	H	H	H	H	L	H	H	H	H	H	H	H	H	H	H	H	H
L	L	L	H	L	L	H	H	H	H	L	H	H	H	H	H	H	H	H	H	H	H
L	L	L	H	L	H	H	H	H	H	H	L	H	H	H	H	H	H	H	H	H	H
L	L	L	H	H	L	H	H	H	H	H	H	L	H	H	H	H	H	H	H	H	H
L	L	L	H	H	H	H	H	H	H	H	H	H	L	H	H	H	H	H	H	H	H
L	L	H	L	L	L	H	H	H	H	H	H	H	H	L	H	H	H	H	H	H	H
L	L	H	L	L	H	H	H	H	H	H	H	H	H	H	L	H	H	H	H	H	H
L	L	H	L	H	L	H	H	H	H	H	H	H	H	H	H	L	H	H	H	H	H
L	L	H	L	H	H	H	H	H	H	H	H	H	H	H	H	H	L	H	H	H	H
L	L	H	H	L	L	H	H	H	H	H	H	H	H	H	H	H	H	L	H	H	H
L	L	H	H	L	H	H	H	H	H	H	H	H	H	H	H	H	H	H	L	H	H
L	L	H	H	H	L	H	H	H	H	H	H	H	H	H	H	H	H	H	H	L	H
L	L	H	H	H	H	H	H	H	H	H	H	H	H	H	H	H	H	H	H	H	L
L	H	X	X	X	X	H	H	H	H	H	H	H	H	H	H	H	H	H	H	H	H
H	L	X	X	X	X	H	H	H	H	H	H	H	H	H	H	H	H	H	H	H	H
H	H	X	X	X	X	H	H	H	H	H	H	H	H	H	H	H	H	H	H	H	H

H = high level, L = low level, X = irrelevant

(b)

Figure 5.14   *The 74154 4-to-16 line decoder (a) Logic circuit and (b) Function table*

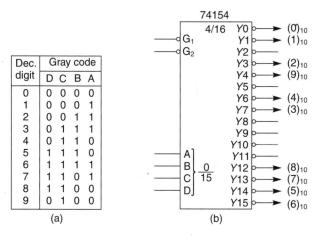

Dec.	Gray code			
digit	D	C	B	A
0	0	0	0	0
1	0	0	0	1
2	0	0	1	1
3	0	1	1	1
4	0	1	1	0
5	1	1	1	0
6	1	1	1	1
7	1	1	0	1
8	1	1	0	0
9	0	1	0	0

(a)   (b)

Figure 5.15 *Gray code to decimal conversion with 74154 4-to-16 line coder (a) Code tabulation (b) Implementation*

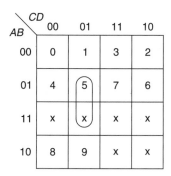

Figure 5.16 *K-map for NBCD/ decimal conversion*

There are two types of NBCD/decimal decoders available. First, the decoder that implements the minterm corresponding to decimal digit five, so that $(5)_{10} = \bar{A}B\bar{C}D$. For the second type, the digit $(5)_{10}$ is simplified by combining the minterm $\bar{A}B\bar{C}D$ with the adjacent 'can't happen' term $AB\bar{C}D$ so that $(5)_{10} = B\bar{C}D$, as illustrated on the K-map in Figure 5.16. A typical example of a 4-to-10 line NBCD/decimal decoder in the TTL family is the 7442 which rejects all false data since it is implemented without minimisation. For the decimal decoder that employs simplification techniques the appearance at the input terminals of the minterm $AB\bar{C}D$ will be recognised as $(5)_{10}$ or $(9)_{10}$ depending upon the simplification.

5.10 Decoder networks

When a large decoding network is required it cannot be implemented in a single MSI package because of the large number of pins needed. For example, a 6-to-64 line decoder requires seventy pins for input and output in addition to those for enabling the package and the voltage supply. The decoding range can be extended by interconnecting decoder chips. Two possible schemes are available, (a) *tree decoding* and (b) *coincident* or *2-dimensional decoding*.

A block diagram of a 4-to-16 line tree decoder is shown in Figure 5.17. It consists of the interconnection of five 2-to-4 line decoders. This requires three dual 2-to-4 line chips with four interconnections between the two levels of decoding where the select lines in the second level of decoding are commoned. An extension of the scheme would require extra levels of decoding and a 6-to-64 line decoder is obtained by the addition of one extra level of decoding consisting of a bank of sixteen 2-to-4 line decoders. The alternative approach of coincident decoding is also illustrated in Figure 5.17. Here the 4-to-16 line decoder consists of two 2-to-4 line decoders, available on a single

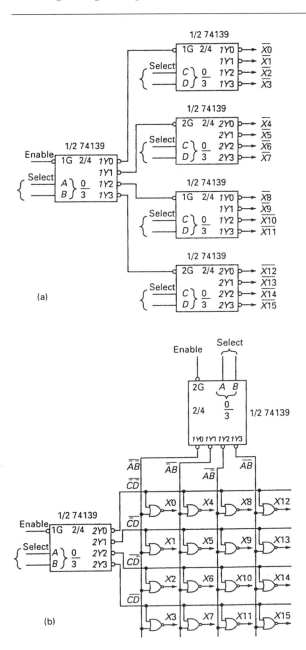

Figure 5.17 *(a) 4-to-16 line tree decoder (b) 4-to-16 line coincident decoder*

chip, and, in addition, sixteen NOR gates. Assuming that four 2-input NOR gates are available on a single chip, a total of five chips are needed for the implementation of this scheme. The coincidence scheme clearly requires more chips and more interconnections than the corresponding tree decoder and as the number of inputs increases, the superiority of the tree decoder becomes more marked.

In spite of this disadvantage, coincident decoders are widely used in conjunction with memory arrays because the NOR gates can be incorporated in the array. The choice of a coincident decoder in this application leads to a significant reduction in the number of lines to be taken to the memory array. In this example (see Figure 5.17) the tree decoder requires 16 lines, while for the coincident arrangement only 8 lines are needed and the gap widens significantly as the number of decoder inputs increases. For 10 inputs, the tree decoder needs 1024 lines compared with 64 for the coincident decoder.

5.11 The decoder as a minterm generator

The 16 outputs of a 4-to-16 line decoder such as the 74154 each correspond to the inverse of one of the sixteen minterms of four Boolean variables. If $A = B = C = D = 0$ the output 0 of the decoder is active low while all other outputs are 1, and can be identified as m_0. The decoder can generate the inverse of the 16 minterms and can be used in conjunction with one NAND gate to implement a Boolean function of four variables. As an example of this application the four variable function

$$f(A, B, C, D) = \sum 0, 1, 5, 8, 10, 12, 13, 15$$

is implemented in Figure 5.18.

The Boolean function can also be expressed in the following alternative minterm forms:

$$f = m_0 + m_1 + m_5 + m_8 + m_{10} + m_{12} + m_{13} + m_{15}$$
$$f = \overline{m_2 + m_3 + m_4 + m_6 + m_7 + m_9 + m_{11} + m_{14}}$$
$$f = \bar{m}_2 \cdot \bar{m}_3 \cdot \bar{m}_4 \cdot \bar{m}_6 \cdot \bar{m}_7 \cdot \bar{m}_9 \cdot \bar{m}_{11} \cdot \bar{m}_{14}$$

The first two forms of these equations require a decoder whose output is active high while the third form needs an active low decoder. The gates required at the decoder outputs in these three cases are OR, NOR and AND respectively, and their configuration will depend entirely on the available gate fan-in. If, for example, the maximum available fan-in for OR gates is four, then an interconnection of three gates

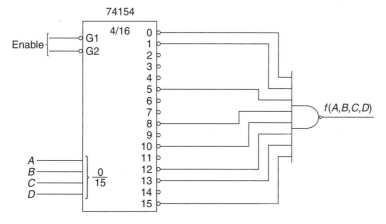

Figure 5.18 *Decoder implementation of a 4-variable Boolean-function*

would be needed and such a complication may well be regarded as uneconomic and an alternative implementation of the function would then have to be sought.

5.12 Display decoding

Many devices in everyday use such as calculators, digital watches, car radios and a wide range of measuring instruments have an illuminated decimal display. *Light emitting diode (LED)* or *liquid crystal display (LCD)* segments provide the illuminated display output. A single display element consists of seven segments arranged in the configuration shown in Figure 5.19. In the same diagram the numerical allocations and resultant displays for each of the 10 decimal combinations of four binary digits, after decoding, are also shown.

LEDs emit light energy when the anode of the device is positive with respect to its cathode. There are two possible connections, *common anode* and *common cathode*. For the common anode connection illustrated in Figure 5.20 the seven anodes are connected to a common voltage supply while the cathodes are controlled individually. To illuminate a segment an active low signal is required at the cathode. For the common cathode connection (see Figure 5.20) an active high signal on the individually controlled anodes is required in order to activate the LEDs.

There are two types of LCD in use. *Reflective* LCDs use ambient light such as sunlight or normal room light to activate the device. *Back-lit* LCDs use the light generated by part of the display. These devices have gained in popularity because of their low power consumption and are eminently suitable for battery powered displays. However it should be pointed out that LEDs always provide a much brighter display.

A typical example of a display decoder is the 7449 BCD/seven-segment decoder (see Figure 5.21) which consists of an array of seven AOI gates, one for each segment. The AOI gate is an AND/OR circuit followed by an inverter. For this reason the inverse function for each segment appears at the outputs and can be obtained by plotting the 0's tabulated in the truth table. For segment a the 0's in the column headed a are plotted on a K-map and simplified. This then gives the inverse Boolean function for the a segment:

$$a = BD + \bar{A}C + A\bar{B}\bar{C}\bar{D}$$

The inverse functions for the remaining six segments can be determined using the same method. When the blanking input is held at 1 the four NAND gates are enabled. If this input is held at 0 the NAND gates are disabled and the segment outputs are inhibited.

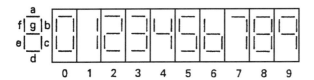

Figure 5.19 *Segment identification and resultant numerical displays*

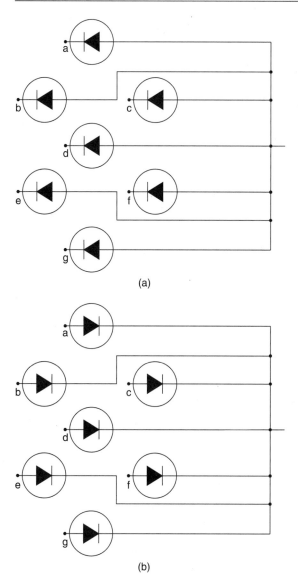

Figure 5.20 *LED segment connections (a) common anode (b) common cathode*

5.13 Encoder circuit principles

An encoder performs the inverse operation to that of a decoder. For an encoder having 2^n inputs there will be n outputs. Hence, for $n = 2$ there are four input lines and two output lines.

A typical example of the use of an encoding circuit is illustrated in Figure 5.22 where a number of *peripherals* P_0, P_1, P_2 and P_3 are serviced by a *central processing unit*. Each peripheral can generate a *flag* when it wishes to be serviced by the CPU. The flags from all the peripherals are ORed to generate a *master flag*. This signal requests the CPU to interrupt its current activity and jump to the service routine of the interrupting peripheral. It is then the function of the encoder to identify the peripheral whose flag has been raised.

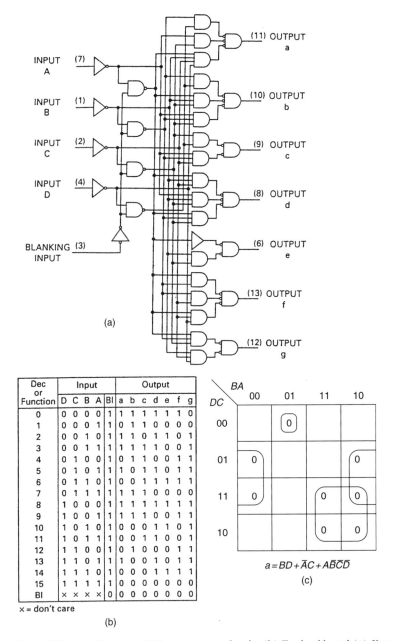

(a)

Dec or Function	Input					Output						
	D	C	B	A	BI	a	b	c	d	e	f	g
0	0 0 0 0				1	1	1	1	1	1	1	0
1	0 0 0 1				1	0	1	1	0	0	0	0
2	0 0 1 0				1	1	1	0	1	1	0	1
3	0 0 1 1				1	1	1	1	1	0	0	1
4	0 1 0 0				1	0	1	1	0	0	1	1
5	0 1 0 1				1	1	0	1	1	0	1	1
6	0 1 1 0				1	0	0	1	1	1	1	1
7	0 1 1 1				1	1	1	1	0	0	0	0
8	1 0 0 0				1	1	1	1	1	1	1	1
9	1 0 0 1				1	1	1	1	0	0	1	1
10	1 0 1 0				1	0	0	0	1	1	0	1
11	1 0 1 1				1	0	0	1	1	0	0	1
12	1 1 0 0				1	0	1	0	0	0	1	1
13	1 1 0 1				1	1	0	0	1	0	1	1
14	1 1 1 0				1	0	0	0	1	1	1	1
15	1 1 1 1				1	0	0	0	0	0	0	0
BI	× × × ×				0	0	0	0	0	0	0	0

x = don't care

(b)

$a = BD + \overline{A}C + A\overline{B}\,\overline{C}\overline{D}$

(c)

Figure 5.21   *(a) The 7449 BCD/seven segment decoder (b) Truth table and (c) K-map for segments*

The encoder truth table (see Figure 5.22) allocates one of the four combinations of the address variables A and B to each of the peripherals. The equations for the address variables and the master flag are

$$A = f_2 + f_3$$
$$B = f_1 + f_3$$
$$MF = f_0 + f_1 + f_2 + f_3$$

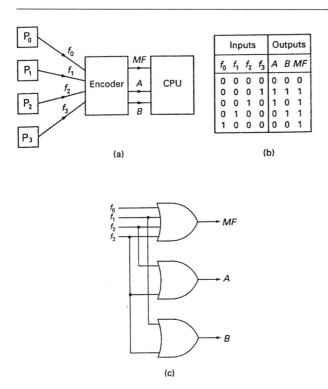

Inputs				Outputs		
f_0	f_1	f_2	f_3	A	B	MF
0	0	0	0	0	0	0
0	0	0	1	1	1	1
0	0	1	0	1	0	1
0	1	0	0	0	1	1
1	0	0	0	0	0	1

Figure 5.22 *(a) Block diagram for a 4-input encoding system (b) Truth table (c) Implementation*

The implementation of these equations is shown in Figure 5.22.

In this arrangement the encoder is designed to identify one, and only one, of the peripherals at any given instant. However, in practice, there is nothing to prevent two or more peripherals requesting service at the same time. To deal with this situation a system of priorities can be attached to the peripheral flags. When more than one flag is raised, the CPU services the peripheral whose flag has the highest priority. When it has been serviced, the flag is turned to the off condition and the peripheral having the next highest priority is serviced.

A truth table for a *priority encoder* and the circuit implementation are shown in Figure 5.23. The truth table assumes that the higher the subscript of the interrupting flag, the higher its priority. The following equations are obtained from the truth table:

$$A = f_3 + \bar{f_3}f_2 = f_3 + f_2$$
$$B = f_3 + \bar{f_3}\bar{f_2}f_1 = f_3 + \bar{f_2}f_1$$
$$MF = f_0 + f_1 + f_2 + f_3$$

5.14 Available MSI encoders

Two MSI encoder packages are available in the '74 series'. The 74147 has nine active low inputs, one for each of the decimal digits $(1)_{10}$ to $(9)_{10}$ inclusive, and encodes them to four active low outputs D, C, B and A. A block diagram and truth table for the package are shown in Figure 5.24. It should be observed that the digit $(0)_{10}$ is available when all the inputs are high and one immediate practical application of the device is

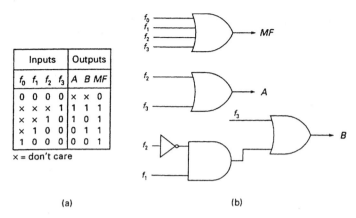

Inputs				Outputs		
f_0	f_1	f_2	f_3	A	B	MF
0	0	0	0	x	x	0
x	x	x	1	1	1	1
x	x	1	0	1	0	1
x	1	0	0	0	1	1
1	0	0	0	0	0	1

x = don't care

(a)　　　　　　　　　　　　(b)

Figure 5.23　*(a) Truth table for a 4-input priority encoder (b) Implementation*

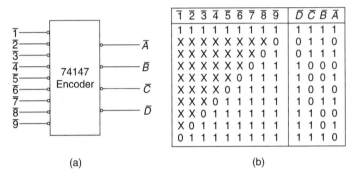

(a)

$\bar{1}$	$\bar{2}$	$\bar{3}$	$\bar{4}$	$\bar{5}$	$\bar{6}$	$\bar{7}$	$\bar{8}$	$\bar{9}$	$\bar{D}$	$\bar{C}$	$\bar{B}$	$\bar{A}$
1	1	1	1	1	1	1	1	1	1	1	1	1
X	X	X	X	X	X	X	X	0	0	1	1	0
X	X	X	X	X	X	X	0	1	0	1	1	1
X	X	X	X	X	X	0	1	1	1	0	0	0
X	X	X	X	X	0	1	1	1	1	0	0	1
X	X	X	X	0	1	1	1	1	1	0	1	0
X	X	X	0	1	1	1	1	1	1	0	1	1
X	X	0	1	1	1	1	1	1	1	1	0	0
X	0	1	1	1	1	1	1	1	1	1	0	1
0	1	1	1	1	1	1	1	1	1	1	1	0

(b)

Figure 5.24　*The 74147 decimal/NBCD priority encoder (a) Block diagram (b) Truth table*

the conversion of the ten decimal digits to the inverted form of the NBCD code. A second practical application is when the device is used in conjunction with a keyboard where the individual decimally identified keypads would generate an inverted NBCD output corresponding to each decimal digit. If two keys are pressed simultaneously the one having the higher decimal digit takes precedence.

Another practical example of an encoding circuit is the 74148 8-to-3 line priority encoder. A logic circuit diagram and the corresponding truth table are shown in Figure 5.25. The circuit consists of an array of four AOI gates with chip-enabling facilities provided by an active low enable input signal *EI*. A group select signal *GS* and an enable output signal *EO* are also provided when the encoder is to be operated in conjunction with other encoders. This situation will arise when the number of signals to be encoded is greater than eight. An active low signal at the *EI* input ensures that all the AND gates in the AOI array are enabled. The enable output equation is

$$EO = \overline{\overline{EI} \cdot 0 \cdot 1 \cdot 2 \cdot 3 \cdot 4 \cdot 5 \cdot 6 \cdot 7}$$

and this signal is active low, provided all the requesting signals are high and the chip has been enabled. The group select equation is

$$GS = EI + 0 \cdot 1 \cdot 2 \cdot 3 \cdot 4 \cdot 5 \cdot 6 \cdot 7 \cdot \overline{EI}$$

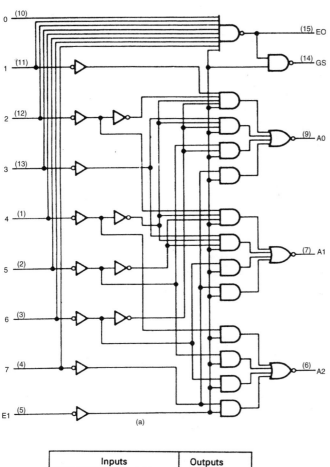

(a)

Inputs									Outputs				
EI	0	1	2	3	4	5	6	7	A2	A1	A0	GS	EO
1	×	×	×	×	×	×	×	×	1	1	1	1	1
0	1	1	1	1	1	1	1	1	1	1	1	1	0
0	×	×	×	×	×	×	×	0	0	0	0	0	1
0	×	×	×	×	×	×	0	1	0	0	1	0	1
0	×	×	×	×	×	0	1	1	0	1	0	0	1
0	×	×	×	×	0	1	1	1	0	1	1	0	1
0	×	×	×	0	1	1	1	1	1	0	0	0	1
0	×	×	0	1	1	1	1	1	1	0	1	0	1
0	×	0	1	1	1	1	1	1	1	1	0	0	1
0	0	1	1	1	1	1	1	1	1	1	1	0	1

× = don't care

(b)

Figure 5.25   *The 74148 8-to-3 line priority encoder (a) Logic diagram (b) Function table*

and this signal is active low provided the chip is enabled and at least one of the requesting signals is active low.

5.15 Encoding networks

Figure 5.26 illustrates the interconnection of nine 8-to-3 line encoders, four 8-input NAND gates and 3 NOT gates, to form a 64-to-6 line encoder network. Since the

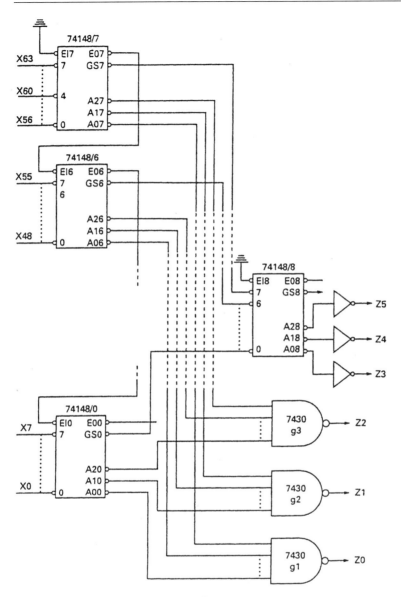

Figure 5.26   *A 64-to-6 line encoder network*

enable inputs $EI7$ and $EI8$ of encoders 7 and 8 are connected to ground, these chips are permanently enabled. Assuming that $X60 = 0$, then $EO7 = 1$ and encoder 6 and all subsequent encoders at the first level of encoding are disabled. In general, the encoding chips at this level are enabled until a chip is reached in the chain where at least one of the requesting signals $X_i = 0$, then all subsequent encoders are disabled.

Assume that for encoder 7, $X60 = 0$ so that the input pin labelled 4 is active low and consequently $A27 = 0$ and $A17 = A07 = 1$. Since all other first level encoders are disabled, their $A0$, $A1$ and $A2$ outputs are all 1. The $A27$ input to g3 $= 0$ while all other inputs to that gate are 1, hence $Z2 = 1$; for gates g2 and g1 all inputs are 1, hence $Z1 = Z0 = 0$.

The group select signal $GS7 = 0$ since encoder 7 is enabled and $X60 = 0$. All other group select signals on the first level encoders are 1. The input pin labelled 7 on encoder 8 is active low, hence $A28 = A18 = A08$ and after inversion of these signals $Z5 = Z4 = Z3 = 1$. It follows that if $X60 = 0$ the encoder network output is $111100 = (60)_{10}$.

Since one of the input signals to encoder 8 is active low, its group select signal $GS8$ is low, having made a transition from 1 to 0. This group select signal will therefore give an indication that at least one of the 64 requesting signals is active low.

If there are no active low requesting signals the 3-line outputs of all the eight first-level encoders are high; consequently all the inputs to g1, g2 and g3 are high so that $Z0 = Z1 = Z2 = 0$. Additionally, all first-level group select signals are high so that after inversion $Z3 = Z4 = Z5 = 0$ and the network output is 000000.

5.16 Parity generation and checking

When data is transmitted from one location to another it is desirable to know at the receiving end whether the received data is free of error. A simple form of error detection can be achieved by adding an extra bit to the transmitted word. This additional bit is called the *parity bit*.

The two different systems currently in use are the *even* and *odd* parity systems. In the even parity system the parity bit added to the word to be transmitted is chosen so that the number of 1's in the modified word are even. This is illustrated in the following example where the 7-bit ASCII code for the decimal digit $(9)_{10}$ is 0111001. An additional 0 in the most significant place is required to give even parity in the modified word which is now written

$(9)_{10} = 0$0111001

┗➤added parity bit

Alternatively, in an odd parity system, the added parity bit ensures that the modified code word contains an odd number of 1's. For the ASCII code for $(9)_{10}$ the modified codeword which will be transmitted is

$(9)_{10} = 1$0111001

┗➤parity bit

The truth table for a 3-bit even/odd parity generator is shown in Figure 5.27 where D_2, D_1 and D_0 represent the data to be transmitted, and p_o and p_e represent the odd and even parity bits to be generated by the parity generation circuit. The Boolean equation for p_e extracted from the truth table is

$$p_e = \bar{D}_2\bar{D}_1D_0 + \bar{D}_2D_1\bar{D}_0 + D_2\bar{D}_1\bar{D}_0 + D_2D_1D_0$$

and this equation can be manipulated algebraically to give

$$p_e = D_2 \oplus D_1 \oplus D_0$$

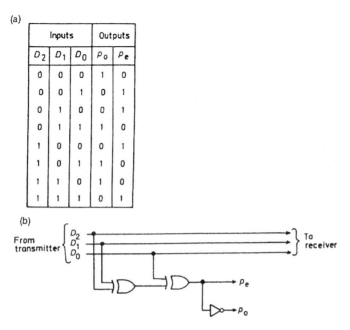

(a)

Inputs			Outputs	
D_2	D_1	D_0	p_0	p_e
0	0	0	1	0
0	0	1	0	1
0	1	0	0	1
0	1	1	1	0
1	0	0	0	1
1	0	1	1	0
1	1	0	1	0
1	1	1	0	1

Figure 5.27 *(a) Truth table for parity generator (b) Implementation of parity generator*

From an examination of the truth table it can be seen that p_0 is the inverse of p_e so that

$$p_0 = \overline{D_2 \oplus D_1 \oplus D_0}$$

The implementation of these equations is shown in Figure 5.27. The addition of extra data bits simply adds extra XOR terms to the above equations.

When the transmitted data arrives at the receiving end, a logic circuit is used to check the modified data. In even and odd parity checking the output is 1 when an error has been detected. A truth table for the two types of parity checking functions and the implementation of these functions is shown in Figure 5.28. The parity checking functions extracted from the truth table for the even and odd parity systems are:

$$F_e = D_2 \oplus D_1 \oplus D_0 \oplus p_e$$

and

$$F_o = \overline{D_2 \oplus D_1 \oplus D_0 \oplus p_0}$$

As a general rule in a digital system where the transmission link is relatively short, it may be assumed that the probability of a single-bit error is small and that of a 2-bit error and higher order errors is extremely small. The parity checking system just described will detect any odd number of errors, but it cannot detect an even number of errors because such errors will not destroy the parity of the transmitted group of bits.

A practical example of a 9-bit parity generator/checker is the 74180 MSI gate circuit shown in Figure 5.29 with its associated truth table. A tree structure of XOR and XNOR gates is used on this package and it is left to the reader to show that the output of the last XNOR gate in the tree is the complement of the XORing of the eight inputs A to H. The pair of AOI gates at the output provides the facility of an additional extra bit input which can be utilised in either the parity generation or checking modes.

Inputs				Outputs	
D_2	D_1	D_0	p_o or p_e	F_e	F_o
0	0	0	0	0	1
0	0	0	1	1	0
0	0	1	0	1	0
0	0	1	1	0	1
0	1	0	0	1	0
0	1	0	1	0	1
0	1	1	0	0	1
0	1	1	1	1	0
1	0	0	0	1	0
1	0	0	1	0	1
1	0	1	0	0	1
1	0	1	1	1	0
1	1	0	0	0	1
1	1	0	1	1	0
1	1	1	0	1	0
1	1	1	1	0	1

(a)

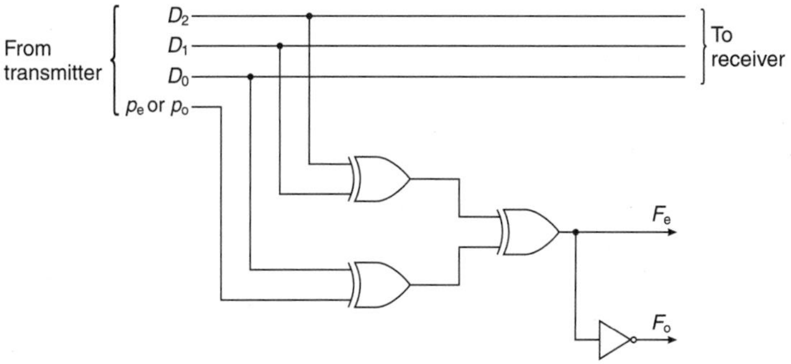

(b)

Figure 5.28 *(a) Truth table for parity-checking circuit (b) Implementation of parity checking function*

The 74280 (see Figure 5.30) is a more recent 9 bit parity generator/checker which utilises AOI gates rather than XOR or XNOR gates for implementing the parity function and has a shorter propagation time than the 74180. For the tree structure of the 74180 the number of logic levels required for implementation is greater than the number required for the 74280, giving a significant difference in the propagation times of the two devices.

Parity generation and checking for longer word lengths may also be achieved by cascading chips. For example, the two 74280s shown in cascade in Figure 5.31 will provide parity generation for a 17-bit word while ten 9-bit 74280s arranged in two levels can be used for parity generator/checkers for word lengths of up to 81 bits.

5.17 Digital comparators

The basic comparison element is the XNOR coincidence gate. The output of the gate is high, provided both inputs are either low or, alternatively, high. This is indicated in Figure 5.32 where $f = 0$ if $A = B = 0$ or, alternatively, if $A = B = 1$. In practice, comparators may be required to indicate more than equality. There are three possible

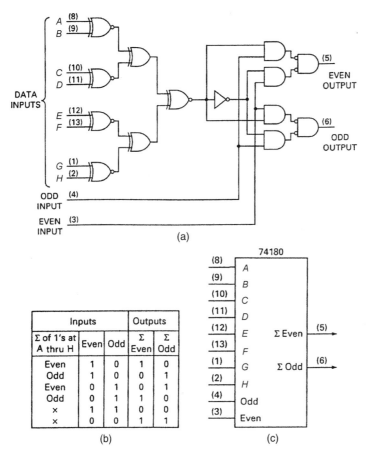

Figure 5.29 *The 74180 9-bit odd/even parity generator/checker (a) Gate circuit (b) Function table (c) Logic symbol*

conditions at the output of a comparator circuit, one for $A > B$, a second for $A = B$ and a third for $A < B$. A suitable NAND implementation for a single bit comparator which gives an output for all the three specified conditions is illustrated in Figure 5.33.

The usual problem for a comparator is the comparison of two multi-digit words such as $A = A_3 A_2 A_1$ and $B = B_3 B_2 B_1$. To compare two such words it is necessary to develop an algorithm which can be used as the basis of a hardware implementation. Such an algorithm is:

1. Examine the most significant pair of digits. If $A_3 > B_3$ then $A > B$; if $A_3 < B_3$ then $A < B$; if $A_3 = B_3$ no decision can be made about the relative magnitude of the two words and the next pair of digits must be examined.
2. If $A_2 > B_2$ and $A_3 = B_3$ then $A > B$; if $A_2 < B_2$ and $A_3 = B_3$ then $A < B$; if $A_3 = B_3$ and $A_2 = B_2$ no conclusion can yet be drawn regarding the relative magnitudes of the two words and the last pair of digits must be examined.
3. If $A_1 > B_1, A_2 = B_2$ and $A_3 = B_3$ then $A > B$; if $A_1 < B_1, A_2 = B_2$ and $A_3 = B_3$ then $A < B$; if $A_3 = B_3, A_2 = B_3$ and $A_1 = B_1$ then $A = B$.

If the most significant pair of digits are equal, then

$$E_3 = \bar{A}_3 \bar{B}_3 + A_3 B_3$$

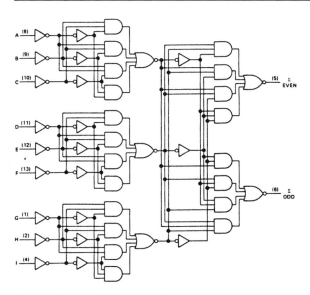

Figure 5.30 *The 74280 9-bit odd/even parity generator/checker*

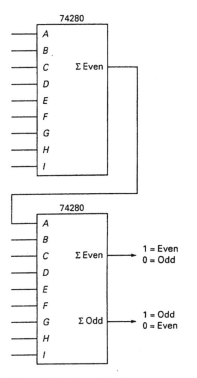

Figure 5.31 *Cascading of parity generators/ checkers*

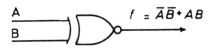

Figure 5.32 *The coincidence gate*

If the next most significant pair and the least significant pair are equal, then

$$E_2 = \bar{A}_2\bar{B}_2 + A_2B_2 \quad \text{and} \quad E_1 = \bar{A}_1\bar{B}_1 + A_1B_1$$

If $A = B$ then

$$E = E_1E_2E_3 = 1$$

The equation for determining whether $A > B$ is

$$A > B = A_3\bar{B}_3 + E_3A_2\bar{B}_2 + E_3E_2A_1\bar{B}_1$$

The first term in this equation $A_3\bar{B}_3 = 1$ if $A > B$, and if that is the case, then $A > B$. The second term $E_3A_2\bar{B}_2 = 1$ if $A_3 = B_3$ and $A_2 > B_2$ and if those two conditions exist then $A > B$. Finally, the third term $E_3E_2A_1\bar{B}_1 = 1$ if $A_3 = B_3$, $A_2 = B_2$ and $A_1 = B_1$. If those three conditions are satisfied then $A > B$.

The equation for determining whether $A < B$ is

$$A < B = \bar{A}_3B_3 + E_3\bar{A}_2B_2 + E_3E_2\bar{A}_1B_1$$

This equation has the same form as the equation for $A > B$ and can be developed using the same line of reasoning. Alternatively

$$A < B = \overline{E + A > B}$$
$$= \bar{E} \cdot \overline{A > B}$$

The implementation of a 3-bit comparator based on the single bit comparator of Figure 5.33 and using the equations developed above is shown in

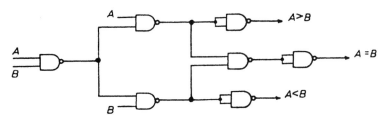

Figure 5.33   *The single-bit comparator*

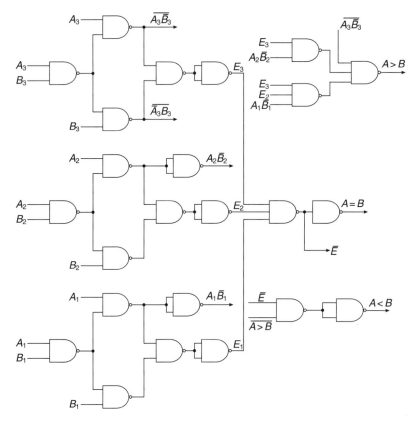

Figure 5.34   *The 3-bit comparator*

Figure 5.34. If an *Identity Comparator* is all that is required then the tree structure illustrated in Figure 5.35 will suffice.

There are a number of comparators available in the TTL family. The 7485 is a 16-pin 4-bit magnitude comparator providing three outputs, $A = B$, $A > B$ and $A < B$. Facilities are provided for cascading comparator chips so that words of greater length can be compared. The gate level circuit for the 7485 is shown in Figure 5.36 along with its function table. The logic of this circuit is based on the equations developed for the 3-bit comparator.

For each pair of input variables an **XNOR** gate generates the individual equality functions. The four equality functions and any input equality signal are ANDed to provide the $A = B$ output. The $A > B$ and $A < B$ functions are generated by two 6-wide AOI gates.

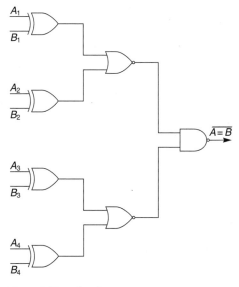

Figure 5.35   *4-bit identity comparator*

A number of different 8-bit comparators are available in the 74 TTL family. Some, such as the 74688, are identity comparators and generate an active low $P = Q$ output when comparing the magnitudes of two 8-bit words. Others, such as the 74682, produce active low outputs for $P = Q$ and $P > Q$. The third output $P < Q$ is obtained by applying the $P = Q$ and $P > Q$ outputs to the inputs of a NAND gate, as illustrated in Figure 5.37. The small triangle pointing in the direction of signal flow denotes that the 74682 package has amplification, that is, its output has a higher output current available than is usual in the MSI TTL series. The hysteresis symbol at the inputs denotes a package whose input characteristics exhibit hysteresis as would be the case with Schmitt Trigger circuits. Open-collector and totem-pole outputs are also available in the comparator group and some chips such as the 74886 have either one or two enable pins.

5.18 Iterative circuits

An iterative network consists of a number of identical cells interconnected in a regular manner as shown in Figure 5.38. $X_1, X_2, \ldots\ldots\ldots X_n$ are termed the *primary input* signals while $Z_1, Z_2, \ldots\ldots\ldots Z_n$ are termed the *primary output* signals. $a_1, a_2, \ldots\ldots\ldots a_{n+1}$ are termed the *secondary inputs* or *outputs* depending on whether these signals are entering or leaving a cell. The structure of an iterative circuit may be defined as one which receives the incoming primary data in parallel form where each cell processes the incoming primary and secondary data and generates a secondary output signal which is transmitted to the next cell. Secondary data is transmitted along the chain of cells and the time taken to reach steady state is determined by the delay times of the individual cells and their interconnections. The disadvantage of this design method is the amount of hardware required and the space it occupies. However, with the introduction of MSI and LSI circuits, the length of the interconnections has been reduced quite dramatically.

Magnitude comparison is a possible choice for an iterative design. It will be assumed that the two words to be compared, A and B, are to be scanned from the most significant end to the least significant end of the words. A block diagram for the ith cell is shown in Figure 5.39. There are two secondary signals, x_i and y_i, at the cell input and x_{i+1} and y_{i+1} at its output. Two secondary signals are required since there are three possible pieces of information to be transmitted along the chain which are defined by the following combinations of those signals:

1. $x_i y_i = 00$ $A = B$ up to cell i
2. $x_i y_i = 01$ $A > B$ up to cell i
3. $x_i y_i = 10$ $A < B$ up to cell i

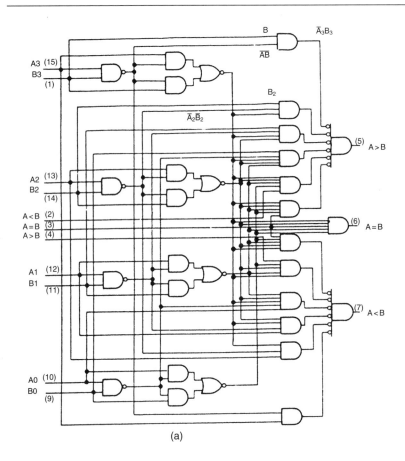

(a)

Comparing Inputs				Cascading Inputs			Outputs		
A3, B3	A3, B3	A3, B3	A0, B0	A > B	A < B	A = B	A > B	A < B	A = B
A3 > B3	×	×	×	×	×	×	1	0	0
A3 < B3	×	×	×	×	×	×	0	1	0
A3 = B3	A2 > B2	×	×	×	×	×	1	0	0
A3 = B3	A2 < B2	×	×	×	×	×	0	1	0
A3 = B3	A2 = B2	A1 > B1	×	×	×	×	1	0	0
A3 = B3	A2 = B2	A1 < B1	×	×	×	×	0	1	0
A3 = B3	A2 = B2	A1 = B1	A0 > B0	×	×	×	1	0	0
A3 = B3	A2 = B2	A1 = B1	A0 < B0	×	×	×	0	1	0
A3 = B3	A2 = B2	A1 = B1	A0 = B0	1	0	0	1	0	0
A3 = B3	A2 = B2	A1 = B1	A0 = B0	0	1	0	0	1	0
A3 = B3	A2 = B2	A1 = B1	A0 = B0	0	0	1	0	0	1

(b)

(c)

Figure 5.36   *The 7485 4-bit magnitude comparator (a) Gate level circuit (b) Function table (c) Logic symbol*

The K-map plotted in Figure 5.39 summarises the logical behaviour of the ith cell for all possible combinations of the primary and secondary input signals. For example, if $x_i y_i = 00$ and $A_i = 0$ and $B_i = 1$ then the secondary output signal combination $x_{i+1} y_{i+1} = 10$ which will indicate to the $(i + 1)$th cell that $A < B$.

Two separate maps for x_{i+1} and y_{i+1} are plotted and after simplification the following equations are obtained:

$$x_{i+1} = x_i + \bar{y}_i \bar{A}_i B_i$$
$$y_{i+1} = y_i + \bar{x}_i A_i \bar{B}_i$$

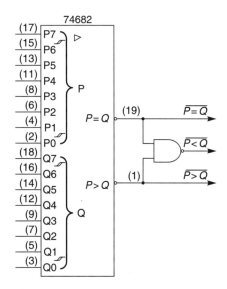

Figure 5.37 *The 74682 8-bit comparator connected to give the three outputs P=Q, P>Q and P<Q*

The first cell can be designed by assuming that the two digits preceding the most significant digits A and B are both 0's, hence $x_1 y_1 = 00$ and

$$x_2 = \bar{A}_1 B_1$$
$$y_2 = A_1 \bar{B}_2$$

In order to give magnitude comparison, an output circuit is required. The implementation of the ith cell and the magnitude output circuit are shown in Figure 5.39.

The gate count for the 7485 MSI 4-bit comparator and the iterative magnitude comparators is identical. However, the number of logic levels needed for the iterative comparator is more than twice the number needed for the 7485 and consequently there would be a significant difference in the propagation

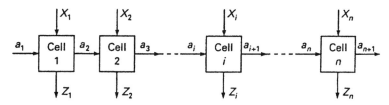

Figure 5.38 *Structure of an iterative circuit*

times for the two circuits due to the ripple effect in the iterative circuit. For this reason in practical situations the iterative comparator design is less likely to be used.

It is also possible to use the same technique to design an iterative adder. A typical cell has the two digits to be added as the primary inputs and the sum would appear at the primary outputs. The carry may have to travel the whole length of the chain of single-bit adder cells and a long delay occurs before steady state is reached. To eliminate the delay 4-bit adders are now provided with carry lookahead circuits.

Iteration can, however, be used in a wider sense of the word. If, for example, it is required to compare the magnitude of two 16-bit words, then four 4-bit 7485 comparators have to be connected in cascade. A cumulative delay due to the delays of the individual 4-bit packages will appear at the output of the last package in the cascade. Similarly, 16-bit addition requires four 4-bit 74283s connected in cascade thus forming an iterative array where the delay at the output is the sum of the individual delays of each 4-bit package.

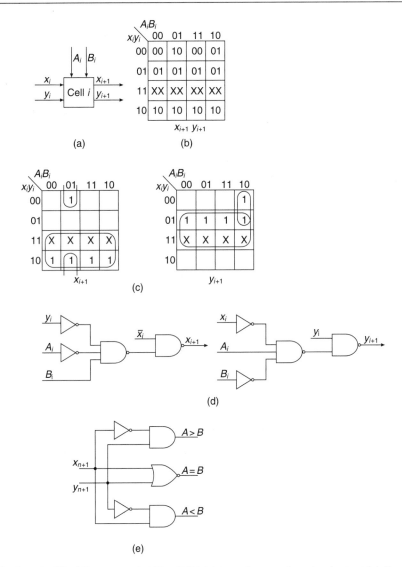

Figure 5.39   *Iterative Word Comparator (a) ith cell (b) joint map for secondary signal output (c) K-map plots for x_{i+1} and y_{i+1} (d) cell implementation (e) magnitude outputs*

Problems

5.1 Implement the following 3-variable Boolean functions using 4-input multiplexers:

(a) $f = \sum 0, 2, 3, 5, 7$, control variables A and B

(b) $f = \sum 1, 3, 4, 6, 7$, control variables B and C

(c) $f = \sum 0, 2, 4, 5, 6, 7$, control variables A and C

5.2 Implement the following 4-variable Boolean functions using 4-input multiplexers and NAND gates:

(a) $f = \sum 0, 1, 3, 5, 6, 8, 9, 11, 12, 13$, control variables A and B

(b) $f = \sum 0, 7, 8, 9, 10, 11, 15$, control variables B and C

(c) $f = \sum 0, 1, 3, 5, 9, 10, 11, 13, 14, 15$, control variables C and D

(d) $f = \sum 1, 8, 9, 12, 13, 14, 15$, control variables A and D

5.3 Implement the following 5-variable Boolean functions using 4-input multiplexers:

(a) $f = \sum 0, 1, 2, 3, 4, 8, 9, 11, 12, 13, 14, 18, 19, 20, 21, 25, 26, 29, 30, 31$

(b) $f = \sum 5, 6, 7, 8, 9, 10, 14, 15, 16, 17, 18, 19, 22, 23, 24, 25, 26, 29, 30, 31$

5.4 Implement the following 6-variable Boolean function using

(a) four-input multiplexers and NAND gates, and

(b) eight-input multiplexers and 4-input multiplexers.

$f = \sum 0, 1, 3, 5, 7, 12, 14, 16, 18, 20, 22, 26, 28, 30, 32, 34, 37, 39, 41, 43, 45,$
$50, 51, 53, 60, 61, 62, 63.$

5.5 Design a circuit for converting from the 8421 code to the 5421 code and implement the design with 4-to-1 multiplexers.

5.6 Design an NBCD to seven-segment decoder which is able to accept decimal information expressed in NBCD and generates outputs which select segments in the seven-segment indicator for displaying the appropriate decimal digit. The arrangement of the seven segments is shown in Figure P5.6(a) and the segmental representation of each decimal digit is shown in Figure P5.6(b). Implement the design using (a) NAND gates, (b) NOR gates and (c) 8-to-1 multiplexers.

(a)

(b)

Figure P5.6   *(a) Arrangement of segments (b) Segmental representation of decimal digits*

5.7 A combinational circuit is defined by the equations

$$f_1 = AB + \bar{A}\bar{B}\bar{C}$$
$$f_2 = A + B + \bar{C}$$
$$f_3 = \bar{A}B + A\bar{B}$$

Design a circuit which will implement these three equations using a decoder and NAND gates external to the decoder.

5.8 The tabulation below gives details of four frequently used codes. Using 4-to-10 line decoders and external logic, design three code converters for converting from 8421 to each of the other three codes.

Binary number	8421	2421	XS3	XS3 Gray
0000	0	0		
0001	1	1		
0010	2	2		0
0011	3	3	0	
0100	4	4	1	4
0101	5		2	3
0110	6		3	1
0111	7		4	2
1000	8		5	
1001	9		6	
1010			7	9
1011		5	8	
1100		6	9	5
1101		7		6
1110		8		8
1111		9		7

5.9   A combinational circuit is defined by the equations

$$f_1 = ABC + \bar{A}\bar{B}C$$
$$f_2 = \bar{A} + \bar{B} + C + D$$
$$f_3 = A + B + \bar{C}D + \bar{A}D$$
$$f_4 = ACD + A\bar{C}\bar{D} + B\bar{C}D + BC\bar{D}$$

Design a circuit which will implement these four equations using a decoder with NAND gates external to the decoder.

5.10 Design a 5-to-32 line decoder. The decoders available are:

(1) 2-to-4 line decoder, active low outputs and a single active low enable.

(2) 3-to-8 line decoder, active low outputs with two active low and one active high enable.

*5.11 Implement the following 4-variable functions using a decoder having active low outputs and NAND gates:

$$f_1 = \sum 0, 1, 3, 9, 12, 14$$
$$f_2 = \sum 5, 9, 10, 12, 13, 15$$
$$f_3 = \prod 0, 3, 8, 11, 12, 15$$
$$f_4 = \prod 1, 2, 7, 8, 11, 12, 14$$

5.12 Develop a 3-to-8 line decoder using NOR gates only, and draw its logic diagram.

5.13 Develop a circuit that resolves priority among eight active low flag inputs f_0 to f_7 where f_7 has highest priority. The address outputs of the encoder should be active high.

5.14 Draw the logic diagram for an 8-to-3 line encoder using just three 4-input NAND gates.

5.15 Develop a set of equations which can be used for implementing a circuit that compares two 4-bit words A and B and gives an active high output for each of the three possible conditions, $A > B$, $A = B$ and $A < B$.

5.16 Design an iterative circuit with the aid of a space state diagram that will give an output $Z = 1$ when three consecutive 0's have occurred in a string of binary digits.

5.17 Design an iterative circuit that will give an output $Z = 1$ when the sequence 010 occurs in a string of 10 binary digits that appear in parallel form.

6 Latches and flip-flops

6.1 Introduction

A digital logic circuit or system is usually made up of combinational elements such as NAND and NOR gates and memory elements which may, for example, be discrete flip-flops or latches. Alternatively, an interconnection of these devices may be found in a shift register, a counter, or in a variety of MSI and LSI packages.

With the introduction of memory elements as components in digital systems, an additional variable, time, has been introduced and must be taken into account when designing digital systems. In effect, logic operations can be performed sequentially, information being stored in a memory element and released at some specified instant later so that it can take part in a controlled combinational operation. Systems operating in this way are called *sequentially* operated systems.

There has always been considerable confusion over the use of the terms *latch* and *flip-flop*. It will be assumed in this book that a flip-flop is a device which changes its state at times when a change is taking place in the clock signal. The flip-flop is said to be either *leading edge* or *trailing edge* triggered, the edges referred to being those of the clock signal. On the other hand an *asynchronous latch*, without a control line, is continuously monitoring the input signals and changes its state at times when an input signal is changing. A *synchronous latch* is also continuously monitoring the input signals but in this case a change of state at the output can only occur when the control signal is active. In both cases the latch is driven by events, but for the synchronous latch the control signal has to be high before the input can be translated into a change at the output.

6.2 The bistable element

By cross coupling a pair of NAND gates which are both connected as inverters, a bistable element is formed. There are two possible states for the element: (a) $Q = 0$, $\bar{Q} = 1$ and (b) $Q = 1$, $\bar{Q} = 0$ (see Figure 6.1). Initially, when the circuit is switched on,

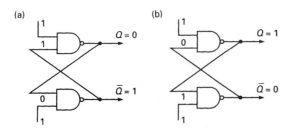

Figure 6.1 *(a) and (b) The two states of a pair of cross-coupled NAND gates*

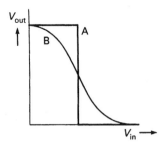

Figure 6.2 *Voltage transfer characteristic of a gate. A Ideal, B practical*

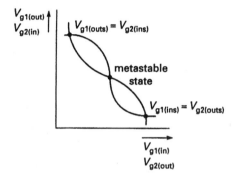

Figure 6.3 *The stable and metastable points of a pair of cross-coupled NAND gates*

the bistable element will take up one of these two states and without external intervention will remain in that state indefinitely, or until the power has been removed. Figure 6.2 shows the ideal voltage transfer characteristic A of a logic gate and it will be seen that there is a sudden change from 0 to 1 midway between logic 0 and logic 1. In practice, the gate characteristic will be similar to that shown in B.

The bistable element shown in Figure 6.1 is made up of two such gates whose characteristics are assumed to be identical. The two characteristics can be plotted on the same axes such that the $V_{g1(out)}$ and $V_{g2(in)}$ axes are coincident. Similarly, the $V_{g1(in)}$ and the $V_{g2(out)}$ axes are also coincident (see Figure 6.3). The two characteristic curves intersect at three points. Two of the points of intersection are the stable states referred to previously and are defined by $V_{g1(ins)} = V_{g2(outs)}$ and $V_{g2(ins)} = V_{g1(outs)}$.

The third point of intersection defines a *metastable state* which lies between the logic 0 and the logic 1 voltages. If the circuit should enter this state it can easily be shown that a small interfering noise voltage will immediately drive the circuit back to one of its two stable states. The state to which it will return depends upon the direction of the noise voltage relative to the metastable voltage. If, on the other hand, a small noise voltage occurs when the circuit is in either of its two stable states, then it will return to its original state.

6.3 The SR latch

The SR latch is shown symbolically in Figure 6.4(a), the set and reset inputs being labelled S and R respectively, and the complementary outputs Q and $\bar{Q}$ respectively. The *state table* for the latch is shown in Figure 6.4(b). In the first three columns of the table all combinations of the present states of S, R, and Q are tabulated, i.e. their states at time t. The fourth column is a tabulation of the next state of the latch, Q at time $t + \delta t$.

Examination of the table shows that a change of the state of the latch occurs in rows 4 and 5 only. In row 4 the latch is being reset or turned off, i.e. its state is changing from 1 to 0 as a consequence of the application of a reset input $R = 1$. In row 5 the latch is being set or turned on, i.e. its state is changing from 0 to 1 as a result of the application of a set input $S = 1$. For rows 1 and 2, $S = R = 0$, and there is no change of state. On row 3, $R = 1$ and this signal in normal circumstances would turn the latch off but $Q = 0$ and the signal $R = 1$ leaves the state unchanged. On row 6, $S = 1$ and this signal would normally turn the latch on, but $Q = 1$ and the latch is already turned on, and consequently there will be no further change of state. Finally, if $S = R = 1$ both

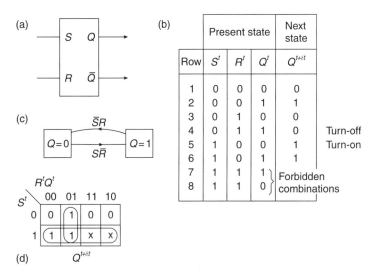

Figure 6.4 *The SR latch (a) Symbolic representation (b) State table (c) External state diagram (d) K-map plot*

outputs Q and $\bar{Q}$ are 1, and in general this is regarded as invalid circuit operation. For this reason the condition $S = R = 1$ is a forbidden input condition, this restriction being expressed algebraically by requiring $SR = 0$.

From the state table the turn-on condition is given by

$$\text{Turn-on} = S\bar{R}\bar{Q}$$

and the turn-off condition is given by

$$\text{Turn-off} = \bar{S}RQ$$

With the aid of these two equations the *external state diagram* can be constructed and is shown in Figure 6.4(c). The transition from $Q = 0$ to $Q = 1$ is made when $S\bar{R} = 1$ and the reverse transition occurs when $\bar{S}R = 1$.

Any change appearing at the output of the latch does so immediately after a change has taken place at the input and is delayed only by the propagation time of the gates that make up the latch. The *characteristic equation* of the latch is obtained by plotting those combinations for which $Q^{t+\delta t} = 1$ in conjunction with the 'can't happen' terms on the K-map shown in Figure 6.4(d). After simplification:

$$Q^{t+\delta t} = (S + \bar{R}Q)^t$$

The implementation of this equation using NAND gates is shown in Figure 6.5(a) and appears in more conventional form in Figure 6.5(b).

It should be observed that the characteristic equation is a Boolean equation but with a difference from the combinational equations that have been seen hitherto. Time has been introduced into the equation and the value of Q on the right-hand side of the equation may well be different from the value of Q on the left-hand side simply because these two values of Q are being observed at different times.

Simplifying the 0's and 'can't happen' terms in Figure 6.4(d) gives the simplest form of the complementary function. From the map:

$$\bar{Q}^{t+\delta t} = (\bar{S}\bar{Q} + R)^t$$

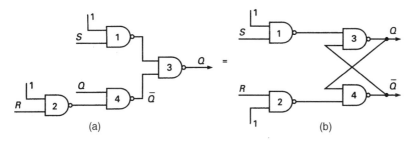

Figure 6.5 *(a) NAND implementation of the SR latch (b) the conventional representation*

and inverting this function

$$Q^{t+\delta t} = (\bar{S}\bar{Q} + R)^t$$

$$Q^{t+\delta t} = [(S + Q)\bar{R}]^t$$

This is the second form of the characteristic equation expressed as a product of sums and is implemented with NOR gates in Figure 6.6(a). The more conventional representation is shown in Figure 6.6(b).

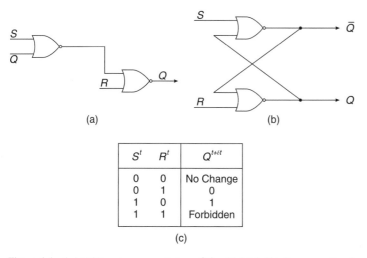

S^t	R^t	$Q^{t+\delta t}$
0	0	No Change
0	1	0
1	0	1
1	1	Forbidden

(c)

Figure 6.6 *(a) NOR gate representation of the SR latch (b) the conventional representation (c) state table*

Q^t	$Q^{t+\delta t}$	S	R
0	0	0	X
0	1	1	0
1	0	0	1
1	1	X	0

Figure 6.7 *The steering table for the SR latch*

The behaviour of the SR latch can be described in a slightly different way by means of the *steering table* shown in Figure 6.7. This table shows every possible output transition which can occur in the first two columns, including $0 \rightarrow 0$ and $1 \rightarrow 1$ both of which are regarded as transitions, while the last two columns give the values of S and R which will produce these transitions. For example, in the first row the $0 \rightarrow 0$ transition will occur providing $S = 0$ and $R = 0$ or 1. Since R can be either 0 or 1 this is indicated in the R column by the symbol X. For the second row the $0 \rightarrow 1$ transition is generated if $S = 1$. Since S and R cannot simultaneously be 1, it follows that $R = 0$. The entries for the other two rows can be determined in a similar fashion.

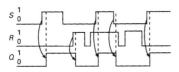

Figure 6.8 *SR latch timing diagram*

An ideal timing diagram for the SR latch is shown in Figure 6.8 where it is assumed that the changes in S, R and Q are instantaneous. Propagation delays are shown on the diagram and also arrows indicating those output transitions caused by a specified input transition. For $S = R = 1$ both Q and $\bar{Q}$ take the same logic level but when one of the inputs is returned to 0 the latch returns to its normal complementary behaviour.

If gates g_1 and g_2 are removed from the gate circuit shown in Figure 6.5(b) the SR latch is modified and becomes an $\overline{SR}$ latch. The stable condition for this latch is $\bar{S} = \bar{R} = 1$, and the forbidden state is $\bar{S} = \bar{R} = 0$. If, in the modified circuit, $Q = 0$, then $\bar{S}$ must make a transition from 1 to 0 to set the latch. Conversely, if $Q = 1$, $\bar{R}$ must make a 1 to 0 transition in order to reset the latch.

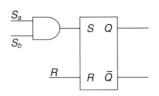

Figure 6.9 *SR latch with AND function for the set inputs*

Latches such as the 74279 are also available with more inputs. A typical example is shown in Figure 6.9 where there is a choice of two set inputs. The set function is equal to the AND of the two inputs S_a and S_b.

6.4 The controlled SR latch

By means of the simple modification shown in Figure 6.10(a), the transparency of the latch can be controlled by the signal G. If $G = 0$, the outputs of gates g_1 and g_2 will always be 1, irrespective of the present values of S and R, or of any changes which may occur in either of these two signals. When G makes a transition from $0 \rightarrow 1$, g_1 and g_2 are enabled and the latch becomes active. The state of the signals S and R at this time, or any subsequent $0 \rightarrow 1$ transitions of these signals during the active period, have an immediate effect on the output of the latch. A timing diagram illustrating this controlled transparency is shown in Figure 6.10(b). The inputs affected by control

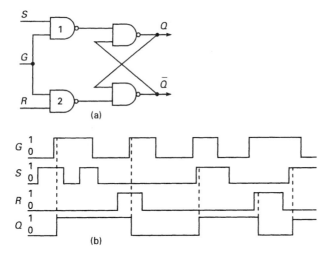

Figure 6.10 *(a) The controlled SR latch and (b) its timing diagram*

signal *G* are termed *synchronous inputs* and a latch operated with this type of control is termed a *gated latch*.

6.5 The controlled D latch

If an inverter is connected from the *S* input line to the *R* input of a controlled SR latch as shown in Figure 6.11(a), the circuit becomes a controlled D latch, and the symbolic representation of this modified latch is shown in Figure 6.11(b). Making this connection results in a modification of the SR state diagram. Since *S* and *R* can never be simultaneously 1 or 0, the first two rows and the last two rows of the SR state table can be deleted. As there is no independent *R* signal, the R column can also be deleted and the S column becomes the D column. The modified state table is shown in Figure 6.11(c) and the characteristic equation may be written as:

$$Q^{t+\delta t} = (D\bar{Q} + DQ)' = D^t$$

When the latch is enabled by the control signal *G*, it takes up the present state of the input signal *D*, delayed only by the propagation time of the latch. The external state diagram is shown in Figure 6.11(d) and a timing diagram for the latch appears in Figure 6.11(e). Latching points are shown at *X*, *Y* and *Z*. At these points the present state of *Q* is latched and cannot change until *G* makes a 0 → 1 transition again.

The controlled D latch has the advantage that it only requires one data input and there is no input condition that has to be avoided. It is also possible to have D latches

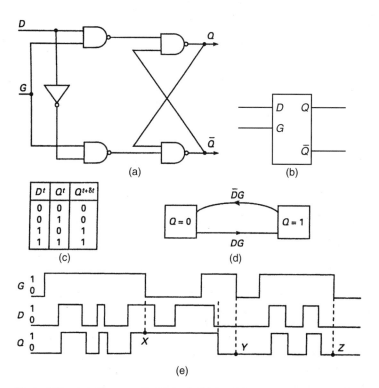

(a) *The controlled D latch* (b) *Symbolic representation* (c) *State table* (d) *External state diagram*

Figure 6.11 (a) *The controlled D latch* (b) *Symbolic representation* (c) *State table* (d) *External state diagram* (e) *Timing diagram*

Present state			Next state
G^t	D^t	Q^t	$Q^{t+\delta t}$
0	0	0	0
0	0	1	1
0	1	0	0
0	1	1	1
1	0	0	0
1	0	1	0
1	1	0	1
1	1	1	1

(a)

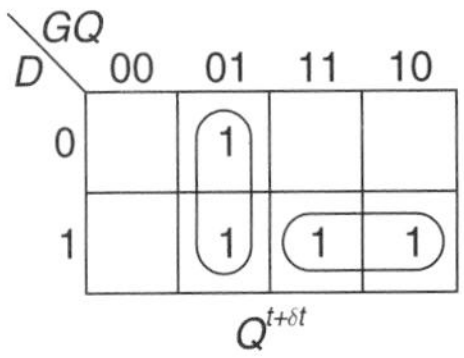

(b)

Figure 6.12   *Controlled D Latch (a) full state diagram (b) K-map plot*

with more inputs and the most common type is the *dual port latch* which has two *D* inputs and two control signals G_1 and G_2. Clearly the two control signals cannot simultaneously be 1.

The state table developed in Figure 6.11(c) does not take into account the action of the control signal *G*. A revised form of this table is shown in Figure 6.12(a) and a K-map plot of $Q^{t+\delta t}$ appears in Figure 6.12(b). This leads to the following modified characteristic equation:

$$Q^{t+\delta t} = (\bar{G}Q + DG)^t$$

Since there are two 1's in adjacent cells on the K-map, a *static hazard* is present in this K-map (see Chapter 9). To eliminate the hazard a third term *DQ* is added to cover the adjacent 1's, and the hazard-free characteristic equation of the controlled D latch follows below:

$$Q^{t+\delta t} = (\bar{G}Q + DG + DQ)^t$$

6.6 Latch timing parameters

There are three important timing parameters to be considered when designing circuits containing controlled latches:

Set-up time, t_{su}. The time interval preceding the deactivating transition of the control signal *G* during which the signal input must be maintained to ensure that it will be latched correctly.

Hold time, t_h. The time interval following the deactivating transition of the control signal *G* for which the input has to be maintained to ensure its latching.

Control pulse width, t_p. The time interval during which the control signal *G* is active.

The above timing parameters are illustrated in Figure 6.13. Other delays to be considered by the designer are the propagation delays defined earlier in Chapter 4.

Satisfying the hold and set-up times ensures that a latch input change provides a stable output state before the next input change occurs. This is of significant importance since latches are used in circuits that operate in the *fundamental mode*. This mode of operation requires that further input changes will not take place until a stable state has been reached.

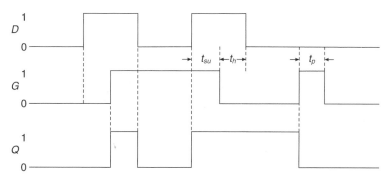

Figure 6.13 *Set-up time and hold time*

6.7 The JK flip-flop

The latch circuits previously described are not suitable for operation in synchronous sequential circuits because of their transparency. For synchronous circuits a clock signal is provided which governs the time at which the outputs of the memory elements are allowed to change state. In a synchronous circuit, flip-flops are used as the basic memory element, a typical example being the JKFF. Unlike latches, they only respond to a transition on a clock input or to a change in an asynchronous input such as Clear.

The symbolic representation of the JKFF is shown in Figure 6.14(a) and the state table describing its logical operation is in Figure 6.14(b). The logical operation of this

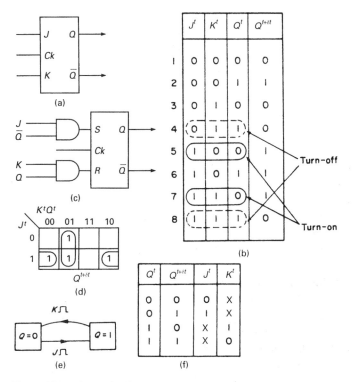

Figure 6.14 *The JK flip-flop (a) symbolic representation (b) state table (c) representation of JK flip-flop by an SR latch and two AND gates (d) K-map plot of $Q^{t+\delta t}$ (e) state diagram (f) steering table*

flip-flop differs in one respect from that of the SR latch in that it is allowable for J and K to be simultaneously equal to 1. When $J = K = 1$, the flip-flop *toggles*, i.e. in row 7 the flip-flop changes state from 0 to 1, while in row 8 the converse action takes place. In rows 4 and 5 normal reset and set operations take place, as described for the SR latch in section 6.3.

An examination of the state table shows that the flip-flop is turned on in rows 5 and 7, while it is turned off in rows 4 and 8. The turn on condition for Q is

$$S = J\bar{K}\bar{Q} + JK\bar{Q}$$
$$= J\bar{Q}$$

The turn off condition for Q is

$$R = \bar{J}KQ + JKQ$$
$$= KQ$$

These two equations indicate that a JK flip-flop may be regarded as an SR latch preceded by two AND gates which implement the turn-on and turn-off functions respectively, as illustrated in Figure 6.14(c).

The characteristic equation of the JK flip-flop is obtained by plotting the present state conditions on the K-map shown in Figure 6.14(d). After simplification, the characteristic equation can be written as

$$Q^{t+\delta t} = (J\bar{Q} + \bar{K}Q)$$

The state diagram describing the terminal behaviour of the flip-flop is shown in Figure 6.14(e). Assuming that the flip-flop is clocked and is presently in the state $Q = 0$ with $J = 1$ and Ck changing from 0 to 1, it makes a transition to the state $Q = 1$. Similarly, in the state $Q = 1$ with $K = 1$ and Ck changing from 0 to 1 it makes a transition to $Q = 0$.

A steering table for the JK flip-flop derived from the state stable is shown in Figure 6.14(f). Comparing the steering table of the SR latch and the JK flip-flop in Figures 6.7 and 6.14(f), it will be noticed that the JK flip-flop has more 'X' or 'don't care' input conditions. In practice, the increased number of 'don't care' terms leads to simpler combinational logic when designing a sequential logic circuit.

A JK flip-flop can be implemented by connecting the output of two AND gates in Figure 6.14(c) to the S and R inputs of the controlled latch shown in Figure 6.10(a). The Q and $\bar{Q}$ outputs of this latch and its clock connections are fed to the inputs of the two AND gates in conjunction with the J and K inputs, as shown in Figure 6.15(a). Notice that the AND gates are formed from two pairs of NAND gates in cascade, namely g_5 and g_7, and g_6 and g_8. Clearly, gates g_7 and g_1, and gates g_8 and g_2, give a double inversion and are redundant, thus reducing the JKFF to an array of four gates only, as shown in Figure 6.15(b).

As in the case of the controlled latches described earlier in this chapter, the flip-flop is disabled when $Ck = 0$ and is active when $Ck = 1$. Unfortunately, the connection shown in Figure 6.15(b) exhibits instability when $J = K = 1$ and $Ck = 1$ due to the feedback of the complementary output signals to the input. The state diagram indicates that under these conditions the Q output is oscillatory and will remain so until such time as the Ck makes a $1 \to 0$ transition when the clock is disabled.

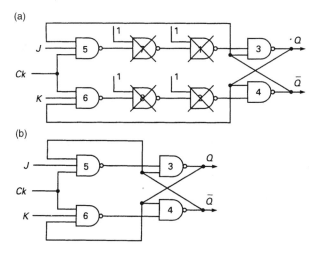

Figure 6.15 *(a) NAND implementation of JK flip-flop and (b) its reduced form*

6.8 The master/slave JK flip-flop

In order to overcome the difficulties described in the previous section, a master/slave JKFF can be used. It consists of two SR latches, the master and the slave, connected in cascade, as shown in Figure 6.16(a). The master is clocked in the normal way, while the clock signal to the slave is inverted. Assuming that changes in the J and K signals are only allowed to occur when the clock is low, the master then being disabled, changes in its output will take place on the rising edge of the clock pulse and these changes are transmitted to the input of the slave. However, no change can occur at the output of the slave until the rising edge of the inverted clock pulse, which is the trailing edge of the

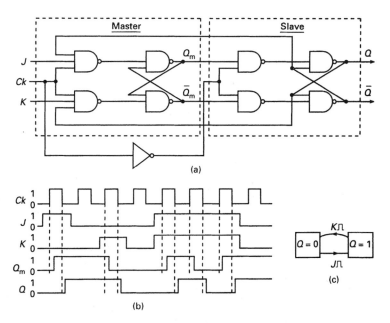

Figure 6.16 *(a) The master/slave JK flip-flop (b) The timing diagram (c) The external state diagram*

clock pulse. Consequently changes in Q and $\bar{Q}$ which are fed back to the input of the master do not take place until the trailing edge of the clock pulse arrives. In terms of the state diagram shown in Figure 6.16(c), when $J = K = 1$, the transition of Q from $0 \to 1$ is made on the trailing edge of the clock pulse. The flip-flop will remain in that state until the trailing edge of the next clock pulse when the reverse transition will take place. The flip-flop is then said to be operating in a *toggling* mode which is analogous to the unstable oscillatory condition described earlier. However, the toggling of Q is now controlled while the condition $J = K = 1$ is maintained, and the flip-flop will toggle on the trailing edge of each successive clock pulse. Timing diagrams for the master/slave JKFF are given in Figure 6.16(b).

6.9 Asynchronous controls (direct preset and clear)

As well as the J, K and Ck inputs a master/slave JKFF may have one or two additional controls which allow both the master and the slave to assume one of their two states irrespective of whether $Ck = 0$ or 1. These asynchronous controls are usually called *preset* and *clear*. Most commercially available flip-flops are provided with a clear control whereas the preset control is not quite as common. The operation of these controls is described in the table shown in Figure 6.17(a) and a circuit including these controls appears in Figure 6.17(b).

With both controls at 0 they are inactive, and the flip-flop is under the control of J, K and Ck. If $Cl = 1$ and $Pr = 0$ both master and slave are cleared so that $Q_m = Q = 0$. If $Cl = 0$ and $Pr = 1$ the flip-flop is preset and $Q_m = Q = 1$. Active high signals on the Pr and Cl lines will override signals on the J and K lines. These signals are normally only used during the asynchronous periods when the clock is low. Typically, the clear control might be used to clear all the flip-flops in an array when the power is first switched on.

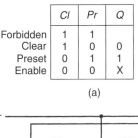

	Cl	Pr	Q
Forbidden	1	1	
Clear	1	0	0
Preset	0	1	1
Enable	0	0	X

(a)

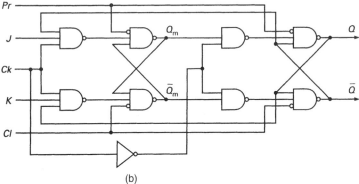

(b)

Figure 6.17 *(a) Pr and Cl control tabulation (b) Master/Slave JK flip-flop with Pr and Cl controls*

6.10 1's and 0's catching

Although the uncontrolled toggling has been eliminated by the master/slave connection, unfortunately the master/slave JK flip-flop exhibits another difficulty which may lead to faulty circuit operation in the presence of noise spikes. This phenomenon is termed *1's or 0's catching*.

In practice the JKFF may be in the hold condition with $J = K = 0$ and the outputs of the master and the slave $Q_m = Q_s = 0$; then, when the clock goes high, the master is enabled. If now a positive going noise spike appears at the J input, Q_m makes a transition from $0 \rightarrow 1$ and on the trailing edge of the same clock pulse Q_s also makes a $0 \rightarrow 1$ transition. This spurious transition is referred to as 1's catching. An example of 1's catching is illustrated in Figure 6.16. Similarly a $1 \rightarrow 0$ transition will be made by Q_s if initially $J = K = 0$ and a positive going spike appears at the K input when the clock is high. Such a transition is referred to as 0's catching, and is also illustrated in Figure 6.18.

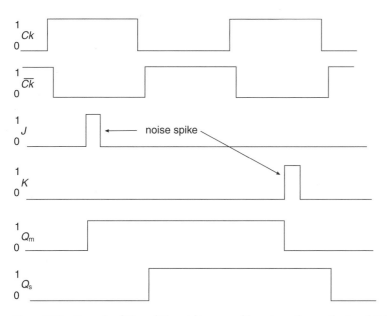

Figure 6.18 *Example of 1's and 0's catching caused by noise spikes on the J and K lines*

6.11 The master/slave SR flip-flop

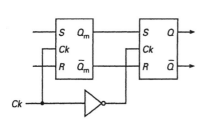

Figure 6.19 *Master/slave SR flip-flop*

It is also possible to construct a master/slave SR flip-flop from two SR latches connected in cascade as shown in Figure 6.19. In practice there is very little application for this device and it has been largely superseded in the manufacturers' catalogues by the master/slave JK flip-flop.

The master/slave SRFF, unlike the SR latch and the controlled SR latch, is no longer transparent. Any change at the output of the slave cannot take place until the trailing edge of a clock pulse. However, like the JKFF, faulty operation may occur due to 1's and 0's catching.

6.12 The edge-triggered D flip-flop

A negative-edge triggered D type master/slave flip-flop consists of a pair of D-latches connected, as shown in Figure 6.20(a). The master follows the D input while the clock is high, and latches the value of the input at the output of the master on the trailing edge of the clock pulse. The master is now disabled and will remain so until the clock goes high again. When the clock goes low the inverted clock signal at the clock input of the slave enables it, and the output of the master is transferred to the output of the slave. When the clock next goes high the slave is disabled and will remain so until the clock goes low again. Edge-triggering is indicated on the symbolic diagram in Figure 6.20(b) by the triangle at the clock input. This triangle is termed a *dynamic input indicator*. Timing diagrams describing the behaviour are shown in Figure 6.20(c).

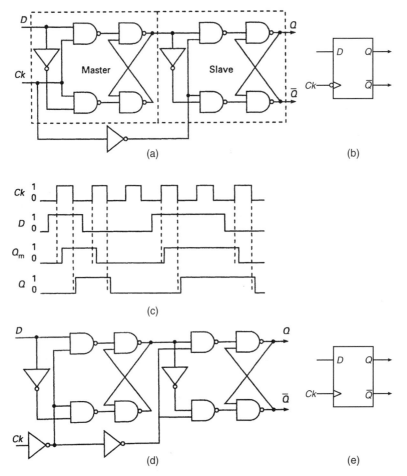

Figure 6.20   *Master/slave D-type flip-flip (a) Negative-edge triggered (b) Symblic representation (c) Timing diagram (d) Positive-edge triggered (e) Symbolic diagram*

The circuit of Figure of 6.20(a) can be modified to provide leading-edge triggering by including a second inverter in the clock line [see Figure 6.20(d)]. The corresponding symbolic diagram is shown in Figure 6.20(e).

The problem of 1's and 0's catching does not arise with this type of flip-flop. Assuming a negative-edge triggered device and that the leading edge of a positive going noise spike occurs when $Q_m = 0$ and the master clock $Ck = 1$, then the master latch will be set to 1. However, on the trailing edge of the spike, the master clock still being high, the master latch will be reset to 0 before the slave latch is enabled by the inverted clock signal.

An alternative configuration of a DFF that can operate in noisy conditions because of data lockout at the input has a wide range of applications. The flip-flop, which is leading-edge triggered, consists of three pairs of cross-coupled NAND gates, each pair constituting a basic $\bar{S}\bar{R}$ latch, of the type shown in Figure 6.21(a). The latch is in a stable state when $\bar{S} = \bar{R} = 1$, $Q = 0$ and $\bar{Q} = 1$. To change the state of the flip-flop $\bar{S}$ must make a $1 \rightarrow 0$ transition and this action will set the flip-flop to $Q = 1$ as shown in the diagram.

The three latches are interconnected as shown in Figure 6.21(b), with g_1 and g_2 comprising one latch while g_3 and g_4 comprise a second latch. The output latch is formed by gates g_5 and g_6. In order to maintain the output latch in a stable state, both $\bar{S}$ and $\bar{R}$ must be held at 1 and this is achieved when the clock $Ck = 0$ since the outputs of g_2 and g_3 are then 1. If additionally $D = 0$, then the remaining signals at different parts of the circuit can easily be determined, and they have been inserted in Figure 6.21(b).

When the data D is changed from $0 \rightarrow 1$ during the asynchronous period then the output of g_4 changes from $1 \rightarrow 0$ which initiates a $0 \rightarrow 1$ transition at the output of g_1 and that change is transferred to the input of g_2 as shown in Figure 6.21(b). The time delay before this change occurs is equal to the sum of the gate delays g_4 and g_1 and is the set-up time for the flip-flop. The inference is that there should be no change in Ck until after the elapse of the set-up time.

After the set-up time, the clock is allowed to go high, and as a consequence the output of g_2, $\bar{S}$, makes a $1 \rightarrow 0$ transition. The change in $\bar{S}$ initiates a change of state in the output latch and Q makes a $0 \rightarrow 1$ transition followed by a $1 \rightarrow 0$ transition in $\bar{Q}$. It should be noticed that there is no change in $\bar{R}$ as a consequence of the clock going high. This is because the lower input of g_3 made a $1 \rightarrow 0$ transition during the set-up time. The time taken for $\bar{S}$ to change $1 \rightarrow 0$ is the hold time and is equal to the gate delay of g_2. It is essential that there should be no change in D during this period. The changes taking place in the circuit after the clock transition from $0 \rightarrow 1$ are recorded in Figure 6.21(c).

If a change in D from $1 \rightarrow 0$ takes place after the hold time has elapsed and while the clock is still high, there will be no further change in the flip-flop output. The consequence of such a transition is that the output of g_4 makes a $0 \rightarrow 1$ transition which is transferred to one of the inputs of both g_1 and g_3 without affecting their outputs. Hence $\bar{S}$ and $\bar{R}$ remain unchanged, as do the outputs Q and $\bar{Q}$ of the flip-flop.

The set-up and hold times represent important timing constraints which have some influence on the maximum clock frequency at which the device can be operated. Additionally, if these two factors are not taken into consideration during the circuit design stage, data may be missed, and required transitions may not take place. It is also possible to have transient outputs, referred to as partial sets and resets. When these occur, a change of state has been initiated, but before it has been completed

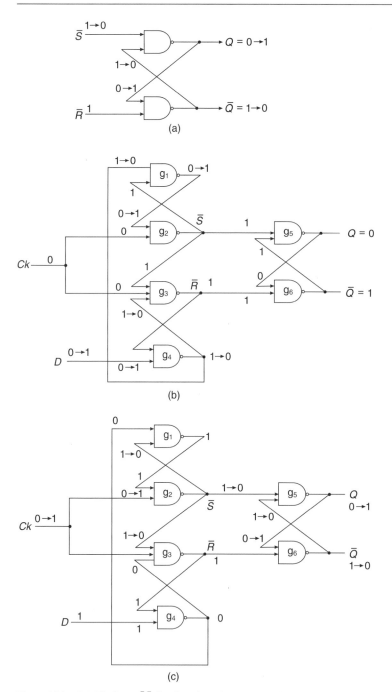

Figure 6.21   *(a) The basic $\bar{S}\bar{R}$ flip-flop (b) Edge triggered flip-flop. The diagram shows the effect of a $0 \rightarrow 1$ transition on the D line (c) Effect of a $0 \rightarrow 1$ transition on the clock line*

the flip-flop returns to its original state. Alternatively, the flip-flop may enter the metastable state and stay there for a time which cannot be precisely defined.

In general edge-triggered DFFs take up less space on a silicon chip than the edge-triggered JKFF and for this reason are the most widely used of the various flip-flops described in this chapter. Furthermore, the DFF having a single data input is easier to program.

6.13 The edge-triggered JK flip-flop

The edge-triggered DFF described in the previous section can be modified to provide an edge-triggered JKFF which eliminates the problem of 1's and 0's catching. The modified circuit is shown in Figure 6.22. In this circuit the outputs of the DFF are fed back to the AND/OR gates preceding the flip-flop, in conjunction with the J and K inputs. If $J = K = 1$ and the complementary output of the flip-flop $\bar{Q} = 1$ then the input to the DFF is $D = 1$ and the Q output becomes 1. If, on the other hand, $J = K = 1$ and the complementary output $\bar{Q} = 0$ then the input $D = 0$ and the output becomes $Q = 0$. Clearly for the condition $J = K = 1$, the flip-flop toggles. It is left to the reader to show that the flip-flop is turned on when $J = 1$, $K = 0$ and $Q = 0$ and that turn-off occurs when $J = 0$, $K = 1$ and $Q = 1$. Additionally, if $J = K = 0$ no change will occur at the output. Combining the above results leads to the steering table for the JKFF developed earlier in this chapter.

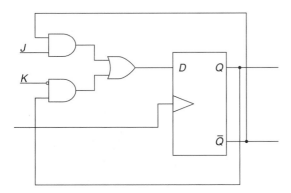

Figure 6.22 *The edge-triggered JK flip-flop*

6.14 The T flip-flop

This flip-flop is symbolically represented by the diagram shown in Figure 6.23(a) and its behaviour is described by means of the state table shown in Figure 6.23(b). It will be noted that if $T^t = 1$ and $Q^t = 0$ a transition is made such that $Q^{t+\delta t} = 1$, and if $T^t = 1$ and $Q^t = 1$ a transition is made such that $Q^{t+\delta t} = 0$. The circuit is said to toggle, and

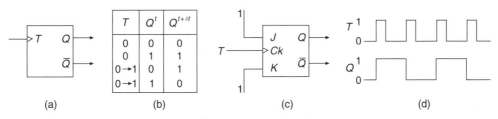

T	Q^t	$Q^{t+\delta t}$
0	0	0
0	1	1
0→1	0	1
0→1	1	0

(a) (b) (c) (d)

Figure 6.23 *The T flip-flop (a) Symbolic representation (b) State table (c) JK master/slave flip-flop connected as a T flip-flop (d) Timing diagram*

indeed, the TFF is frequently called a *toggle circuit*. The equation describing the behaviour of the flip-flop can be extracted from the state table. The equation is obtained by writing down the present state conditions which give a value of $Q^{t+\delta t} = 1$. Hence:

$$Q^{t+\delta t} = (\bar{T}Q + T\bar{Q})^t$$
$$= (T \oplus Q)^t$$

Rather than implementing the above equation, it is a simple matter to develop a T flip-flop from a master/slave JK flip-flop. All that is required is that the J and K inputs should be permanently connected to 1, as illustrated in Figure 6.23(c), and that the toggle signal T should be connected to the clock input. On the rising edge of every T input pulse, the flip-flop will change state, as shown in Figure 6.23(d). The flip-flop is now behaving in a toggling mode in the sense that the Q output is alternately taking up the 0 and 1 states. This circuit is the basis of all counting circuits. It is, in fact, a *scale-of-two counter*.

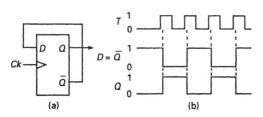

(a)   (b)

Figure 6.24   *(a) D flip-flop connected as a T flip-flop (b) Timing diagram*

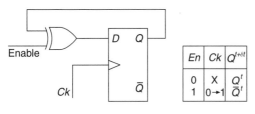

Figure 6.25   *T flip-flop with enable/disable facility*

Additionally, an examination of the timing diagram shows that the frequency of the output waveform Q is half the frequency of the T input and for this reason it is also called a *divide-by-two* circuit.

The D flip-flop can, like the JKFF, be converted to a TFF by connecting the $\bar{Q}$ output to the D-input and the toggling signal T to the clock input. The connections for this modification and a timing diagram are shown in Figure 6.24. An alternative connection, shown in Figure 6.25, uses an XOR gate to provide an enable/disable signal and the behaviour of the circuit is described in the accompanying table. Similar enable/disable arrangements can be provided with the converted JKFF.

6.15 Mechanical switch debouncing

Because of contact bounce it is almost impossible to obtain a clean transition from 5 to 0 V when the switch is moved from position 1 to position 2 in Figure 6.26(a). The voltage bounces between 0 and 5 V for a few milli-seconds before it settles to its steady value of 0 V. A typical voltage waveform is shown in the diagram and the voltage variations occurring are unacceptable in many circuits.

The effects of contact bounce can be eliminated at the output by using an $\bar{S}\bar{R}$ latch as shown in Figure 6.26(b). Assuming that the switch is presently in position 1, then $\bar{S}$ is low and the Q output is high. When the switch is moved to position 2, $\bar{S}$ goes high, $\bar{R}$ goes low and Q goes low a few nanoseconds afterwards. If the connection at position 2 is now broken due to contact bounce, both $\bar{S}$ and $\bar{R}$ are now high and no further voltage change takes place at the Q output. The converse action takes place if the switch is now moved back to position 1.

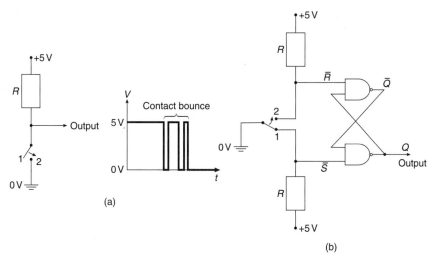

Figure 6.26 *(a) Effect of contact bounce (b) debouncing using an $\bar{S}\bar{R}$ latch*

6.16 Registers

Registers are very important elements in a digital system and the structure of these devices highlights the difference between the behaviour of latches and flip-flops. The basic requirements of a register are that it should be able to store data and that it should also provide the facility for moving data either to the right or left. For example, the 4-bit register storing the data 1011 may be required to shift this data to the right until the contents of the register are 0000. This process should be carried out in an orderly fashion, one bit at a time, so that after the receipt of the first shift pulse the contents of the register should be 0101. After the receipt of the second shift pulse the contents should be 0010, and after the receipt of four shift pulses the contents should be 0000.

An array of D latches is able to store the data 1011, as shown in Figure 6.27(a). On the application of a load signal at the control inputs of the latches, because of the transparency of the latches, the data appears almost immediately at the output of the latches. If the stored data is to be shifted to the right, the array of latches would be

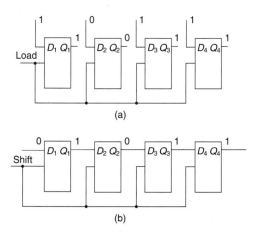

Figure 6.27 *(a) Four D latches store data and (b) the failure to shift data with D latches*

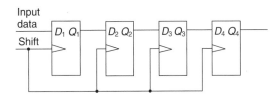

Figure 6.28   *A D flip-flop shift register*

connected, as shown in Figure 6.27(b). Unfortunately on the rising edge of the first shift pulse and because of the transparency of the latches the contents of the register would almost immediately become 0000, delayed only by the propagation times of the individual latches. Clearly, in order for the above scheme to allow a shift of one bit for each shift pulse, some delay must be inserted in the connection between each latch.

In practice, this delay can be provided by the edge-triggered D flip-flops shown in Figure 6.28. On the rising edge of the first shift pulse the data is held at the output of the master latches which are directly connected to the inputs of the slave latches; the slave latches are activated and the latch inputs are transferred to their outputs, which are of course the flip-flop outputs. The data has shifted one place to the right under the control of the first shift pulse. After the arrival of three more shift pulses the contents of the register will be 0000. The use of a flip-flop as a storage element in the register allows the orderly shift of data.

A more detailed account of the structure and operation of registers appears in the next chapter.

Problems

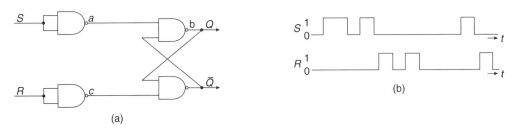

(a)   (b)

Figure P6.1

6.1 An SR latch constructed from NAND gates is shown in Figure P6.1(a). Determine the logic levels at points a, b and c under the following conditions:

    (a) $S = 0$, $R = 0$ and $Q = 0$
    (b) As in (a), but S changes from $0 \rightarrow 1$
    (c) $S = 0$, $R = 0$ and $Q = 1$, and R changes from $0 \rightarrow 1$.

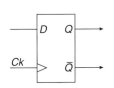

Waveforms for S and R are shown in Figure P6.1(b). Draw the corresponding waveform for Q assuming that the initial value of $Q = 0$.

6.2 A positive edge triggered D-type flip-flop combines two D latches, as shown in Figure P6.2. With the aid of a timing diagram show that the flip-flop senses the input data present at the rising edge of the clock and produces a corresponding output.

Figure P6.2

6.3 A master/slave JK flip-flop is shown in Figure P6.3. Assuming that the initial condition of the flip-flop is $J = K = Q_m = Q_s = 0$, trace the logic levels through the diagram for the following changes. (N.B.: changes in J and K take place in the time intervals between clock pulses.)

(i) $J = 0 \rightarrow 1, K = 0 \rightarrow 0$, Clock pulse 1 applied
(ii) $J = 1 \rightarrow 1, K = 0 \rightarrow 1$, Clock pulse 2 applied
(iii) $J = 1 \rightarrow 0, K = 1 \rightarrow 0$, Clock pulse 3 applied
(iv) $J = 0 \rightarrow 1, K = 0 \rightarrow 0$, Clock pulse 4 applied

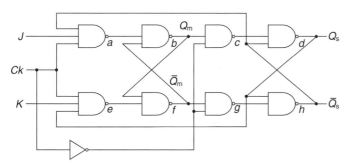

Figure P6.3

Draw a timing diagram displaying the J, K, Q_m and Q_s waveforms for the period of four clock pulses.

Assuming the same initial conditions, determine the final value of Q_s as the inputs are changed in the following order:

(v) $Ck = 0 \rightarrow 1, J = 0 \rightarrow 1, Ck = 1 \rightarrow 0$
(vi) $J = 0 \rightarrow 1, Ck = 0 \rightarrow 1, K = 0 \rightarrow 1, J = 1 \rightarrow 0, Ck = 1 \rightarrow 0$

6.4 With the aid of external logic, show that a D-type flip-flop can be converted to a JK flip-flop. Construct a timing diagram for the JK flip-flop and show that the circuit produces an output which depends only on the input data present at the instant of the rising edge of the clock pulse.

6.5 A JK flip-flop is modified, as shown in Figure P6.5, to form a J'K flip-flop. Draw the state table for this flip-flop and derive its characteristic equation.

6.6 Draw the external state diagram for the flip-flop whose characteristic equations are

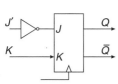

Figure P6.5

(a) $Q^{t+\delta t} = (X \oplus Y \oplus Q)^t$
(b) $Q^{t+\delta t} = (X \odot Y \odot Q)^t$

6.7 The waveforms shown in Figure P6.7(a) are to be applied to the circuit shown in Figure P6.7(b); assuming the initial value of $Q = 0$, determine the Q output.

6.8 Given the S and R waveforms for an SR latch shown in Figure P6.8 and assuming the initial value of $Q = 0$ plot the time variations of the Q output of the latch. How does the Q output vary if the latch is controlled by the G waveform?

6.9 Using timing diagrams analyse the behaviour of the clocked SR flip-flop shown in Figure P6.9.

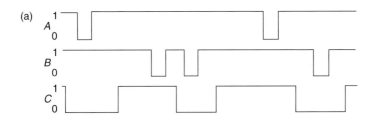

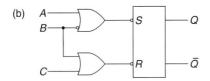

Figure P6.7

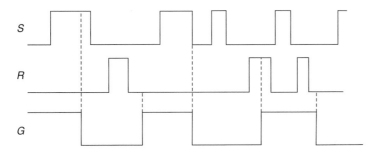

Figure P6.8

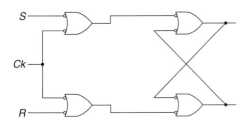

Figure P6.9

7 Counters and registers

7.1 Introduction

A counting circuit composed of memory elements, such as flip-flops and electronic gates, is the simplest form of sequential circuit available. All sequential circuits are of two types, (1) *synchronous* (clock driven) and (2) *asynchronous* (event driven). In synchronous circuits, changes in the circuit state are synchronised to the normally periodic clock pulses, whereas in event driven circuits state changes are governed by events such as, for example, the occurrence of a system fault.

Counting circuits can be in either of the two categories described above. All counter circuits count clock pulses and store the number received in an array of memory elements. In the case of synchronous counters the flip-flops are all clocked at precisely the same instant in time, whereas in an asynchronous circuit only the least significant stage is clocked, and succeeding flip-flops are clocked at later times which depend on the flip-flop propagation times. Design of synchronous counters is generally more complex than that of asynchronous counters.

Counters are fundamental and important components of a digital system and can be used for timing, control or sequencing operations. Alternatively, they can be used for frequency division and in some cases there may be a non-binary count, for example a Gray code counter or a BCD counter. In practice it would be most unusual for the logic designer to design a counter circuit since there are a large number available on MSI chips. Nevertheless, it is important that the reader should be aware of the basic design techniques employed.

7.2 The clock signal

An essential feature of a synchronous system is that flip-flops which are part of the system should all change at the same instant in time. This is achieved by the use of a synchronising signal which is formally known as the *clock*. The clock signal is normally periodic, and there must always be a sufficient time period between adjacent clock pulses to ensure that the combinational logic has reached its steady state condition before the next clock pulse in the sequence arrives.

In general, as shown in Figure 7.1, an idealised form of clock signal will approximate to a square wave, and the period in the cycle when the clock is high is termed the active period. Flip-flop transitions, initiated by the synchronising clock, are arranged to take place on either the leading edge or, alternatively, the trailing edge of a clock pulse, and these two types of flip-flop operation should not be used in the same circuit. The active clock edge will initiate a change of state in a synchronous circuit providing there is no

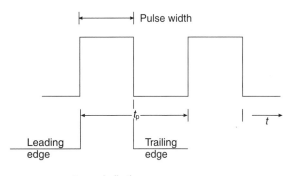

t_p = periodic time

Figure 7.1   *The clock pulse*

other external input to the circuit. It should be recognised that transitions in clock values from low to high and vice versa are never instantaneous, but providing the flip-flop changes take place during the course of the transition, the conditions for synchronous operation are satisfied.

7.3 Basic counter design

The simplest possible counter is the scale-of-two counter which has only two states, 0 and 1. Since the output of the flip-flop can only exist in one of these states at any time the counter can be implemented with a single flip-flop.

One design technique is to draw up a state table in which the first column represents the present state of the counter while the second column gives the next state of the counter after the arrival of a clock pulse, as shown in Figure 7.2(a). The table identifies

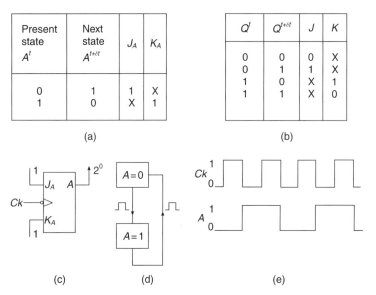

Figure 7.2   *Scale-of-two counter (a) State table (b) Steering table of JK flip-flop (c) Implementation (d) State diagram (e) Timing diagram*

the transitions that have to be made as the counter moves from its present state to the next state. Assuming that the circuit is to be implemented with a JK flip-flop, the inputs required to produce the transitions tabulated in the state table can be obtained from the JK flip-flop steering table shown in Figure 7.2(b). Since the entries in the J and K columns of the state table are all either 'don't cares' or 1 it follows that $J_A = K_A = 1$.

The counter is implemented in Figure 7.2 along with the state diagram and a timing diagram. The state diagram is both the internal and external state diagram since $A = 0$ and $A = 1$ represent the internal state of the circuit as well as being the externally displayed count. Examination of the timing diagram shows the flip-flop toggling continously from 0 to 1 and 1 to 0 but it should be recognised that the timing diagram is idealised since flip-flop delays and rise and fall times of the clock have not been taken into account.

A scale-of-four up counter has four states and requires two flip-flops. The design method used for the scale-of-two counter can be extended to cover the scale-of-four counter and the required flip-flop inputs are $J_A = K_A = 1$ and $J_B = K_B = A$.

For a scale-of-eight counter, the state table is tabulated in Figure 7.3. The design of the A and B stages employs the techniques used for the design of the scale-of-two counter and adding a further stage to the counter in no way alters the design of the

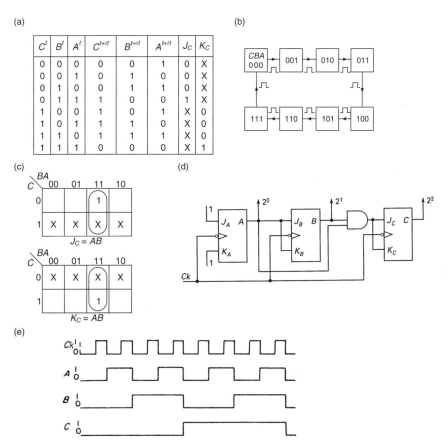

Figure 7.3 *Scale-of-eight counter (a) State table (b) State diagram (c) K-maps (d) Implementation (e) Timing diagram*

earlier stages. Hence $J_A = K_A = 1$ and $J_B = K_B = A$. K-maps for $J_C = K_C$ are plotted in Figure 7.3 and, after simplification, the input signals for the C flip-flop are found to be $J_C = K_C = AB$. Implementation of the counter and the state diagram are also shown in Figure 7.3.

Results for the three flip-flops are tabulated below:

$$J_A = K_A = 1 \qquad J_B = K_B = A \qquad J_C = K_C = AB = J_B B$$

and, by observation of these equations, it is clear that:

$$J_D = K_D = ABC = J_C C \quad \text{and}$$
$$J_N = K_N = ABC \cdots (N-1) = J_{(N-1)}(N-1)$$

	C	B	A	C̄	B̄	Ā
0	0	0	0	1	1	1
1	0	0	1	1	1	0
2	0	1	0	1	0	1
3	0	1	1	1	0	0
4	1	0	0	0	1	1
5	1	0	1	0	1	0
6	1	1	0	0	0	1
7	1	1	1	0	0	0

Figure 7.4 *Using the complementary outputs of a chain of flip-flops to count down*

Synchronous down-counters can also be designed using the techniques employed for upcounters, and the following flip-flop equations are obtained:

$$J_A = K_A = 1 \quad J_B = K_B = \bar{A} \quad J_C = K_C = \bar{A}\bar{B} = J_B\bar{B}$$
$$J_N = \bar{A}\bar{B}\bar{C}\cdots(\overline{N-1}) = J_{(N-1)}(\overline{N-1})$$

It is also possible, in the case of binary counters, to use an up-counter to count down by utilizing the complementary flip-flop outputs. This is illustrated for a scale-of-eight counter in the tabulation shown in Figure 7.4.

7.4 Series and parallel connection of counters

There are two ways of connecting the inputs to successive flip-flops and these are illustrated in Figure 7.5. In the first method, the gates providing the J and K inputs to adjacent flip-flops in the counter are all fed in parallel. As the number of stages increases, the fan-in to the AND gates also increases. However, the gate delay at the input to each flip-flop is identical and equal to t_g, the time delay of a single AND gate.

In the second method, the fan-in for each of the AND gates is always two, but the gate delay at the inputs to the flip-flops increases with the number of stages in the

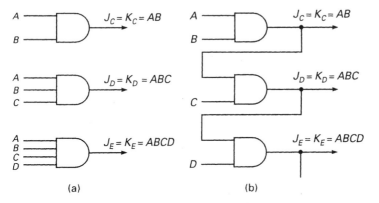

Figure 7.5 *Flip-flop input gates for (a) parallel connection and (b) series connection*

counter. Examination of Figure 7.5(b) shows that the gate delay at the J_C input is t_g, at the J_D input is $2t_g$, and so on. Since longer gate delays are experienced at each successive flip-flop input in the chain if this method of connection is used, it is clear that the upper frequency limit of a counter using this method is lower than one using the parallel connection.

If the switching time of individual flip-flops is t_f, then, for the parallel connection, the upper frequency limit is given by

$$f_u = \frac{1}{t_g + t_f}$$

While for the series connection

$$f_u = \frac{1}{(N - 2)t_g + t_f}$$

where N is the number of stages in the counter.

For the parallel connection, the first two flip-flops are required to drive $N - 2$ gates, the third flip-flop $N - 3$ gates and so on, whereas, for the serial connection, all the flip-flops in the counting chain, except the last one, are required to drive one gate only.

7.5 Scale-of-five up-counter

Often a counter with a scale that is not a power of 2 is required. For example, a scale-of-five counter has five states and requires three flip-flops. This will leave three unused states on the state diagram, as shown in Figure 7.6(a). The state table [Figure 7.6(b)] has been developed using the JK steering table, as illustrated in the case of the scale-of-2 counter (see Figure 7.2(b)). The unused states have been plotted as 'can't happen' terms on the K-maps for the flip-flop input signals J_C, J_B, J_A, K_B. All the entries in the K_A and K_C columns are either 1 or X. Hence $K_A = K_C = 1$.

If, for some reason, the counter enters one of the unused states, for example when the power is switched on, or due to faulty circuit operation, it is interesting to note its subsequent behaviour. This can be determined by examining the flip-flop input signals. For example:

$$(CBA)^t = 101: \quad J_C = AB = 0; \quad K_C = 1; \quad \text{FFC resets}$$
$$J_B = K_B = A = 1; \quad \text{FFB toggles}$$
$$J_A = \bar{C} = 0; \quad K_A = 1; \quad \text{FFA resets}$$

On the receipt of the next clock pulse, $CBA = 010$.

The next states for the unused states 110 and 111 are 010 and 000 respectively. If the counter should enter any one of the unused states it will return to the correct counting sequence after one clock pulse. The transitions that would occur under these circumstances are shown dotted on the state diagram in Figure 7.6(a). The implementation of the counter is shown in Figure 7.6(d).

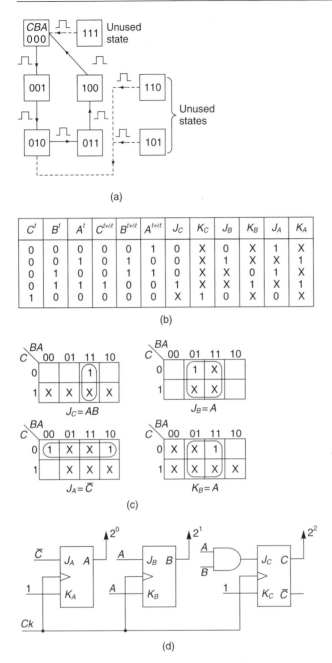

(a)

(b)

(c)

(d)

Figure 7.6   *Scale-of-5 counter (a) State diagram (b) State table (c) K-maps for flip-flop input signals (d) Circuit implementation*

In practice, it would be more logical to return all the unused states directly to the initial state $CBA = 000$ as shown in Figure 7.7(a). The state table for the modified counter is also shown in Figure 7.7 along with the K-maps for determining the flip-flop input signals. Implementation of the modified counter is shown in Figure 7.7(d) and it will be noted that a *Clear* signal has been provided for resetting the flip-flops to the starting state.

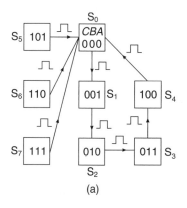

(a)

State	C^t	B^t	A^t	$C^{t+\delta t}$	$B^{t+\delta t}$	$A^{t+\delta t}$	J_C	K_C	J_B	K_B	J_A	K_A
S_0	0	0	0	0	0	1	0	X	0	X	1	X
S_1	0	0	1	0	1	0	0	X	1	X	X	1
S_2	0	1	0	0	1	1	0	X	X	0	1	X
S_3	0	1	1	1	0	0	1	X	X	1	X	1
S_4	1	0	0	0	0	0	X	1	0	X	0	X
S_5	1	0	1	0	0	0	X	1	0	X	X	1
S_6	1	1	0	0	0	0	X	1	X	1	0	X
S_7	1	1	1	0	0	0	X	1	X	1	X	1

(b)

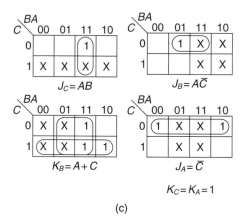

$K_C = K_A = 1$

(c)

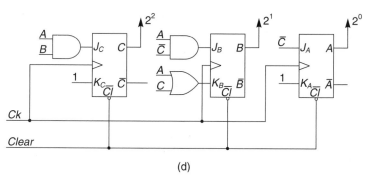

(d)

Figure 7.7 *Modified scale-of-5 counter*

7.6 The design steps for a synchronous counter

The scale-of-eight and the scale-of-five counters have been designed from basic principles in sections 7.3 and 7.5, and it is convenient at this point to summarise the design steps in the form of the following algorithm.

1. Define the count sequence.
2. Construct a state table for the counter where the left hand column is a tabulation of the present state, and the right hand column is a tabulation of the corresponding next states.
3. Any unused states should be tabulated in the present state column of the state table and should have, as their next state, the initial state in the count sequence, usually 000.
4. Select the type (D, JK, or RS) and number of flip-flops to be used, bearing in mind that $2^n \geq p$ where n is the number of flip-flop outputs and p is the magnitude of the count sequence.
5. Tabulate the flip-flop inputs for each change of the state of the counter as specified by the state table.
6. Plot the tabulated FF input signals on K-maps.
7. Simplify the FF input signals wherever possible.
8. Implement the counter, including the Clock and Clear signals.

Although DFFs can be used for the design of synchronous counters, the designer must recognise that the flip-flop transitions are taken directly from the next state entries and consequently there are no 'don't cares' available for simplification of the flip-flop input functions, and this leads to more complex logic.

Using the steps set out in the algorithm, a decade-up counter has been designed. The state diagram, state table, the tabulation of the JK input signals and their corresponding K-maps, and the implementation of the counter are all shown in Figure 7.8. It is suggested that as an exercise in logic design, the reader should check the validity of this design.

The BCD count frequently has to be displayed in decimal form. The simplest possible decimal representation is obtained by using a 4-to-10 line decoder. The ten outputs of the decoder may be active low or active high, depending on the MSI decoder selected. However, this method only gives an indication that a particular decimal digit has been received. More frequently, a BCD/seven segment decoder would be employed, as described in Chapter 5. In this case, the decimal digit received will be displayed as a decimal digit by the seven segments.

7.7 Gray code counters

Consider the transition from state 0001 to 0010 in the decade binary up-counter and assume that FFB changes faster than FFA. The sequence of changes that take place is:

```
D C B A
0 0 0 1
0 0 1 1 (transient state)
0 0 1 0
```

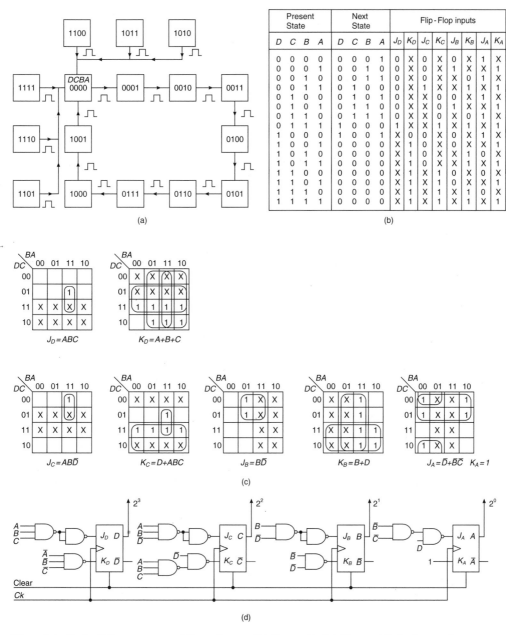

Figure 7.8   *Scale-of-10 up counter (a) State diagram (b) State table (c) K-maps (d) Implementation*

If a 4-to-10 line decoder is being used to convert the binary output of the counter to a decimal representation, a spike will occur on the $(3)_{10}$ output, and this is clearly incorrect circuit operation. This can occur at any point in the counting sequence where more than one flip-flop is required to change state during a transition. Faulty operation of this kind can be eliminated by using a Gray code counter in which only one flip-flop changes state at each transition.

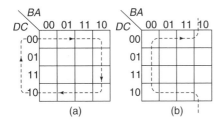

Figure 7.9   *(a) and (b) Alternative Gray code paths through K-maps*

There are a number of Gray codes suitable for decade counting and they can be developed by plotting a closed path on a K-map that consists of ten adjacent cells. Two such examples are shown in Figure 7.9. Both of them are reflected binary codes, the first, in Figure 7.9(a), being reflected about 110, the first combination in this sequence being 0000; while the second, in Figure 7.9(b) is reflected about 100, the first combination in the sequence being 0100.

7.8 Design of decade Gray code up-counter

The counter is to be designed using the Gray code established in Figure 7.9(a). All the unused states are to be returned to the initial count combination 0000. This will ensure that if the counter enters one of the unused states due to faulty operation it will return to the correct count sequence after the receipt of a single clock pulse. The state table gives the transitions for each of the JKFFs as the counter progresses from one state to the next, and with the aid of the JK steering table (see Figure 7.2) the flip-flop input signals J and K can be obtained for each transition. These signals are tabulated on the right-hand side of the state table. Eight K-maps are needed, one for each of the flip-flop input signals, and the J and K inputs are obtained after map simplification. The state diagram, state table, K-maps and counter implementation are all shown in Figure 7.10.

7.9 Scale-of-16 up/down counter

In many applications a counter must be able to count both up and down. For a scale-of-16 up-counter the equations are:

$$J_{Au} = K_{Au} = 1; \quad J_{Bu} = K_{Bu} = A; \quad J_{Cu} = K_{Cu} = AB; \quad \text{and} \quad J_{Du} = K_{Du} = ABC$$

and for a scale-of-16 down counter the equations are:

$$J_{Ad} = K_{Ad} = 1; \quad J_{Bd} = K_{Bd} = \bar{A}; \quad J_{Cd} = K_{Cd} = \bar{A}\bar{B}; \quad \text{and} \quad J_{Dd} = K_{Dd} = \bar{A}\bar{B}\bar{C}$$

Normally, a control signal Z is available for controlling the direction of the count. Counting up takes place when $Z = 1$ and counting down when $Z = 0$. The modified equations for up/down counting are:

$$J_A = K_A = 1$$
$$J_B = Z J_{Bu} + \bar{Z} J_{Bd} = ZA + \bar{Z}\bar{A}$$
$$K_B = Z K_{Bu} + \bar{Z} K_{Bd} = ZA + \bar{Z}\bar{A}$$

Similarly

$$J_C = K_C = ZAB + \bar{Z}\bar{A}\bar{B}$$

and

$$J_D = K_D = ZABC + \bar{Z}\bar{A}\bar{B}\bar{C}$$

The implementation of the counter is shown in Figure 7.11.

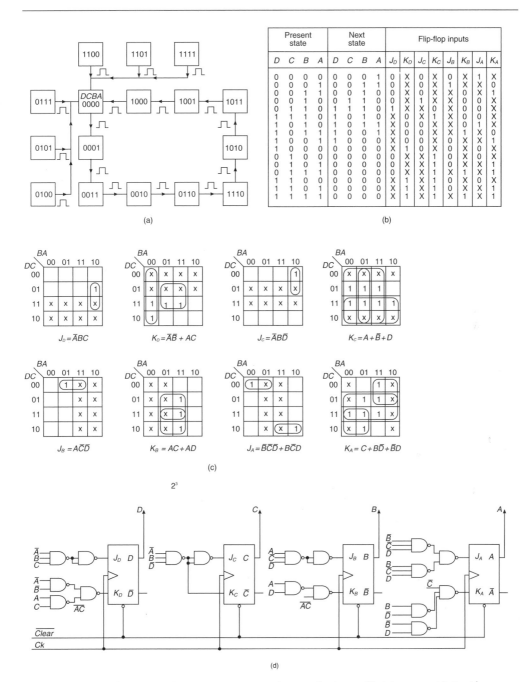

Figure 7.10   *Scale-of-10 Gray code up-counter (a) State diagram (b) State table (c) K-maps (d) Implementation*

7.10 Asynchronous binary counters

The simplest type of counter available is the 'ripple through' or asynchronous counter. For this type of counter the individual FFs are not controlled by a synchronous clock pulse. Withdrawal of clock synchronisation reduces the amount of circuitry required for implementation of the counter. For counts that are powers of 2, the counter

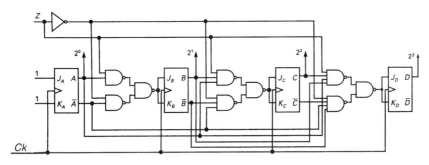

Figure 7.11 *Scale-of-16 up/down counter*

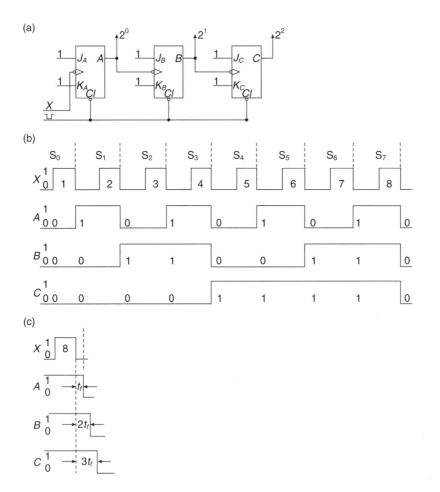

Figure 7.12 *(a) Implementation of, and (b) timing diagrams for a three-stage ripple-through counter (c) The ripple-through effect in the counter*

consists of a cascade of TFFs (JKFFs with $J = K = 1$) as shown in Figure 7.12(a). The output of each flip-flop provides the clock signal for the next one in the chain and the input signal pulses are connected to the clock of the first stage. The time diagrams for a scale-of-eight up-counter are shown in Figure 7.12(b), where all changes of state take place on the trailing edge of the pulses applied to the clock terminals of the three flip-flops A, B and C. Examination of the time diagrams shows that FFA changes state on each trailing edge of the input pulses X.

The output of FFA is used as the clock pulse for FFB, and a change of state of this flip-flop occurs on the trailing edge of the A pulses. Similarly, the output of FFB provides the clock pulse for FFC and this flip-flop changes state on the trailing edge of the B pulses. The various states of the counter, and the binary digits associated with each, state are marked on the time diagrams.

A scale-of-eight counter is also a frequency division circuit. An inspection of the time diagrams shows that the output of FFC produces one pulse for every eight input pulses X. It follows that if the frequency of the input pulses is f then the frequency at the output of FFC is $f/8$. Similarly, the output of FFB is $f/4$ and the output of FFA is $f/2$. Every stage of this counter divides the frequency of the succeeding stage by two.

The idealised behaviour of the counter is shown in Figure 7.12. On the trailing edge of the eighth input pulse, the outputs of the three flip-flops are all shown changing simultaneously from 1 to 0. In practice, these changes ripple through the counter and FFA does not change to 0 until time t_f, the propagation delay of FFA, after the trailing edge of the eighth X pulse. Similarly, FFB and FFC change at times $2t_f$ and $3t_f$, respectively, after the trailing edge of the eighth pulse being counted.

If a ripple-through counter has n stages, then the maximum ripple-through delay of the counter is nt_f. Assuming that the period of the input pulses X is T, then

$$T \geq nt_f$$

and the upper frequency limit of the counter is given by

$$f_u = \frac{1}{T} \leq \frac{1}{nt_f}$$

After modifying the up-counter, shown in Figure 7.12, so that signals $\bar{A}$ and $\bar{B}$ are used as the clock signals for FFB and FFC respectively, the circuit will operate as a scale-of-8 down counter. For up/down counting, a further modification is required. XOR gates are used for transmitting the true or inverted signals from the outputs of FFA and FFB to the succeeding stages of the counter, as shown in Figure 7.13. If the mode control M is set to 0, A and B are transmitted to the clock inputs of FFB and FFC respectively, giving an up-count. For $M = 1$, A and B are both inverted before transmission to the succeeding stages and this initiates a down-counting mode. Initialisation of the counter is provided by the active-low $\overline{Cl}$ inputs.

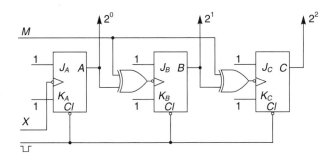

Figure 7.13 *3-stage scale-of-8 asynchronous up/down counter*

7.11 Decoding of asynchronous counters

Decoding problems can occur with asynchronous counters due to the different delay times occurring at the outputs of each of the FFs in the counting chain. Consider, for example, the transition from $CBA = 001$ to $CBA = 010$ in the scale-of-8 up-counter. The sequence of changes that take place is

CBA
0 0 1
0 0 0   transient state
0 1 0

The least significant flip-flop A makes the transition from 1 to 0 before the next most significant flip-flop B changes from 0 to 1. During the transient period a spike or *glitch* will appear at the output of the gate that decodes $(0)_{10}$. The generation of the glitch is shown in Figure 7.14.

At some stages of the count, more than one transient state may occur. Consider the possible sequence of changes that may take place when CBA changes from 011 to 100:

CBA
0 1 1
0 1 0   transient states
0 0 0
1 0 0

In this case the transient states will generate spikes at the outputs of the gates that decode $(0)_{10}$ and $(2)_{10}$.

If the circuits are to be used to give a visual display, the generation of spikes of a very short time duration will not show on the display and are of no consequence. However, if the counter is used to control some other digital circuit, the spikes may initiate false circuit operation and the designer should take steps to eliminate their effect. The problem can be overcome by generating a *strobe* pulse which disables all the decoding gates when the clock goes high. At time $3t_f$, when the three FFs

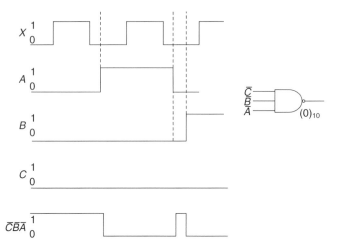

Figure 7.14   *Generation of a decoding spike by an asynchronous counter*

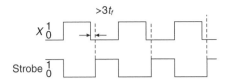

Figure 7.15 *Elimination of decoding spikes using a strobe*

have all reached their final state, the strobe pulse goes high and enables all the decoding gates. The strobe remains high until the leading edge of the next clock pulse arrives. This sequence of events is illustrated in Figure 7.15.

7.12 Asynchronous resettable counters

An asynchronous resettable counter can be used when scales that are not a power of 2 are required. A scale-of-N counter of this type is allowed to count up to the number N, and a logic signal testing for this number is used to clear all the flip-flops in the counter. The state diagram for a resettable scale-of-5 counter is shown in Figure 7.16(a). The counter remains in each of the first five states for one clock period, but on entering S_5, the sixth state (101), a reset signal $r = \overline{A\bar{B}C}$ is generated by a NAND gate. Circuit implementation and the timing diagrams are shown in Figures 7.16(b) and (c).

The reset times for the individual flip-flops in the counter may well be different. For example, in the circuit described, FFA may reset faster than FFC. The negative-going reset signal will cease to exist when FFA is cleared and is simply not wide enough to reset FFC. This problem can be overcome by latching the reset signal until

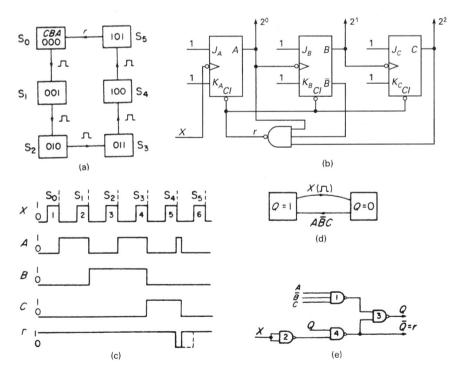

Figure 7.16 *Resettable scale-of-five counter (a) State diagram (b) Implementation (c) Timing diagrams (d) State diagram for the latching circuit (e) Implementation of the latching circuit*

the leading edge of the sixth clock pulse arrives, as indicated by the dotted lines on the timing diagrams. A suitable state diagram for the latching circuit is shown in Figure 7.16(d).

The turn-on condition for Q is $S = A\bar{B}C$

The turn-off condition for Q is $R = X$

This yields

$$Q^{t+\delta t} = (A\bar{B}C + \bar{X}Q)^t$$

This 2-level sum-of-products is shown implemented in Figure 7.16(e). It is, in fact, the implementation of an $\bar{S}\bar{R}$ latch and the output of the gate marked 4 is the complementary output of the latch. In this circuit, the output Q of gate 3 becomes 1 when the counter enters S_5. It then follows that $\bar{Q}$ becomes 0. Hence $\bar{Q} = r$ is used to clear the flip-flops in the counter. The latching circuit remains in this condition until the sixth X pulse arrives. This resets the flip-flop and $\bar{Q} = r$ becomes 1 again. The cycle of operation of the latching circuit is completed when $A\bar{B}C$ is detected again.

7.13 Integrated circuit counters

In practice, synchronous and asynchronous counters can be designed using discrete JK, D and T flip-flops; however, in the type 74 series, counters already packaged on IC chips are readily available. For their use in a digital system, the designer needs to study the manufacturer's data sheet carefully in order to understand the various modes of operation of the circuit.

A typical example of a synchronous presettable counter is the 74ALS560. The logic diagram for the counter is shown, along with its function table, in Figure 7.17. It consists of four DFFs which operate on the leading edge of the clock signal. The flip-flops are provided with tri-state outputs which can be put into the high impedance state when $G = 1$.

The function table shows that when asynchronous clear $\overline{ACLR}$ is low, it overrides all other control inputs and unconditionally clears the four flip-flops. Alternatively, when synchronous clear $\overline{SCLR}$ goes low, the FFs are cleared on the leading edge of the next clock pulse. Data can be loaded into the counter at terminals A, B, C and D when asynchronous load $\overline{ALOAD}$ is low, otherwise if synchronous load $\overline{SLOAD}$ goes low, then on the leading edge of the next clock pulse the data will be loaded into the four DFFs. The count enable signals ENP and ENT are set high for counting. ENT also provides the additional function of enabling the ripple carry output (RCO) gate. An alternative carry output is provided by the clocked carry output (CCO) gate, which, unlike RCO, is free of glitches. Cascading of counters is achieved by connecting either RCO or CCO to the ENT terminal of the next counter in the chain.

The 74176 is an example of a presettable asynchronous counter. A logic diagram for this device, along with a function table, is shown in Figure 7.18. It consists of four trailing edge triggered flip-flops, two of them being TFFs and the other two being JKFFs. The logic for clearing, loading and counting is identical for each of the four flip-flops and is

$$\overline{Pr} = \overline{\bar{R} \cdot D \cdot \overline{C/\bar{L}} \cdot \bar{R}} = \bar{D} + C/\bar{L} + R$$

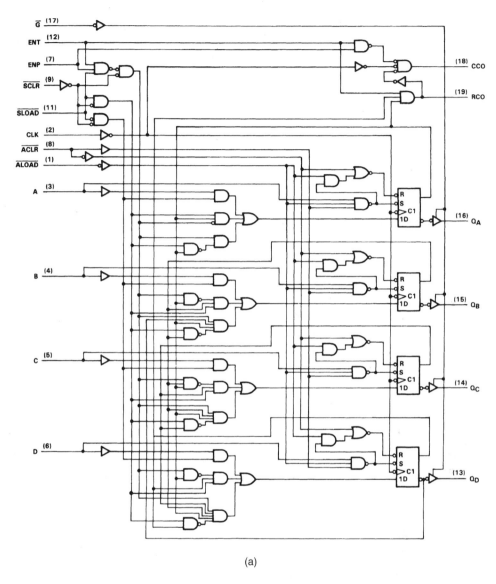

(a)

Inputs								Operation
$\bar{G}$	ACLR	ALOAD	SCLR	SLOAD	ENT	ENP	CLK	
H	X	X	X	X	X	X	X	Q outputs disabled
L	L	X	X	X	X	X	X	Asynchronous clear
L	H	L	X	X	X	X	X	Asynchronous load
L	H	H	L	X	X	X	↑	Synchronous clear
L	H	H	H	L	X	X	↑	Synchronous load
L	H	H	H	H	H	H	↑	Count
L	H	H	H	H	L	X	X	Inhibit counting
L	H	H	H	H	X	L	X	Inhibit counting

(b)

Figure 7.17 *The 74ALS560 synchronous 4-bit counter with tri-state outputs (a) logic diagram (b) function table*

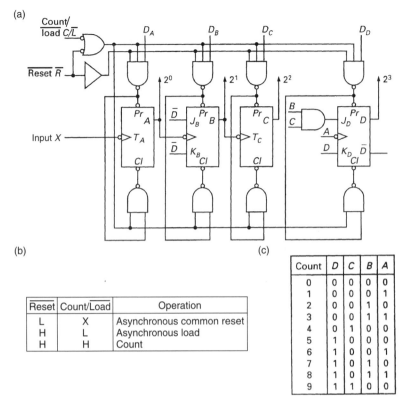

Figure 7.18   *(a) The 74176 operation as a decade counter (b) function table (c) mode 2 count sequence*

and the equation for clearing the individual flip-flops is:

$$\overline{Cl} = \overline{(\overline{C/\bar{L}} + R) \cdot (\bar{D} + C/\bar{L} + R)}$$

which, after manipulation, reduces to:

$$\overline{Cl} = \bar{R}(C/\bar{L} + D)$$

In order to clear all the FFs in the counter, $\overline{RESET}$ must be low, and for counting $\overline{RESET}$ and $COUNT/LOAD$ both must be high.

Assuming all the FFs are cleared, the count follows the normal binary sequence up to, and including, the count of nine. On the trailing edge of the tenth input pulse X, FFA makes a transition from $1 \rightarrow 0$, which would normally induce a transition in FFB, changing its state from $0 \rightarrow 1$. However, at this instant, $J_B = K_B = \bar{D} = 0$ and consequently FFB remains in the reset condition. At the same instant it is also necessary to clear FFD. Now $J_D = BC = 0$ and $K_D = D = 1$, hence when A makes a $1 \rightarrow 0$ transition at the trailing edge of the tenth input pulse X, FFD is reset. All the FFs are now reset to 0 and are awaiting the arrival of the next input pulse.

The 74176 has three modes of operation:

1. To operate as a decade counter, an external interconnection has to be made from A to the clock input of FFB, the incoming count being connected to the clock pin of FFA.

2. For the count tabulated in Figure 7.18(c), the output D is externally connected to the clock pin of FFA and the input count is applied at the clock pin of FFB.
3. To operate as a scale-of-2 and scale-of-5 counter, no external connections are required. FFA provides the scale-of-2 count with the input count applied at its clock pin. FFs B, C, and D are used as the scale-of-5 counter, the input count being fed to the clock pin of FFB.

A second asynchronous counter, the 74290, is available as an IC package. It consists of two parts: a single flip-flop acting as a scale-of-2 counter, and three other flip-flops acting as a scale-of-5 counter. In order to use the package effectively, it is not essential to have a detailed knowledge of the circuit. However, the digital designer must be

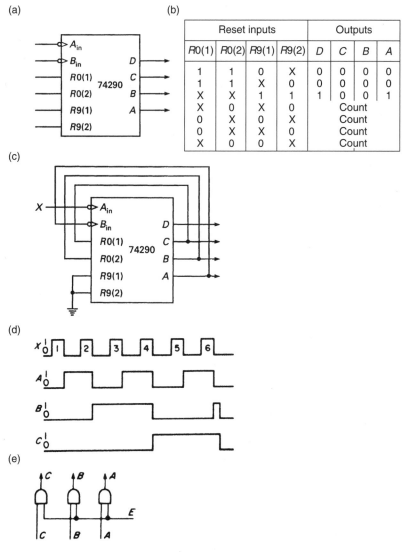

	Reset inputs				Outputs			
	$R0(1)$	$R0(2)$	$R9(1)$	$R9(2)$	D	C	B	A
	1	1	0	X	0	0	0	0
	1	1	X	0	0	0	0	0
	X	X	1	1	1	0	0	1
	X	0	X	0	Count			
	0	X	0	X	Count			
	0	X	X	0	Count			
	X	0	0	X	Count			

Figure 7.19 *(a) Chip connections for the 74290 (b) Truth table for the reset inputs (c) The 74290 connected as a scale-of-six counter (d) The 74290 scale-of-six counter timing diagrams (e) Elimination of spikes with the enable signal, E*

familiar with the package connections and in order to use it intelligently, must understand the basic principles of counting. For the 74290, the important chip connections are shown in Figure 7.18(a). They are:

1. Four outputs, D, C, B and A, where D is the most significant digit
2. Input terminal A_{in} where the input count is connected
3. Input terminal B_{in} which is connected to output A when the counter is operating in the decade mode. Otherwise the input count can be connected to B_{in} when in the scale-of-5 mode.
4. $R0(1)$ and $R0(2)$, which are direct clear terminals. Both must be held at 1 to clear all the FFs.
5. $R9(1)$ and $R9(2)$, which set a count of nine in the counter if they are both held at 1.

One other operating rule must be observed, and that is, for normal counting, at least one of the $R0$ terminals and one of the $R9$ terminals must be held at 0. A function table defining the operation of the reset terminals is given in Figure 7.19(b).

Having become familiar with the chip connections, it is now possible to make use of the package. If a scale-of-6 counter is required, the package is connected as shown in Figure 7.19(c). In this configuration the chip is acting as a resettable ripple counter. When the output combination $B = C = 1$ and $A = 0$ is reached, terminals $R0(1)$ and $R0(2)$ make a transition to 1 and all the flip-flops are cleared.

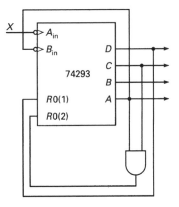

Figure 7.20   *The 74293 used as a scale-of-13 ripple counter*

The timing diagrams for this connection are shown in Figure 7.19(d) and it will be seen that after a count of five the output of FFB becomes 1 for a very short period of time, leading to a spike output on the B line. If the output data is to be decoded, it is desirable that this should be done during clearly defined time intervals in order that the spikes of this type can be eliminated. This can be achieved by means of a strobe which only enables the output gates at appropriate times. The method is illustrated in Figure 7.19(e).

The closely related 74293 package consists of a scale-of-2 counter along with a scale-of-8 counter. An example of this chip connected as a scale-of-13 resettable counter is illustrated in Figure 7.20.

7.14 Cascading of IC counter chips

If two counter chips, such as the 74290 and the 74293, are cascaded and a frequency of 320 kHz is applied at the input terminal of the 74290, as shown in Figure 7.21, the frequency of the signal appearing at the output of the 74293 will be 2 kHz.

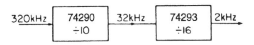

Figure 7.21   *Two IC counters connected in cascade and dividing the frequency input by 160*

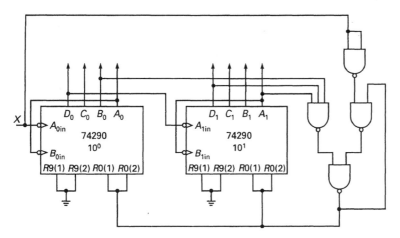

Figure 7.22 *Two 74290 chips in cascade forming a scale-of-92 counter*

When frequency division by a large number is required, the only practical way of achieving this is to use a cascade of counter chips.

Another example of the cascading of counter chips is shown in Figure 7.22. If a scale-of-92 counter is required, this can be achieved by cascading two 74290s. The most significant digit output D_0 of the first chip is connected to terminal A_{1in} of the second chip and acts as the clock signal for it. For every ten X pulses there is one D_0 pulse, and on the tenth X pulse the chip labelled 10^0 makes a transition from 1001 to 0000 and the chip labelled 10^1 makes a transition from 0000 to 0001. The counter is allowed to count up to 92 when the signal representing this number is fed back via the latch circuit to the clear inputs of the two chips. The latching circuit eliminates problems that may be caused by the flip-flops having different resetting times.

7.15 Shift registers

A shift register is a sequential logic device which consists of a cascade of FFs contained in a single IC package. The output of each FF in the cascade is connected to the input of the succeeding FF, and data can be shifted from left to right or vice versa by the clock which is frequently referred to as the *shift pulse*. A basic 4-stage register is shown in Figure 7.23 along with a series of timing diagrams. The register consists of four trailing edge triggered master/slave JKFFs which, alternatively, could be either master/slave SR or D flip-flops. The timing diagrams illustrate the serial movement of 1 bit of data from the input of the register to its output. This operation requires four clock pulses, the data moving from one FF in the cascade to the succeeding one on the receipt of the next clock pulse.

Shift registers can be classified into four distinct groups.

1. *Serial-in/serial-out (SISO)*, in which data can be moved serially in and out of the register, one bit at a time.
2. *Serial-in/parallel-out (SIPO)*, in which the register is loaded serially, one bit at a time, and when an output is required the data stored in the register can be read in parallel form.

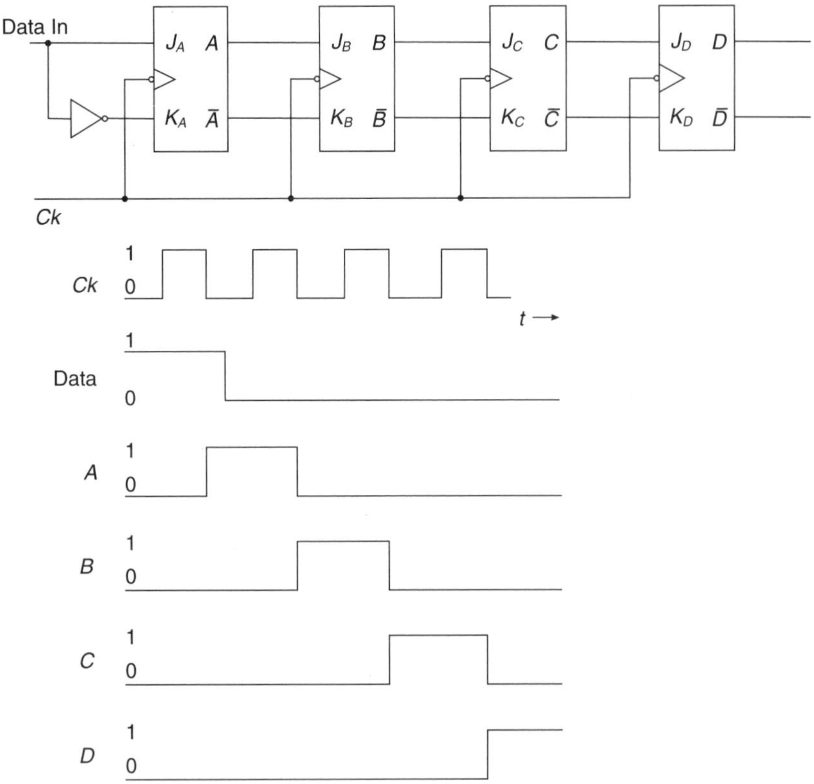

Figure 7.23 *Basic 4-bit shift register with timing diagrams*

3. *Parallel-in/serial-out (PISO)*, in which all the flip-flops are loaded simultaneously and when an output is required, the data stored is removed serially from the register one bit at a time under clock control.
4. *Parallel-in/parallel-out (PIPO)*, in which all the flip-flops in the register are loaded simultaneously, and when an output is required the flip-flops are read simultaneously.

Additional to the input and output terminals, a shift register will have an asynchronous clear terminal which is used to drive all the FFs in the register to logic 0. For those shift registers having parallel data inputs, an asynchronous preset or load is required for entering the data pattern into the register. A clock input is also required for shifting data through the register.

It is also possible to classify shift registers according to their input arrangements:

1. *Double-rail input.* For this type of register there are two input terminals for either the *J* and *K* inputs or, alternatively, the *S* and *R* inputs.
2. *Single-rail input*, as illustrated in Figure 7.23. Here the first flip-flop in the cascade has been converted into a DFF by placing an inverter between the *J* and *K* input lines.

There can also be double-rail output, where the true and complementary outputs of the last flip-flop in the register are brought out to separate pins, or, alternatively, there can be a single-rail output where only the true output of the last flip-flop is made available at a pin.

Data can be transferred by shift registers in either serial or parallel form. Serial transfer between two 4-bit registers will require four clock pulses and one interconnection, while parallel transfer between two registers needs four interconnections. The type of transfer to be used depends upon the distance between the sending and receiving registers. For registers which are near to one another, parallel transfer will be faster, even if more interconnections are needed, but for registers some distance apart, the large number of interconnections required would prove to be uneconomic both in terms of cost and space.

7.16 The 4-bit 7494 shift register

This register, shown in Figure 7.24, can be operated in the serial-in/serial-out mode or, alternatively, as a dual source parallel-in/serial-out register. The register consists of four SR master/slave flip-flops, four AOI gates and four inverter drivers. In order to prepare the register for operation, the FFs are set to logic 0 by applying logic 1 at the CLEAR input. Data can now be loaded asynchronously into each stage of the register by setting the corresponding preset enable inputs, PE1 or PE2, high. For serial operation, the true and inverted data are set up at the R and S inputs of the first flip-flop in the register. On the trailing edge of the clock pulse the data is entered into the master and appears at the slave input. When the leading edge of the clock pulse arrives, the data is transferred to the output of the slave.

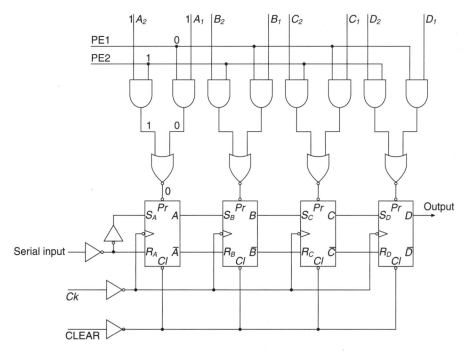

Figure 7.24   *The 7494 shift register with serial and parallel loading*

7.17 The 4-bit 7495 universal shift register

A typical example of a versatile shift register is the 7495 and its logic diagram is shown in Figure 7.25. It has facilities for parallel loading and parallel output, serial loading and serial output, and, additionally, it has shift-left and shift-right facilities. This is, in effect, a universal shift register which can operate in all the four modes previously described, besides having the facility of bi-directional shifting.

The mode control (MC) signal controls whether data inputs are serial or parallel. With $MC = 0$, the AND gates marked 1 are enabled. In this mode data is serially entered under the control of $Ck1$. Alternatively, with $MC = 1$, the AND gates marked 2 are enabled. In this case the input data is entered in parallel and appears at the data outputs after the $1 \rightarrow 0$ transition of $Ck2$. Shift right takes place on the $1 \rightarrow 0$ transition of $Ck1$ and the shift left operation takes place on the $1 \rightarrow 0$ transition of $Ck2$ when $MC = 1$ by connecting the output of each flip-flop to the parallel input of the previous flip-flop, as shown in Figure 7.25.

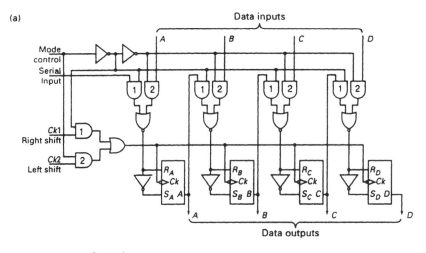

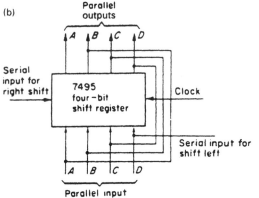

Figure 7.25   *(a) The 7495 universal shift register (b) The 7495 connections for bi-directional shifting*

7.18 The 74165 parallel loading 8-bit shift register

An example of an 8-bit shift register is the 74165 (see Figure 7.26) which can be operated as a SISO or a SIPO. It consists of eight SRFFs with parallel access which is enabled when the Shift/Load signal is low. The data is loaded asynchronously into the eight flip-flops on a $1 \rightarrow 0$ transition of the Shift/Load signal. When loading, the two gates associated with Clock and Clock Inhibit are disabled, and shifting cannot take place. Serial transmission of data is also inhibited when Clock Inhibit and Shift/Load are high, but on returning Clock Inhibit to logic 0, shifting from left to right can take place.

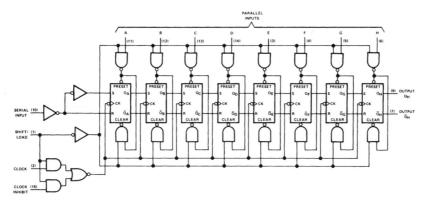

Figure 7.26 *The 74165 parallel load 8-bit shift register*

7.19 The use of shift registers as counters and sequence generators

An alternative method of designing digital counters or sequence generators is to use a shift register chip. A typical shift register counter configuration is shown in Figure 7.27. The individual flip-flops form part of an N-stage shift register and the connections between individual flip-flops are internal to the chip. The output of each stage and its complement are both available, and they may be used to drive combinational feedback logic which provides the J and K inputs to the least significant stage of the register. Such a circuit can be used to generate specified binary sequences or, alternatively, it can operate as a scale-of-M counter, where $M \leq 2^N$.

The input-output relationships for each stage of the counter, shown in Figure 7.27, are defined by the following set of equations:

$$A^{t+\delta t} = f(A, \bar{A}, B, \bar{B}, \ldots, N, \bar{N})$$
$$B^{t+\delta t} = A^t, \quad C^{t+\delta t} = B^t, \quad \text{etc.}$$

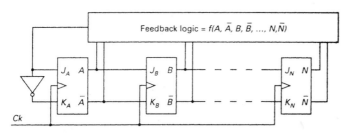

Figure 7.27 *Basic configuration of a feedback shift register*

The feedback circuit produces either a 1 or a 0 which is fed to the input of FFA where it determines the next state of *A* on receipt of the next clock pulse. For example, assuming that the *N*-stage shift register is in the state *N*...*CBA* = 0...001, the next state of the shift register will be either 0...010 or 0...011, depending on whether the feedback logic provides a 0 or a 1 at the *J*-input of FFA.

7.20 The universal state diagram for shift registers

The transition table for a two-stage left-shift register is shown in Figure 7.28(a). If the shift register is initially in the state 00 there are two possible next states. These are 00, if the *J*-input to the least significant stage of the register is a 0, or 01, if the *J*-input is a 1. Similarly, if the initial state of the shift register is 01 then the two possible next states are either 10 or 11.

The transition table can be translated into the *universal state diagram* shown in Figure 7.28(b), which is also called the *De Bruijn diagram*. It will be noted that the shift register is permanently 'locked' in the state 00 if the feedback signal is a 0 and similarly it is 'locked' in the state 11 if a 1 is provided by the feedback logic.

A similar transition table can be developed for a 3-stage shift register, and this can be translated into a universal state diagram, as shown in Figure 7.29. The universal state diagram for a 4-stage register shown in Figure 7.30 has been developed in the same way, and clearly as the number of stages in the register increases the complexity of this type of diagram increases rapidly.

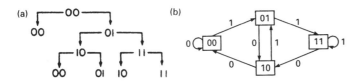

Figure 7.28   *Two-stage shift register (a) Transition table (b) Universal state diagram*

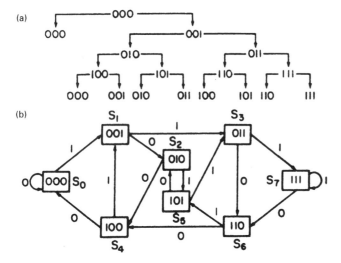

Figure 7.29   *Three-stage shift register (a) Transition table, (b) Universal state diagram*

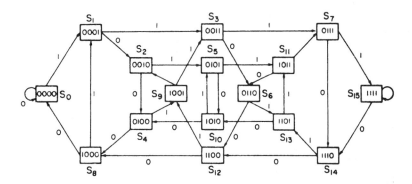

Figure 7.30 *Universal state diagram for a four-stage shift register*

The universal state diagram is a departure from the kind of state diagram that defines a single count sequence. All possible internal states of the register and all possible transitions between states are shown on the universal state diagram. The logic designer may choose a suitable sequence of states on the diagram and design the feedback logic that will allow the register to cycle through the chosen sequence.

7.21 The design of a decade counter

The first step in the design is to choose a ten-state sequence on the universal state diagram for a 4-stage register. One possible sequence is:

S_0-S_1-S_2-S_5-S_{11}-S_6-S_{13}-S_{10}-S_4-S_8-S_0

It should be noted that this is not the only ten state sequence available on the universal state diagram.

The second step in the design is to draw up the state table, as shown in Figure 7.31(a) in order to determine the logical value of the feedback function for each change of state. For example, in going from S_0 to S_1, the output of FFA must change from 0 to 1, and hence the required input to this flip-flop, $J_A = 1$. This is the logical value of the feedback function required for this change of state, and it is entered in the right hand column of the state table.

The feedback function and the unused states are plotted on a 4-variable K-map (see Figure 7.31(b)). It should be noted that S_{15} is an unused state and it appears that the S_{15} cell on the K-map should have been marked with an X. However, a general rule that should be followed when designing this type of counter is that the entry in the S_{15} cell should always be a 0, and that in the S_0 cell should always be a 1, irrespective of whether these two states are in the counting sequence. This ensures that the counter will never be locked in either the 0000 or 1111 states.

Minimising, the feedback function is found to be

$$f = J_A = B\bar{D} + AC\bar{D} + \bar{A}\bar{C}\bar{D}$$

If the counter enters an unused state due to faulty circuit operation, it will return to the correct sequence after a maximum of five clock pulses. The return to the correct

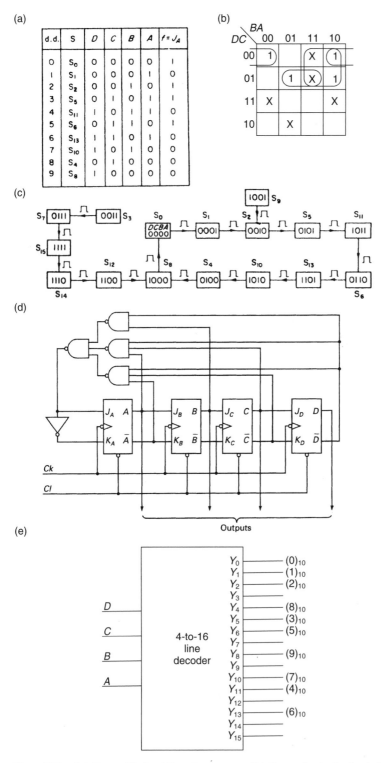

Figure 7.31   *(a) State table for shift register counter (b) K-map for the feedback function $f = J_A$ (c) Full state diagram (d) Implementation of shift register counter (e) Count sequence used to generate a circulating ring of ten 1's*

sequence when faulty operation occurs is illustrated in the full state diagram shown in Figure 7.31(c), and implementation of the counter is shown in Figure 7.31(d). If a decimal display is required, then the counter, in conjunction with the appropriate combinational logic, can be used to drive a seven segment indicator. Alternatively, the flip-flop outputs can be fed to the input terminals of a 4-to-16 line decoder (see Figure 7.31(e)), whose outputs will be either active low or active high depending on the MSI package selected. If, for example, $DCBA = 1101$ the corresponding decimal output is $(6)_{10}$ and will appear at the decoder output terminal marked Y_{13}. It will be observed that if the decoder outputs are active high they will produce a continuously circulating ring of ten pulses which could be used to initiate operations in other sequential logic circuits.

7.22 The ring counter

The simplest type of shift register counter is the ring counter, where feedback from the last stage of the register feeds the input of the first stage, as shown in Figure 7.32(a). The register has ten stages and it can be used as a decimal counter since the number of stages is equal to the number of states. The data contained in each stage is shifted to the next stage on the receipt of a clock pulse, and the counter circulates a 1 which is initially preset in the least significant stage of the register, all other stages

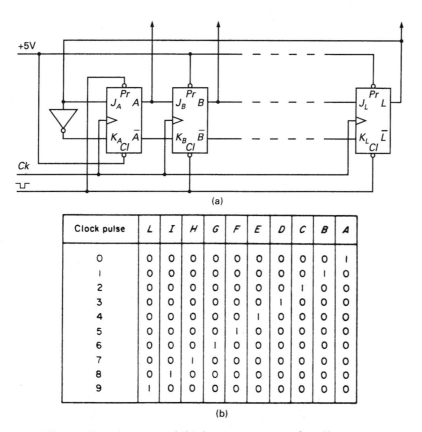

Clock pulse	L	I	H	G	F	E	D	C	B	A
0	0	0	0	0	0	0	0	0	0	1
1	0	0	0	0	0	0	0	0	1	0
2	0	0	0	0	0	0	0	1	0	0
3	0	0	0	0	0	0	1	0	0	0
4	0	0	0	0	0	1	0	0	0	0
5	0	0	0	0	1	0	0	0	0	0
6	0	0	0	1	0	0	0	0	0	0
7	0	0	1	0	0	0	0	0	0	0
8	0	1	0	0	0	0	0	0	0	0
9	1	0	0	0	0	0	0	0	0	0

(b)

Figure 7.32 *(a) The ring counter and (b) the counting sequence for a 10-stage ring counter*

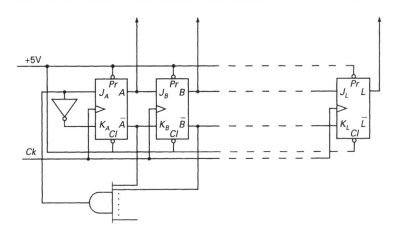

Figure 7.33   *The self-starting self-correcting ring counter*

being simultaneously cleared. The count sequence of the register is tabulated in Figure 7.32(b).

The ten outputs of the ring counter can be used directly as decimal outputs without the need for a decoding network. Alternatively, the circulating 1 can be used to enable a group of logic circuits sequentially. The number of stages required in this case will be equal to the number of circuits to be enabled.

An obvious advantage of the ring counter is its simplicity. Additionally, it has spike-free outputs since decoding logic is not required. However, it has the disadvantage of not having a binary readout and its counting sequence is radically changed if, through faulty circuit operation, it enters one of the many unused states.

A binary counter, synchronous or asynchronous, having ten stages will have $2^{10} = 1024$ counting states and can count up to 1023, whereas the decimal ring counter only has 10 counting states and it follows that there are $2^{10} - 10 = 1014$ unused states. If the counter, for some reason, enters one of these states it enters a forbidden counting sequence, of which there are many, and it will never again re-enter the correct counting sequence unless forced to do so.

The circuit of Figure 7.32(a) can be modified so that it becomes both *self-starting* and *self-correcting*. The required modification is shown in Figure 7.33. The input to FFA is:

$$J_A = \bar{A}\bar{B}\bar{C}\bar{D}\bar{E}\bar{F}\bar{G}\bar{H}\bar{I}$$

and this can only be 1 provided $A = B = C = D = E = F = G = H = I = 0$.

Clearly, if any section of the counter, except the last one, contains a 1, $J_A = 0$ and the counter will now enter the required sequence within a maximum of 9 clock pulses. If, for some reason, the counter enters a false state, the counter is also self-correcting and will return to the correct sequence after, at most, 9 clock pulses.

7.23 The twisted ring or Johnson counter

As the name implies, the difference between the twisted ring counter and the ordinary ring counter is that the feedback is taken from the complementary output of the last stage in the register and is connected to the *J*-input of the first stage, while the inverted

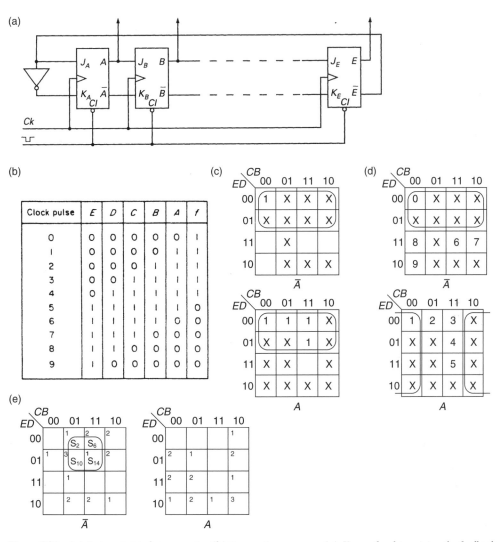

Figure 7.34 *(a) 5-stage twisted ring counter (b) its counting sequence (c) K-map for determining the feedback function (d) K-map for determining the decode logic (e) K-map for determining self-correction function*

form of the feedback is fed to the K-input. If all the flip-flops are initially preset to the same state, either 0 or 1, then the number of stages in the count sequence is equal to twice the number of stages in the shift register. Hence, a decade counter can be constructed from a 5-stage shift register, as shown in Figure 7.34(a). The counting sequence of the circuit, assuming that initially all the flip-flops are cleared to zero, is tabulated in Figure 7.34(b).

This is a 10-state sequence which could have been selected from the universal state diagram of a 5-stage shift register. The feedback logic could have been developed by first tabulating the required value of the feedback function in the column headed f in the count sequence tabulation. The function is then plotted on the K-map shown in Figure 7.34(c). Simplifying, the K-map plot gives $f = J_A = \bar{E}$.

For this circuit, decoding logic is required to obtain a decimal count. This logic is obtained from a 5-variable K-map on which the decimal equivalent (corresponding to the clock pulse numbering in Figure 7.34(b)) for each of the states in the counting cycle has been marked as shown in Figure 7.34(d). The simplifying adjacencies for $(0)_{10}$ and $(1)_{10}$ have also been marked with X's on the map, and if the reader cares to continue the process of simplification it can be seen that it is always possible to combine seven unused states with each of the decimal entries. The resulting decimal decode logic, after simplification, is tabulated below:

$(0)_{10} = \bar{A}\bar{E}$ $(5)_{10} = AE$

$(1)_{10} = A\bar{B}$ $(6)_{10} = \bar{A}B$

$(2)_{10} = B\bar{C}$ $(7)_{10} = \bar{B}C$

$(3)_{10} = C\bar{D}$ $(8)_{10} = \bar{C}D$

$(4)_{10} = D\bar{E}$ $(9)_{10} = \bar{D}E$

There are also three other undesired and independent count sequences for the Johnson counter. They are:

1. $S_2\text{-}S_5\text{-}S_{11}\text{-}S_{23}\text{-}S_{14}\text{-}S_{29}\text{-}S_{26}\text{-}S_{20}\text{-}S_8\text{-}S_{17}\text{-}S_2$

2. $S_4\text{-}S_9\text{-}S_{19}\text{-}S_6\text{-}S_{13}\text{-}S_{27}\text{-}S_{22}\text{-}S_{12}\text{-}S_{25}\text{-}S_{18}\text{-}S_4$

3. $S_{10}\text{-}S_{21}\text{-}S_{10}$

If the counter should enter any one of these sequences due to faulty circuit operation or when switching on, it will remain in that sequence unless arrangements are made to return the counter to the required sequence.

The unwanted sequences are shown plotted on the K-map in Figure 7.34(e), cells marked with a 1 being in unwanted sequence 1, and so on. It will be observed that the four adjacent states S_2, S_6, S_{10} and S_{14} are all in one of the three unwanted sequences. If the Boolean function that represents these four states, $f = \bar{A}B\bar{E}$, is used to clear the five stages of the counter, then within a maximum of ten clock pulses the counter will enter the Johnson count sequence. The reader should note that there are alternative combinations that will achieve the same effect.

The Johnson counter has an even-numbered cycle length of $2N$ where N is the number of stages in the register. However, with a suitable modification of the feedback it is possible to achieve an odd-numbered cycle length of $(2N - 1)$. For example, if the state 00000 is omitted, the counting cycle becomes that shown tabulated in Figure 7.35(a)

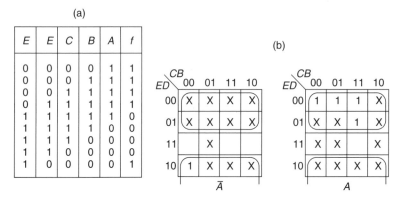

Figure 7.35   *(a) Counting sequence of an odd-numbered cycle length Johnson counter (b) Determination of the feedback function for the Johnson counter of cycle length (2N−1)*

and the values of the feedback function required to generate this sequence appear in the last column of this table. Plotting this function in conjunction with the unused states on the 5-variable K-map (see Figure 7.35(b)) and minimising, leads to the revised feedback function $f = \bar{D} + \bar{E}$. It is left to the reader to show that if the state 11111 is omitted rather than the 00000 state, the modified feedback function will be $f = \bar{D}\bar{E}$.

7.24 Series and parallel interconnection of Johnson counters

For a modulo-10 count, a 5-stage Johnson counter is required with its associated decode logic. As the modulus increases, the number of stages required also increases, and beyond modulo-12 the use of a single Johnson counter is no longer economic. However, by means of a series or parallel interconnection it is possible to use two Johnson counters of smaller moduli to generate a larger modulus.

In Figure 7.36, two mod-4 counters are connected in series to generate the tabulated mod-16 count. The AND gate provides the clock pulses for the right-hand counter. Each time the left-hand counter is in the state $BA = 10$, a clock pulse is generated for the right-hand counter. If required, a 4-to-16 line decoder can be used to decode the outputs.

A pair of parallel connected Johnson counters is shown in Figure 7.37, both of them being clocked by the same signal. The two moduli chosen, mod-3 and mod-5, are both prime numbers. Odd numbered counting sequences are obtained for both counters by using the appropriate feedback signals. For the mod-3 counter $BA = 11$ is removed from the count sequence, and for the mod-5 counter $EDC = 000$ is removed. The feedback signal for the mod-3 counter is $J_A = \bar{B}\bar{A}$ and for the mod-5 counter it is $J_C = \bar{D} + \bar{E}$. The initial state of the tabulated count sequence is $EDCBA = 10000$ and the counter re-enters this state after 15 clock pulses. Two parallel connected modified Johnson counters have combined to form a mod-15 counter.

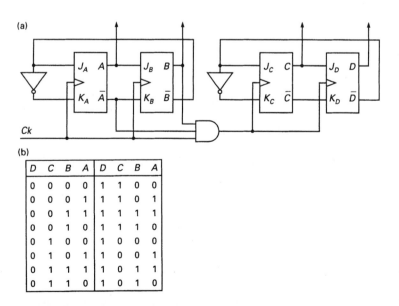

(b)

D	C	B	A	D	C	B	A
0	0	0	0	1	1	0	0
0	0	0	1	1	1	0	1
0	0	1	1	1	1	1	1
0	0	1	0	1	1	1	0
0	1	0	0	1	0	0	0
0	1	0	1	1	0	0	1
0	1	1	1	1	0	1	1
0	1	1	0	1	0	1	0

Figure 7.36 *(a) A series-connected pair of mod-4 Johnson counters providing a mod-16 count and (b) the count sequence*

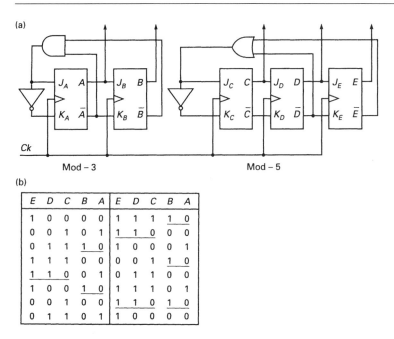

Figure 7.37   *(a) A parallel-connected pair of Johnson counters providing a mod-15 count and (b) the count sequence*

It is left to the reader to show that if two moduli which are not prime, such as 4 and 6, are selected, the two counters return to the initial state after twelve clock pulses. A mod-12 counter has been obtained rather than a mod-24. The reader should observe that the modulus obtained is the lowest common multiple (LCM) of the individual moduli.

7.25 Shift registers with XOR feedback

The 4-stage shift register shown in Figure 7.38(a) has XOR feedback from stages C and D such that the input to the first stage $J_A = C \oplus D$. To determine the sequence of states for the register, it is assumed initially that the shift register is in the state $D = 0, C = 0, B = 0$ and $A = 1$, in which case $J_A = 0 \oplus 0$, and on receipt of the next clock pulse the register enters the state $D = 0, C = 0, B = 1$ and $A = 0$. The complete sequence of states for the register is tabulated in Figure 7.38(b), the value of the feedback function for each state appearing in the right-hand column of the tabulation.

In all, there are 15 states, and this is the maximum number a 4-stage register having XOR feedback can have. This sequence is termed the *maximum length sequence* (MLS). $S_0 = 0000$ is not included in the sequence since this is a *lock-in* state. If the register enters this state $J_A = 0 \oplus 0 = 0$; it is unable to leave it when the next and subsequent clock pulses arrive. In general, the maximum length sequence for such a circuit is given by $l = 2^N - 1$ where N is the number of stages in the shift register.

Not all XOR connections result in a maximum length sequence. The table in Figure 7.39 gives the feedback functions which will give the maximum length sequence for values of N up to and including 18.

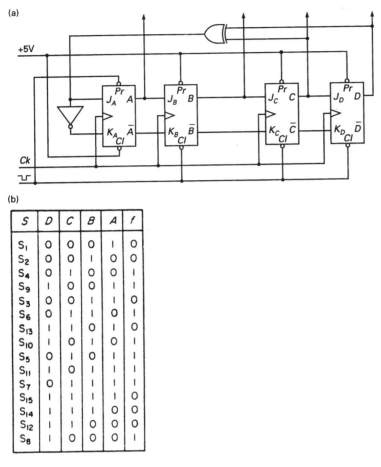

(a)

(b)

S	D	C	B	A	f
S_1	0	0	0	1	0
S_2	0	0	1	0	0
S_4	0	1	0	0	1
S_9	1	0	0	1	1
S_3	0	0	1	1	0
S_6	0	1	1	0	1
S_{13}	1	1	0	1	0
S_{10}	1	0	1	0	1
S_5	0	1	0	1	1
S_{11}	1	0	1	1	1
S_7	0	1	1	1	1
S_{15}	1	1	1	1	0
S_{14}	1	1	1	0	0
S_{12}	1	1	0	0	0
S_8	1	0	0	0	1

Figure 7.38 *(a) Four-stage MLS shift register generator (b) MLS for four-stage shift register*

No of stages, N	Feedback equation	No of stages, N	Feedback equation
1	A	10	$G \oplus J$
2	$A \oplus B$	11	$I \oplus K$
3	$B \oplus C$	12	$F \oplus H \oplus K \oplus L$
4	$C \oplus D$	13	$I \oplus J \oplus L \oplus M$
5	$C \oplus E$	14	$D \oplus H \oplus M \oplus N$
6	$E \oplus F$	15	$N \oplus O$
7	$F \oplus G$	16	$D \oplus M \oplus O \oplus P$
8	$D \oplus E \oplus F \oplus H$	17	$N \oplus Q$
9	$E \oplus I$	18	$K \oplus R$

Figure 7.39 *Feedback functions for maximum-length sequences*

Other maximum length sequences are available with the same register length. For example, if the inverse of the XOR function $C \oplus D$ is used as feedback, then an alternative maximum length sequence is obtained and is tabulated in Figure 7.40. Furthermore, an examination of the feedback equations in Figure 7.39 shows that

(a)

S	D	C	B	A	$f = C \odot D$
S_1	0	0	0	1	1
S_3	0	0	1	1	1
S_7	0	1	1	1	0
S_{14}	1	1	1	0	1
S_{13}	1	1	0	1	1
S_{11}	1	0	1	1	0
S_6	0	1	1	0	0
S_{12}	1	1	0	0	1
S_9	1	0	0	1	0
S_2	0	0	1	0	1
S_5	0	1	0	1	0
S_{10}	1	0	1	0	0
S_4	0	1	0	0	0
S_8	1	0	0	0	0
S_0	0	0	0	0	1

(b)

S	D	C	B	A	$f = A \oplus D$
S_1	0	0	0	1	1
S_3	0	0	1	1	1
S_7	0	1	1	1	1
S_{15}	1	1	1	1	0
S_{14}	1	1	1	0	1
S_{13}	1	1	0	1	0
S_{10}	1	0	1	0	1
S_5	0	1	0	1	1
S_{11}	1	0	1	1	0
S_6	0	1	1	0	0
S_{12}	1	1	0	0	1
S_9	1	0	0	1	0
S_2	0	0	1	0	0
S_4	0	1	0	0	0
S_8	1	0	0	0	1

(c)

S	D	C	B	A	$f = A \odot D$
S_1	0	0	0	1	0
S_2	0	0	1	0	1
S_5	0	1	0	1	0
S_{10}	1	0	1	0	0
S_4	0	1	0	0	1
S_9	1	0	0	1	1
S_3	0	0	1	1	0
S_6	0	1	1	0	1
S_{13}	1	1	0	1	1
S_{11}	1	0	1	1	1
S_7	0	1	1	1	0
S_{14}	1	1	1	0	0
S_{12}	1	1	0	0	0
S_8	1	0	0	0	0
S_0	0	0	0	0	1

Figure 7.40   *(a) The MLS for a four-stage shift register with feedback $C \odot D$ (b) $A \oplus D$ and (c) $A \odot D$*

one of the digits in the equation is always the Nth digit in the register, and the other digit (or digits) is obtained by looking back down the register. For example, for $N = 4$ the Nth digit is D, and the other digit in the equation, C, is the $(N - 1)$th digit. Two alternative maximum length sequences for a 4-stage register can be obtained by looking forward to the $(N + 1)$th digit which, in this case, is A. Hence the other two maximum length sequences are obtained by using the feedback $A \oplus D$ and $A \odot D$, and these sequences are shown tabulated in Figure 7.40.

Clearly, the circuit shown in Figure 7.38(a) can be used as a binary sequence generator, the output sequence being taken directly from the output of one of the flip-flops in the register. In this case, the binary output sequence appearing at the output of FFD is 000100110101111. This kind of generator is sometimes referred to as a *pseudo-random binary sequence generator* because the digits in the sequence are in apparently random order. However, the randomness repeats itself every $2^N - 1$ clock pulses. For a given clock frequency, the periodicity of the randomness increases very rapidly with the number of stages in the register.

If $N = 10$, $(2^N - 1) = 1023$

and if the clock frequency is 1 MHz the sequence repeats itself every 1.023 ms.

If $N = 20$ $2^N - 1 = 1048575$

and the period of the sequence is 1.05 s.

If $N = 30$ $2^N - 1 = 1073741823$

and the period of the sequence is 1073.74 s.

The design of pseudo-random sequence generators is based on the theory of *finite fields* developed by the French mathematician Evariste Galois. The algebra associated

S	D	C	B	A	f	S	D	C	B	A	f	S	D	C	B	A	f
S_1	0	0	0	1	0	S_3	0	0	1	1	1	S_6	0	1	1	0	1
S_2	0	0	1	0	1	S_7	0	1	1	1	1	S_{13}	1	1	0	1	1
S_5	0	1	0	1	0	S_{15}	1	1	1	1	0	S_{11}	1	0	1	1	0
S_{10}	1	0	1	0	0	S_{14}	1	1	1	0	0						
S_4	0	1	0	0	0	S_{12}	1	1	0	0	1						
S_8	1	0	0	0	1	S_9	1	0	0	1	1						

Figure 7.41 *Non-maximum length sequences generated by a four-stage shift register with feedback $B \oplus D$*

with finite field theory is frequently referred to as *Galois field algebra*. This type of binary sequence generator has a number of applications. Typical of these is the generation of repetitive noise for test circuits and also in the process of *encrypting* serial transmissions to ensure message security.

Non-maximum length sequences can be generated with a 4-stage register if an alternative XOR feedback is used. For example, if the feedback function is $B \oplus D$, one of the sequences tabulated in Figure 7.41 will be generated. The form which the sequence takes will depend on the initial state of the register.

The basic MLS generator shown in Figure 7.38 is not necessarily self-starting, since on switching on the initial state of the generator may be 0000. As the circuit stands, there is no way in which it can leave this state. With a slight modification to the feedback circuit it is possible to make the generator self-starting. The required modification is the logical addition of the term $\bar{A}\bar{B}\bar{C}\bar{D}$ to the feedback equation so that it becomes:

$$f = C \oplus D + \bar{A}\bar{B}\bar{C}\bar{D}$$

This function is plotted on the K-map shown in Figure 7.42(a) and, after simplification, it reduces to:

$$f = C \oplus D + \bar{A}\bar{B}\bar{D}$$

The implementation of the self-starting generator is shown in Figure 7.42(b).

It is also possible to generate non-maximum length sequences by using a jump technique. The method of approach is to start with an MLS generator using XOR feedback and then reduce the length of the sequence by introducing additional feedback. The method will be described for the 4-stage shift register generator shown in Figure 7.43.

It will be assumed that initially the generator is in the state $DCBA = 0011$ (S_3). If, when in this state, the feedback is a 0, then the next state of the generator will be $DCBA = 0110$ (S_6). Examination of the state table for the 4-stage MLS generator in Figure 7.38 shows that $C \oplus D = 0$ when the generator is in state S_3, and the next state is S_6. If the feedback is modified to a 1 then the next state of the generator is S_7.

The state diagram for the MLS generator having four stages is shown in Figure 7.43(a), and it can be seen that by modifying the feedback, the states S_6, S_{13},

(a)

(b)

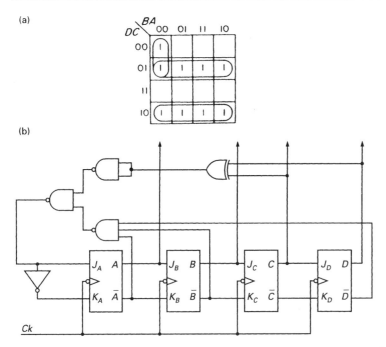

Figure 7.42 *(a) K-map plot for a self-starting MLS generator (b) implementation of self-starting generator*

S_{10}, S_5 and S_{11} will be omitted from the sequence, thus reducing its length from 15 to 10 states.

The modified sequence for the generator is shown in the state table in Figure 7.43(b) and the new value of the feedback function in state S_3 is shown encircled. The feedback function in conjunction with the unused states and the 'lock-in' state S_0 are plotted on a K-map and then simplified (see Figure 7.43(c)). This gives a modified feedback function of

$$f = C \oplus D + AB\bar{D} + \bar{A}\bar{B}\bar{D}$$

and the implementation of this self-starting non-maximum length sequence generator is shown in Figure 7.43(d).

7.26 Multi-bit rate multipliers

It is, on occasions, desirable to have a counter that is capable of generating a variety of count sequences under the control of a variable combination of inputs, termed *rate constant* inputs. Consider, for example, the scale-of-eight counter shown in Figure 7.44(a) having the normal count sequence shown in Figure 7.44(b) but in this case under the control of the binary rate inputs W, X and Y.

If the binary rate inputs are $W = 1$, $X = 0$ and $Y = 0$ the output $Z = W\bar{Q}_A Ck$, or when expanded:

$$Z = W(\bar{Q}_C\bar{Q}_B\bar{Q}_A + \bar{Q}_C Q_B\bar{Q}_A + Q_C\bar{Q}_B\bar{Q}_A + Q_C Q_B\bar{Q}_A)Ck$$

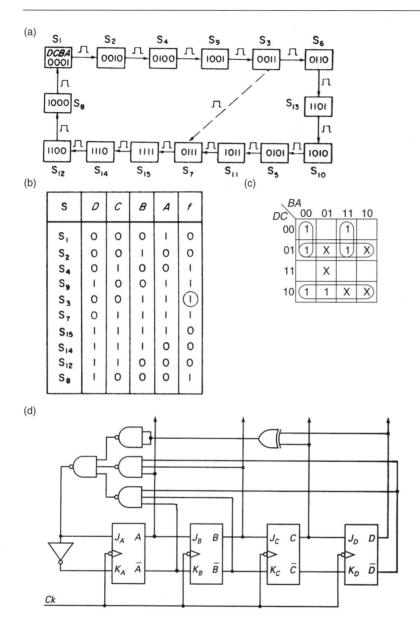

Figure 7.43 *(a) State diagram of the four-stage MLS generator with modified feedback showing the jump (b) Modified MLS sequence (c) K-map plot of the feedback function f (d) Implementation of an MLS generator employing the 'jump' technique*

And a sequence of four pulses appears at the output Z rather than the normal scale-of-eight count. For a binary rate input of $W = 0$, $X = 1$ and $Y = 0$ the output $Z = XQ_A\bar{Q}_BCk$, or when expanded:

$$Z = X(Q_C\bar{Q}_BQ_A + \bar{Q}_C\bar{Q}_BQ_A)Ck$$

In this case a sequence of two pulses appears at the output Z while the counter cycles through the scale-8 count, and it is clear that if the rate inputs are $W = 0$,

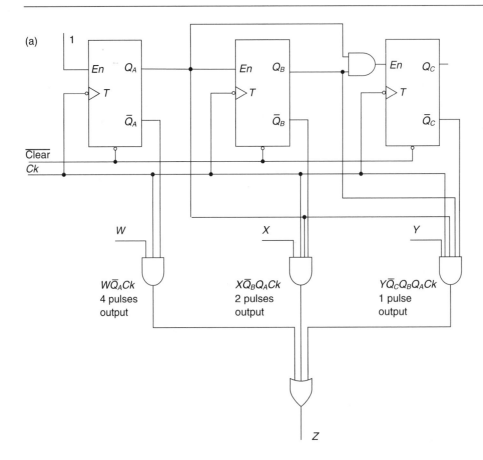

Figure 7.44   *(a) Basic 3-bit binary rate multiplier (b) Scale-of-8 count sequence*

$X = 0$ and $Y = 1$, the output Z will consist of a single pulse. Clearly, any combination of the rate constant inputs can be chosen, and this will lead to a variety in the number of pulses appearing at the output. A timing diagram for the rate multiplier is shown in Figure 7.45.

Typical examples of rate multipliers in the type 74 series are the 7497, a 6-bit binary rate multiplier, and the 74167 decade rate multiplier. The 7497 has a basic count cycle of 64 clock pulses and the maximum number of pulses appearing at the output Z during one complete count cycle is 63 when all the rate inputs are high. A circuit diagram of the device is shown in Figure 7.46 along with the truth table. The 7497

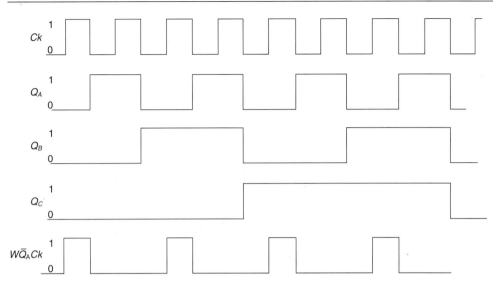

Figure 7.45 *Timing diagram for 3-bit rate multiplier with W = 1*

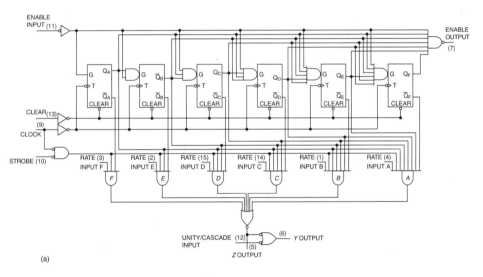

(a)

INPUTS					OUTPUTS			
			BINARY RATE				LOGIC LEVEL OR NUMBER OF PULSES	
CLEAR	ENABLE	STROBE	F E D C B A	NUMBER OF CLOCK PULSES	UNITY / CASCADE	Y	Z	ENABLE
H	X	H	X X X X X X	X	H	L	H	H
L	L	L	L L L L L L	64	H	L	H	1
L	L	L	L L L L L H	64	H	1	1	1
L	L	L	L L L L H L	64	H	2	2	1
L	L	L	L L L H L L	64	H	4	4	1
L	L	L	L L H L L L	64	H	8	8	1
L	L	L	L H L L L L	64	H	16	16	1
L	L	L	H L L L L L	64	H	32	32	1
L	L	L	H H H H H H	64	H	63	63	1
L	L	L	H H H H H H	64	L	H	63	1
L	L	L	H L H L L L	64	H	40	40	1

(b)

Figure 7.46 *(a) Type 7497 6-bit binary rate multiplier (b) truth table*

features buffered clock, clear and enable inputs to control the device operation. The strobe input is used to enable or inhibit the rate inputs. To enable the multiplier, the clear, strobe, and enable inputs are low and the output frequency is given by the equation

$$f_{out} = \frac{M \times f_{in}}{64}$$

where $M = F \times 2^5 + E \times 2^4 + D \times 2^3 + C \times 2^2 + B \times 2^1 + A \times 2^0$. If, for example, $F = 1$ and $E = 1$ while D, C, B, and $A = 0$

$$f_{out} = \frac{32 + 16}{64} \times f_{in} = \frac{3}{4} \times f_{in}$$

The unity/cascade input allows the range of the binary rate multiplier to be extended to 12 bits or more.

Problems

7.1 Design a synchronous modulo-12 counter using NAND gates and

(a) T flip-flops,
(b) SR flip-flops,
(c) JK flip-flops,
(d) D flip-flops.

Develop the decode logic for the counters.

7.2 Design a cyclic generator for the following sequence using *JK* flip-flops and NAND gates:

Clock pulse	C	B	A
1	0	0	1
2	1	0	0
3	0	1	0
4	1	0	1
5	1	1	0
6	0	1	1

Examine the behaviour of this circuit in its unused states and show that one of the unused states is a 'lock-in' state. Suggest a way of avoiding a 'lock-in'.

7.3 Convert the binary code in the tabulation shown below to its corresponding Gray code, and design a counter using JK flip-flops and NAND gates to generate this new counting sequence. Assume unused counts are 'don't care' states.

D	C	B	A
0	0	0	0
0	1	1	1
0	1	1	0
0	1	0	1
0	1	0	0
1	0	1	1
1	0	1	0
1	0	0	1
1	0	0	0
1	1	1	1

7.4 The operational characteristics of a PQ flip-flop are as follows:

$PQ = 00$ the next state of the flip-flop is 1, irrespective of its present state.
$PQ = 01$ the next state of the flip-flop is the complement of the present state, irrespective of its present state.
$PQ = 10$ the next state of the flip-flop is the same as the present state, irrespective of its present state.
$PQ = 11$ the next state of the flip-flop is 0, irrespective of its present state.

Using the above information, obtain the steering table of the PQ flip-flop and develop the input equations for a scale-of-eight binary counter which uses this flip-flop.

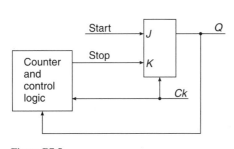

Figure P7.5

7.5 The circuit shown in figure P7.5 is to be used to generate an output pulse Q having a time duration equal to 14 clock periods. Draw a timing diagram showing the principal circuit waveforms.

7.6 A five-stage ripple-counter uses flip-flops having a delay time of 30 ns and a decode time of 50 ns. Determine the maximum frequency of operation of the counter. If the counter is operating at this frequency, draw a timing diagram for each of the flip-flops as the count advances from 01111 to 10000.

Assuming that the counter is operated now at a frequency of 8.33 MHz, draw timing diagrams showing the behaviour of the flip-flops between the fifteenth and sixteenth clock pulses.

7.7 Draw the timing diagrams for the following asynchronous counters:

(a) a 4-bit binary down-counter
(b) a 4-bit binary up-counter

assuming that the flip-flops used in the counting array trigger on the leading edge of the pulse applied to the clock terminal.

7.8 The contents of a serial-in/serial-out shift register are $DCBA = 0101$, where A is the least significant digit of the register. A serial input 10011 is moved into the shift register, from left to right, most significant bit first, by five successive clock pulses. Draw time diagrams showing how the outputs of the four flip-flops vary with time during the period of the five clock pulses.

7.9 Design a modulo-12 counter using a shift register and feedback logic. Develop the decode logic required to give a decimal output.

7.10 Using a shift register and combinational logic, design a sequence generator which will generate the binary sequence 0-1-0-0-1-0-1-1-1-0-1.

7.11 Develop the state diagrams for the following shift register generators which employ exclusive-OR feedback:

(a) A four-stage shift register. Feedback function $f = B \oplus C$.
(b) A five-stage shift register. Feedback function $f = D \oplus E$.

7.12 A three-stage shift register is to be used to generate two sequences of length 7 and 5, respectively. When a control signal $m = 1$, it generates a sequence of length 7, and when the control signal $m = 0$ it generates a sequence of length 5. Design a shift register generator using exclusive-OR feedback to implement the above specification.

7.13 A three-stage shift register ABC having exclusive-OR feedback $B \oplus C$, where A is the least significant stage of the register, is to be used to produce a repeating sequence of binary coded decimal digits for e (2.718282) on four output lines P, Q, R and S.

Determine the sequence developed by the generator and develop the combinational logic required to generate the sequence for e.

7.14 Draw a timing diagram for a four-stage twisted ring counter for a period of eight clock pulses. Display the outputs of each of the flip-flops on the timing diagram.

If the counting sequence is to be reduced from eight to seven by the omission of the 1111 state, determine the modification of the feedback logic that is required.

8 Clock-driven sequential circuits

8.1 Introduction

In this chapter a design procedure will be established for the design and implementation of clock-driven sequential circuits. Such circuits have many applications in the digital field and consist of both combinational and memory elements. For an SSI design, members of one of the commonly used logic families would be employed in conjunction with either JK or D flip-flops. In this field JK flip-flops would probably be selected since their use normally leads to simpler circuit implementation. However, in recent years, enormous advances in technology have led to the introduction of a variety of large scale programmable devices (PLDs) and the flip-flops used as memory elements on these devices are likely to be D flip-flops.

8.2 The basic synchronous sequential circuit

A block diagram of a basic synchronous sequential circuit is shown in Figure 8.1. The circuit is controlled by the synchronising clock signal and the memory is realised with edge-triggered flip-flops, changes taking place on either the leading or trailing edge of a clock pulse. If there are n flip-flops in the memory, for storing the state of the circuit, there are 2^n possible states, not all of which need be used in the design of the circuit. The state of the circuit can only change on a transition of the clock signal. Relationships between the various quantities specified in the diagram may be expressed in the form of state tables or state diagrams.

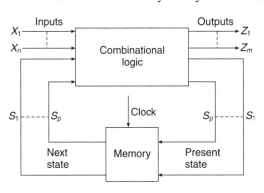

Figure 8.1 *Basic synchronous sequential circuit*

8.3 Analysis of a clocked sequential circuit

The logic diagram shown in Figure 8.2 is that of a clocked sequential circuit having two inputs, X and clock, and one output Z. The memory elements used are two edge-triggered D flip-flops which define the four possible internal states of the circuit, $AB = 00, 01, 10,$ and 11.

(a)

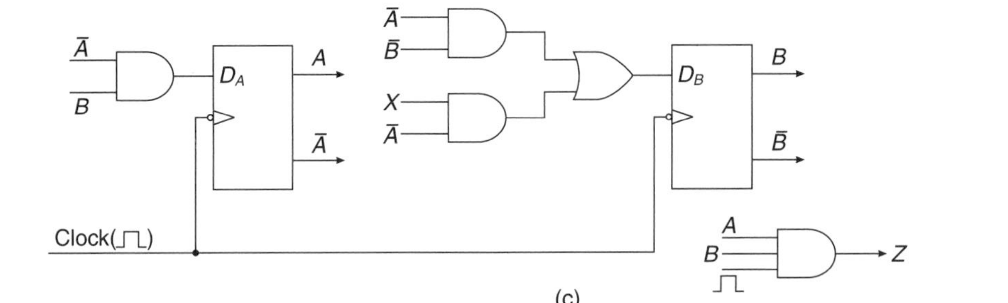

(b)

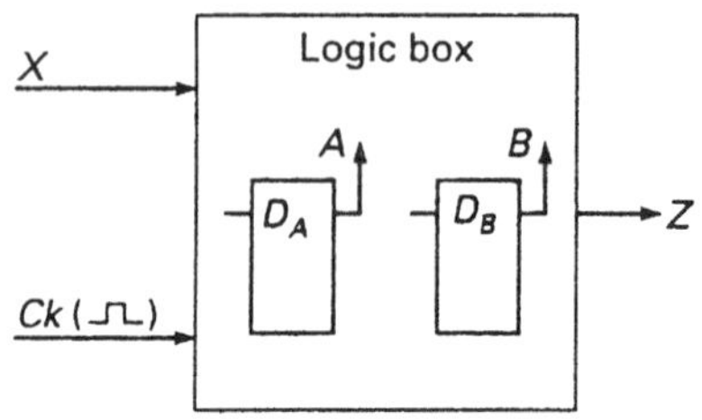

(c)

Present state	Next state		Output	
	$X=0$	$X=1$	$X=0$	$X=1$
AB	AB	AB	Z	Z
S_0 00	01	01	0	0
S_1 01	10	11	0	0
S_2 11	00	00	1	1
S_3 10	00	00	0	0

(d)

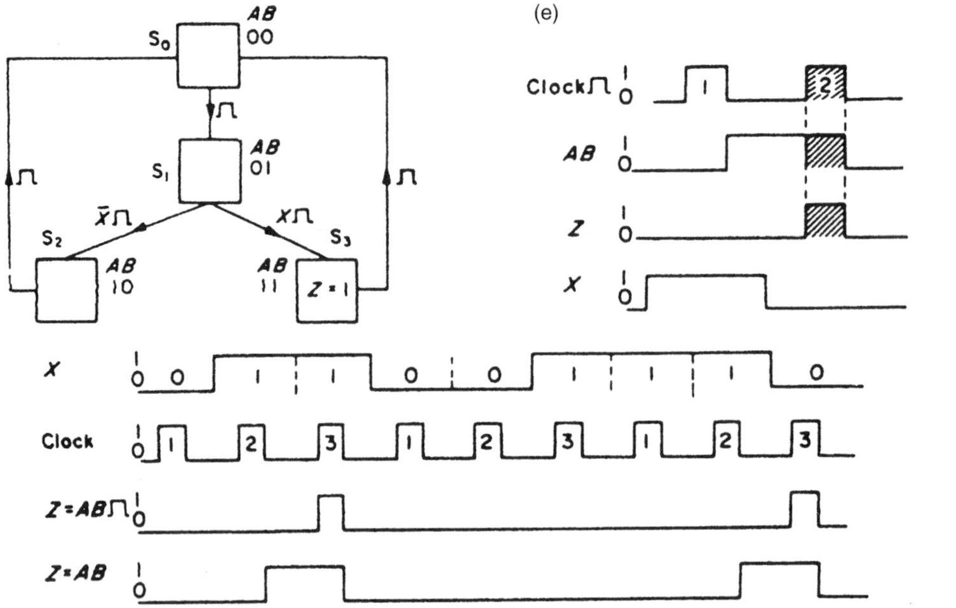

(e)

(f)

Figure 8.2 *(a) Sequential circuit to be analysed (b) Block diagram for circuit (c) State table for circuit to be analysed (d) State diagram of the circuit (e) Generation of output signals (f) Timing diagram for the circuit*

An alternative way of representing this circuit is by means of the block diagram shown in Figure 8.2(b). This diagram depicts a logic box which contains the combinational logic as well as the two flip-flops, A and B, whose output combinations define the four internal states of the circuit. The input equations for the flip-flops A and B can be obtained directly from Figure 8.2(a).

$$D_A = \bar{A}B \qquad D_B = \bar{A}\bar{B} + X\bar{A}$$

In Chapter 6 it was shown that the next state output for an edge-triggered D flip-flop is given by the equation:

$$Q^{t+\delta t} = D^t$$

By substituting D_A and D_B in this equation, the next state outputs of the two flip-flops A and B are obtained:

$$A^{t+\delta t} = (\bar{A}B)^t$$

and

$$B^{t+\delta t} = (\bar{A}\bar{B} + X\bar{A})^t$$

With the aid of these equations, it is now possible, for given present state values of A and B, and for a given value of the input signal X, to determine the next state values of A and B. For example, if $A = 0, B = 0$ and $X = 0$, then $A^{t+\delta t} = 0$ and $B^{t+\delta t} = 1$.

It is now possible to determine the output Z of the circuit for all possible combinations of X, A, and B. This requires a knowledge of the output equation which is obtained directly from Figure 8.2(a):

$$Z = AB\Pi$$

The interpretation of this equation is that the output $Z = 1$, when the present state of the circuit is $A = 1, B = 1$, and if $X = 0$ or 1 is received at the input in conjunction with a clock signal. For all other combinations of A and B the output $Z = 0$, irrespective of the value of X or the presence of the clock. Further, the equation indicates that the time duration of $Z = 1$ at the output can never be greater than the time duration of the clock pulse.

It is now possible to construct a table showing the present state, the next state and the output. This table, shown in Figure 8.2(c), may be regarded as the state table of the circuit where the various states have been designated as S_0, S_1, S_2 and S_3. With the aid of this state table the internal state diagram can now be constructed and it is shown in Figure 8.2(d), each of the four rectangles representing one of the four states. A transition from one state to the next is represented by a straight line with an arrowhead which indicates the direction of the transition. The transition signal is placed at the side of the arrowhead. In order to make a transition from S_1 to S_3, the circuit needs to receive $X = 1$ and clock ($X\Pi$). Since the flip-flops used are negative edge-triggered D flip-flops, a transition between states will always take place on the trailing edge of a clock pulse. It follows that the transition from S_1 to S_3 will take place on the trailing edge of the clock pulse which forms part of the transition signal $X\Pi$.

The output $Z = 1$ has been entered in the rectangle marked S_3. This output allocation should be interpreted as follows. $Z = 1$ if the circuit is in internal state $AB = 11$, and a clock pulse is received. The generation of the output signal Z is illustrated in Figure 8.2(e). The circuit enters state S_3 when $X = 1$, and on the trailing edge of the clock pulse marked 1. The circuit remains in state S_3 until the trailing edge of the clock pulse marked 2, which initiates the transition from S_3 to S_0. The output $Z = AB\Pi$ is formed by ANDing AB and the clock pulse marked 2. Both of these signals are logic 1 in the shaded region shown on the diagram, and in this region the output $Z = 1$. It should be observed that $Z = 1$ during the time duration of the shaded region irrespective of whether $X = 0$ or 1.

What can be deduced about the function of this circuit from the state diagram? An initial observation indicates that there are two distinct paths through the state diagram starting with S_0. The first branch is via S_3 and back to S_0, and the second branch is via S_2 and then back to S_0. Secondly, it is clear that no matter which of these two paths is taken from S_0, there are always three transitions before returning to S_0. This implies that strings of three binary digits arriving on the X line are being examined by the circuit.

Certain combinations of three digits will result in an output for the duration of the third clock pulse. This will only occur if the path taken through the state diagram is via S_3. Other combinations of the three digits result in the path via S_2 being selected, and in this case there is no output on the Z line during the third clock pulse.

The first transition in the sequence, from S_0 to S_1, is initiated by a clock pulse and takes place irrespective of whether $X = 0$ or 1. In order for the transition from S_1 to S_3 to take place $X = 1$, and finally the transition from S_3 back to S_1 on the third consecutive clock pulse will take place irrespective of whether $X = 0$ or 1. Clearly there are four combinations of three binary digits that will generate an output of $Z = 1$. They are 010, 011, 110 and 111. The remaining four combinations, 000, 001, 100 and 101 will be associated with the alternative path through the state diagram, and for this path the output $Z = 0$.

The timing diagram for three different strings of three binary digits, 011, 001 and 110, is shown in Figure 8.2(f). The X signal is synchronised to the clock and it is assumed that changes in this signal always take place between clock pulses. For the combinations 011 and 110 there is an output $Z = 1$ which lasts for the duration of the clock pulses marked 3, while for the combination 001 the output $Z = 0$.

The last waveform in the timing diagram is for $Z = AB$, the clock signal having been removed from the output equation, and as a consequence the output becomes $Z = 1$ on the trailing edge of clock pulse 2 and terminates on the trailing edge of clock pulse 3. It is interesting to note that in the case where $Z = AB$, the output goes high before the third digit has arrived. This is satisfactory in this case since having once recognised what the second digit is by entering the state S_3, it is now irrelevant whether the third digit is 0 or 1, and it does not have to be recognised. However, in the case where $Z = AB\sqcap$, the output does not go high until the leading edge of the third clock pulse. The circuit has to recognise this clock pulse before an output can occur.

8.4 Design steps for synchronous sequential circuits

The analysis of the sequential circuit in section 8.3 has identified some of the processes required for the design of small scale synchronous circuits having a limited number of inputs. An orderly design process can be carried out using the following 10 steps:

Step 1 Receive the problem specification.
Step 2 Draw up a block diagram for the proposed design which displays all the inputs and the required outputs.
Step 3 Make an attempt to construct a basic state diagram using information obtained from steps 1 and 2.
Step 4 Using the basic state diagram construct a state table and check for redundant states.
Step 5 Reconstruct the state diagram if redundancy has occurred, using the information obtained in step 4.

Step 6 Make a state assignment.
Step 7 Draw up a new state table, excluding any redundancies, and using the state assignment of step 6.
Step 8 Select the flip-flops, D, T or JK, to be used as memory elements.
Step 9 Using the reduced state table derive the logic equations for the next state inputs to the selected flip-flops.
Step 10 Develop the output logic with the aid of the reduced state diagram.

Step 1: Problem specification

For relatively simple sequential circuits the specification will usually consist of a verbal statement of the problem and, in particular, details of the inputs available and the outputs required. The specification of the problem in completely unambiguous terms is not always straightforward and may require several discussions between designer and user. If the ambiguities are not resolved at this stage, a circuit implementation will be reached that does not satisfy the user's requirements and the design process will have to be repeated.

Step 2: Problem block diagram and timing diagram

Having studied the problem specification, construct a block diagram showing the sources of all the inputs and the required outputs. Additionally, draw up a timing diagram displaying the outputs specified by the problem.

Step 3: The internal state diagram

In this step the verbal statement of the problem should be expressed in terms of the internal states of the circuit in the form of a state diagram. There are no rules for constructing state diagrams, and the ability to draw them can only be acquired by experience. For example, the inexperienced designer will almost certainly not, in the first instance, produce the state diagram shown in Figure 8.2(d) for the circuit analysed in Section 8.3. To construct the state diagram for that problem, the designer might have been given the following verbal statement of the problem.

'A logic circuit is to receive binary data serially on an input line, which is synchronised with an external clock signal. Non-overlapping strings of three successive bits of the input data are to be examined by the logic circuit, and if the combinations 010, 011, 110 and 111 are detected, a 1 will appear at the output. The output must occur when the third bit of the string is present and the third clock pulse is high.'

In practice, the inexperienced designer may well develop the tree-like structure of states shown in Figure 8.3. The method of approach used to arrive at this diagram would be to commence in the arbitrarily selected state S_0. This internal state of the circuit can be left by two separate transition paths, one associated with the transition signal X leading to S_1, and the other associated with the transition signal $\bar{X}$ leading to S_2, where X represents logic high level at the input. Each of the states S_1 and S_2 can be left by a pair of transition paths, one associated with the transition signal X, and the other with the transition signal $\bar{X}$. These four paths lead to the four states S_3, S_4, S_5 and S_6. For each of these four states there are two exit paths, but the next transition

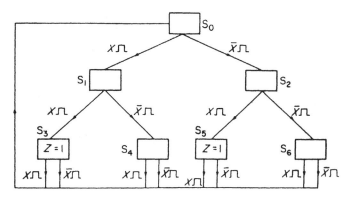

Figure 8.3   *Internal state diagram for a combination detector*

is the third one and consequently all eight exit paths must lead back to the starting state.

The combinations 111 and 110 take the path $S_0 \rightarrow S_1 \rightarrow S_3 \rightarrow S_0$ through the state diagram, and the output $Z = 1$ in state S_3. Similarly, the combinations 011 and 010 take the path $S_0 \rightarrow S_2 \rightarrow S_5 \rightarrow S_0$ through the state diagram and the output $Z = 1$ in state S_5. The other two paths through the state diagram deal with those combinations that do not have to be detected.

In developing this diagram, no short-cuts have been taken. Each combination of three bits of input data appear explicitly on the diagram. However, this version of the state diagram requires eight states compared to the four states in Figure 8.2(d). In terms of hardware, this means that a circuit implementation developed from Figure 8.3 would require three JK or D flip-flops and an additional amount of combinational logic.

Developing the state diagram from the problem specification is the most interesting and rewarding part of any digital design. Beginners are likely to experience difficulties at this stage. Their aim should be to produce a state diagram that contains no redundant states. A beginner, for example, might well have produced the state diagram shown in Figure 8.3 from the given problem specification. Clearly this diagram contains redundant states and after drawing up a state table, methods of state reduction (see Step 5) should be applied to generate the state diagram shown in Figure 8.2.

Step 4: State table

The state table corresponding to the state diagram shown in Figure 8.3 appears in Figure 8.4(a). The table has a row for every state of the circuit and a column for every combination of the input signals. In this case there is only one input signal and this only requires two columns, one for $X = 1$, and one for $X = 0$. In each of the cells formed by the intersection of the rows and columns the next state of the circuit is entered along with the output Z. If, for example, $X = 0$ when in the state S_0, the next state is S_2. Alternatively, if $X = 1$ when in state S_0, the next state is S_1.

Step 5: State reduction

The more states there are in the state diagram, the more hardware is required for the circuit implementation. For this reason, it is in the interests of the designer to reduce

(a)

X	0	1
S_0	S_2, $Z=0$	S_1, $Z=0$
S_1	S_4, $Z=0$	S_3, $Z=0$
S_2	S_6, $Z=0$	S_5, $Z=0$
S_3	S_0, $Z=1$	S_0, $Z=1$
S_4	S_0, $Z=0$	S_0, $Z=0$
S_5	S_0, $Z=1$	S_0, $Z=1$
S_6	S_0, $Z=0$	S_0, $Z=0$

(b)

X	0	1
S_0	S_2, $Z=0$	S_1, $Z=0$
S_1	S_{46}, $Z=0$	S_{35}, $Z=0$
S_2	S_{46}, $Z=0$	S_{35}, $Z=0$
S_{35}	S_0, $Z=1$	S_0, $Z=1$
S_{46}	S_0, $Z=0$	S_0, $Z=0$

(c)

X	0	1
S_0	S_{12}, $Z=0$	S_{12}, $Z=0$
S_{12}	S_{46}, $Z=0$	S_{35}, $Z=0$
S_{35}	S_0, $Z=1$	S_0, $Z=1$
S_{46}	S_0, $Z=0$	S_0, $Z=0$

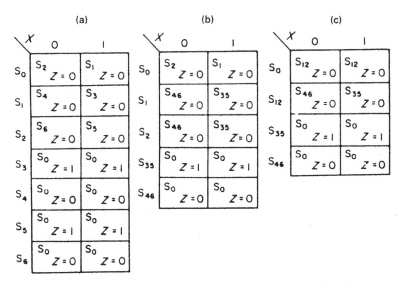

Figure 8.4 *Combination detector (a) State table (b) Reduced state table (c) Minimal state table*

the number of states if possible. The process of state reduction in sequential circuit design corresponds to the process of minimisation in combinational circuit design.

State reduction can be done systematically with the aid of the state table and by using Caldwell's merging procedure which depends upon proving that two states are equivalent. Equivalence is defined by the following statement:

Two states a and b are equivalent if (1) both have identical next states and (2) both have identical outputs.

For the table shown in Figure 8.4(a) the rows headed S_4 and S_6 satisfy this definition as do the rows headed S_3 and S_5. After states S_4 and S_6 have been merged, the state formed is designated S_{46} and wherever S_4 and S_6 appear in the state table they are replaced by S_{46}. Similarly, S_3 and S_5, when merged, form an equivalent state S_{35} which replaces S_3 and S_5 wherever they appear in the state table.

Figure 8.5 *State diagram for the combination detector*

Using Caldwell's merging procedure, the state table of Figure 8.4(a) can be reduced to that shown in Figure 8.4(b) which also has two rows, S_1 and S_2, that are equivalent and can be merged to form the equivalent state S_{12}. The table of Figure 8.4(b) can now be replaced by the state table shown in Figure 8.4(c) and no further reduction is now possible. The reduced state diagram that can now be constructed from the reduced state table is identical to that shown in Figure 8.2(d) and is repeated here in Figure 8.5.

The best situation in practice is one in which the number of states n is a power of two. There is little point in reducing the number of states below 2^n unless it is to a lower power of two since this would lead to a number of unused states. For example, if N is the number of states after reduction so that $2^{n-1} < N < 2^n$, then the number of unused states is $2^n - N$. Unused states create additional difficulties for the digital designer. An unused state can be entered at 'power on' or, alternatively, due to faulty circuit operation. It is the responsibility of the designer to specify the behaviour of the circuit if it should, by chance, enter an unused state, otherwise a 'lock-in' may occur. If a 'lock-in' occurs it means there is no exit from the unused state and the circuit will remain in that state for an indefinite period. It should also be stressed that unused states are not 'can't happen' states, and for this reason they should not be used for simplification of the circuit equations.

Step 6: State assignment

Having obtained the minimum state table, the next step the designer must take is to allocate secondary variables to the various states. The number of secondary variables required to define a state is governed by the total number of states in the diagram. In this case there are four states in Figure 8.5 and two secondary variables are required to define each state uniquely.

For this problem a state assignment has been selected which conforms to the state assignment in Figure 8.2(c). Clearly there are other possible allocations of these variables and consequently there are a number of different circuit solutions to this problem, some of which may lead to more economical circuitry. However, it is rarely worthwhile to search for a minimal solution since this can be a very time consuming process.

The number of secondary variables needed to define a state is equal to the number of flip-flops required to implement the design. In the state diagram of Figure 8.5 there are four states and two secondary variables A and B define these states; consequently two flip-flops will be required for the circuit implementation.

Step 7: The revised state table

The reduced state table of Figure 8.4(c) is now tabulated in terms of the secondary variables as shown in Figure 8.6(a). This table gives every possible transition of these variables for both $X = 0$ and $X = 1$.

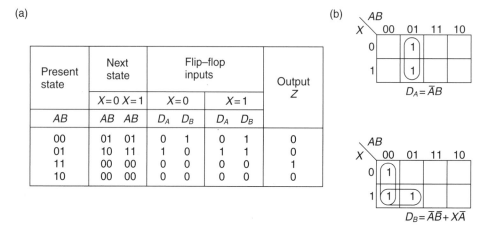

(a)

Present state	Next state		Flip–flop inputs				Output Z
	$X=0$	$X=1$	$X=0$		$X=1$		
AB	AB	AB	D_A	D_B	D_A	D_B	
00	01	01	0	1	0	1	0
01	10	11	1	0	1	1	0
11	00	00	0	0	0	0	1
10	00	00	0	0	0	0	0

(b)

$D_A = \bar{A}B$

$D_B = \bar{A}\bar{B} + X\bar{A}$

Figure 8.6 *(a) State table and flip-flop input tabulation for the combination detector (b) K-map plots for the flip-flop inputs*

Step 8: Flip-flop selection

To complete this design D flip-flops have been selected to implement the next state equations.

Step 9: The next state equations

The technique used for determining the next state equations consists of tabulating the D-inputs for every transition on the state table. The D-inputs for the various transitions are then mapped on a K-map and are simplified where that is possible.

For a D flip-flop it will be recalled that $Q^{t+\delta t} = D^t$, i.e. the next state of the flip-flop is given by the present state of the D-input. It follows that the D_A and D_B columns are identical to the next state entries for A and B. For example, if $X = 0$, the four next state entries for A are 0, 1, 0 and 0 and consequently the corresponding entries in the D_A column for $X = 0$ will be 0, 1, 0 and 0.

Maps can now be plotted for D_A and D_B. These are shown in Figure 8.6(b) and the next state equations derived from these maps are:

$$D_A = \bar{A}B \quad \text{and} \quad D_B = \bar{A}\bar{B} + X\bar{A}$$

As might be expected, these equations are identical to the flip-flop input equations for the circuit shown in Figure 8.1(a).

Step 10: The output

A single column is tabulated in the state table shown in Figure 8.6(a) for the output Z. The output $Z = AB$ occurs when the circuit is in the state $AB = 11$, and $Z = 1$ is entered in the output columns opposite this state. If the entry in the state had been clock ($\sqcap$) then the output would have been $Z = AB\sqcap$.

8.5 The design of a sequence detector

Step 1: Problem definition

Serial binary data is received on the X-input line of a logic circuit, each bit being synchronised with the clock signal. An output signal is generated at the output Z each time the sequence 101 is detected. Overlapping sequences are permitted. A block diagram for the proposed circuit is shown in Figure 8.7(a) in conjunction with a stream of input data X and the output Z.

Step 2: The internal state diagram

A suitable state diagram, consisting of three states, for detecting the sequence 101 is shown in Figure 8.7(b). The reader should note that if $X = 0$ is received when the circuit is in state S_0, it will remain in that state and will continue to do so until the signal $X = 1$ arrives. Similarly if, after making a transition from $S_0 \rightarrow S_1$ on the signal $X = 1$, a succession of 1's is received the circuit will remain in S_1 until such time as $X = 0$ arrives at the input, when a transition will be made to S_2. To define three states, two secondary variables A and B are required. Since there are four combinations of these variables there is one unused state S_3. If, due to faulty operation of the circuit, it should

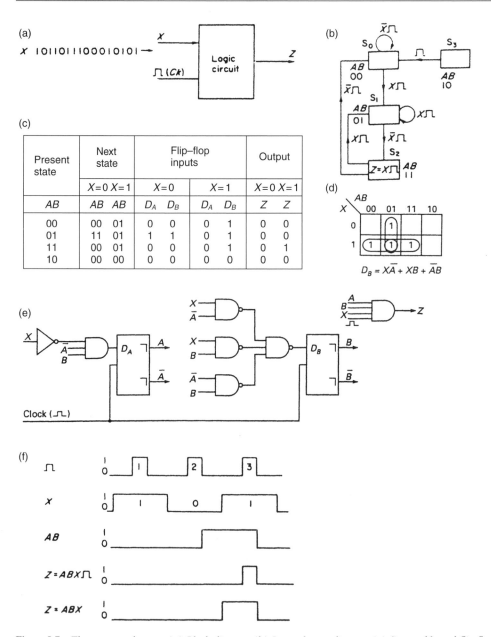

Figure 8.7   *The sequence detector (a) Block diagram (b) Internal state diagram (c) State table and flip-flop inputs tabulation (d) K-map plot for D_B (e) Circuit implementation (f) Timing diagrams*

enter this state, it might be desirable to return to the main sequence of states as soon as possible. This can be achieved by returning S_3 to S_0 via a transition which is initiated by the first clock pulse that occurs after the entry into S_3.

Step 3: State reduction

An examination of the state table in Figure 8.7(c) shows that state reduction is not possible since there are no rows having the same next state entries and outputs in corresponding columns of the table.

Step 4: Development of the next state equations

Since there are only four states in the state diagram, which have been arbitrarily assigned, just two D flip-flops are required to implement the detector. The flip-flop inputs, D_A and D_B, and the output Z for each entry on the state table are tabulated alongside those entries. The D flip-flop input entries in the state table are simply a repeat of the next state entries. As there is only one 1 entry in the two columns for D_A, the equation for the D-input of FFA may be taken directly from the state table and is:

$$D_A = \bar{X}\bar{A}B$$

A K-map plot of the D-input for FFB is shown in Figure 8.7(d), and after simplification

$$D_B = X\bar{A} + XB + \bar{A}B$$

The output Z is read directly from the state table and is:

$$Z = ABX \underline{\hspace{0.1em}}\sqcap\underline{\hspace{0.1em}}$$

The circuit implementation for the detector is shown in Figure 8.7(e) and the timing diagrams for a 101 sequence of bits is shown in Figure 8.7(f). Postponed output DFFs are used in this design, so the circuit enters state $AB = 11$ on the trailing edge of clock pulse 2 and leaves on the trailing edge of clock pulse 3. If the output is defined as $Z = ABX$ then it will go high when the circuit recognises the leading edge of the input bit associated with the clock pulse numbered 3. If, on the other hand, the output is defined as $Z = ABX \underline{\hspace{0.1em}}\sqcap\underline{\hspace{0.1em}}$ it does not go high until the leading edge of clock pulse number 3 is recognised. By using $Z = ABX \underline{\hspace{0.1em}}\sqcap\underline{\hspace{0.1em}}$ the possibility of contact bounce on the signal X being propagated to the output Z is avoided, whereas using $Z = ABX$ gives a longer detection pulse if contact bounce is known not to be present on X.

8.6 The Moore and Mealy state machines

There are two types of synchronous sequential machines. The first of these machines has an output that depends only on its present state and is referred to as the *Moore machine*. The behaviour of the machine is defined by the equations:

Next State $= f($*Present State, Inputs*$)$

    Output $= g($*Present State*$)$

The configuration of the machine is shown in Figure 8.8(a).

In the second type of machine the output depends on both its present state and also its inputs. This type of machine is referred to as the *Mealy machine* and its behaviour is defined by the following equations:

Next State $= f($*Present State, Inputs*$)$

    Output $= g($*Present State, Inputs*$)$

The general structure of the Mealy machine is shown in Figure 8.8(b).

The circuit developed in Figure 8.7(e) is an example of a Mealy circuit, since the output $Z = ABX \underline{\hspace{0.1em}}\sqcap\underline{\hspace{0.1em}}$ depends not only on the state of the circuit but also on the input X, whilst its time duration is limited by the width of the clock pulse. A slight modification to the state

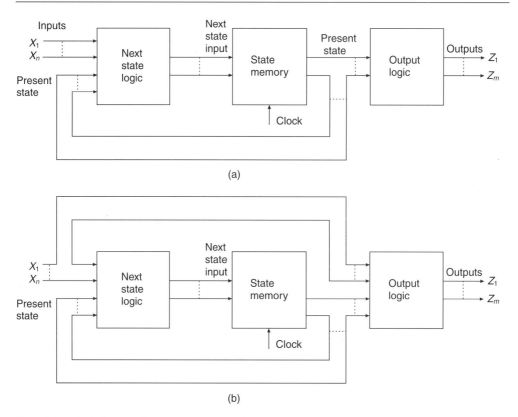

Figure 8.8 *Block diagrams for (a) the Moore machine and (b) the Mealy machine*

diagram shown in Figure 8.7(b) will convert the Mealy circuit to a Moore circuit. This modification is illustrated in Figure 8.9(a). An additional state, S_3, has been introduced and this is now used as the output state of the circuit so that $Z = A\bar{B}$. The timing diagrams corresponding to an input sequence $X = 10100$ are shown in Figure 8.9(b).

The state table and the tabulation of the flip-flop inputs for the Moore circuit are shown in Figure 8.9(c) and the K-map plots for the D flip-flops are shown in Figure 8.9(d).

After simplification, the equations for the D inputs are found to be

$$D_A = \bar{X}\bar{A}B + \bar{X}A\bar{B} + XAB$$

and

$$D_B = X\bar{A} + \bar{A}B + A\bar{B}$$

These equations are sometimes referred to as the *excitation equations*. Using the above equations and the output equation $Z = A\bar{B}$, the Moore implementation of the sequence detector is shown in Figure 8.9(e). It is left to the reader to show that if the states had been allocated such that $S_2 = A\bar{B} = 10$ and $S_3 = AB = 11$ much simpler excitation equations would have been obtained leading to a much simpler circuit implementation.

As a further example of the Mealy and Moore representations, consider the following problem. 'A logic circuit receives binary information on the input line X. Non-overlapping strings of three successive digits are to be examined by the circuit. If the last two digits in the group are both 1's the output Z will be 1'.

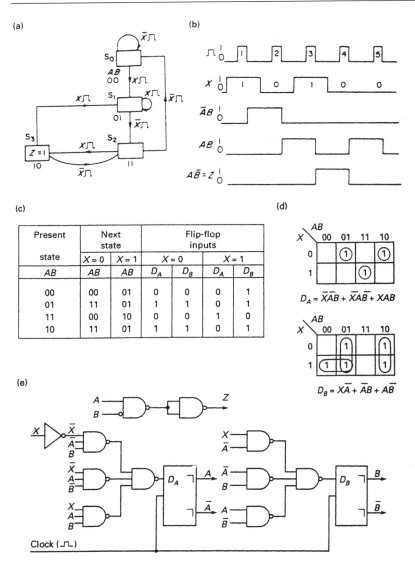

(a)

(b)

(c)

Present	Next		Flip-flop			
state	state		inputs			
	$X = 0$	$X = 1$	$X = 0$		$X = 1$	
AB	AB	AB	D_A	D_B	D_A	D_B
00	00	01	0	0	0	1
01	11	01	1	1	0	1
11	00	10	0	0	1	0
10	11	01	1	1	0	1

(d)

$$D_A = \bar{X}\bar{A}B + \bar{X}A\bar{B} + XAB$$

$$D_B = X\bar{A} + \bar{A}B + A\bar{B}$$

(e)

Clock ($\sqcap\!\sqcap$)

Figure 8.9 *The sequence detector (a) Moore representation state diagram (b) Timing diagrams (c) State table and flip-flop inputs tabulation (d) K-map plots for D_A and D_B (e) Circuit implementation*

The block diagram for the problem is shown in Figure 8.10(a) and a possible state diagram is shown in Figure 8.10(b). For this problem the entry in states S_3 and S_5 is $Z = X\sqcap$. This indicates that an output will occur in those two branches when the last digit is 1. The state table is shown in Figure 8.10(c) and it is apparent on inspection that states S_3 and S_5 are equivalent and can be merged to form one state S_{35}. Furthermore, S_4 and S_6 are also equivalent and can be merged to form the state S_{46}. After merging, the reduced state table is tabulated in Figure 8.10(d). From an inspection of this table it is evident that a further reduction is possible since S_1 and S_2 are equivalent and can be merged to form the one state S_{12}. The final state table is shown in Figure 8.10(e), no further reduction being possible.

The reduced state diagram shown in Figure 8.10(f) has been obtained from the information tabulated in Figure 8.10(e). This state diagram will lead to a Mealy-type

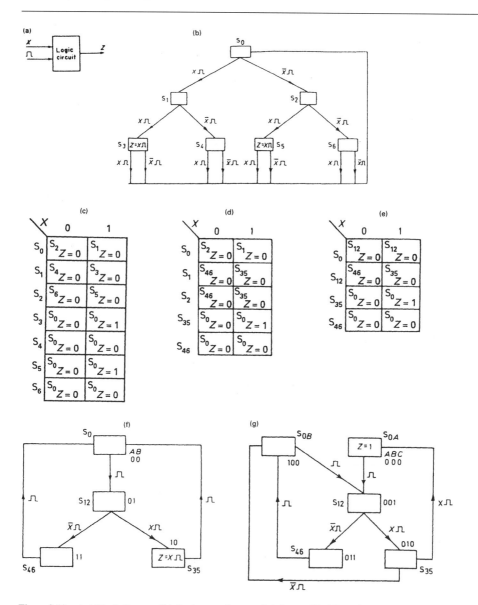

Figure 8.10 *(a) Block diagram (b) Basic state diagram (c) State table (d) Reduced state table (e) Further reduced state table for the word scanner (f) State diagram for the Mealy type representation (g) State diagram for the Moore diagram representation*

circuit since the output $Z = A\bar{B}X\sqcap$ depends upon the present state and the input signal X. To convert the state diagram to one which will lead to a Moore-type circuit the state S_0 is split into two states, S_{0A} and S_{0B}, as shown in Figure 8.10(g). The output Z now appears in state S_{0A} and is dependent only on the state so that $Z = \bar{A}\bar{B}\bar{C}$. If it is required that the output should be time limited by the clock, the output would be written $Z = \bar{A}\bar{B}\bar{C}\sqcap$. The conversion to a Moore state diagram increases the number of states from four to five. Since one of the states has to be set aside for the output it is clear that Moore-type circuits will require more states than Mealy-type circuits.

In the case of the Moore circuit there are three unused states, and it is desirable that the behaviour of the circuit is predictable if a fault condition arises in the circuit such that it enters one of the unused states. The best solution is to ensure that the circuit returns to the initial state S_{0B}. Additionally, it may be desirable to raise an alarm and disable the circuit by, for example, stopping the clock. The precautions taken by the designer will depend on the requirements of the design specification.

8.7 Analysis of a sequential circuit implemented with JK flip-flops

JK flip-flops are also used for implementing sequential circuits. They have the disadvantage of having two separate inputs compared to the single input of the D flip-flop. However, there are four 'don't care' terms available in the JK steering table and this will normally result in simpler next state equations.

The logic diagram of a JK sequential circuit is shown in Figure 8.11(a). The circuit has a single input signal m and a synchronizing clock signal. Since there are two JKFFs the circuit has four internal states. The next state equations for the A and B flip-flops are:

$$J_A = \bar{B}m \quad J_B = A$$
$$K_A = 1 \quad K_B = A + \bar{m}$$

The characteristic equation of a JK flip-flop developed in chapter 6 is:

$$Q^{t+\delta t} = (J\bar{Q} + \bar{K}Q)^t$$

By substituting J_A, K_A and J_B, K_B in this equation, the next state functions of the two flip-flops are obtained. They are:

$$A^{t+\delta t} = (\bar{A}\bar{B}m)^t$$

and

$$B^{t+\delta t} = (A\bar{B} + \bar{A}Bm)^t$$

With the aid of these equations it is now possible, for given present state values of A and B, and for a given value of the input signal m, to determine the next state values of A and B. For example, if $A = 0, B = 0$ and $m = 1$, then $A^{t+\delta t} = 1$ and $B^{t+\delta t} = 0$. Similarly, next state values can be obtained for the other seven combinations of A, B and m.

The output equation taken directly from the logic circuit is:

$$Z = A\bar{B}\sqcap$$

and is independent of the input signal m, so this is an example of a Moore-type circuit.

With the aid of the output equation and the next state functions it is now possible to develop the state table for this circuit [see Figure 8.11(b)]. For example, if the present state is $AB = 01$ and the input signal $m = 0$, on the trailing edge of the next clock pulse the circuit will enter the state $AB = 00$. The transition details

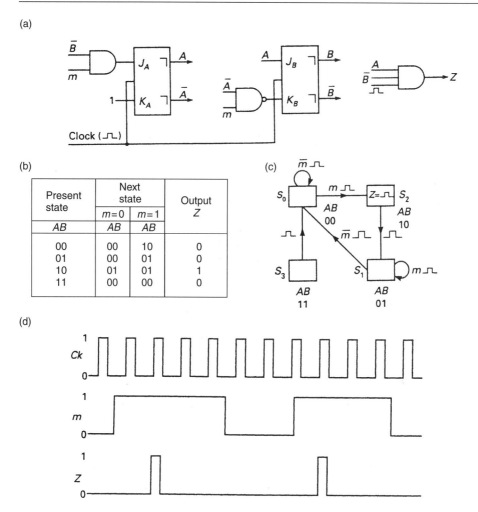

Figure 8.11   *(a) Sequential circuit to be analysed (b) State table (c) State diagram (d) Timing diagrams*

provided by the state table have been used to develop the state diagram in Figure 8.11(c).

The function of this sequential circuit can be deduced from the state diagram and is illustrated in the timing diagrams shown in Figure 8.11(d). When in the initial state S_0, with $m = 0$, the circuit will remain in that state. If the input signal m changes from $0 \rightarrow 1$, then on the trailing edge of the next clock pulse, a transition will be made to state S_2. On the receipt of the next clock pulse the output Z is generated and on the trailing edge of that clock pulse the circuit makes a transition to state S_1. The output, it will be noted, is a single clock pulse. The circuit now remains in state S_1 until the input signal returns to $m = 0$ and on the trailing edge of the next clock pulse after this event has occurred the circuit returns to the initial state S_0. State S_3 is an unused state and if the circuit should at some instant enter that state it will return to the initial state S_0 on the trailing edge of the next clock pulse. This circuit is called a 'one-shot' and has practical application where it is required to slow down high speed operations to manual speeds.

8.8 Sequential circuit design using JK flip-flops

Step 1: Problem definition

Serial NBCD codes arrive on line X, most significant bit first, each bit of the 4-bit code being synchronised with a clock pulse. Develop a circuit that will give an output when an invalid NBCD code is received.

Step 2: The internal State Diagram

In this example, a logical approach has been adopted to develop the state diagram shown in Figure 8.12(a). For example, the path $S_0 \rightarrow S_1 \rightarrow S_2 \rightarrow S_3$ is associated with the first eight combinations of the code, $0000 \rightarrow 0111$ inclusive, all of which can be identified by the most significant digit 0. These are all valid code combinations. A second path, $S_0 \rightarrow S_4 \rightarrow S_7 \rightarrow S_3$, is associated with the remaining two valid

(a)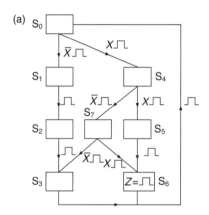

(b)

Present state	Next state	
	$X=0$	$X=1$
S_0	S_1	S_4
S_1	S_2	S_2
S_2	S_3	S_3
S_3	S_0	S_0
S_4	S_7	S_5
S_5	S_6	S_6
S_6	S_0	S_0
S_7	S_3	S_6

(c)

Q	$Q^{t+\delta t}$	J	K
0	0	0	X
0	1	1	X
1	0	X	1
1	1	X	0

(d)

Present state	Next state		Flip-flop inputs										
	$X=0$	$X=1$	$X=0$		$X=1$		$X=0$		$X=1$		$X=0$		$X=1$	
CBA	CBA	CBA	J_C	K_C	J_C	K_C	J_B	K_B	J_B	K_B	J_A	K_A	J_A	K_A
S_0 000	001	100	0	X	1	X	0	X	0	X	1	X	0	X
S_1 001	011	011	0	X	0	X	1	X	1	X	X	0	X	0
S_2 011	010	010	0	X	0	X	X	0	X	0	X	1	X	1
S_3 010	000	000	0	X	0	X	X	1	X	1	0	X	0	X
S_4 100	110	101	X	0	X	0	1	X	0	X	0	X	1	X
S_5 101	111	111	X	0	X	0	1	X	1	X	X	0	X	0
S_6 111	000	000	X	1	X	1	X	1	X	1	X	1	X	1
S_7 110	010	111	X	1	X	0	X	0	X	0	0	X	1	X

Figure 8.12 *The invalid code detector (a) The state diagram (b) The state table (c) JK flip-flop steering table (d) Tabulation of flip-flop inputs (e) K-maps for flip-flop inputs (f) Circuit implementation*

(e)

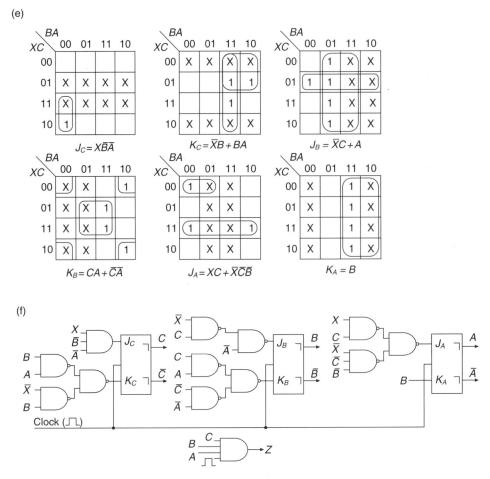

Figure 8.12   (*Continued*)

combinations 1000 and 1001. The invalid code combinations 1100 → 1111 inclusive are covered by path $S_0 \to S_4 \to S_5 \to S_6$, and the invalid combinations 1010 and 1011 take the path $S_0 \to S_4 \to S_7 \to S_6$. The output for invalid code combinations $Z = \sqcap$ is entered in the state S_6. There are eight states in all, and these can be defined by three secondary variables A, B and C. Each state has been arbitrarily allocated one of the eight combinations of these variables.

Step 3: State reduction

The state table, showing present and next states, is tabulated from the information provided by the state diagram, and is shown in Figure 8.12(b). Examination of this table shows that rows S_3 and S_6 have identical next state entries; however, they cannot be merged since the only output entry appears in present state 6 and consequently these two states are not equivalent.

Step 4: Development of the next state equations

It is at this point where design with JK flip-flops differs from a design using D flip-flops. For a JK flip-flop there are two inputs, J and K, which have to be determined from every transition recorded in the state table, and this is achieved with the help of the JK flip-flop steering table developed in Chapter 6 and presented again for convenience in Figure 8.12(c).

Since three secondary variables are required to uniquely define each of the eight states, three JK flip-flops will be required for the circuit implementation. In Figure 8.12(d) the state table has been redrawn, each state now being represented by the combination of three secondary variables allocated to it on the state diagram. Alongside the state table are twelve columns in which the flip-flop input signals J_C, K_C, J_B, K_B, J_A, and K_A are tabulated. The entries in these columns are obtained from the steering table. For example, if the present state is $CBA = 000$ and $X = 1$, the next state $CBA = 100$; hence, $C^t = 0$ and $C^{t+\delta t} = 1$. Thus, from the steering table, the entry for a $0 \rightarrow 1$ transition is $J_C = 1$ and $K_C = X$. For both B and A the transitions recorded are $0 \rightarrow 0$ and from the steering table the entries for the B and A inputs are $J_B = J_A = 0$ and $K_B = K_A = X$.

K-maps for each of the input signals are now plotted. These are shown in Figure 8.12(e) and the next state equations, derived from these maps after simplification, are

$$J_C = X\bar{B}\bar{A} \qquad\qquad K_C = \bar{X}B + BA$$
$$J_B = \bar{X}C + A \qquad\qquad K_B = CA + \bar{C}\bar{A}$$
$$J_A = XC + \bar{X}\bar{C}\bar{B} \qquad K_A = B$$

The output equation is taken directly from the state diagram and is:

$$Z = ABC\,\text{⊓}$$

The implementation of the invalid code detector is shown in Figure 8.12(f).

8.9 State reduction

There are three methods available for determining equivalent states in a completely specified state table. They are:

1. Inspection
2. Partitioning
3. The implication table

A method of state reduction by inspection has already been introduced in Section 8.4. In practice, all methods of state reduction depend upon the principle of equivalence defined earlier in Section 8.4. However, two states S_p and S_q may also be deemed to be equivalent if, and only if, every possible input sequence produces identical output sequences, irrespective of whether S_p or S_q is the initial state. One method of determining state equivalence would therefore be to apply all possible input sequences and tabulate the corresponding output sequences of the circuit, assuming each of the states of the circuit to be in turn the initial state of the circuit. This would clearly be a tedious process for a circuit having a number of input signals and a number of states. Fortunately there are two other simple and non-tedious techniques available for state reduction. They are (1) partitioning and (2) by implication table. The method of partition will be discussed next.

Partitioning

It will be assumed that the state table shown in Figure 8.13(a) has been obtained from a state diagram relating to a problem in which there is a single input X and a single output Z. A first partition is made by placing all those present states in the same section of the partition if the outputs generated are identical for all possible inputs. For example, if the present state is S_0, the two possible inputs are $X = 0$ and $X = 1$ for which the outputs are $Z = 0$ and $Z = 1$. Similarly, if the present states are either S_3 or S_5 then for $X = 0$ and $X = 1$ the outputs are $Z = 0$ and $Z = 1$. The three states S_0, S_3 and S_5 are said to be 1-equivalent. From a further inspection of the table it is clear that S_1 and S_4 are 1-equivalent and that S_2 and S_6 are also 1-equivalent. Hence, the first partition is

$$P_1 = (S_0, S_3, S_5)(S_1, S_4)(S_2, S_6)$$

The partition has been obtained by the application to the circuit of an input sequence of length one.

The second partition, P_2, is obtained using the following procedure. In the first section of P_1, for $X = 0$ the next states for S_0, S_3 and S_5 are all in the same section of P_1. However, for $X = 1$ the next states for S_0, S_3 and S_5 are S_4, S_6 and S_4 respectively, and the next state of S_3 lies in a different section of the partition. The first section of the partition P_1 is now split into two sections, the first one containing S_0 and S_5, and the second containing S_3 only. The procedure is now repeated for the second section of the first partition. With $X = 0$, S_1 and S_4 have next state entries both in the same section of the first partition, whilst with $X = 1$ the next state entries are both S_2, which is also in the same section of the first partition, and hence no split of this section is required. An examination of the third section of the original partition shows that no splitting of this section is required.

Hence:

$$P_2 = (S_0S_5)(S_3)(S_1S_4)(S_2S_6)$$

(a)

Present state	Next state	
	$X=0$	$X=1$
S_0	S_0 $Z=0$	S_4 $Z=1$
S_1	S_4 $Z=1$	S_2 $Z=0$
S_2	S_0 $Z=1$	S_3 $Z=1$
S_3	S_5 $Z=0$	S_6 $Z=1$
S_4	S_1 $Z=1$	S_2 $Z=0$
S_5	S_5 $Z=0$	S_4 $Z=1$
S_6	S_0 $Z=1$	S_3 $Z=1$

(b)

Present state	Next state	
	$X=0$	$X=1$
S_{05}	S_{05} $Z=0$	S_{14} $Z=1$
S_{14}	S_{14} $Z=1$	S_{26} $Z=0$
S_{26}	S_{05} $Z=1$	S_3 $Z=1$
S_3	S_{05} $Z=0$	S_{26} $Z=1$

Figure 8.13 *(a) State table for the partitioning example (b) Reduced state table after partitioning*

This partition has been obtained by the application of an input sequence of length two. The procedure described above is used again to determine P_3, but in this case no further partitioning is possible and $P_3 = P_2$. It follows that the individual sections of P_2 contain the equivalent states of the circuit and the reduced state table is shown in Figure 8.13(b).

The implication table

The final method of state reduction available to the designer employs the implication table. A state table for a synchronous sequential circuit is shown in Figure 8.14(a). An implication table can be constructed by listing all the states vertically except the first one, and all the states horizontally except the last one, as illustrated in Figure 8.14(b). The implication table displays all possible combinations of state pairs, and the individual cells in the table represent the testing ground for the equivalence of a state pair. For example, the top left-hand cell at the intersection of S_0 and S_1 is where these two states are tested for equivalence.

One of the conditions for equivalence is that the next state outputs of a pair of states must be identical if the two states are equivalent. On the implication table, all the cells that cannot possibly be equivalent are marked with a cross. For example, S_0 and S_1 cannot be equivalent states since the next state outputs are 0,0 and 1,0 respectively, and the cell situated at the intersection of S_0 and S_1 is marked with a cross. Similarly, all the other cells for non-equivalent state pairs are marked with a cross in Figure 8.14 (c).

The next step is to place in the empty cells the implications required to make the pair of states associated with a particular cell equivalent, by having identical next states. For example, the cell at the intersection of S_0 and S_2 contains the implication that both S_0 and S_2 must be equivalent to S_5 in order that they will be equivalent. The remaining equivalent implications are entered in the empty cells in Figure 8.14(d).

If the pairs implied in any of the cells in Figure 8.14(d) contain only those states defined by the cell, or, alternatively, if the next states of the two states defining the cell are the same state for a given input, then the two states defining the cell are equivalent and are marked with a tick. The first part of this rule applies to two cells in Figure 8.14(d), the first at the intersection of S_0 and S_7, and the second at the intersection of S_2 and S_5. These two cells have been marked by a tick.

An examination of the state table indicates that S_2 and S_5 are a pair of 'lock-in states'. S_5 can be entered from S_2 on the receipt of a clock pulse and vice versa, but there is no other exit from these two states. Clearly, these two states can be merged, and on the receipt of a clock pulse the circuit will stay in the merged state. To leave this 'lock-in' state, a reset signal is required. A similar argument can also be applied to states S_0 and S_7.

The next step is to examine the implication table row by row, beginning with the bottom right-hand cell. A cross can be entered into any cell containing implied pairs if either of the implied pairs have previously been crossed out. The first cell qualifying for a cross is at the intersection of S_4 and S_6 since the cell associated with the implied pair S_6 and S_7 has already been crossed out. This procedure is repeated until no further cells can be crossed out and leads to the final form of the implication table shown in Figure 8.14 (e).

The states are now listed in reverse order, as shown in Figure 8.14(f) and the implication table is examined, column by column, from right to left, to determine

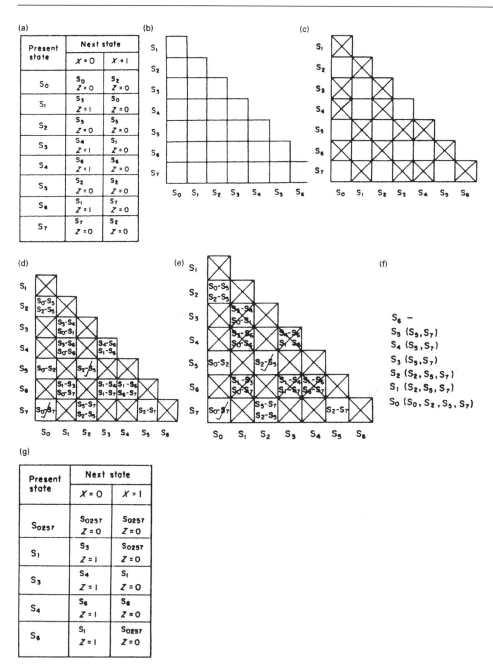

(a)

Present state	Next state	
	X = 0	X = 1
S0	S0, Z = 0	S2, Z = 0
S1	S3, Z = 1	S0, Z = 0
S2	S3, Z = 0	S3, Z = 0
S3	S4, Z = 1	S1, Z = 0
S4	S4, Z = 1	S4, Z = 0
S5	S2, Z = 0	S2, Z = 0
S6	S1, Z = 1	S7, Z = 0
S7	S7, Z = 0	S2, Z = 0

(f)

S_6 —
$S_5 (S_5, S_7)$
$S_4 (S_5, S_7)$
$S_3 (S_5, S_7)$
$S_2 (S_2, S_5, S_7)$
$S_1 (S_2, S_5, S_7)$
$S_0 (S_0, S_2, S_5, S_7)$

(g)

Present state	Next state	
	X = 0	X = 1
S0257	S0257, Z = 0	S0257, Z = 0
S1	S3, Z = 1	S0257, Z = 0
S3	S4, Z = 1	S1, Z = 0
S4	S6, Z = 1	S6, Z = 0
S6	S1, Z = 1	S0257, Z = 0

Figure 8.14 *(a) State table to be reduced by the implication table (b) Implication table (c) Elimination of non-identical outputs (d) Insertion of implied pairs (e) Completed implication table (f) The partition listing (g) The reduced state table*

whether there are any cells that have not been crossed out, since such cells define pairs of equivalent states. In the first column the single cell is crossed out, and there is no entry opposite S_6 in the partition listing. In the second column the pair S_5 and S_7 have not been crossed out; it follows that they are equivalent states and are entered opposite S_5 in the listing. There are no uncrossed entries in columns 3 and 4 and the

($S_5 S_7$) entry is repeated against S_4 and S_3 in the listing. In the fifth column there are two uncrossed cells which define two equivalent state pairs, ($S_2 S_7$) and ($S_2 S_5$). Now the transitivity law states:

$$(S_p S_q)(S_p S_r) \rightarrow (S_p S_q S_r)$$

and using this rule the entry opposite S_2 becomes ($S_2 S_5 S_7$). Remaining entries in the listing are found using the same procedure and the final partition of states is found to be

$$P = (S_0 S_2 S_5 S_7)(S_1)(S_3)(S_4)(S_6)$$

The reduced state table resulting from this partition is shown in Figure 8.14(g).

8.10 State assignment

In all the design problems dealt with in this chapter, a perfectly arbitrary state assignment has been adopted. For example, in the 101 sequence detector, designed in Section 8.5, the state assignment selected was $S_0 = 00, S_1 = 01, S_2 = 11, S_3 = 10$. It is clear that other state assignments could have been selected and they would have led to different circuit solutions.

The number of different ways of choosing N states out of a possible 2^n states is given by:

$$\frac{2^n!}{N!(2^n - N)!}$$

and there are $N!$ ways to assign each different choice of N states; hence the number of possible state assignments N_{SA} is given by:

$$N_{SA} = \frac{2^n!}{N!(2^n - N)!} \times N!$$

where n is the number of state variables. If $N = 5$ then the number of state variables required is $n = 3$ and the number of possible state assignments is 6720.

The number of state assignments for a given number of states are tabulated in Figure 8.15 and it is clear that the number of assignments increases very rapidly with the number of states. For the designer, the criterion for a well-chosen state assignment is that it should lead to a simple circuit implementation.

Number of states to be assigned N	Number of flip-flops n	Number of state assignments N_{SA}
1	0	1
2	1	2
3	2	24
4	2	24
5	3	6720
6	3	20 160
7	3	40 320
8	3	40 320

Figure 8.15 *Number of state assignments*

Simpler circuits will mean that fewer gates are required, and this in turn means that a smaller number of chips are required. If the designed circuit is to be manufactured in large numbers there may be a significant reduction in manufacturing costs. A simpler circuit realisation will also result in a reduced number of interconnections, and finally there may also be a significant saving of space.

More recently, advances in technology have resulted in the development of program-mable logic sequencers. The essential features of these devices are on-chip AND and OR arrays and also a number of single-bit memory elements. When a state machine, either synchronous or asynchronous, is implemented by programming a logic sequencer, the need for an efficient state assignment is no longer of the same importance.

The need for a well-chosen assignment when designing with MSI and SSI circuits will be demonstrated by randomly selecting three different assignments for the invalid code detector designed in Section 8.8. The three assignments chosen are tabulated in Figure 8.16(a). Using the state diagram shown in Figure 8.12(a) the next state equations for each of the flip-flops A, B and C and the output equations for each of the state assignments are found to be:

Assignment 1

$$J_C = X\bar{B}\bar{A} \qquad\qquad J_B = X\bar{C} + \bar{C}A + \bar{X}C\bar{A} \qquad J_A = XC + \bar{X}C\bar{B}$$
$$K_C = \bar{X}\bar{B} + \bar{B}\bar{A} \qquad K_B = \bar{X}C + \bar{C}\bar{A} + CA \qquad K_A = C\bar{B} + \bar{C}B$$
$$Z = C\bar{B}A\,\square$$

Assignment 2

$$J_C = X\bar{B}\bar{A} \qquad\qquad J_B = A + \bar{X}C \qquad\qquad J_A = \bar{X}\bar{C} + C\bar{B} + \bar{C}B$$
$$K_C = \bar{X}B + B\bar{A} \qquad K_B = \bar{C}A + C\bar{A} \qquad K_A = X + \bar{C} + \bar{B}$$
$$Z = CB\bar{A}\,\square$$

Assignment 3

$$J_C = X\bar{B}\bar{A} \qquad\qquad J_B = A + \bar{X}C \qquad\qquad J_A = XC + \bar{X}\bar{C}\bar{B}$$
$$K_C = \bar{X}B + BA \qquad K_B = CA + \bar{C}\bar{A} \qquad K_A = B$$
$$Z = CBA\,\square$$

A comparison of the number of gates required to implement the design for each of the three assignments is shown in Figure 8.16(b). Inspection of this table reveals that assignment 3 requires the least hardware. However, rather than use a random process, it is possible with the aid of two simple rules to choose a state assignment which will with some certainty lead to a simpler circuit implementation.

The state table for the invalid code detector is shown in Figure 8.16(c) and the rules for obtaining a good assignment follow:

Rule 1: Present states which lead to identical states for a given input should be given state assignments that differ in one digit place only, i.e. the present states should be logically adjacent, and separated by a Hamming distance of 1.

Referring to Figure 8.16(c), for input $X = 0$, present states S_3 and S_6 have the same next state S_0 and present states S_2 and S_7 have the same next state S_3. In each case these state pairs should be given logically adjacent state assignments. For $X = 1$, present states S_5 and S_7 have the same next state S_6 and on applying rule 1 should have logically adjacent state assignments.

Rule 2: States which are the next states of the same present state should be given logically adjacent assignments.

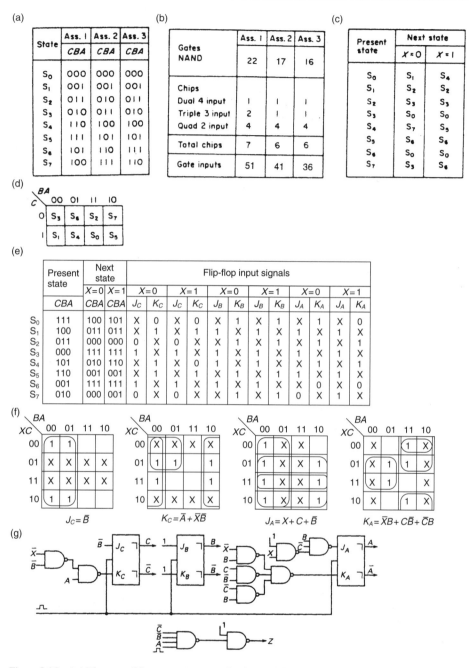

Figure 8.16 *(a) Three possible state assignments for the invalid code detector (b) Gate and chip comparison for the three randomly selected state assignments (c) State table for the invalid code detector (d) State assignment map (e) State table for the state assignment obtained using rules 1 and 2 (f) K-maps for the invalid code detector (g) Implementation of the invalid code detector*

There is a corollary to this rule which states that the assignments to the next states should be given logically adjacent assignments corresponding to the branching variable(s). An example of the application of the corollary to Rule 2 is shown in Figure 8.17. The assignment obtained is referred to as the *reduced input dependency* assignment.

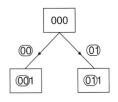

Figure 8.17 *An example of reduced state dependency*

Summarizing, the sets of adjacent states determined by using rules 1 and 2 are tabulated below:

Rule 1: (S_3, S_6) (S_2, S_7) (S_5, S_7)
Rule 2: (S_1, S_4) (S_5, S_7) (S_3, S_6)

A suitable state assignment is shown plotted in Figure 8.16(d), in which the above adjacencies are satisfied. If it is not possible to satisfy all the adjacencies obtained using these rules without conflict, then the adjacencies obtained from the first rule should have priority.

The state table for the state assignment shown in Figure 8.16(d) is tabulated in Figure 8.16(e), together with the flip-flop inputs. K-maps for simplifying the flip-flop input signals are shown in Figure 8.16(f). Note that it is not necessary to plot J_B and K_B since all the entries in their tabulations are either 1 or X. Hence $J_B = K_B = 1$.

The next state inputs obtained from the maps are:

$$J_C = \bar{B} \qquad J_B = 1 \qquad J_A = X + C + \bar{B}$$
$$K_C = \bar{A} + \bar{X}\bar{B} \qquad K_B = 1 \qquad K_A = \bar{X}B + C\bar{B} + \bar{C}B$$
$$Z = \bar{C}\bar{B}A \sqcap$$

The implementation of the invalid code detector is shown in Figure 8.16(g). Ten NAND gates and three JK flip-flops are required for the circuit implementation. The gates needed are:

Dual 4-input	1
Triple 3-input	1
Quad 2-input	2
Total chips	4
Gate inputs	24

It is clear that this state assignment requires less hardware than any of the other three randomly selected assignments shown in Figure 8.16(a).

An additional provision must also be made by the circuit designer to ensure that the invalid code detector is switched to the state S_0 on power-up ready to receive the first bit of an NBCD code. Otherwise the circuit will be out of synchronism with the data.

A number of state assignment procedures have been developed for determining an optimal or near optimal state assignment. An optimal assignment for one type of flip-flop may not necessarily be optimal for another type of flip-flop. For example, the JKFF has a number of 'don't care' terms in the JK steering table and for this reason is more likely to provide an optimal assignment than D or T flip-flops.

8.11 Algorithmic state machine charts

An alternative method of designing sequential circuits utilises the algorithmic state machine (ASM) chart rather than a state diagram. When this technique is used the state diagram is constructed in the form of a flowchart. The chart describes a sequence of events which are designed to initiate a set of state transitions and outputs from a set of data inputs. The basic elements of the ASM chart are illustrated in Figure 8.18.

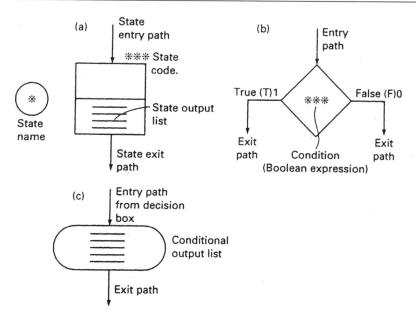

Figure 8.18 *ASM chart components (a) State box (b) Decision box (c) Conditional output box*

State Box: A machine remains in a state box for a period of one state time which may be for one clock period or for an integral number of clock periods in a clock-driven machine. The state is identified by a code which is a unique combination of the state (secondary) variables and is defined when state assignment takes place. A state output is active while the machine remains in the state, and is present for the period of the state time unless its time duration is constrained by the clock signal. States are frequently identified by a number or a mnemonic. There is one entry path and one exit path for each state and the exit path may lead directly to another state box or, alternatively, to one or more decision boxes.

Decision Box: This box contains a Boolean expression that can be regarded as a condition expression which involves the machine inputs. If the logical value of the condition is 1 the true exit path is taken, while if it is 0 the false exit path is taken. These two paths can be identified by the letters T and F. Exit paths may lead to state boxes, conditional output boxes, or to other decision boxes.

Conditional output box: The input path to a conditional output box always comes from a decision box and it specifies the condition required to generate an active output. A conditional output depends upon the state of the machine as well as one or more of the machine inputs; consequently, it is a Mealy type output. For a Mealy machine all the outputs appear in conditional output boxes, while for a Moore machine they appear in the state boxes.

Each state in the ASM chart is associated with an ASM block which may contain the other two basic elements. The ASM block illustrated in Figure 8.19 has one input path and three exit

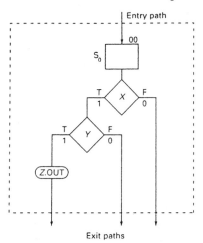

Figure 8.19 *An ASM block*

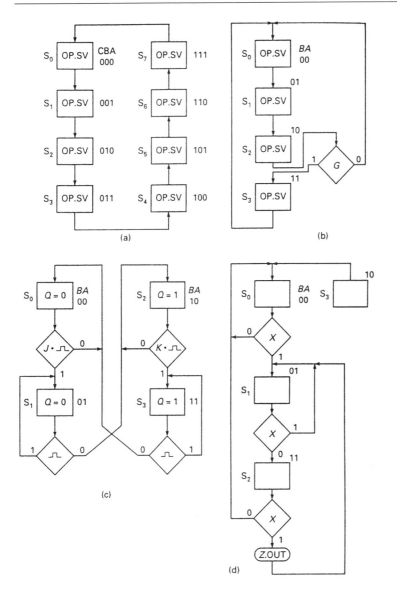

Figure 8.20   *ASM charts (a) Scale-of-8 counter (b) Counter controlled by signal G having scale-of-3 or scale-of-4 count (c) JK master/slave flip-flop (d) 101 sequence detector*

paths which will link it to other state boxes which in turn are associated with their own ASM blocks.

Some examples of ASM charts are shown in Figure 8.20. The first example in Figure 8.20(a) is the ASM chart for a scale-of-8 counter. It consists of state boxes and is almost identical to the state diagram for a scale-of-8 counter shown in Figure 7.3(d). The outputs OP.SV appear in each of the state boxes and are the state variables; for example, when in state S_3 the output is CBA $= 011$. Figure 8.20 shows the ASM chart for a counter which can operate as either a scale-of-3 or scale-of-4 counter, depending on the value of the condition variable G. A more complex ASM chart is that for the master/slave JK flip-flop shown in Figure 8.20 (c). When the flip-flop is in state S_0, i.e. $Q = 0$, it is linked to the decision box containing the condition expression $J \cdot \square$. If $J \cdot \square = 0$ the flip-flop remains in S_0, but if $J \cdot \square = 1$ it makes a transition to S_1 where $Q = 0$. This state is linked to a second decision box containing the condition variable $\square$. If the clock line is high, the flip-flop remains in S_1 but at the trailing edge of the clock signal it makes a transition to S_2 where $Q = 1$ and the flip-flop has been set. The right-hand half of the chart covers the reset process which can be described in a similar manner to the set process. The last of these examples shown in Figure 8.20(d) is for the Mealy representation of the 101 detector designed earlier in this chapter.

8.12 Conversion of an ASM chart into hardware

A typical example of an ASM chart is shown in Figure 8.21(a). The machine represented by the chart is a word scanner which provides an output Z_2 when the last two bits in consecutive 3-bit words are ones, and a second output Z_1 which identifies the start of each 3-bit word. Each of the state boxes has been coded with the state variables B and A and has been assigned identifying letters P, Q, R and S. The output Z_1 is in state box P and depends on the state only, while the second output Z_2 has its own conditional output box, associated with state box S, indicating that it is a Mealy-type output which is dependent on the input signal d. Since there are four states, two flip-flops are required, and DFFs have been selected for the implementation.

The state table in Figure 8.21(b) details all the state transitions and because D flip-flops have been selected for the implementation, the flip-flop input tabulation is a repeat of the next state tabulation. In this simple example the next state equations for the two flip-flops can be read directly from the tabulation, and they are:

$$D_B = \bar{B}A$$
$$D_A = \bar{B}\bar{A} + d\bar{B}A$$

Simplifying the equation for D_A using the consensus theorem gives:

$$D_A = \bar{B}\bar{A} + d\bar{B}$$

The two outputs Z_1 and Z_2 are taken directly from the ASM chart. They are:

$$Z_1 = \bar{B}\bar{A} \qquad Z_2 = dBA$$

In the case of Z_2 it is worth noting that the equation has been derived using the principle that a decision box is a part of the preceding state box.

If it is required to limit the outputs to the time duration of the clock, these equations would be written:

$$Z_1 = \bar{B}\bar{A}\square \qquad Z_2 = dBA\square$$

The implementation of the scanner is shown in Figure 8.21(c).

Alternatively, the next state equations can be implemented using 4-to-1 multiplexers. Selecting the flip-flop outputs B and A as the control variables, the next state equations may be written:

$$D_B = \bar{B}\bar{A}(0) + \bar{B}A(d + \bar{d}) + B\bar{A}(0) + BA(0)$$

And the multiplexer inputs are:

$$D_0 = 0 \qquad D_1 = 1 \qquad D_2 = 0 \qquad D_3 = 0$$

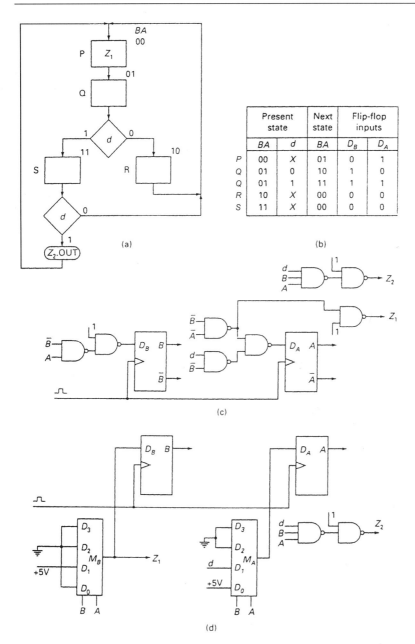

The state table (b)

	Present state		Next state	Flip-flop inputs	
	BA	d	BA	D_B	D_A
P	00	X	01	0	1
Q	01	0	10	1	0
Q	01	1	11	1	1
R	10	X	00	0	0
S	11	X	00	0	0

Figure 8.21   *The 3-bit word scanner (a) The ASM chart (b) The state table (c) Gate and flip-flop implementation (d) Multiplexer and flip-flop implementation*

Also

$$D_A = \bar{B}\bar{A}(d + \bar{d}) + \bar{B}A(d) + B\bar{A}(0) + BA(0)$$

And the inputs to the second multiplexer are:

$$D_0 = 1 \qquad D_1 = d \qquad D_2 = 0 \qquad D_3 = 0$$

The multiplexer implementation of the word scanner is shown in Figure 8.21(d).

8.13 The 'one-hot' state assignment

Sequential circuits described by ASM charts may be implemented using a 'one-hot' state assignment with the intention of reducing design time. The number of states required by the machine is defined by the ASM chart. In this type of assignment only one flip-flop will be high at any given instant of time. If the chart has n states then n flip-flops are required, one for every state. For an 8-state machine eight flip-flops are required, whilst using the state assignment technique described earlier in this chapter only three flip-flops are needed.

The technique provides an alternative method of implementation which in the following example employs one DFF per state. When using the technique, encoding of states is not needed and the problems associated with state assignment do not arise. However, a slightly different method of tabulation will be used.

The ASM chart for a 4-state machine is shown in Figure 8.22 along with the tabulation of the present and next states. For each of the state transitions, the corresponding transition signal is tabulated. For example, if the present state of the machine is S_1 and the transition (input) signal is $XY = 1$ then the machine will make

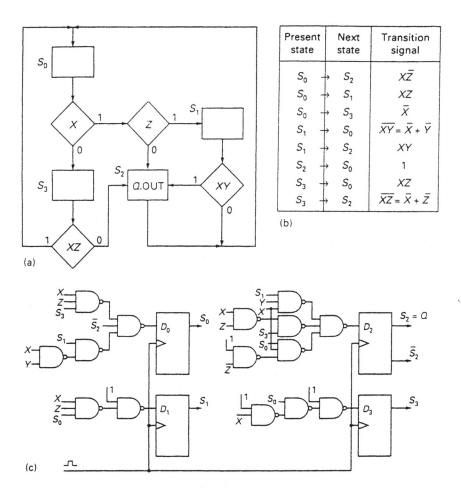

Present state		Next state	Transition signal
S_0	$\rightarrow$	S_2	$X\bar{Z}$
S_0	$\rightarrow$	S_1	XZ
S_0	$\rightarrow$	S_3	$\bar{X}$
S_1	$\rightarrow$	S_0	$\overline{XY} = \bar{X} + \bar{Y}$
S_1	$\rightarrow$	S_2	XY
S_2	$\rightarrow$	S_0	1
S_3	$\rightarrow$	S_0	XZ
S_3	$\rightarrow$	S_2	$\overline{XZ} = \bar{X} + \bar{Z}$

(b)

(a)

(c)

Figure 8.22 *'One-hot' implementation technique (a) ASM chart (b) Transition table (c) Machine implementation*

the transition from S_1 to S_2. The remaining two terms in the equation for S_0 are obtained in a similar manner. The next-state equations are:

$$S_0 = (\bar{X} + \bar{Y})S_1 + S_2 + XZS_3$$
$$S_1 = XZS_0$$
$$S_2 = XYS_1 + (\bar{X} + \bar{Z})S_3 + X\bar{Z}S_0$$
$$S_3 = \bar{X}S_0$$

Implementation of the machine is shown in Figure 8.22.

8.14 Clock skew

When designing a synchronous circuit, all the flip-flops should normally be synchronised by the same clock signal. The flip-flops used in the design should all be of the same type, either leading-edge or trailing-edge triggered. Furthermore, the clock signal must be routed by the shortest possible path on the circuit board to avoid delays caused by the finite speed of electrical signals along the connecting wires and which lead to the problem of clock skew.

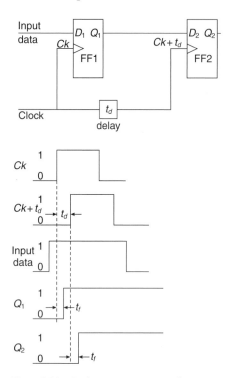

As an example of faulty circuit operation due to *clock skew* consider the first two stages of the serial-in shift register shown in Figure 8.23. The two edge-triggered DFFs are both fed from the same clock source, but the arrival of $Ck + t_d$ at the clock input of FF2 is delayed by an amount t_d relative to the clock input Ck at the clock input of FF1. The input data I to FF1 is transferred to its output Q_1 at a time t_f after the rising edge of Ck. Q_1 is also the data input to FF2, and if $t_d > t_f$ it follows that the input data is transferred to the output of FF2 at a time t_f after the rising edge of $Ck + t_d$. The input data has been transmitted through two stages of the shift register on the receipt of a single clock pulse. Since there is no combinational logic in between each stage of a shift register it is clear that the problem of clock skew is of particular importance in shift register design and operation.

There are a number of reasons why unacceptable clock skew may occur in a large digital system implemented with edge-triggered flip-flops:

Figure 8.23 *Faulty circuit operation due to clock skew*

1. Proper attention has not been paid to the layout of the circuit board, and consequently the clock connection to some of the devices on the board may take inordinately long paths. Two possible methods of path routing are shown in Figure 8.24. Path delay increases as the clock connection is taken to each flip-flop in turn in Figure 8.24(a). In practice, a more realistic clock routing is shown in Figure 8.24(b)

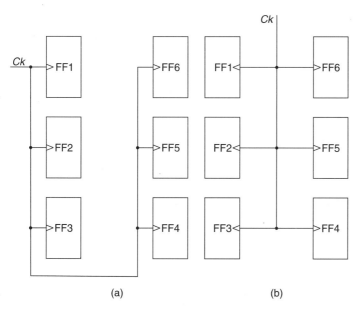

Figure 8.24 *(a) Clock routing leading to clock skew (b) Clock routing designed to minimise possibility of skew*

where a tree-like structure is used for the clock connection to the array of six flip-flops.

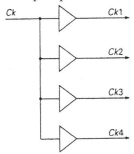

Figure 8.25 *Clock Buffering*

2. A single clock output may not be capable of driving all the flip-flops in the system and it is then necessary to provide a number of identical clock signals from the same source. This can be achieved by *buffering* the clock (see Figure 8.25), but the buffers selected should have approximately the same propagation delay.

3. In the case of multiple clock signals generated by a single source, one clock may be be more heavily loaded than the others and this can lead to significant clock skew. Equalisation of clock loading should be the aim of the logic designer.

8.15 Clock timing constraints

Most users wish to run the clock at the maximum possible frequency; moreover, a common second requirement is that a precise clock frequency should be generated. In order to satisfy this stringent requirement a crystal controlled oscillator is used.

The maximum allowable clock frequency is constrained by a number of circuit parameters. A typical situation is illustrated in Figure 8.26. The output of FFA changes at some time t_{ff} after the leading edge of *Ck*. This change is transmitted via the combinational logic to the input of FFB with a time delay t_{comb}. The timing diagram showing these various circuit transitions are shown in Figure 8.26.

For satisfactory operation of the circuit, any change occurring at the input of FFB should do so at a time $> t_{su}$ before the arrival of the leading edge of the next clock pulse, where t_{su} is the flip-flop set-up time. Hence, the maximum allowable clock period T_{Ck} is given by:

$$T_{Ck} = t_{su} + t_{ff} + t_{comb}$$

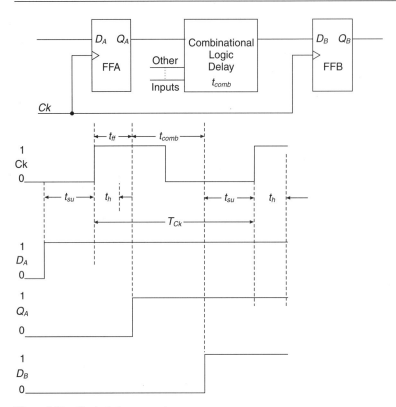

Figure 8.26 *Clock timing constraints*

And the maximum allowable clock frequency f_{Ck} is

$$f_{Ck} = 1/T_{Ck} = 1/(t_{su} + t_{ff} + t_{comb})$$

If the leading edge of the clock pulse at FFB is skewed, then the maximum allowable frequency is:

$$f_{Ck} = 1/(t_{su} + t_{ff} + t_{comb} + t_{skew})$$

8.16 Asynchronous inputs

There are many digital systems that receive asynchronous inputs from external sources and it is essential that these asynchronous signals should be synchronised with the system clock. There may also be cases where the incoming signal has a short time duration by comparison with the sampling period of the system clock. A typical example of such an occurrence is shown in Figure 8.27 and it is clear that the asynchronous input misses the sampling edge of the clock, in this case its trailing edge.

The problem can be overcome by the use of a catcher cell which is in effect an $\overline{S}\overline{R}$ latch whose set signal is the asynchronous

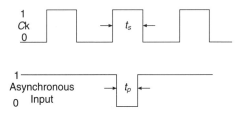

Figure 8.27 *The asynchronous signal that was never sampled*

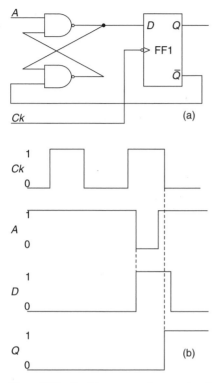

Figure 8.28 *Catching an asynchronous signal (a) the circuit (b) the timing diagrams*

input with the reset signal coming from the complementary output of FF1. The catcher cell consists of the gates connected to the system flip-flop FF1, as shown in Figure 8.28. If it is assumed that $Q = 0, D = 0$ and $A = 1$, the logic levels for this condition are indicated at appropriate points of the circuit. On the arrival of the asynchronous signal, A makes a $1 \rightarrow 0$ transition and D makes a $0 \rightarrow 1$ transition. The outputs of the $\overline{SR}$ latch are now 1 and 0 respectively. The catcher cell input A makes a return transition from $0 \rightarrow 1$ before the trailing edge of the clock pulse. FF1 is now set on the trailing edge of the clock pulse and the complementary output $\bar{Q}$ is fed back to the lower input of the $\overline{SR}$ latch, resetting the catcher cell to its original condition before the arrival of the next asynchronous input. The catcher cell has been used to synchronise the asynchronous signal to the system clock.

Rather than using the simple catching cell shown in Figure 8.28, an edge-triggered DFF can be used as the synchroniser as shown in Figure 8.29. Each individual asynchronous input signal requires its own synchroniser and if the asynchronous input has to be routed to a number of different points of the system, the synchronisation should take place at one point only and any delays must be matched carefully. Furthermore, it is always advisable to precede any combinational logic with the synchronisation process because of differing combinational delays.

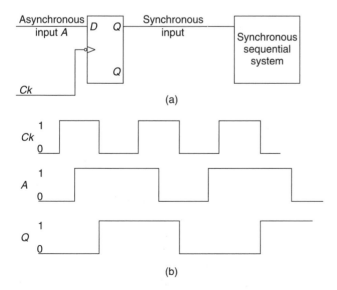

Figure 8.29 *Basic synchroniser circuit (a) logic diagram (b) timing diagram*

Synchronisation failure may occur intermittently and upset the operation of the system. Such failures occur because the asynchronous signal arrives at the input of the synchroniser at any instant of time and may breach its set-up and hold requirements. This leads to the possibility that the synchroniser may enter the metastable state, and the time it occupies that state cannot be defined precisely.

8.17 The handshake

Data frequently has to be transferred, for example, from some external system to a processor. This has led to the widely used *handshaking* transaction which involves the acknowledgement of the receipt of the data and simultaneously defines the time at which the transfer was complete. Conceptually, the handshake mode is analogous to the despatch of an invitation with the letters RSVP attached. The person sending the invitation does not know that it has been received until the acknowledgement in response to the letter's RSVP has been returned.

A state diagram describing a transfer in the handshake mode is given in Figure 8.30. When in the quiescent state (QS) the sender indicates that data is available by sending a data available signal (DAV) signal to the processor, thus initiating a transition to state S_1. While in this state, the sender waits for the processor to acknowledge the DAV signal, which it does by returning the data acknowledge (DAA) signal to the sender. The DAA signal initiates a transfer to S_2, and in this state the data transfer takes place; DAV is set low by the sender. On completion of the transfer, the processor sets DAA $= 0$, and a return to S_0, the quiescent state, occurs.

Handshaking transactions similar to this are the basis of the 'bus systems' used for transferring information between different interface cards inside a computer, or between computers and various peripheral equipment such as printers, scanners, and so on. Many different specifications for bus systems have been drawn up alongside the recent rapid development of computers, but most may be classified as either (1) internal bus systems, for use only inside and within a self-contained computer or processor unit, or (2) external bus systems, for use in transferring information between two or more self-contained pieces of equipment.

Examples of internal systems include the old S-100 computer 'backplane' bus, and the various bus systems developed and specified for use inside IBM-compatible computers. Generally speaking, internal bus systems transfer data in parallel, i.e. several

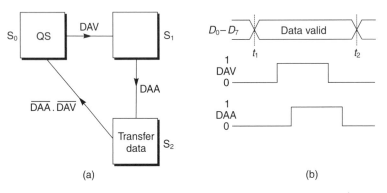

Figure 8.30   *(a) State diagram for the handshake mode of operation (b) Timing diagram for the handshake*

bits at a time – typically 8 or 16 bits in old bus systems like S-100 or ISA, and up to 32 bits or more simultaneously for newer bus systems such as PCI. They also include provision for specifying memory addresses.

Well-known examples of external bus systems include RS-232C, a system where data is transferred serially, one bit at a time, and IEEE-488 where data is transferred in parallel, 8 bits at a time.

RS-232C is most often encountered where relatively slow data transfer can be tolerated, for example from a mouse or other pointing device to a computer, as its serial data transfer imposes serious speed limitations compared to fast parallel bus systems. IEEE-488 was originally developed as a system allowing a computer to act as a controller and data collector for peripheral devices such as measurement instruments, and it is still widely used in such experimental applications.

The RS-232C bus system (also known as EIA-232 or V24) utilises special multi-way plugs and sockets known as 'D connectors', since they are roughly shaped like an elongated 'D', and uses voltage levels that are slightly different from the usual TTL levels. Logic 1 is indicated by a voltage greater than $+3$ V (typically between $+5$ V and $+15$ V) and logic 0 is indicated by a voltage less than -3 V (typically the same magnitude as the voltage indicating logic 1). One reason for this voltage choice is that since 0 V cannot normally occur on a correctly operating line, it is easy for interface circuits to detect an unconnected line.

If no data is being transmitted by a connected line, i.e. the 'quiescent' or 'idle' state, then the line remains at logic level 1. To transmit an ASCII character, the 7-bit ASCII code is preceded by one or more 'Start bits' that are always 0. The seven ASCII bits are then transmitted, followed by one parity bit, corresponding either to even or odd parity, and one or more 'Stop bits' that are always 1. Therefore, each ASCII character is typically represented by a total of 10 (or more) bits. The total number of bits per second transmitted by RS-232C is usually referred to as the 'baud rate', and rates of up to 9600 baud are widely encountered. This rate therefore corresponds to 960, or slightly fewer, ASCII characters per second. The maximum rate allowable in RS-232C is normally 19200 baud.

In the RS-232C system, after the start bit is sent, the rest of the bits, data and parity until the next stop bit, are sent one after the other at exactly the predefined and agreed baud rate, so that the receiver must examine the received bit stream at precisely the same baud rate to avoid errors. The full specification of RS-232C defines a total of 25 connection lines between the two pieces of equipment to be connected, but in practice very rarely are all of these lines used and it is normal to use a subset of the complete bus system. For example, commonly a mouse is connected to a computer using a special cut-down 9-line version of RS-232C. The minimum configuration for bidirectional data transfer uses just three connections, two plus earth, with no handshaking; for more reliable communications, handshaking is used. In the RS-232C bus the 'Request to send' or RTS line has a function similar to the DAV signal, and the 'Clear to send' or CTS line has a function similar to the DAA signal.

Special ICs known as UARTs (universal asynchronous receiver/transmitters), or alternatively ACIAs (asynchronous communications interface adaptors) or ACEs (asynchronous communications elements), are available that convert parallel data to and from the format needed to operate the RS-232C bus directly with the correct handshaking. More recent versions of the serial interface specification include RS-422 and RS-423, which allow greater connecting cable lengths to be used and greater baud rates to be transmitted and received, but for serial communications RS-232C remains the most common wire connection specification. For faster communications, parallel bus systems may represent a suitable alternative.

The IEEE-488 (also known as **GP-IB** or **HP-IB**) standards specify a total of 24 lines used for 8-bit parallel data transfer between the 'Controller', usually a computer, and a maximum of 14 other peripheral devices. Each device is connected in the same way to all the bus lines using special 'stackable' 24-way connectors. At any one time, one of the devices connected to the bus will be 'talking' (placing data onto the bus) and the others will be 'listening' (receiving data from the bus). The 'talker' will specify for which 'listener' the data is intended; this is done by a control on each IEEE-488 device that sets it to a unique code number or 'primary address' that is usually not changed unless there is a conflict with another device connected to the same bus having the same address number.

Here, 'data' can mean either actual measurement or output data, or commands in a format that a 'listener' can interpret. Therefore, some instruments are capable of being both 'listeners', acting on commands or storing measurements, and 'talkers', transmitting measurements or issuing commands, at different times. In the IEEE-488 bus the DAV line is active-low and the function of the DAA signal is similar to the active-low NDAC (No Data Accepted) line. Data transfers are always undertaken using handshakes, and eight bits at a time are transferred on eight separate data lines plus ground. There are also five lines specified by IEEE-488 that control the operation of the bus; for example, IFC, Interface clear, for resetting the bus to an initialised state, and others.

Problems

8.1 For the sequential circuit shown in Figure P8.1 find

    (a) the state table,
    (b) the internal state diagram, and
    (c) the function of the circuit.

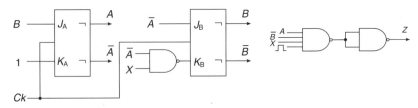

Figure P8.1

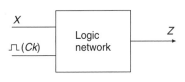

Figure P8.2

8.2 Serial binary data X, synchronised with the clock, is fed to the logic network shown in Figure P8.2. An output 1 will occur on the Z line of the network whenever the string of digits 1101 is received (an output 1 will occur for overlapping strings). Develop a synchronous sequential circuit using D-type flip-flops and NAND gates to implement the above specification.

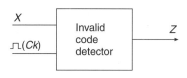

Figure P8.3

8.3 XS3 information is received serially, most significant bit first, and in synchronism with the clock, by the logic network shown in Figure P8.3. The function of the network is to generate an output signal $Z = 1$ when an invalid code combination has been received. Using JK flip-flops and NAND gates, develop a synchronous sequential logic circuit that will perform this function.

8.4 A clock signal X is to be gated on and off by a signal m. The gating signal must be arranged so that the circuit produces complete clock pulses only. A timing diagram for the network is shown in Figure P8.4. Develop a synchronous sequential circuit for implementing the above specification.

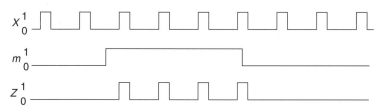

Figure P8.4

8.5 A circuit is to be designed in which a single clock pulse Z is to be selected by a push button control S. The push button is pushed at random intervals and the time duration for which the push button contact is on is long in comparison with the periodic time of the clock. A typical timing diagram is shown in Figure P8.5. Construct an ASM chart and determine an implementation that will satisfy the given specification using D-type flip-flops.

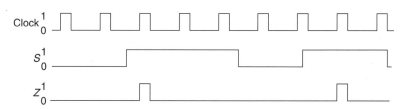

Figure P8.5

8.6 A sequential network has two inputs, X and clock, and one output Z. Incoming data are examined in consecutive groups of four digits and the output $Z = 1$ if any of the three input sequences 1010, 0110 or 0010 should occur. Develop a state diagram and implement the circuit using JK flip-flops and NOR gates.

8.7 A sequential logic network is to be used for determining the parity of a continuous string of binary digits. If an even number of 1's has been received the output of the network $Z = 1$, provided two consecutive 0's have never been received. If two consecutive 0's are received the circuit should return to its initial state and recommence the parity determination. Draw an ASM chart and hence design a circuit to satisfy the specification. Implement the design with D-type flip-flops and NAND gates.

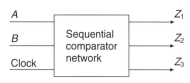

A

B

Clock

Sequential comparator network

Z_1

Z_2

Z_3

Figure P8.8

8.8 Four-bit binary numbers, $A_3A_2A_1A_0$ and $B_3B_2B_1B_0$, are fed to a sequential comparator circuit most significant bit first as shown in Figure P8.8. Design the synchronous sequential circuit whose outputs are $Z_1 = 1$ if $A > B$, $Z_2 = 1$ if $A = B$, and $Z_3 = 1$ if $A < B$.

8.9 A synchronous counter is controlled by two signals, A and B. If $A = 0$ and $B = 0$, the counter is non-operative, if $A = 0$ and $B = 1$ the counter operates as a scale-of-four counter, and if $A = 1$ and $B = 0$ the counter operates as a scale-of-eight counter. Draw an ASM chart and hence design a circuit to satisfy the specification. Implement the design with JK flip-flops and NAND gates.

8.10 Find a minimal state table for the synchronous sequential machines whose state tables are given below, by

(a) Caldwell's merging rules:
(b) partitioning; and
(c) the implication table.

Present state	Next state		Present state	Next state		Present state	Next state	
	$X = 0$	$X = 1$		$X = 0$	$X = 1$		$X = 0$	$X = 1$
S_0	S_1 $Z = 0$	S_2 $Z = 0$	S_0	S_1 $Z = 0$	S_8 $Z = 0$	S_8	S_9 $Z = 0$	S_{12} $Z = 0$
S_1	S_3 $Z = 0$	S_4 $Z = 0$	S_1	S_2 $Z = 0$	S_5 $Z = 0$	S_9	S_{10} $Z = 0$	S_{11} $Z = 0$
S_2	S_5 $Z = 0$	S_6 $Z = 0$	S_2	S_3 $Z = 0$	S_4 $Z = 0$	S_{10}	S_0 $Z = 0$	S_0 $Z = 0$
S_3	S_0 $Z = 1$	S_0 $Z = 0$	S_3	S_0 $Z = 0$	S_0 $Z = 0$	S_{11}	S_0 $Z = 0$	S_0 $Z = 0$
S_4	S_0 $Z = 0$	S_0 $Z = 0$	S_4	S_0 $Z = 0$	S_0 $Z = 1$	S_{12}	S_{13} $Z = 0$	S_{14} $Z = 0$
S_5	S_0 $Z = 0$	S_0 $Z = 0$	S_5	S_6 $Z = 0$	S_7 $Z = 0$	S_{13}	S_0 $Z = 0$	S_0 $Z = 0$
S_6	S_0 $Z = 0$	S_0 $Z = 1$	S_6	S_0 $Z = 0$	S_0 $Z = 0$	S_{14}	S_0 $Z = 0$	S_0 $Z = 1$
			S_7	S_0 $Z = 0$	S_0 $Z = 1$			

8.11 The 2-4-2-1 self-complementing code is received serially, most significant bit first, and in synchronism with the clock. An output signal $Z = 1$ is generated when an invalid code combination is received. Draw an ASM chart and design a circuit that will detect an invalid code. Implement the design with D-type flip-flops and multiplexers.

8.12 A sequential circuit has a single input x and a single output z. The input signal x can occur in groups of 1, 2 and 3 pulses.

If $x = 1$ for one clock period, the output z will be 1 for three clock periods before returning to the starting state.

If $x = 1$ for two clock periods, the output z will be 1 for two clock periods before returning to the starting state.

If $x = 1$ for three clock periods, the output z will be 1 for a single clock period before returning to the starting state.

Construct a state diagram and implement your design with DFFs. The circuit when designed acts as a pulse width adjuster.

8.13 A sequential circuit has two inputs, x and s, and a single output z. The input x is a train of high frequency pulses. It is required to output every fourth input pulse when $s = 0$, and every third input pulse when $s = 1$.

Draw a state diagram and develop a state table. Implement your design with trailing edge triggered master/slave JK flip-flops.

8.14 A sequential circuit has two inputs, x and s. Input x is a train of high frequency pulses and the control signal s selects whether the x train of pulses will appear on one of the two output lines z_1 and z_2. If $s = 0$ the output z_1 is activated and when $s_1 = 1$ output z_2 is activated. Design the circuit. The circuit is operating as a pulse train switch.

8.15 A sequential circuit waveform generator has four possible input waveforms selected by the control signals x_1 and x_2.

The waveform selected by $x_1 = x_2 = 0$ has a period of three clock cycles.

The waveform selected by $x_1 = 0$ and $x_1 = 1$ has a period of four clock cycles.

The waveform selected by $x_1 = x_2 = 1$ has a period of four clock cycles.

The waveform selected by $x_1 = 1$ and $x_2 = 0$ has a period of two clock cycles.

Develop an ASM chart and construct a state table for the generator and implement your design with D-type flip-flops.

8.16 A sequential circuit has an input x which consists of a chain of intermittently occurring pulses. The x pulses, when they appear, do so midway between a pair of successive clock pulses. The output z will occur for the period of the time interval between a pair of successive clock pulses providing an x pulse occurred in the preceding interval between clock pulses. Develop an ASM chart and implement your design with D-type flip-flops.

9 Event driven circuits

9.1 Introduction

Some sequential circuits are driven by events rather than by a train of clock pulses. For example, a digital alarm will be activated by the event that raised the alarm. In this example it is the event that drives the logic, and since the events are frequently irregular occurrences, such a circuit is referred to as an asynchronous sequential circuit or, perhaps more meaningfully, as an event driven circuit.

Asynchronous circuits are also called fundamental mode circuits. The main characteristic of this type of circuit is that only one input is allowed to change at any given instant. Simultaneous changes are forbidden as, indeed, are changes that may take place before the circuit reaches a stable condition after the preceding change. This is clearly different from the behaviour of a synchronous sequential circuit, where inputs changing at arbitrary times are allowed and state changes are activated by the repetitive clock signal.

There are two conditions in which an asynchronous circuit may exist, namely stable and unstable. The total state of the circuit at a given time is defined by the logical values of the inputs and the present state of the circuit. If the next state is the same as the present one the circuit is in a stable condition. If, however, an input changes, the circuit may move to an unstable condition and at some later time the state variables will have taken on their new values such that the next state has become the present state, and stability has been restored.

When designing asynchronous circuits, the designer has to eliminate the possibility of the occurrence of static hazards, dynamic hazards, essential hazards and races, in order to avoid circuit malfunction. These problems, with the exception of static hazards, do not exist in synchronous circuits since they are always designed to reach a steady-state condition before the next clock pulse arrives. Bearing in mind the design difficulties, perhaps the main advantage of asynchronous circuits is that they can work at their own speed and are not constrained to work within the time limits imposed on them by a repetitive clock signal.

9.2 Design procedure for asynchronous sequential circuits

The design procedure for asynchronous sequential circuits is similar in many respects to that developed for synchronous circuits in Chapter 8. The aim of the design is to produce hazard-free next state equations and output functions. The steps in the design procedure are summarised below:

1. *Problem definition*: An unambiguous statement is required by the designer.
2. *Basic state table and internal state diagram*: A basic state table should be constructed from the information given in step 1 above. In many cases the designer

may find it helpful to produce a state diagram first, and then develop the basic state table from the information provided on the state diagram.

3. *Reduction of the basic state table*: If possible by using Caldwell's merging rules or a merging diagram, reduce the number of rows in the table, thus reducing the number of states. In some cases it may be necessary to use an implication chart to reduce the number of states.

4. *State assignment*: Secondary variables are assigned to the states, care being taken to avoid races.

5. *Equations for the state variables*: The equations for the variables assigned to the states can be obtained using a sequential equation, such as $Q^{t+\delta t} = (S + \bar{R}Q)^t$, as developed in Chapter 6. This will lead to a gate implementation of the equations and steps should be taken to ensure that they are hazard-free. Alternatively, the equations can be implemented using latches and the next state equations for their inputs may be determined from the reduced state table.

9.3 Stable and unstable states

The equation of the SR latch developed in Chapter 6 can be written as follows:

$$Q^{t+\delta t} = (S + \bar{R}q)^t$$

where $Q^{t+\delta t}$ is the next state while q^t is its present state. The gate circuit for the latch is shown in Figure 9.1(a), and if the feedback path is removed it can be regarded as a purely combinational circuit in which the condition $S = R = 1$ is not allowed. A K-map for those combinations of the variables that are allowed is plotted in Figure 9.1(b).

For the condition $SRq = 000, q = Q = 0$, and if the feedback path is reconnected, the state of the latch will remain unchanged. This is a stable state which is indicated by ringing the entry on the K-map.

For the condition $SRq = 011, q = 1$ and $Q = 0$. In this case, if the feedback path is reconnected, there will be a change of state to $SRq = 010, q = 0$ being the next present state. $SRq = 011$ is an unstable state and is not ringed on the K-map.

On the K-map in Figure 9.1(b) all the stable states have been ringed leaving the remaining unstable states not ringed. In Figure 9.1(c) the states have been defined numerically and the unstable states are given the same number as the adjacent stable state having the same values of S and R. If the latch is in state 2 and R makes the

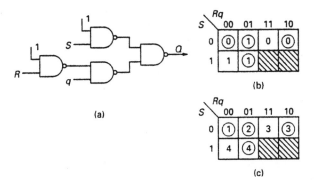

Figure 9.1 *(a) SR latch gate circuit (b) next state map for Q (c) stable and unstable states*

transition $0 \rightarrow 1$ it enters the unstable state 3 before finally settling in stable state 3. These changes can be summarised as follows:

SRq 001 $\rightarrow$ 011 $\rightarrow$ 010
         Stable $\rightarrow$ Unstable $\rightarrow$ Stable
         2        $\rightarrow$ 3 $\rightarrow$ 3

9.4 Design of a lamp switching circuit

Step 1: Problem definition

An asynchronous sequential circuit is to be designed to ensure that a correct manual switching procedure is carried out by the operator for part of an electrical key operating mechanism. If switch X is made, followed by switch Y, then a red lamp L_R is to be turned on, indicating that the incorrect switching procedure has been followed. If switch Y is made followed by switch X, then a green lamp L_G is to be illuminated, indicating that the correct switching procedure has been adopted.

A block diagram for the problem is shown in Figure 9.2(a). The two inputs X and Y generated when the switches are made are referred to as the primary or input variables. The circuit has two outputs, one of which drives the red lamp and the other, the green lamp.

Step 2: Internal state diagram and basic state table

The internal state diagram for the problem is shown in Figure 9.2(b). S_0 may be regarded as the quiescent state in that it represents the condition that both switches are off. The path taken through the state diagram for the correct switching procedure is $S_0 \rightarrow S_3 \rightarrow S_4$, and for the incorrect procedure the path is $S_0 \rightarrow S_1 \rightarrow S_2$. A number of options are available for the switching procedure in the reverse direction. The ones available for this design are:

(a) From green light on, either $S_4 \rightarrow S_3 \rightarrow S_0$ or $S_4 \rightarrow S_1 \rightarrow S_0$
(b) From red light on, either $S_2 \rightarrow S_1 \rightarrow S_0$ or $S_2 \rightarrow S_3 \rightarrow S_0$

A basic state table can now be drawn up from the information appearing on the state diagram (see Figure 9.2(b)). The table has five rows, one for each possible present state, and four columns, one for each of the possible combinations of the input variables.

The entry in the top left-hand cell is $\widehat{S_0}$ and is ringed, indicating a stable state. The entry in the top right-hand cell is S_1, indicating an unstable state. This implies that if, in the quiescent state S_0 where $XY = 00$, X changes from 0 to 1, then a new unstable state S_1 is defined by the total state S_0XY before the circuit settles into a new stable state defined by S_1XY. This condition defines the fourth cell on the second row where the entry is $\widehat{S_1}$.

It will be noted that there is only one stable state per row, and that each unstable state is preceded and succeeded by a stable state. A further examination of the basic state table shows that some cells do not have an entry at all. For example, in the first row where the present state is S_0, there is no entry in the cell corresponding to the input combination $XY = 11$. The basic state table shows that the state S_0 is entered from either S_1 or S_3 with an input signal $XY = 00$, and to enter the cell on the first row where $XY = 11$ would now require a simultaneous change of the input variables. Such a change is not allowable for a circuit operating in the fundamental mode. Cells with no entries in them are marked with a '–'.

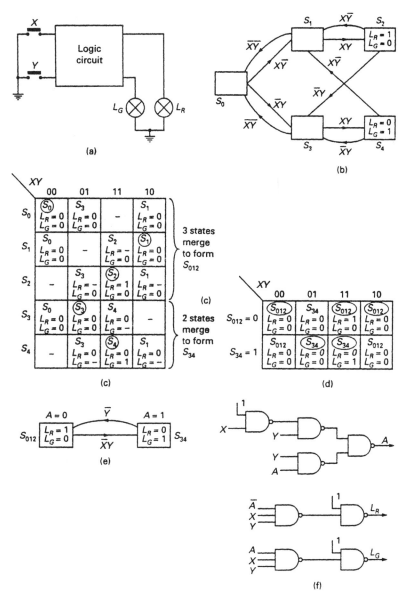

Figure 9.2 *(a) Block diagram for lamp switching circuit (b) the internal state diagram (c) primitive state table (d) reduced state table (e) reduced state diagram (f) circuit implementation*

State tables which contain cells marked with a '–' are referred to as *incompletely specified tables*. In the table shown in Figure 9.2(c) the cells marked with a '–' correspond to forbidden input combinations. These cells can be regarded as 'can't happen' conditions, and may enable a simplification of the table which did not at first sight seem possible. The justification of the allocation of a 'can't happen' condition in the state table is the same as for 'can't happen' conditions in combinational logic problems. If an event cannot happen, the designer 'doesn't care' what the circuit would do in response to it.

Entries are made in each cell of the table for the outputs, and two possible situations can arise:

1. The output entries are identical in the states immediately preceding and succeeding an unstable state. For this situation, the output entries in an unstable state should be identical to those in the immediately preceding and succeeding states.
2. The output entries in the immediately preceding and succeeding states are different. For this situation, the entries in the intervening unstable state can be '–', indicating a 'don't care' entry which can be used in the simplification of the output function.

Step 3: Reduction of the basic state table

When simplifying an incompletely specified table, it is possible to assign a next state and output to a cell containing a '–' in such a way as to make the row in which the '–' occurs identical to a second row. The states at the head of these two rows are then identical, since all the next state entries and outputs in corresponding cells on the two rows are the same, and the rows can be merged.

An examination of the table in Figure 9.2(c) shows that the rows headed S_0, S_1 and S_2, and those headed S_3 and S_4 are identical and can be merged using Caldwell's merging rules. S_0, S_1 and S_2 are merged to form a new state S_{012} and states S_3 and S_4 are merged to form a new state S_{34}. The reduced state table is shown in Figure 9.2(d) and the reduced state diagram in Figure 9.2(e) is constructed from the information in the reduced state table.

Step 4: State assignment

Since there are only two states in the reduced state diagram, just one state variable A is required to define them. For the state S_{012}, $A = 0$, and for the state S_{34}, $A = 1$. As there is only one state variable in this case, the problem of races does not arise.

Step 5: Equations for the state variable and the outputs

The equation for the state variable can now be obtained with the aid of the NAND sequential equation $Q^{t+\delta t} = (S + \bar{R}Q)^t$, where S is defined as the turn-on condition for Q and R is defined as the turn-off condition for Q. The turn-on and the turn-off conditions for the secondary variable A can be obtained directly from the reduced state diagram:

Turn-on condition for $A = \bar{X}Y$
Turn-off condition for $A = \bar{Y}$
Hence $A^{t+\delta t} = (\bar{X}Y + YA)^t$

And the outputs may be written

$$L_G = AXY \quad \text{and} \quad L_R = \bar{A}XY$$

The implementation of the circuit is shown in Figure 9.2(f).

9.5 Races

When the state variables were allocated to the internal states of a clock-driven sequential circuit, the criterion for the allocation was that it should lead to a minimum hardware implementation. It was pointed out in the previous chapter that there is no known method for the allocation of the state variables that will lead to minimum hardware implementation, although guidelines were presented which, when used, lead to a simple, if not the simplest, circuit. The criterion for the allocation of state variables in event-driven circuits is somewhat different, and in this section those factors which govern this allocation will be examined.

An alternative state diagram for the light-switching problem is shown in Figure 9.3(a). An extra state S_5 has been introduced to allow an extra return path from the 'light-on' states S_2 and S_4 to the quiescent state S_0. Using the techniques described in section 9.4, the reduced state table and state diagram have been obtained. Four combinations of two state variables A and B have been arbitrarily allocated, one to each of the four states, and it will be noticed that when a transition is made from S_5 to S_0 on the signal XY, both of the state variables have to change. There are three possible cases to consider:

1. A and B change simultaneously: a direct transition is made from S_5 to S_0.
2. A changes before B: the circuit makes a transition to S_0 via the route $S_5 \rightarrow S_{34} \rightarrow S_0$.
3. B changes before A: the circuit makes2 a transition to S_0 via the route $S_5 \rightarrow S_{12} \rightarrow S_0$.

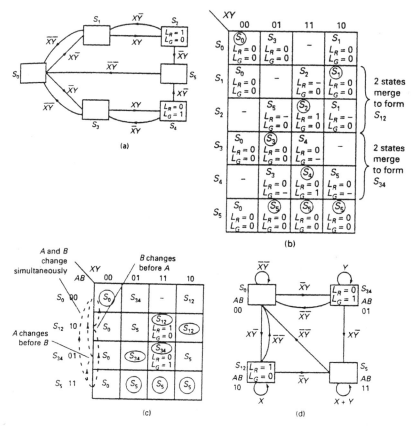

Figure 9.3   *(a) Modified light switching state diagram (b) primitive state table (c) reduced state table (d) reduced state diagram exhibiting a non-critical race*

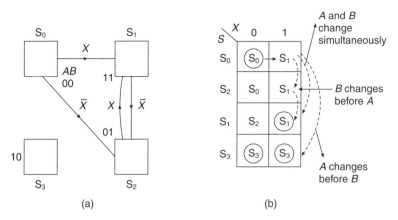

Figure 9.4   *(a) State diagram for a circuit exhibiting critical races (b) State table illustrating a critical race*

In all three cases the circuit enters a stable state S_0 and remains there until a further change of input variables occurs. The various transitions between states for the three conditions described above are illustrated in the reduced state table shown in Figure 9.3(c).

From the foregoing remarks it may be concluded that whenever two state variables change in response to a change in an input variable, a race condition exists. The condition has its origin in the different delays when the A and B signals are generated. In the case described above the races identified are both non-critical races since, irrespective of the transition made, the circuit always ends up in the same stable state.

However, there are races that can occur in event-driven circuits in which the final state reached depends upon the order in which the state variables change. Such races are termed critical races. For example, the internal state diagram of a state machine and its corresponding state table are shown in Figure 9.4. It will be assumed that the machine is in the state defined by $AB = 00$, and $X = 0$. If the input X is now changed to 1, the machine will make a direct transition to the stable state defined by $X = 1$ and $AB = 11$ (S_1) providing A and B change simultaneously. Alternatively, if A changes before B, the machine will make a transition to the state defined by $X = 1$ and $AB = 10$. Since this is a stable state, the circuit will remain there, and in fact a quick glance at the state diagram shows that the circuit remains locked in that state indefinitely because of the absence of an output path from the state. However, B may change before A and then the circuit will make a transition to the state defined by $X = 1$ and $AB = 01$. This state is unstable and the circuit makes a further transition to the state defined by $X = 1$ and $AB = 11$. The transitions described can clearly lead to faulty circuit operation. Critical races occur in this circuit because it is possible to end up in one of two stable states, depending on the order in which the state variables change. The various transitions which can take place in this circuit are indicated on the state table shown in Figure 9.4(b).

9.6 Race free assignments

If critical races are to be avoided, it is necessary to provide a race-free assignment of the state variables on the state diagram. In effect, this means that when a transition is made from one state to the next, only one state variable should be allowed to change.

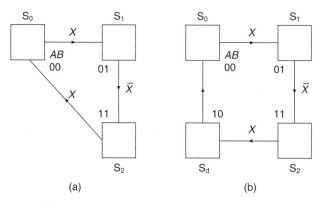

Figure 9.5 *(a) State machine requiring race-free secondary assignment (b) Inclusion of a dummy state to give race-free assignment*

In some cases it is not possible to satisfy this requirement without making modifications to the state diagram. For example, the three-state diagram shown in Figure 9.5(a) requires two state variables to define the three states. An arbitrary state assignment has been made on the diagram, but inspection reveals that on making a transition from S_2 to S_0 both secondary variables must change. Unfortunately, it is impossible to find a race-free assignment for a three-state diagram if transitions are required between each pair of states.

Furthermore, two state variables can define four states, which implies that for the three-state diagram of Figure 9.5(a) there is one unused state which has been omitted from the diagram. If there is no exit from the unused state it can become a 'lock-in' state as described in section 9.5.

These two problems are overcome by incorporating the unused state $AB = 10$ (S_d) in the modified state diagram shown in Figure 9.5(b). This modification allows the circuit to return unconditionally from this dummy state to state S_0.

The four-state diagram in Figure 9.6 is structured in such a way that there are no race problems providing adjacent states are allocated state variables that differ in one variable only. If, however, the state diagram for the machine includes transitions between two states that are not adjacent, for example $S_3 \rightarrow S_1$ in the state diagram shown in Figure 9.7(a), then a race-free assignment is not possible with two state variables. The state diagram reveals that with the same state assignment as the one shown in Figure 9.6 there is a double change in state variables when the transition $S_3 \rightarrow S_1$ is made. No matter how the state variables are allocated, there will always be at least one transition which will result in a double change of the state variables, and this implies that a race-free assignment can only be achieved by using three state variables.

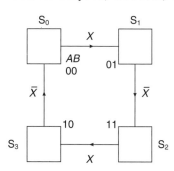

Figure 9.6 *State diagram for a four-state machine with transitions between adjacent states*

A race-free assignment can most easily be obtained from a K-map of the three state variables, as shown in Figure 9.7(b). It is a property of the K-map that adjacent cells differ in one digit position only, and consequently two states allocated to adjacent cells will have state assignments that differ in one digit place.

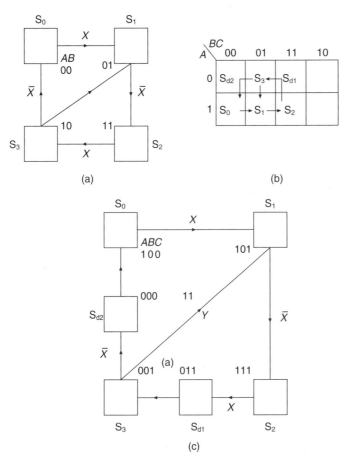

Figure 9.7 *(a) State diagram for a four-state machine with one diagonal transition (b) K-map for determining a race-free assignment (c) Race-free state diagram for a four-state machine having a diagonal transition*

Four of the states have been allocated to cells such that S_0 is adjacent to S_1, S_1 to S_2, and S_1 to S_3. However, for a race-free assignment, S_2 should be adjacent to S_3, and so should S_0. The K-map shows that with three state variables such adjacencies are impossible, and the transitions $S_2 \rightarrow S_3$ and $S_3 \rightarrow S_0$ have been made via the dummy states S_{d1} and S_{d2} respectively. The modified state diagram consists of six states, two of which are dummies, as shown in Figure 9.7(c). Each transition on this diagram has only one change of state variable, and hence the assignment is race-free. An event driven circuit will now be designed which requires the inclusion of a dummy state.

9.7 The pump problem

Step 1: Problem definition

Water is pumped into a water tank by two pumps, p_1 and p_2. Both pumps are to turn on when the water goes below level 1 and they are to remain on until the water reaches level 2, when pump p_1 turns off and remains off until the water is below level 1 again. Pump p_2 remains on until level 3 is reached when it also turns off and remains off until the water falls below level 1 again. Level sensors are used to provide level detection signals as follows:

Signal $a = 1$ when the water is at or above level 1, otherwise $a = 0$
Signal $b = 1$ when the water is at or above level 2, otherwise $b = 0$
Signal $c = 1$ when the water is at or above level 3, otherwise $c = 0$

The aim is to develop an event driven circuit to control pumps p_1 and p_2 according to the specification given above.

A schematic diagram of the water tower is shown in Figure 9.8(a), and a block diagram of the proposed circuit is shown in Figure 9.8(b).

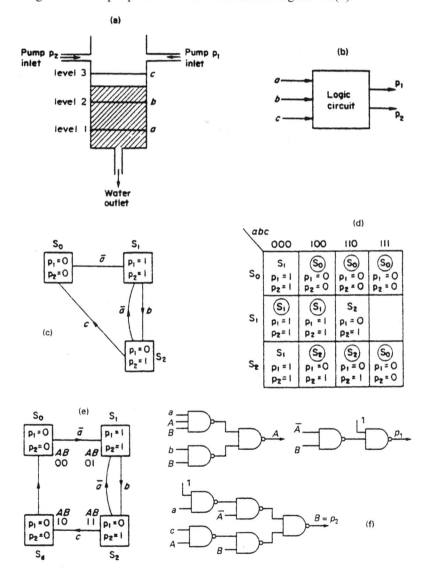

Figure 9.8   *(a) Diagram of the water pump problem (b) Block diagram of pump controller (c) Basic internal state diagram for the pump problem (d) State table (e) Modified state diagram (f) Circuit implementation of the pump controller*

Step 2: The state diagram

A suitable state diagram is shown in Figure 9.8(c), in which the state S_0 is related to the condition when the water is above level 3 and both pumps are off. As the tank empties, the water level falls until it is below level 1 and a transition is then made to S_1, since $\bar{a} = 1$. In state S_1, both pumps are on. If the water continues to rise and reaches level 2, a transition is made to S_2 and pump p_1 is then turned off. In state S_2, two options are available. If the water level falls below level 1 again a transition will be made back to S_1 on the signal $\bar{a} = 1$. Alternatively, if the water continues to rise, when level 3 is reached a transition is made to S_0 and both pumps are turned off.

Step 3: The state table

The state table for the pump problem is shown in Figure 9.8(d). It should be observed that input conditions $abc = 001$, 010, 011 and 101 are missing from the table since these combinations can only exist under fault conditions.

Two state variables A and B are required to define three states. Because there are transitions between each pair of states, a race-free assignment of the state variables is not possible. To overcome this problem an additional dummy state S_d is added to the state diagram. The modified state diagram is shown in Figure 9.8(e).

Step 4: Development of the circuit equations

Turn-on condition for $A = bB$

Turn-off condition for $A = \bar{B} + B\bar{a} = \bar{B} + \bar{a}$

Turn-on condition for $B = \bar{a}\bar{A}$

Turn-off condition for $B = cA$

$$\begin{aligned}
\text{Hence } A^{t+\delta t} &= [bB + \overline{(\bar{B} + \bar{a})A}]^t \\
&= [bB + aAB]^t \\
\text{and} \quad B^{t+\delta t} &= [\bar{a}\bar{A} + \overline{(cA)}B]^t \\
&= [\bar{a}\bar{A} + (\bar{c} + \bar{A})B]^t \\
\text{Also} \quad p_1 &= \bar{A}B \\
\text{and} \quad p_2 &= \bar{A}B + AB = B
\end{aligned}$$

Step 5: Circuit implementation

The circuit implementation of the pump controller is shown in Figure 9.8(f).

9.8 Design of a sequence detector

In this section a further example of the design of an event driven circuit will be studied to emphasise some of the problems faced by the designer when developing this type of circuit. The opportunity will also be taken to look at various methods of implementation and the construction of an ASM chart for this problem. The design to be studied concerns a sequence detector which has two inputs X_1 and X_2, and one output Z, as shown in Figure 9.9(a), and which is required to give an output $Z = 1$ when the sequence of input signals $X_1 X_2 = 00$, 10, 11 has occurred.

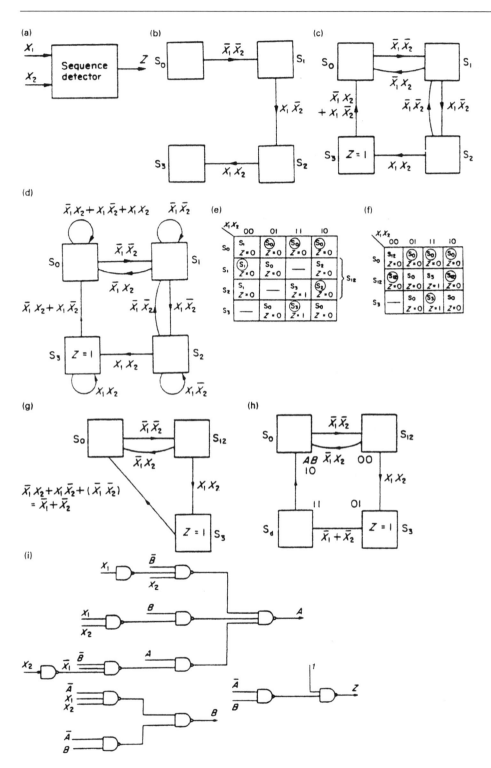

Figure 9.9 *(a) Block diagram of a sequence detector (b) Basic elements of the internal state diagram (c) Internal state diagram for the sequence detector (d) Complete internal state diagram including slings (e) State table (f) Reduced state table (g) Reduced state diagram (h) State diagram including dummy state and secondary assignment (i) Implementation of the sequence detector*

One method of approach open to the designer is to develop the state diagram. In this type of problem a good beginning to the state diagram is to insert the required sequence, as shown in Figure 9.9(b). This requires four states, connected via three transitions, initiated by the transition signals $\bar{X}_1\bar{X}_2, X_1\bar{X}_2$ and X_1X_2 respectively.

To complete the state diagram it is now necessary to insert the additional transition paths that may originate at each of the states. For example, the machine enters state S_1 on the transition signal $\bar{X}_1\bar{X}_2$. Since the machine to be designed will be operating in the fundamental mode, there cannot be a simultaneous change in the input variables when state S_1 is entered and the state can only be left on the transition signals $X_1\bar{X}_2$ or $\bar{X}_1X_2$. The transition signal $X_1\bar{X}_2$ represents the second combination of the input signals in the required sequence, and initiates the transition from S_1 to S_2. Alternatively, a change in X_2 from $0 \rightarrow 1$ results in an input signal $\bar{X}_1X_2$ and the machine should be designed to return to the state S_0 to await the arrival of the first signal in the sequence, $\bar{X}_1\bar{X}_2$. The completed state diagram is shown in Figure 9.9(c). In this diagram, the output $Z = 1$ appears in state S_3 at the completion of the required sequence. If, when in this state, the input signals $\bar{X}_1X_2$ or $X_1\bar{X}_2$ are received, the machine will return to S_0, where it will await the next occurrence of the signal $\bar{X}_1\bar{X}_2$, the first combination of the required sequence.

Some designers insert slings (arrows indicating a "transition" to the same state) on the state diagram, and an example of the use of slings has already appeared in Figure 9.3(d). In this problem, if the machine enters state S_1 on the signal $\bar{X}_1\bar{X}_2$, it will stay there as long as this signal still exists, and this can be indicated by a sling originating from and terminating on S_1, as shown in Figure 9.9(d). This diagram includes all possible slings, and it will be observed that when in S_0 the sling signal is $\bar{X}_1X_2 + X_1\bar{X}_2 + X_1X_2$. This means that if the machine entered S_0 on either of the signals $\bar{X}_1X_2$ or $X_1\bar{X}_2$ it would be possible to get a change of input signal from either $\bar{X}_1X_2$ to X_1X_2 or, alternatively, from $X_1\bar{X}_2$ to X_1X_2. If such a sequence of events occurs, the machine will remain in state S_0 and will only leave the state if the input signals X_1 and X_2 change in either of the following two sequences:

1. $11 \rightarrow 01 \rightarrow 00$
2. $11 \rightarrow 10 \rightarrow 00$

The state table is constructed from the information given on the state diagram, and is shown in Figure 9.9(e). Examination of the table shows that rows S_1 and S_2 are mergeable and the table can be reduced to three rows. At first sight this may appear to be a disadvantage for the following two reasons. First, it leads to the presence of an unused state, and second, since the state diagram will now only consist of three states, a race-free assignment is not possible. However, the unused state can be reintroduced as a dummy state having an unconditional transition to the next state. The presence on the state diagram of an unconditional transition will lead to simpler turn-on and turn-off conditions and a simpler logic implementation.

The reduced state table is shown in Figure 9.9(f) and it will be noticed that there is one unoccupied cell on this diagram on the row headed S_3. This is effectively a 'can't happen' condition. If the present state is S_3, then a transition signal $\bar{X}_1\bar{X}_2$ is forbidden. Since this signal cannot occur when the machine is in state S_3 it may be used as an optional term added into the Boolean equation for the $S_3 \rightarrow S_0$ transition, as shown in the reduced state diagram of Figure 9.9(g). In this case the optional term leads to a simplification of the transition signal.

The state diagram, including the dummy state and with a suitable state assignment, is shown in Figure 9.9(h). The turn-on and turn-off equations are taken directly from this diagram:

Turn-on condition for $A = \bar{B}\bar{X}_1 X_2 + B(\bar{X}_1 + \bar{X}_2)$
Turn-off condition for $A = \bar{B}\bar{X}_1 \bar{X}_2$

$$A^{t+\delta t} = [\bar{B}\bar{X}_1 X_2 + B(\bar{X}_1 + \bar{X}_2) + (\overline{\bar{B}\bar{X}_1\bar{X}_2})A]^t$$
$$= [\bar{B}\bar{X}_1 X_2 + B(\bar{X}_1 + \bar{X}_2) + (B + X_1 + X_2)A]^t$$

Turn-on condition for $B = \bar{A}X_1 X_2$

Turn-off condition for $B = A$

$$B^{t+\delta t} = (\bar{A}X_1 X_2 + \bar{A}B)^t$$

The output Z is given by $Z = S_3 = \bar{A}B$, and the machine implementation is shown in Figure 9.10(i).

The simplest form of the equations for the next state of the state variables $A^{t+\delta t}$ and $B^{t+\delta t}$ can be obtained directly from a pair of K-map plots. The state table compiled from the information given in the state diagram of Figure 9.9(h) is shown in Figure 9.10(a) and in this diagram, assignment of A^t and B^t has been placed alongside the states. However, it is more convenient to rearrange the rows of this table so that the secondary variables appear in normal K-map order. At the same time the state entries in the cells in Figure 9.10(a) are replaced by the state variables that define them, as shown in Figure 9.10(b). This table can be regarded as a plot of the next states of the state variables, $A^{t+\delta t}$ and $B^{t+\delta t}$, for every possible combination of the total present state $(X_1 X_2 \, AB)^t$.

The K-maps for $A^{t+\delta t}$ and $B^{t+\delta t}$ have been separated out in Figure 9.10(c). Using the normal simplification techniques gives the following equations:

$$A^{t+\delta t} = (\bar{X}_1 X_2 + B\bar{X}_2 + AX_1)^t$$
$$B^{t+\delta t} = (\bar{A}X_1 X_2 + \bar{A}B)^t$$

and

$$Z = S_3 = \bar{A}B$$

The implementation of these three functions is shown in Figure 9.10(d).

An ASM chart for the sequence detector is given in Figure 9.10(e). The chart has been constructed from the information extracted from the state table in Figure 9.10(a). For example, the decision box immediately below the state box S_0 contains the Boolean term $\bar{X}_1\bar{X}_2$. If $\bar{X}_1\bar{X}_2 = 0$ the machine remains in S_0, but if $\bar{X}_1\bar{X}_2 = 1$ a transition is made to the state S_{12}. The condition for the machine to remain in state S_{12} is $\bar{X}_1 X_2 + X_1 \bar{X}_2 = \bar{X}_2 = 1$, i.e. $X_2 = 0$. On leaving state box S_{12} the first decision box contains the variable X_2, and if $X_2 = 0$ the path from this decision box returns the machine to state S_{12}. If, however, $X_1 = 0$ and $X_2 = 1$ the machine takes the path back to S_0 while if $X_1 = 1$ and $X_2 = 1$ the path on the chart leads to state box S_3. The remainder of the chart is constructed using the information obtained from the last two rows of the state table.

An alternative method of implementing the design of the sequence detector would be to use SR latches and combinational logic. This requires the development of the next state equations for the two latches, A and B. The tabulation of the next state functions,

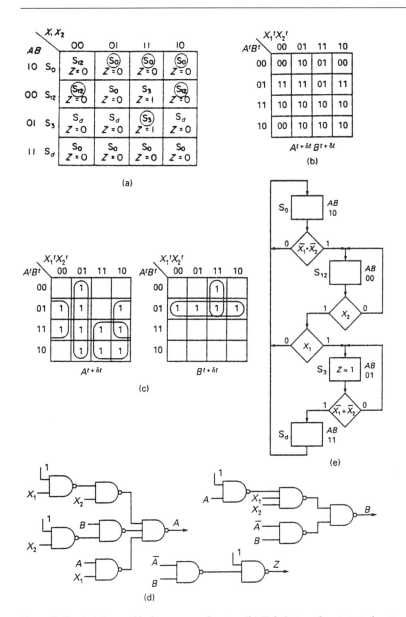

Figure 9.10   *(a) State table for sequence detector (b) Tabulation of next state functions $A^{t+\delta t}$ and $B^{t+\delta t}$ (c) K-maps for the next state functions (d) Alternative implementation of sequence detector (e) The ASM chart for the sequence detector*

repeated again for convenience in Figure 9.11(a), may be regarded as the next state map for the two latches which, in conjunction with the steering table for the SR latch shown in Figure 9.11(b), enables the designer to obtain the K-maps for the S and R inputs to both latches. For example, when the present total state of the machine is $ABX_1X_2 = 0001$ and the input combination $X_1X_2 = 01$ is received, the next total state is $ABX_1X_2 = 1001$. Latch A has made a $0 \rightarrow 1$ transition which requires $S_A = 1$ and $R_A = 0$, while latch B has made a $0 \rightarrow 0$ transition which requires $S_B = 0$ and $R_B = X$.

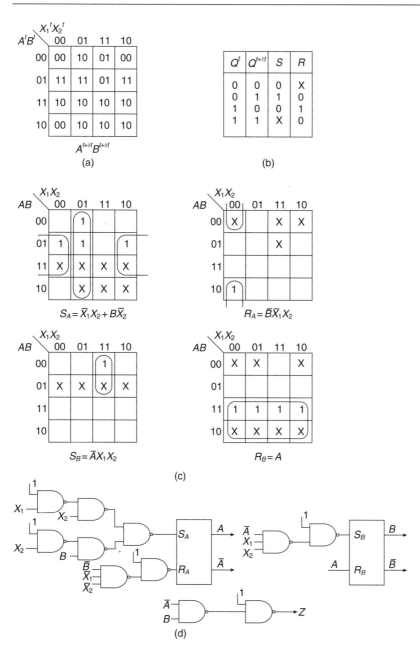

Figure 9.11 *(a) Latch excitation table (b) Steering table for an SR Latch (c) K-maps for the latch input signals (d) Implementation of sequence detector using SR latches*

The K-maps for the latch input signals are shown in Figure 9.11(c) and, after simplification, the following equations are obtained for the set and reset signals:

$$S_A = \bar{X}_1 X_2 + B\bar{X}_2 \qquad S_B = \bar{A} X_1 X_2$$
$$R_A = \bar{B}\bar{X}_1 \bar{X}_2 \qquad R_B = A$$

The output is given by $Z = S_3 = \bar{A}B$ and the machine implementation is shown in Figure 9.11(d).

9.9 State reduction for incompletely specified machines

In a completely specified machine there is an entry for the next state and output in every cell of the state table. For incompletely specified machines the outputs and the next states may not be specified for some combinations of the present states and inputs. Unspecified next states and outputs can be regarded as 'can't happen' conditions and can be specified in any way the designer may choose. Because of this freedom of choice it is possible to have more than one state reduction for an incompletely specified machine. A state table for an incompletely specified machine and two possible state reductions for the machine are shown in Figure 9.12.

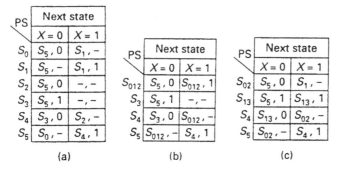

Figure 9.12 *(a) State table for an incompletely specified machine (b) and (c) Two possible state reductions*

9.10 Compatibility

In the previous chapter, when dealing with completely specified state tables, state reduction was achieved by combining equivalent states to form a single state. Equivalent states were defined as those states where next-state entries and outputs were identical for each input condition. When dealing with incompletely specified tables such as the one shown in Figure 9.12, state reduction is achieved by finding compatible states.

As an example of compatibility, consider the two states S_2 and S_3 tabulated below which appear in the state table of an incompletely specified machine:

PS	NS	
	$X = 0$	$X = 1$
S_2	$S_5, 0$	$-, -$
S_3	$-, -$	$S_4, 1$

The unspecified next state and outputs are regarded as 'can't happen' conditions, and entries in the tabulation can be made where the $-$'s occur which will enable the two states to be combined to form a new state S_{23} where:

PS	NS	
	$X = 0$	$X = 1$
S_{23}	$S_5, 0$	$S_4, 1$

The two states have formed a *compatible pair*. If the output conditions for S_2 and S_3 had been conflicting for either of the two input conditions they would then have been incompatible. Alternatively, if the next state entries for either of the two input conditions had been different, the two states would have been incompatible.

As a further example of compatibility, the two rows tabulated below have been taken from an incompletely specified table:

PS	NS	
	$X = 0$	$X = 1$
S_0	$S_1, 0$	$S_3, -$
S_1	$S_1, -$	$S_6, 1$

The two rows can be made output consistent by inserting a 1 in place of the '−' on the first row, and a 0 in place of the '−' on the second. However, after these two insertions, S_0 and S_1 can only form a compatible pair providing S_3 and S_6 are also compatible.

Additionally, the compatibility relationship is not transitive. It is possible for S_i to be compatible with S_j and S_j may also be compatible with S_k, but it does not follow that S_i will be compatible with S_k. This point is demonstrated by the following example:

PS	NS	
	$X = 0$	$X = 1$
S_1	$-, 0$	$S_6, 0$
S_2	$S_5, 0$	$-, -$
S_3	$-, -$	$S_4, 1$

Clearly S_1 and S_2 form a compatible pair. Similarly, S_2 and S_3 are compatible. However, S_1 and S_3 are incompatible since they are not output consistent.

Summarising, the conditions for the compatibility of two states S_i and S_j are:

1. The outputs on the rows headed by S_i and S_j must be identical for each possible input condition.
2. The next-state entries of S_i and S_j must be compatible when both are specified for each possible input.

A set of output consistent states in which every pair within the set is compatible is called a *compatibility class* while a *maximal compatibility class* is defined as a set of compatible states which are output consistent but are not a subset of any other class. For example, if (S_1, S_5) is a compatibility class, it is not a maximal compatibility class if it is a subset of a maximal compatibility class (S_1, S_4, S_5, S_7).

9.11 Determination of compatible pairs

To determine the compatible pairs for the incompletely specified machine whose state table is shown in Figure 9.13(a), the implication chart described in the previous chapter is used. For an incompletely specified machine, each cell in the chart represents the testing ground for the compatibility of a state pair. The top left-hand cell of the chart is the testing ground for the compatibility of states A and B. In order that the two states should be output consistent, all the output '−'s in the two rows must be replaced by 1's. To satisfy the second condition, states A and C must be compatible, and this implication is entered in the cell.

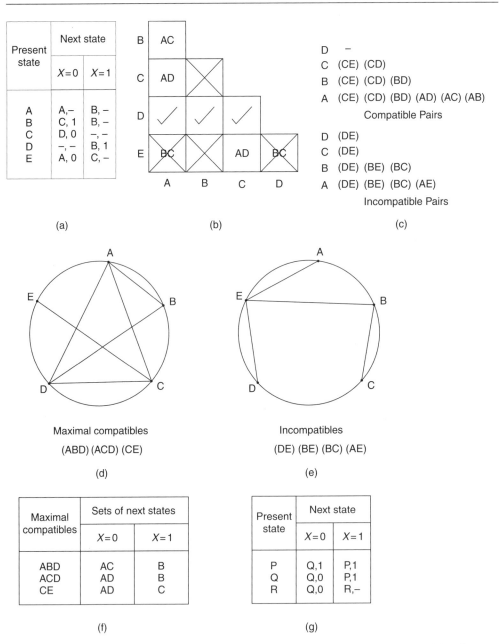

Figure 9.13 *(a) State table for incompletely specified machine (b) implication table (c) compatible and incompatible pairs (d) and (e) merger diagrams (f) closure table (g) reduced state table*

It is clear from an examination of the state table that states B and C cannot be compatible since they are not output consistent and the cell at the intersection of these two states is marked with an X. On the other hand, states A and D can be made output consistent by replacing the output '−'s, on the rows headed by A and D, with 1's. The second condition is satisfied if the next state '−' for $X = 0$ in the row headed D is replaced by A. The two states are then compatible, and the cell that identifies them is marked with a $\sqrt{}$. Every cell on the chart is examined in this way, and the appropriate entry is made in each cell, as shown in Figure 9.13(b). Finally, the chart must be examined systematically to see if any of the implications involve a pair of states that have already been found to be incompatible. For example, the entry BC in the cell at the intersection of D and E is an incompatibility, and this cell must be marked with an X. Similarly, the entry BC at the intersection of A and E is an incompatibility and must also be marked with an X. The states are now listed in reverse order as shown in Figure 9.13(c), and the implication table is examined column by column from right to left to determine first the compatible and then the incompatible pairs. The compatible pairs are:

(CE)(CD)(BD)(AD)(AC)(AB)

and the incompatible pairs are:

(AE)(BE)(BC)(DE)

9.12 The merger diagram

The next step in the state reduction process is to find the maximal compatibles, and this process can be assisted by the construction of a merger diagram. In this diagram the original states of the machine can be represented by dots equally spaced round a circle as shown in Figure 9.13(d). A line is then used to connect each of the compatible pairs. The maximal sets of compatible states can be obtained from the merger diagram by noting those sets of states in which every state is connected to every other state by a line. A typical example of a maximal compatible in Figure 9.13(d) is (ACD) and it will be observed that it is impossible to add any other state to this triangular grouping. The remaining maximal compatibles on the merger diagram are (ABD) and (CE). The maximal incompatibles can also be found on the incompatible merger diagram shown in Figure 9.13(e). They are (AE), (BC), (BE) and (DE).

9.13 The state reduction procedure

The maximal compatibles are now selected to provide a reduced state table which will represent the behaviour of the incompletely specified machine. When making the selection, three conditions must be satisfied:

1. *Completeness*: The chosen set of maximal compatibles must contain all the states in the original machine.
2. *Consistency*: The set of chosen maximal compatibles must be closed. This condition is satisfied if the implied next states of each selected maximal compatibility is contained by another maximal compatibility within the selected set.
3. *Minimality*: The smallest number of maximal compatibles required for a minimal realisation.

The process of selecting a set of maximal compatibles to represent the machine, whose incompletely specified state table is shown in Figure 9.13(a), is one of trial and error. In this problem, all three maximal compatibles will be selected in order to satisfy completeness and consistency. Hence the reduced state table will consist of three states, P = ABD, Q = ACD and R = CE. The reduced state table is shown in Figure 9.13(g).

9.14 Circuit hazards

One cause of malfunction in combinational and sequential circuits can be traced to the presence of race hazards. The designer should have a clear understanding of the mechanism that produces such hazards and should also be aware of their effects on circuit performance.

There are four types of hazard which can occur in digital systems:

1. Static hazards
2. Dynamic hazards
3. Function hazards
4. Essential hazards

Static hazards are due to a momentary change in output caused by an input change that does not affect the steady-state output. They may be present in both combinational circuits and gate-implemented asynchronous circuits. Dynamic hazards occur when, due to a single output change, the output changes several times before reaching its steady state value. Function hazards occur when more than one input variable change takes place at the same time, while essential hazards are peculiar to fundamental mode sequential circuits and they cannot be eliminated without controlling the delays in the circuit.

9.15 Gate delays

If a two-input NAND gate is used as an inverter in a combinational network, as illustrated in Figure 9.14, there will be a finite time delay t_g before any change at the input to the gate produces the required change at the output. This delay is demonstrated in the timing diagrams, where the change in A from 0 to 1 is followed by a change in $\bar{A}$ from 1 to 0, t_g seconds later. Similarly, when A changes from 1 to 0, the corresponding change in $\bar{A}$ from 0 to 1 also occurs later (see Chapter 4).

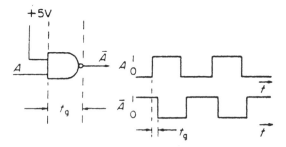

Figure 9.14   *The effect of gate delays when inverting a signal A*

9.16 The generation of spikes

If the signal A and its complement $\bar{A}$, generated by the NAND gate shown in Figure 9.14, are both fed to the inputs of a two-input AND gate as shown in Figure 9.15, then according to the laws of Boolean algebra the output of the gate should be $A \cdot \bar{A} = 0$ at all times. However, it will be observed from an examination of the timing diagrams that in the time periods that have been shaded, A and $\bar{A}$ are simultaneously equal to 1, so that

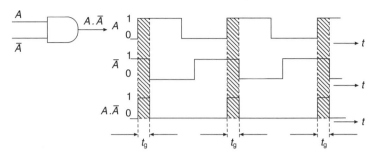

Figure 9.15   *Generation of spikes by an AND gate*

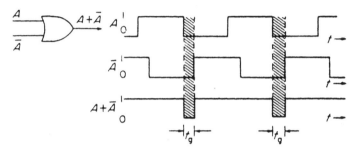

Figure 9.16   *Generation of spikes by an OR gate*

during these periods the gate output is $A \cdot \bar{A} = 1$. The output of the gate, $A\bar{A}$, consists of a series of positive going spikes which are initiated when A is changing from 0 to 1, each of time duration t_g, the gate delay of the inverter shown in Figure 9.14. The circuit used to generate the signal $A \cdot \bar{A}$ is said to exhibit a *static 0-hazard* because the output signal, which should be permanently 0, goes to 1 for a short transient period.

Alternatively, if the signals A and $\bar{A}$ are applied to the inputs of a two-input OR gate as shown in Figure 9.16 then the output of the gate is $A + \bar{A}$, which, according to the laws of Boolean algebra, should be 1 at all instants of time. The waveforms of A and $\bar{A}$ (see Figure 9.16) show that during the shaded time periods, they are both simultaneously equal to 0. In these shaded time periods, which are of short time duration, the output goes to 0. The circuit is said to exhibit a *static 1-hazard* because its output, which is normally 1, goes to 0 for short time periods. It will be observed that for the OR gate, the negative-going spikes are initiated at the instant when A is changing from 1 to 0.

The generation of spikes by NAND and NOR gates is illustrated in Figure 9.17. Negative-going spikes are generated by a NAND gate at the instant when A is

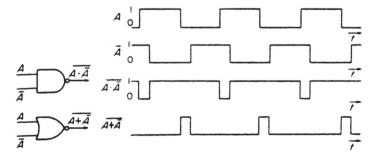

Figure 9.17   *Generation of spikes by NAND and NOR gates*

changing from 0 to 1. The circuit exhibits a static 1-hazard. In the NOR circuit, positive-going spikes are generated at the instant when A is changing from 1 to 0. This circuit exhibits a static 0-hazard.

9.17 The generation of static hazards in combinational networks

When an input to a combinational network is changing, spikes may be generated at the output of the circuit. The spikes, when they occur, are due to different path lengths in the network which introduce different time delays. For example, the Boolean function

$$f = AB + \bar{A}C$$

may be implemented by NAND gates, as shown in Figure 9.18. There are two paths through the circuit, the first via g_1, g_2 and g_3 and the second via g_4 and g_3. If it is assumed that all gates have exactly the same time delay, then it is apparent that the delay through the first path is greater than the delay through the second path.

The changes taking place in the circuit are illustrated in Figure 9.18 for the circuit condition $B = 1$, $C = 1$ and A, changing from 1 to 0. For this change in A, the output of g_4 changes from 0 to 1 and produces a change in the output of g_3 from 1 to 0. For the other path through the circuit, the output of g_1 first changes from 0 to 1, followed by the output of g_2 changing from 1 to 0, thus producing a change in the output of g_3 from 0 to 1. Because the g_4, g_3 path has the shorter time delay, it is clear that the change in output propagated along this path occurs earlier in time than the change propagated along the alternative path.

Since it has been assumed that $B = C = 1$, the network equation reduces to $f = A + \bar{A}$. When a circuit equation, under certain specified input conditions, reduces to this form, a static 1-hazard will be generated. In the example chosen here, the timing diagrams shown in Figure 9.18 reveal that due to the inverter delay, for a short period of time both A and $\bar{A}$ are equal to 0, and $A + \bar{A} = 0$. Providing the condition $B = C = 1$ is maintained and the input signal consists of a train of positive-going pulses, a series of negative-going spikes will be generated. The presence of the negative-going spikes confirms the earlier deduction, made by following the signal changes through the circuit diagram, that the output changes are $1 \rightarrow 0 \rightarrow 1$.

The dual function of $f = AB + \bar{A}C$ is:

$$f_d = (A + B)(\bar{A} + C)$$

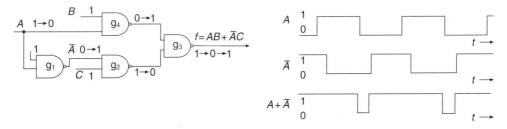

Figure 9.18   *The production of a static 1-hazard in a combinational network*

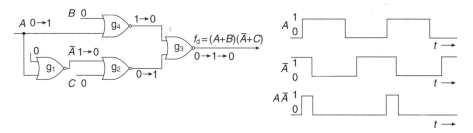

Figure 9.19 *The production of a static 0-hazard in a combinational network*

The implementation of this function using NOR gates is shown in Figure 9.19. When $B = C = 0$ the circuit equation reduces to $f_d = A \cdot \bar{A}$. Under the conditions specified a static 0-hazard will be generated when A is changing from 0 to 1. The production of the static 0-hazard is illustrated in Figure 9.19. Immediately after A changes from 0 to 1 both A and $\bar{A}$ are simultaneously 1, hence $A \cdot \bar{A} = 1$. The output remains at this value until $\bar{A}$ falls to 0 when $A \cdot \bar{A}$ resumes its value of 0 again.

Signal changes are also illustrated in Figure 9.19, where it has been assumed that $B = C = 0$ and that A consists of a stream of positive-going pulses. If all the gates have the same time delay, then path g_4, g_3 has the shortest time delay and the change in output due to A changing from 0 to 1 will propagate along this path faster than along path g_1, g_2, g_3. This results in the output changing from 0 to 1. When the corresponding change arrives at the output along the alternative path, the output changes back to zero again.

A similar analysis can be carried out for both the AND/OR and OR/AND configurations, and this will show that the AND/OR circuit implementing the function $f = AB + \bar{A}C$ will generate a static 1-hazard. Similarly, for the OR/AND circuit implementing the function $f_d = (A + B)(\bar{A} + C)$, it can be shown that a static 0-hazard will be generated.

9.18 The elimination of static hazards

The equation of the NAND circuit shown in Figure 9.18 is $f = AB + \bar{A}C$. The consensus product for this equation is BC, and this can be added to the original equation without altering its value. Thus:

$$f = AB + \bar{A}C + BC$$

and for the condition $B = C = 1$ the equation reduces to $f = A + \bar{A} + 1$, and even if A and $\bar{A}$ are, for a short period of time, simultaneously equal to 0, the value of the function f remains at 1.

The effect of adding the consensus product can be studied by examining the K-map plot of the function before and after the addition of the consensus product. The original function is shown plotted in Figure 9.20(a) and the plot of the function, after the inclusion of the consensus product, is shown in Figure 9.20(b). Comparison of the two plots shows that before the addition of the consensus product, there are two 1's in adjacent cells not covered by the same prime implicant. On covering these two adjacent 1's by the same prime implicant, as in Figure 9.20(b), the hazard is removed from the circuit.

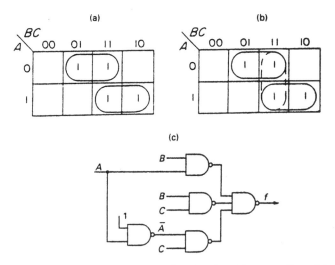

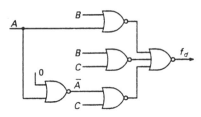

Figure 9.20   *(a) Plot of f* $= AB + \bar{A}C$ *(b) Plot of f* $= AB + \bar{A}C + BC$ *(c) Implementation of the hazard-free function f* $= AB + \bar{A}C + BC$

It follows that static 1-hazards can be detected by looking for adjacent 1's on a K-map plot of the function that are not covered by the same prime implicant. They can then be removed at the design stage by including additional prime implicants which cover adjacent 1's not otherwise covered by the same prime implicant.

The hazard-free circuit for the Boolean function $f = AB + \bar{A}C$ is shown in Figure 9.20(c), and it will be observed that an additional NAND gate has been introduced for generating the required consensus product BC.

For the NOR circuit of Figure 9.19, $f_{\mathrm{d}} = (A + B)(\bar{A} + C)$. The consensus term for this equation is $(B + C)$, and this can be included in the above equation without altering its value, so that:

$$f_{\mathrm{d}} = (A + B)(\bar{A} + C)(B + C)$$

If $B = C = 0$ then:

$$f_{\mathrm{d}} = A \cdot \bar{A} \cdot 0$$

With the inclusion of the consensus sum, the value of the function is always 0, irrespective of whether A and $\bar{A}$ are simultaneously equal to 1.

The static 0-hazard is eliminated by the inclusion of the consensus term $(B + C)$, and the resulting hazard-free circuit is shown in Figure 9.21. Elimination of the hazard requires the inclusion of an additional gate which generates the inverse of the consensus sum.

When looking for a static 0-hazard, a K-map plot of the function which identifies those combinations of the variables that cause the function value to be 0 is required. To obtain a plot of the 0-terms, the inverse of the function f_{d} must be plotted. The equation of the circuit is:

Figure 9.21   *Implementation of the hazard-free function* $f_{d} = (A + B)(\bar{A} + C)(B + C)$

$$f_{\mathrm{d}} = (A + B)(\bar{A} + C)$$

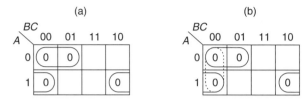

Figure 9.22 *(a) Plot of $\bar{f} = \bar{A}\bar{B} + A\bar{C}$ (b) Plot of $\bar{f}$ including consensus term for removing the hazard*

Inverting:

$$\overline{f_d} = \bar{A}\bar{B} + A\bar{C}$$

The inverse function is shown plotted in Figure 9.22(a) and it will be noticed that the two 0's in the adjacent cells 000 and 100 are not covered by the same prime implicant. The function containing the additional prime implicant $\bar{B}\bar{C}$ becomes:

$$\overline{f_d} = \bar{A}\bar{B} + A\bar{C} + \bar{B}\bar{C}$$

Inverting, $f_d = (A + B)(\bar{A} + C)(B + C)$ which is the hazard-free function obtained previously by introducing the consensus term to the function equation.

The algorithm for finding static 0-hazards follows:

Step 1: Plot the inverse function.
Step 2: Look for adjacent 0's not covered by the same prime implicant.
Step 3: Insert additional prime implicants to cover all adjacent 0's that are not covered by the same prime implicant.
Step 4: Modify the inverse equation by including the additional prime implicants.
Step 5: Re-invert the equation to obtain the hazard-free form of the function.

9.19 Design of hazard-free combinational networks

In this section the function represented by the equation

$$f = \sum 2, 5, 6, 7, 10, 13, 15$$

will be implemented in hazard-free form using (a) NAND gates, and (b) NOR gates. A fan-in limitation of three will be imposed.

For the NAND implementation, the circuit has to be free of static 1-hazards. The first step in the design is to plot the K-map of the function and simplify in the normal way (see Figure 9.23). The plot is now examined to see if there are any 1's in adjacent cells not covered by the same prime implicant. In this case a pair of such cells are 0111 and 0110, and an additional prime implicant is added to the plot to eliminate the uncovered adjacency. The 1's that constitute the added prime implicant are enclosed by dotted lines on the K-map plot.

Reading from the map, the hazard-free function is:

$$f = BD + \bar{A}C\bar{D} + \bar{B}C\bar{D} + \bar{A}BC$$

To meet the fan-in restriction, the equation can be factorised and then:

$$f = C\bar{D}(\bar{A} + \bar{B}) + BD + \bar{A}BC$$

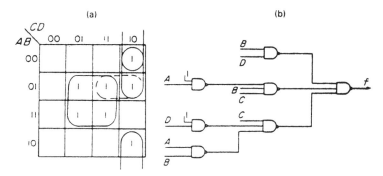

Figure 9.23   *(a) Plot of f = Σ 2, 5, 6, 7, 10, 13, 15 (b) NAND hazard-free implementation*

The factorisation of an equation in this way does not reintroduce hazards. In this problem the hazard would have occurred when $A = 0$, $B = 1$ and $C = 1$, with D changing from 1 to 0. Insertion of these conditions in the factorised equation gives:

$$f = \bar{D}(1 + 0) + D + 1$$
$$= \bar{D} + D + 1$$

which is the required condition for the removal of the hazard. The NAND implementation of the hazard-free function is shown in Figure 9.23(b).

To obtain the hazard-free NOR realisation, the inverse function is plotted and simplified. The inverse plot is derived from Figure 9.23(a) by marking the vacant cells on the map with 0's, as shown in Figure 9.24. The presence of 0's in adjacent cells not covered by the same prime implicant indicates that the simplified function will produce a static 0-hazard under certain prescribed conditions. In this case there are two such pairs of adjacent cells, (a) 0000 and 0001, and (b) 1000 and 1001. The introduction of an additional prime implicant $\bar{B}\bar{C}$, enclosed by dotted lines on the map, covers the uncovered adjacencies and eliminates the static 0-hazard. Reading the inverse function from the map:

$$\bar{f} = \bar{C}\bar{D} + \bar{B}D + \bar{B}\bar{C} + AB\bar{D}$$

and factorising to satisfy the fan-in restriction gives:

$$\bar{f} = \bar{C}\bar{D} + \bar{B}(\bar{C} + D) + AB\bar{D}$$

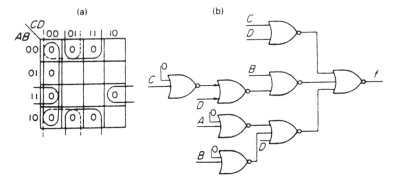

Figure 9.24   *(a) The 0-plot of f = Σ 2, 5, 6, 7, 10, 13, 15 (b) NOR hazard-free implementation*

and re-inverting:

$$f = (C + D)(B + C\bar{D})(\bar{A} + \bar{B} + D)$$

The implementation of this hazard-free function with NOR gates is shown in Figure 9.24(b).

9.20 Detection of hazards in an existing network

The network shown in Figure 9.25 is to be analysed to see if it has any static 0- or static 1-hazards. The equation of the network is :

$$f = AB\bar{C} + (A + B)(\bar{A} + \bar{D})$$

which may be expanded into the following form:

$$f = AB\bar{C} + A\bar{A} + A\bar{D} + \bar{A}B + B\bar{D}$$

This expression contains the term $A\bar{A}$ which, under normal circumstances, would be removed since, by the laws of Boolean algebra, its value is 0. Since the variables A and $\bar{A}$, in combinational networks can be simultaneously 1, they are treated as independent

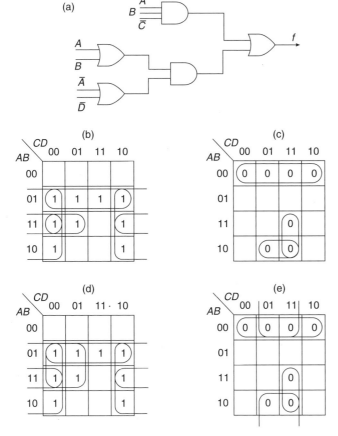

Figure 9.25 *(a) Circuit for the function* $f = AB\bar{C} + (A + B)(\bar{A} + \bar{D})$ *(b) K-map plot of the function (c) Plot of the inverse function (d) The resimplification of the function (e) Hazard-free plot of the inverse function*

variables in this equation which may be regarded as the equation which holds for transient conditions.

When deriving the transient equation of a circuit, some of the theorems of Boolean algebra may not be used. Those which make use of the identities $A\bar{A} = 0$ and $A + \bar{A} = 1$ may not be used to manipulate the equation into its transient form. For example, the expression $A + \bar{A}B = (A + \bar{A})(A + B) = (A + B)$ cannot be used as the reduction depends upon the identity $(A + \bar{A}) = 1$. Earlier in this chapter, it was shown that A and $\bar{A}$ may be simultaneously equal to zero, and in that case $A + \bar{A} \neq 1$, hence the above reduction is not valid for all instants of time.

The hazards can be detected by examining the expanded equation to see whether it reduces to either of the forms $X\bar{X}$ or $X + \bar{X}$ under defined input conditions, where X and $\bar{X}$ may represent any one of the four variables in the equation. For example, if $B = 0$ and $D = !$, the equation reduces to $f = A\bar{A}$. Hence for these input conditions, a static 0-hazard occurs when A is changing from 0 to 1. Additionally, if $B = 1, C = 0$ and $D = 1$, the transient equation reduces to $f = A + A\bar{A} + \bar{A}$ and a static 1-hazard occurs when A is changing from 1 to 0. It should be noted that since A is changing from 1 to 0, $A\bar{A} = 0$ since it can only have a value 1 when A is changing from 0 to 1. If, however, $B = 1, C = 0$ and $D = 0$, the transient equation reduces to $f = A + A\bar{A} + A + \bar{A} + 1$. In this case, irrespective of the instantaneous values of A and $\bar{A}$, $f = 1$, and hence there is no static hazard.

Alternatively, the static 1-hazard can be detected by plotting those values of the variables that make the value of the function $f = 1$, as shown in Figure 9.25(b). Examination of this K-map shows that the two 1's in the adjacent cells 1101 and 0101 are not covered by the same prime implicant. The introduction of the prime implicant $B\bar{C}$ will ensure the coverage of these two cells by the same prime implicant and will remove the static 1-hazard.

To detect the possibility of a static 0-hazard, the circuit function has first to be inverted, and then plotted on a K-map. The inverse of the circuit function $f = AB\bar{C} + (A + B)(\bar{A} + \bar{D})$ is:

$$\bar{f} = \bar{A}\bar{B} + A\bar{B}D + ACD + A\bar{A}D$$

Note that the fourth term (the transient term) cannot be represented on the map.

It is clear from an examination of the K-map (Figure 9.25(c)) that the 0s in cells 1001 and 1011 are adjacent to the 0s in the cells 0001 and 0011 and are not covered by the same prime implicant, and a static 0-hazard is present in the circuit shown in Figure 9.24(a). By introducing the prime implicant $\bar{B}D$ to cover these four cells, the static 0-hazard can be removed.

Poorly designed circuits may generate both kinds of static hazard. In practice, it would be a more satisfactory solution to redesign the circuit, shown in Figure 9.25(a), using the K-map plot of Figure 9.25(b) which, for convenience, is repeated in Figure 9.25(d). On this map the function has been simplified in such a way that the function is free of static 1-hazards. The hazard-free function is:

$$f = \bar{A}B + B\bar{D} + A\bar{D} + B\bar{C}$$

If an AND-OR-INVERT configuration is to be used, all that has to be done is to examine the plot of the inverse function for static 0-hazards. The inverted function is:

$$\bar{f} = \bar{A}\bar{B} + \bar{B}D + ACD$$

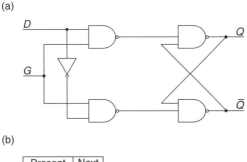

(a)

(b)

Present state			Next state
G^t	D^t	Q^t	$Q^{t+\delta t}$
0	0	0	0
0	0	1	1
0	1	0	0
0	1	1	1
1	0	0	0
1	0	1	0
1	1	0	1
1	1	1	1

(c)

Figure 9.26 *(a) The controlled D-latch (b) state table (c) K-map plot*

and is shown plotted in Figure 9.25(e). Since there are no adjacent 0's not under the same prime implicant there are no static 0-hazards present.

A practical example of the possibility of a static hazard in a controlled D latch was referred to in section 6.5 of Chapter 6. For convenience, the circuit configuration is shown again in Figure 9.26 along with the state table and the K-map plot of the characteristic equation.

The characteristic equation read from the map is:

$$Q^{t+\delta t} = (\bar{G}Q + DG)^t$$

It will be observed that there are two 1's in adjacent cells not covered by the same prime implicant, and consequently a static 1-hazard is present. To eliminate the hazard, an extra prime implicant, DQ, enclosed by dotted lines on the map, is added in Figure 9.26(c), and the modified characteristic equation is:

$$Q^{t+\delta t} = (\bar{G}Q + DG + DQ)^t$$

It is left to the reader to show that the implementation of the latch shown in Figure 9.26(a) does, in fact, correspond to this hazard-free equation.

9.21 Hazard-free asynchronous circuit design

A gate-implemented asynchronous circuit with feedback is, in essence, a group of one or more combinational circuits which, under certain conditions, may generate static hazards. In practice, the designer should examime the design for hazards and then eliminate them using the techniques described earlier in this chapter. To demonstrate the occurrence of hazards in asynchronous circuits, the design of a hazard-free T-type flip-flop will be undertaken.

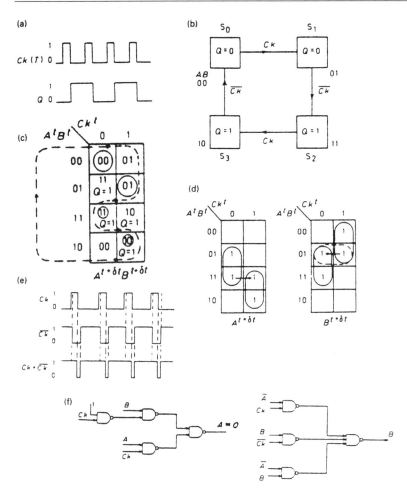

Figure 9.27   *(a) Timing diagram for T-type flip-flop (b) Internal state diagram (c) State table (d) Individual simplified state tables for $A^{t+\delta t}$ and $B^{t+\delta t}$ including hazard-removing prime implicant (e) Timing diagram for the Ck signal (f) Implementation of hazard-free flip-flop*

The timing diagram of a trailing edge triggered TFF is shown in Figure 9.27(a), the output toggling on the trailing edge of successive clock pulses. The state diagram is shown in Figure 9.27(b) and it reveals that the circuit completes a cycle of operation after four changes of the clock signal. It should be noted that in this example the clock transitions can be regarded as events which are able to initiate state transitions.

Since there are four states, two state variables A and B are required, and since this is an asynchronous design, a race-free state assignment has been used. The state table corresponding to the state diagram is shown in Figure 9.27(c), and the path traversed through the state table as one cycle of operation of the flip-flop takes place is illustrated by the dotted line.

In Figure 9.27(d), the state table has been separated into two distinct maps, one for $A^{t+\delta t}$ and one for $B^{t+\delta t}$. After simplification of these two functions it is clear that in both cases two 1's in adjacent cells are not covered by the same prime implicant, and there is a real possibility that a static hazard may be generated in both the A and B circuits. Arrows have been inserted on both maps indicating the direction of the state transitions between the relevant cells.

The equation for $A^{t+\delta t}$ is:

$$A^{t+\delta t} = (B \cdot \overline{Ck} + A \cdot Ck)^t$$

If $A = B = 1$, the equation reduces to:

$$A^{t+\delta t} = (\overline{Ck} + Ck)^t$$

and this condition indicates the possibility of the generation of a static 1-hazard. However, an examination of the timing diagrams for Ck, $\overline{Ck}$ and $(Ck + \overline{Ck})$ in Figure 9.27(e) shows that a static 1-hazard will only occur if Ck is making a $1 \rightarrow 0$ transition. The arrow-head on the $A^{t+\delta t}$ map reveals that the transition concerned is from total state $ABCk = 110$ to $ABCk = 111$; that is, the clock signal is changing from $0 \rightarrow 1$, and it follows that a static 1-hazard can never be generated in the A circuit.

The equation for $B^{t+\delta t}$ is:

$$B^{t+\delta t} = (\bar{A} \cdot Ck + B \cdot \overline{Ck})^t$$

If $\bar{A} = B = 1$ this equation reduces to:

$$B^{t+\delta t} = (Ck + \overline{Ck})^t$$

In this case, the arrow-head on the $B^{t+\delta t}$ map shows that the B circuit makes a transition from total state $ABCk = 011$ to $ABCk = 010$; that is, Ck is making a $1 \rightarrow 0$ transition, and consequently a static 1-hazard will be generated in the B circuit. To eliminate the hazard, the additional prime implicant $\bar{A}B$ is added to the equation for $B^{t+\delta t}$ which now reads:

$$B^{t+\delta t} = (\bar{A} \cdot Ck + B \cdot \overline{Ck} + \bar{A} \cdot B)^t$$

Also from the state diagram:

$$Q = S_3 + S_2$$
$$= AB + A\bar{B} = A$$

The NAND implementation of the hazard-free T flip-flop is shown in Figure 9.27(f).

9.22 Dynamic hazards

A second type of hazard that can occur in gate networks is referred to as a *dynamic hazard*. The output changes normally expected by the circuit designer are either $0 \rightarrow 1$ or alternatively $1 \rightarrow 0$. If, in practice, the output transitions are $1 \rightarrow 0 \rightarrow 1 \rightarrow 0$ then a dynamic hazard has occurred. Similarly, if an output designed to change from $0 \rightarrow 1$ has the change pattern $0 \rightarrow 1 \rightarrow 0 \rightarrow 1$, then a dynamic hazard is present. In either case there is a minimum of three changes appearing at the output as illustrated in Figure 9.28.

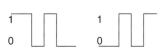

Figure 9.28   *Dynamic hazards*

This type of hazard occurs as a result of the factorisation of a Boolean function, necessary, because of fan-in restrictions, which leads to different path lengths through a circuit. Alternatively, the gates in the circuit configuration may have different time delays, and it is also possible to have differing time delays in the interconnecting leads.

(a)

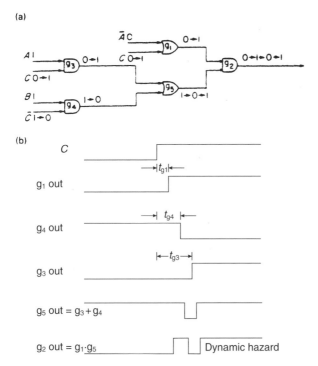

(b)

Figure 9.29   *(a) Network with a dynamic hazard (b) Occurrence of a dynamic hazard in the network*

Consider the function:

$$f = (AC + B\bar{C})(\bar{A} + C)$$

implemented with AND and OR gates as shown in Figure 9.29(a). There are three different paths through this network for the variable C and consequently there is a possibility that a dynamic hazard exists in the network. The three paths through the network are

1. via gates g_1 and g_2
2. via gates g_3, g_5 and g_2, and
3. via gates g_4, g_5 and g_2.

There are eight possible starting combinations of the variables A, B, and C. Since, in this circuit, the dynamic hazard is caused by multiple paths taken by the signal representing variable C, in each case only the next change in C need be examined. For the four combinations starting with $A = 0$, the output of OR gate g_1 remains at 1 irrespective of C, and the output of AND gate g_3 remains at 0. Hence, for these combinations, changes in C take the path g_4–g_5–g_2 only and so no dynamic hazards are present.

However, in the case $A = 1$, $B = 1$, and $C = \uparrow$, the upper input of gate g_2 changes from 0 to 1. The other input to g_2 is $AC + B\bar{C} = C + \bar{C}$ in this case, and if $t_{g4} < t_{g3}$ then there will be a static 1-hazard present at the output of gate g_5. Then, if this static hazard itself occurs after the change at the output of g_1, i.e. if $t_{g1} < t_{g4} < t_{g3}$, there will be a dynamic hazard present at the output of gate g_2 as illustrated in Figure 9.29(b).

A similar analysis shows that if $t_{g1} > t_{g4} > t_{g3}$, then there is a dynamic hazard produced in the case $A = 1$, $B = 1$, and $C = \downarrow$. In the two remaining cases,

$A = 1$, $B = 0$, and $C = $ X, there are no further dynamic hazards as signal C takes only two paths through the network. Using the terminology to be introduced in Chapter 13, all three possible paths for C are *sensitised* only when $A = 1$ and $B = 1$.

It is worth noting that providing AND/OR sum-of-products circuits or if OR/AND product-of-sum circuits have been designed such that there are no static hazards present, then these circuits will have no dynamic hazards.

9.23 Function hazards

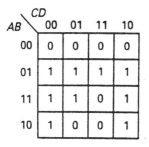

Figure 9.30 *K-map plot used for illustrating function hazards*

This type of hazard, which can be either a static 1- or static 0-hazard, occurs when it is specified that two circuit input variables change at the same time. In practice, it is extremely unlikely that two variables will change at precisely the same time but if this should happen to occur it can lead to the presence of a hazard during a transition.

Consider the K-map plot of a 4-variable function shown in Figure 9.30. If the initial condition of the input variables is $ABCD = 1000$ and circuit operation specifies that the variables B and D change simultaneously, one of three possibilities may occur:

1. B and D change simultaneously:
   $ABCD = 1000 \rightarrow 1101$
   $\quad f = 1 \quad\;\; \rightarrow 1$
2. B changes before D:
   $ABCD = 1000 \rightarrow 1100 \rightarrow 1101$
   $\quad f = 1 \quad\;\; \rightarrow 1 \quad\;\; \rightarrow 1$
3. D changes before B:
   $ABCD = 1000 \rightarrow 1001 \rightarrow 1101$
   $\quad f = 1 \quad\;\; \rightarrow 0 \quad\;\; \rightarrow 1$

If D changes before B a function static 1-hazard is present. Alternatively, if the initial condition of the input variables is $ABCD = 0000$ and a simultaneous change in the variables A and D should occur, one of the following three possibilities may arise:

1. A and D change simultaneously
   $ABCD = 0000 \rightarrow 1001$
   $\quad f = 0 \quad\;\; \rightarrow 0$
2. A changes before D
   $ABCD = 0000 \rightarrow 1000 \rightarrow 1001$
   $\quad f = 0 \quad\;\; \rightarrow 1 \quad\;\; \rightarrow 0$
3. D changes before A
   $ABCD = 0000 \rightarrow 0001 \rightarrow 1001$
   $\quad f = 0 \quad\;\; \rightarrow 0 \quad\;\; \rightarrow 0$

If A changes before D a function static 0-hazard occurs.

A situation may also arise where it is specified that three variables should change at the same time, and in that case there is the possibility that a *function dynamic hazard*

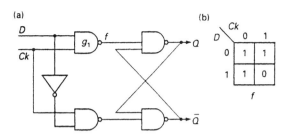

Figure 9.31 *(a) Controlled D-type latch (b) K-map plot for output of gate g_1*

may occur. In practice, function hazards can be avoided at the design stage by ensuring that only one variable can change at any one time.

A function hazard can occur at the input NAND gate of a synchronising latch. It will be assumed that the input D is asynchronous data and that Ck is the synchronising signal, as described in Chapter 8. In this situation there is no way of ensuring that the asynchronous input changes at the same time as the synchronising signal. The K-map for a 2-input NAND gate is shown in Figure 9.31. If the initial condition is $DCk = 10$ and both signals happen to change simultaneously, then the steady-state output of the gate will remain at 1. In practice, they are unlikely to change simultaneously and a spurious output can occur, which in the case of a synchroniser circuit, is referred to as a *runt pulse*. This pulse may not be sufficient to cause the synchroniser to switch from one stable state to another and the latch may enter the metastable state where it will stay for a period which cannot be precisely defined.

9.24 Essential hazards

This type of hazard is peculiar to asynchronous circuits and is caused by a race between an input signal and a state variable. The state diagram for an asynchronous circuit having a race-free state assignment is shown in Figure 9.32. Assuming that the circuit is in state S_0 and a change in the value of X from 0 to 1 occurs, a transition from S_0 to S_1 should take place and, on arriving in S_1, the circuit should remain in that state.

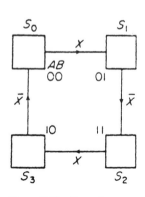

Figure 9.32 *State diagram for a machine which can have an essential hazard*

However, correct operation of the circuit as described above will depend upon the relative values of the inversion time t_i for the input signal X and and the turn-on time t_t for the state variable B. If the circuit arrives in the state S_1 before the value of $\bar{X}$ has changed from 1 to 0, a further transition to S_2 will be made. Since $X = 1$ when the circuit arrives in state S_2, it follows that a further transition will take place to state S_3, where the circuit will now remain, provided the change in $\bar{X}$ has now occurred. Hence, if $t_i > t_t$, incorrect circuit operation will occur as a consequence of the race between the inversion of the input signal X and the turn-on of the state variable B.

An examination of the equation for the state variable A reveals more clearly the origin of the hazard. The turn-on condition for $A = B\bar{X}$, the turn-off condition for $A = \bar{B}\bar{X}$, and

$$A^{t+\delta t} = (B\bar{X} + (\overline{\overline{B}\overline{X}})A)^t$$
$$= (B\bar{X} + (B + X)A)^t$$

The first term of this equation provides the turn-on signal for A when the circuit is in state S_1. If B changes to 1 before $\bar{X}$ changes to 0, the value of $B\bar{X} = 1$ and the state variable A is turned on.

The method of dealing with this type of hazard is to insert a delay in the output line of the circuit generating the state variable B. This will ensure that the change in B does not arrive at the input to the circuit generating the state variable A until the value of $\bar{X}$ has changed.

Problems

9.1 A double-sequence detector has two inputs, X_1 and X_2, and one output Z. For an input sequence $X_1 X_2 = 00$, 10, 11 the output Z becomes 1, and when the reverse sequence is received the output Z returns to 0. A typical timing diagram for the detector is shown in Figure P9.1. Develop:

(1) A state diagram

(2) An ASM chart

for the detector and obtain a state table. If possible, reduce the state table and implement the design with NAND gates.

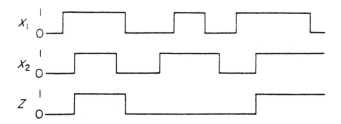

Figure P9.1

9.2 Develop an event-driven circuit to implement a trailing-edge triggered JK flip-flop and draw a timing diagram for the flip-flop.

9.3 X_1 and X_2 are the two inputs to an asynchronous circuit which has two outputs, Z_1 and Z_2. When $X_1 X_2 = 00$ the output $Z_1 Z_2 = 00$. If a $0 \rightarrow 1$ change in X_1 precedes a $0 \rightarrow 1$ change in X_2, then the output of the circuit is $Z_1 Z_2 = 01$. Alternatively, if a $0 \rightarrow 1$ change in X_2 precedes a $0 \rightarrow 1$ change in X_1, then the output of the circuit is $Z_1 Z_2 = 10$. In both cases the outputs remain at 01 and 10, respectively, until $X_1 X_2 = 00$ again. Draw the state diagram for this system.

9.4 Develop an asynchronous circuit that will give an output clock pulse (Z) after every second data pulse arrives on the X input line. The arrival of the data pulses is purely random and it is to be assumed that the minimum time for a pair of consecutive data pulses is greater than the periodic time of the clock. A typical timing diagram is shown in Figure P9.4.

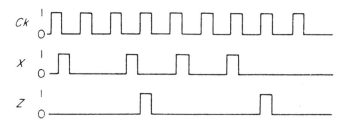

Figure P9.4

9.5 A logic circuit has two asynchronous inputs, X_1 and X_2, and also a synchronous clock signal. The circuit is to be designed so that the first complete clock pulse that occurs after X_1 and X_2 have become 1, in that order, is output on the line marked Z in Figure P9.5. After the output of the clock pulse the circuit must return to its quiescent state when $X_1 X_2 = 00$.

Design a circuit that satisfies this specification and implement the design using NAND gates.

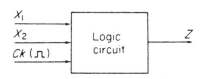

Figure P9.5

9.6 Analyse the fundamental mode circuit shown in Figure P9.6:

(a) Determine the state table.
(b) Determine the state diagram.
(c) Use the state table to determine the output response to the input sequence $X_1 X_2 = 00$, 01, 11, 10, 11, 01, 00, 10, 00, 01. Initial conditions $X_1 = X_2 = A = 0$.

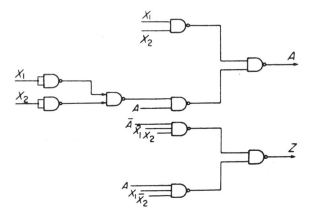

Figure P9.6

9.7 Analyse the circuit shown in Figure P9.7:

(a) Determine the state table.
(b) Determine the state diagram.
(c) Use the state table to determine the output response to the input sequence $X_1 X_2 = 00, 01, 11, 10, 00, 01, 11, 01, 11, 10, 00$. Assume the initial conditions are $X_1 = X_2 = 0$ and $A = B = 0$.

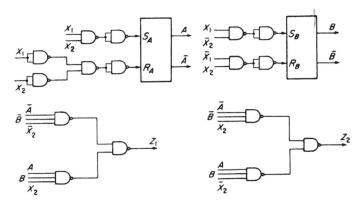

Figure P9.7

9.8 The internal state diagram for a four-state digital machine is shown in Figure P9.8. Construct a state table for the machine and identify all races that will occur if the machine is implemented from the given state diagram, stating whether they are critical or non-critical. For each race, give all the state transitions which may occur.

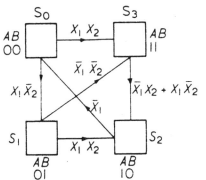

Figure P9.8

9.9 Plot the K-map of the functions

(a) $f(A, B, C, D) = \sum 0, 2, 4, 5, 6, 8, 9, 11, 12, 14, 15$, and

(b) $f(A, B, C, D) = \sum 3, 4, 5, 6, 11, 12, 13, 14, 15$

and determine hazard-free implementations in both cases, using NAND gates.

9.10 Find all the static hazards in the two networks shown in Figures P9.10(a) and (b). Specify the input conditions that must exist for the hazards to occur and draw the logic diagram for modified networks that are hazard-free.

9.11 Design a hazard-free, D-type flip-flop using asynchronous circuit design techniques. It may be assumed that the output will take on the value of the input on the trailing edge of a clock pulse.

9.12 An incompletely specified table is shown in Figure P9.12. With the aid of an implication chart, find the compatible state pairs. Using a merger diagram obtain the maximal compatibles and construct a reduced state table.

9.13 An electrical system is protected by a fault detector. If a fault occurs within the system a fault signal activates an alarm buzzer. The green light that indicates fault

(a)

(b)

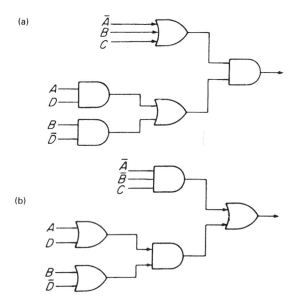

Figure P9.10

X_1X_2	00	01	11	10
S_0	$-,-$	$S_2,1$	$S_4,1$	$S_2,1$
S_1	$S_4,-$	$-,-$	$-,-$	$-,-$
S_2	$S_5,-$	$S_5,1$	$-,-$	$-,-$
S_3	$-,-$	$-,-$	$S_1,1$	$-,-$
S_4	$-,-$	$S_5,0$	$S_0,0$	$S_3,1$
S_5	$S_2,0$	$-,-$	$S_1,0$	$S_2,1$

Figure P9.12

free operation is switched off by the fault signal and a red light is switched on. When the fault is acknowledged by the system controller the alarm buzzer is turned off. After the fault has been cleared the green light is switched on and the red light is turned off. A test signal is to be provided to check the operation of the fault detector.

Develop an appropriate state diagram and implement your design with the aid of the NAND characteristic equation.

9.14 An asynchronous circuit is to be used to control the gates and a red flashing light at a railway level crossing. The gates are to be closed and the red flashing light is to be turned on when a train enters a defined section of track from either direction. When the train is in a further defined section of track which straddles the crossing the gates must remain closed and the red light must remain flashing. After a train has passed through the crossing the gates are opened and the flashing red light is turned off.

Develop an ASM chart, convert it to a state diagram and implement your design using the NAND sequential equation.

9.15 Design an asynchronous lock operated by five input buttons labelled A, B, C, D and R (the reset button). The unlocking operation can only take place if only one button is activated at a time and in the order B, D, A, C. Draw a state diagram and develop a gate-implemented circuit.

9.16 Using asynchronous circuit design techniques, design a hazard free D-type flip-flop whose output takes up the value of the input on the trailing edge of a clock pulse.

10 Instrumentation and interfacing

10.1 Introduction

Very many systems designed today use digital logic components alongside sub-systems based upon analogue electronics, and also sub-systems based upon mechanical components. This allows designers the flexibility to use several design techniques in order to produce the most useful systems as a whole. From one point of view it is true that a digital system is merely a special case or subset of a general (analogue) electronic system where the signals involved always happen to fall into two well-defined voltage or current levels rather than being unconditionally variable between upper and lower limits. Hence, many basic considerations of design, such as response time, current requirements, and so on, are similar in both analogue and digital design. However, as observed elsewhere in this text, the consequences of the basic differences between analogue and digital design are far-reaching as far as the design methodology adopted is concerned, and so it is usual to regard the two systems as separate. Therefore, there are usually components or sub-systems of various types employed at the interface between the two types of systems.

10.2 Schmitt trigger circuits

The 'Schmitt trigger' circuit is one of the simplest of interfaces between the analogue world and the digital world. A full analysis of Schmitt-type trigger circuits is beyond the scope of this text and is covered in texts on analogue electronics, but the important results are summarised in this section.

In a Schmitt trigger circuit, *positive* feedback is applied to an analogue differential amplifier from its output to its non-inverting input to give a circuit having the following characteristics:

1. only two output voltages are possible, almost equal to the two supply rail voltages, and with suitable choice of supply voltages these correspond to the two digital logic levels 0 and 1; and
2. the trigger circuit switches between its two possible output voltage levels according to the voltage applied to the single input of the trigger circuit.

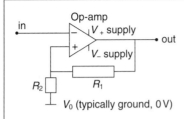

Figure 10.1   *Simple form of Schmitt trigger circuit*

The Schmitt trigger circuit in its simplest form (see Figure 10.1) consists of an operational amplifier with a resistor R_1 connected from its output to its non-inverting input, and a resistor R_2 connected from the non-inverting input to a voltage V_0. An operational amplifier is an analogue amplifier which produces an output voltage proportional to the difference voltage between its two inputs, and approaches an ideal of having infinite differential gain, infinite input impedance at both inputs, and zero output impedance at its single output. An ideal operational amplifier also has negligible limitations regarding its input offset, bandwidth, slew-rate, latch-up, and noise, although in practice, these aspects must often be considered. Typically V_0 is ground, $0\,\text{V}$, or another constant voltage between the positive supply rail V_+ and the lower supply rail V_- to the amplifier. The inverting input of the amplifier is the input of the trigger circuit as a whole, and so the trigger circuit has a high (ideally infinite) input impedance.

Disregarding the unlikely possibility that the input voltage is precisely equal to the voltage applied by the resistor chain R_1 and R_2 to the non-inverting input, there will always be some voltage difference between the two amplifier differential inputs. Therefore, the high amplifier gain would imply a large voltage at the output, but in practice, the amplifier output stage will saturate, and simply give an output voltage almost equal to either its positive supply or its negative supply, dependent upon the polarity of the difference voltage at the inputs. Note that the applied *positive* feedback in this circuit is the opposite of that required to give well-controlled linear operation, which needs *negative* feedback. In the Schmitt trigger circuit, whatever the value of input voltage, the amplifier will always be overdriven, so that its output voltage can only be at either one or the other supply rail value. In the context of Digital Systems, of course, the two possible output voltages are arranged to be equal to two voltage levels recognised by the digital IC technology used in the subsequent parts of the circuit, usually $0\,\text{V}$ and $5\,\text{V}$. Since the trigger circuit has two possible output voltages, controlled by the single input, it follows that at a certain trigger voltage applied to the input, the circuit output changes its state from one of the possible output voltages to the other possible output voltage.

In fact in a Schmitt trigger the analogous transition in the reverse direction takes place at a slightly different input voltage, so that the circuit shows 'hysteresis' by having an 'overlap' region of input voltages within which the output voltage depends upon the direction from which the input voltage entered the 'overlap' region between the two threshold voltages. To quantify this, when a *low* voltage is applied to the circuit input, the amplifier will be overdriven such that its output will be at the voltage of the *high* supply rail, V_+. Therefore, the voltage applied to the non-inverting input will be $V_0 + (V_+ - V_0)R_2/(R_1 + R_2)$ and the input voltage must be *greater* than this in order to change the state of the output voltage. Once this occurs, however, the amplifier will be overdriven in the opposite direction, so the output voltage will be the same as that of the other supply rail, V_-. So, now the voltage applied to the non-inverting input will be $V_0 + (V_- - V_0)R_2/(R_1 + R_2)$. Therefore, to change the output voltage back to its first value, the input voltage must be reduced below this new threshold value. Since V_- is less than V_+, the new threshold voltage is less than the first by the difference of $(V_+ - V_-)R_2/(R_1 + R_2)$, which is the *hysteresis* of the circuit.

10.3 Schmitt input gates

The internal circuitry of a 'Schmitt input gate' is based upon the well-known 'Schmitt trigger' circuit described in section 10.2 above. A 'Schmitt inverter gate' may be regarded as a Schmitt trigger circuit giving outputs at the correct voltage levels to drive subsequent digital logic gates correctly. A 'Schmitt input buffer gate' is similar but has an extra stage of logic inversion prior to the output. In each case, varying analogue voltages, not necessarily corresponding to the logic level specifications for the logic family concerned, may be applied to the Schmitt input. The output will take the appropriate logic level (0 or 1) according to whether the input is above or below the relevant threshold voltage fixed by the manufacturer. Other types of Schmitt input gates, with more complex logic functions, are also available. The threshold voltage is usually fixed between the normal logic level voltages, so that a Schmitt input gate will operate correctly if it is driven by a conventional logic gate rather than an analogue voltage source.

On circuit diagrams, a Schmitt input gate is indicated by the gate symbol for the corresponding conventional logic gate, but with the addition of the special symbol $\varPi$ drawn adjacent to an input or centrally inside the gate symbol as appropriate. This special symbol has a stylised derivation from the letter 'S' and from a diagram indicating hysteresis between input and output voltages. Examples of circuit symbols for typical Schmitt input gates are shown in Figures 10.2(a) and (b).

One typical use of such a Schmitt input gate is shown in Figure 10.2(c). In this circuit, an *RC* network is used to convert an input logic waveform into a waveform with *slow edges*, that is, a signal where the rising and falling edges are governed by the usual exponential law with time constant $\tau = RC$. It is bad practice to apply such a waveform to a conventional logic gate because these slow edges can potentially cause severe problems such as oscillations at the output or out-of-specification output voltages with gates not specifically designed to handle slow edges. However, Schmitt input gates are expressly designed and intended to be able to handle slow edges, and so the output state changes after a delay time determined by the precise value of the threshold voltage and the time constant $\tau = RC$.

The Schmitt delay circuit can be developed into a simple oscillator circuit for producing a repetitive waveform, as shown in Figure 10.3(a). In this circuit, a logic transition at the output of the inverter is fed back to the input of the inverter, but the *RC* network connected to the Schmitt input causes a delay before the complementary transition occurs at the output. Because of the non-zero hysteresis at the gate input, the gate input voltage varies exponentially between the two threshold voltages, and the output oscillates between logic low and high levels indefinitely, as shown in Figure 10.3(b). The frequency of the waveform produced depends upon the values of R and C determining the delay time. However, a serious disadvantage of this circuit is that the exact oscillation frequency also depends upon the precise input threshold voltages and the difference between them, which are usually not precisely known.

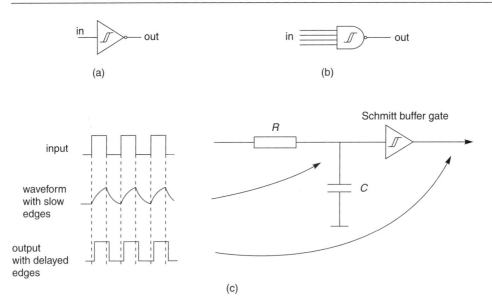

Figure 10.2 *(a) Circuit symbol for a Schmitt inverter (e.g. as in IC type 74LS19) (b) Circuit symbol for a Schmitt NAND gate (e.g. as in IC type 74LS18) (c) Simple digital signal delay generator using a Schmitt input gate*

A simple oscillator circuit that overcomes this difficulty to some extent is shown in Figure 10.3(c). Gate G1 must be a Schmitt input gate; G2 need not have a Schmitt input, although in practice it will usually be part of the same IC package as gate G1 and so will also have a Schmitt input. The voltage waveforms in this circuit are shown

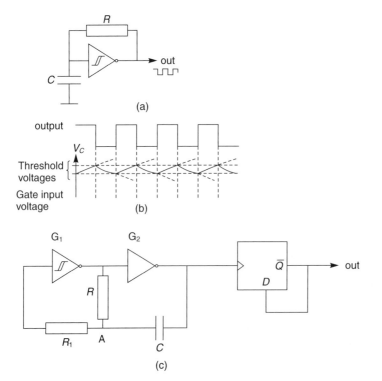

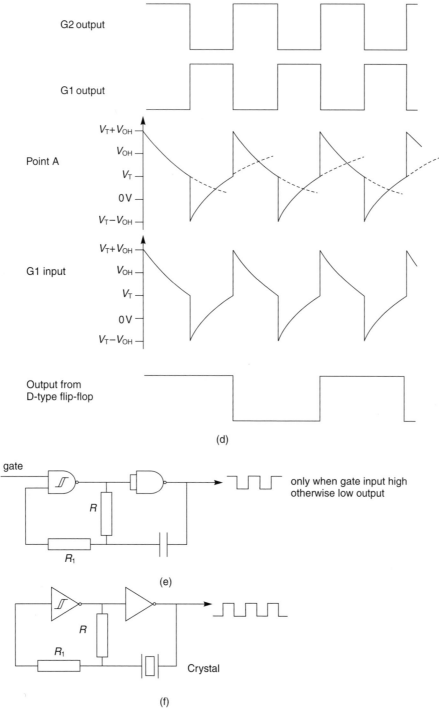

Figure 10.3   *(a) Simple low-precision RC oscillator circuit using a single Schmitt inverter (b) Voltage waveforms in the single-inverter oscillator circuit (c) Simple oscillator circuit using two Schmitt inverters (d) Voltage waveforms in the double-inverter oscillator circuit (e) Gated oscillator circuit (f) Precision oscillator using a resonant quartz crystal*

in Figure 10.3(d). Suppose that the output voltage of gate G2 is currently at logic high level, V_{OH}; because of the inversion in gate G2, this requires that the output of gate G1 is at logic low level. Therefore, the voltage at point A will fall, according to the usual exponential decay law, until the input voltage to gate G1 falls below the threshold value. For simplicity it will be assumed here that gate G1 has a high input impedance (much greater than R_1) so that it draws no current from the RC network. It follows that the voltage at its input is the same as the voltage at point A, and the precise value of resistor R_1 is unimportant. It will also be assumed here for simplicity that the Schmitt threshold voltages, V_T, for rising and falling edges, are exactly half of the logic high voltage level, with zero hysteresis, and also that the logic low voltage level $V_{OL} = 0V$. At the instant where the input of gate G1 falls below its threshold voltage, the voltage across the capacitor will be half of the logic high voltage, or $\frac{1}{2}V_{OH}$. Then the output of gate G1 goes high and so the output of gate G2 goes low. At this point, the voltage at point A is now *negative*, with value $-\frac{1}{2}V_{OH}$, and it starts rising towards the logic high value with time constant RC. When the voltage at the input to gate G1 rises above its threshold value $V_T = \frac{1}{2}V_{OH}$, then the output of gate G1 changes to logic low, the output of gate G2 changes to logic high, and the voltage at point A is immediately $+V_{OH} + V_T = 3V_{OH}/2$ (i.e., the output of gate G2 plus the capacitor voltage of $\frac{1}{2}V_{OH}$). The whole cycle can now repeat indefinitely. It is straightforward to calculate the time taken for the first half of the cycle, during which the voltage at point A rises from $-\frac{1}{2}V_{OH}$ to the threshold value $+\frac{1}{2}V_{OH}$ with an aiming voltage of $+V_{OH}$ and time constant RC; the result is $t_0 = RC \ln(3)$. The time taken for the other half-cycle of the operation is the same, so that the oscillation frequency is given by $f = 1/[2RC \ln(3)]$. The oscillation frequency is now relatively insensitive to the actual threshold voltage values. However, the precise mark-to-space ratio of the signal at the output of gate G2 depends upon the exact threshold voltage values, and so if it is important to have a 1:1 mark-to-space ratio then the output of gate G2 can be taken through a divide-by-2 circuit as shown in Figure 10.3(c).

In practice, a Schmitt input NAND gate might be used in place of one of the inverters, as shown in Figure 10.3(e), to produce a 'gated oscillator'. This only produces a repetitive waveform when the additional input to the first NAND gate is held at logic high level. When the gating input is held low, the oscillator output is held low. Since the precise oscillation frequency still depends to some extent upon the voltage threshold values, this type of RC oscillator is only suitable for applications where the utmost frequency stability and accuracy is not required. A similar circuit, as shown in Figure 10.3(f), using a quartz crystal which resonates at a frequency precisely specified by its manufacturer, will usually be employed in cases where excellent frequency stability or precision is of paramount importance.

10.4 Digital-to-analogue conversion

Digital-to-analogue conversion (abbreviated to D/A, D-A, or D-to-A conversion) is frequently required in a digital system used to control some external analogue circuitry. The D/A converter (or DAC) gives a controlled analogue output voltage or, in certain specialist applications, a controlled analogue output current or another circuit parameter such as resistance, whose value corresponds to an input digital word. Here it will be assumed that the digital input is a conventional positive base 2 integer.

If the digital input is in some other numerical format, it can be converted to base 2 as described elsewhere in this text. If the output is required to be bipolar, that is, the numerical input may have either positive or negative polarity to produce either positive or negative output voltages or currents, then this may be most easily handled by two separate converters, one for each of the output polarities and only one of which is allowed to be active at any one time. Alternatively, there are some bipolar D/A converters available commercially.

Most D/A converters are based upon a precision resistor network containing a network of standard resistor values each of which can be switched into or out of circuit according to which bits are set in the input binary word. In the popular 'binary weighted' resistor network shown in outline in Figure 10.4, the resistors have values of R, $2R$, $4R$, $8R$, and so on, in multiples of powers of 2, and each resistor is switched into circuit as its own associated bit is set equal to 1. When the input bit associated with any resistor is equal to 0, that resistor is not switched into the circuit but instead is replaced by a short circuit of zero resistance. To produce an analogue output voltage, all that is then necessary is to drive a certain standard but constant current through the variable resistor network, and the voltage dropped across the entire resistor network is then the analogue output voltage required.

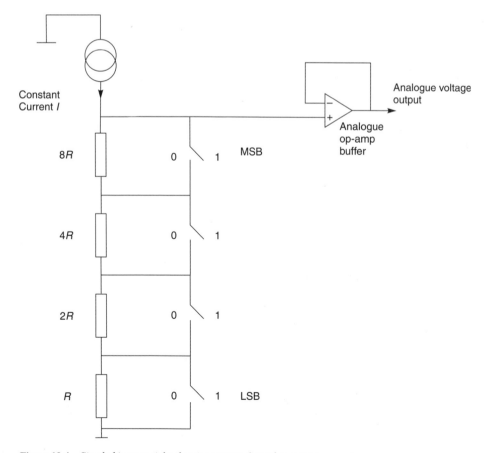

Figure 10.4 *Simple binary weighted resistor network used as a D/A converter*

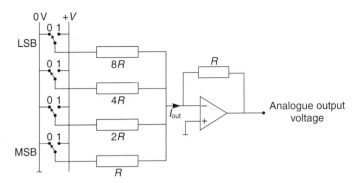

Figure 10.5   *Alternative design of simple binary weighted resistor network used as a D/A converter*

The output voltage may be *scaled*, or multiplied by a constant factor so that its greatest and least values are within the limits required for the particular application intended by adjusting the value of the constant current used. In practice, this output voltage is usually subsequently buffered, using an analogue 'voltage follower' circuit which may include a small amount of extra gain for further scaling purposes, so that the circuit is more tolerant of whatever circuitry is connected to the output of the D/A converter.

Alternatively, the resistors may be connected, as shown in Figure 10.5, together with a stabilised voltage source $+V$. Here the network produces an analogue output current I_{out} equal to the sum of binary-weighted contributions, and an operational amplifier buffer circuit is used to give an analogue voltage output.

The *resolution* or *precision* of such a D/A converter is defined as the smallest output increment possible, divided by the difference between the maximum and minimum output values. The *accuracy* or *linearity* of the converter is defined as the difference between the actual output and the expected output value, measured with any specified digital input value. Typical values of accuracy for commercial A/D converters are of the same order as the output corresponding to the least significant bit.

This binary weighted resistor network suffers from the disadvantage that the resistor with the largest weighting in the network must be manufactured to a precision such that the likely error in its value is comfortably less than the significance of the resistor with the smallest weighting in the network, if the conversion is to be accurate around the values where the most significant resistor is being switched in and out. For example, suppose an 8-bit D/A converter is to be designed using a resistor chain with a smallest resistor value of 10Ω. The other resistors in the chain will take values of 20Ω, 40Ω, 80Ω, 160Ω, 320Ω, 640Ω, and 1280Ω. Further, suppose that the 10Ω resistor is manufactured to a tolerance of $\pm10\%$ or $\pm1\Omega$. To match this precision, all the other resistors must also be manufactured to a precision of $\pm1\Omega$ as otherwise there is little point in making the 10Ω resistor this precise. The most stringent requirement is therefore placed upon the 1280Ω resistor which must be manufactured to a precision of $\pm(1/1280) \times 100\% \approx \pm0.08\%$. Such precision is extremely expensive to achieve. Other designs of resistor networks can be used to circumvent this difficulty in precision D/A converters.

The conversion rate of such converters is limited only by the bandwidth of the analogue parts of the circuit and the response time of the digital parts of the circuit, and so in principle may be very fast. Although Figures 10.4 and 10.5 show mechanical switches for simplicity, normally solid-state analogue switches would be used, as these operate much faster and are more reliable than mechanical reed switches. In practice, D/A converters with limited resolution that convert only small numbers of digital input bits are available with conversion rates up to 1GSa/s or even faster (where Sa/s stands for 'Samples per second', indicating the D/A conversion rate). A D/A converter with a conversion rate of 1GSa/s must produce an analogue output voltage that can change from one value to another within approximately 1ns. This time, akin to the 'rise time' of a pulse circuit, is known as the 'settling time' of the D/A converter. D/A converters with precisions of up to 20 bits or even greater are available at lower conversion rates, corresponding to output changes within around 20 μs.

10.5 Analogue-to-digital conversion

The Schmitt input gate (section 10.3) may be regarded as a special case of the reverse conversion, that is, analogue-to-digital conversion, producing only a single bit output (0 or 1) in response to an analogue varying voltage at its input. A more sophisticated analogue-to-digital converter (or ADC, also called an A/D, A-D, or A-to-D converter) extends this principle to produce a binary integer, typically of 8, 16 or another number of bits in parallel form at its output in response to an analogue voltage at its analogue input. In many ways, an A/D converter resembles a rudimentary digital voltmeter, although its input impedance is not likely to be especially high, there will often only be one voltage conversion range or, at best, external components must be used to change the range of allowable input voltages, and there is no display, apart from what may be added externally. The manufacturer of the A/D converter specifies the maximum and minimum analogue voltages that may be applied to the analogue input for correct conversion to occur. Usually the maximum input voltage will correspond to, and will be converted to, the largest binary integer that can be expressed with the number of bits available at the converter output. The minimum voltage will usually be either 0 V, or a negative voltage of the same magnitude as the maximum allowable input voltage, and the corresponding digital outputs will therefore usually be either binary 0 or a 2s complement integer indicating negative values.

A number of techniques has been developed for designing A/D converters in practice, each of which has advantages and disadvantages. A detailed examination of all the various techniques is outside the scope of this text, but the three main types are described in the following sections. Many of these techniques use *analogue comparators* (see Figure 10.6) which have some characteristics in common with both Schmitt triggers and with conventional operational amplifiers. The usual type is an analogue *voltage* comparator with one output and two inputs, ideally having high input impedance like those of an operational amplifier. Within the limits specified by the manufacturer, any analogue voltages may be applied to the two inputs. The comparator gives a logic 1 output if its *non-inverting* input is at a *greater* voltage than that at its *inverting* input, and gives a logic 0 output if its *inverting* input is at a *greater* voltage than that at its *non-inverting* input. The comparator differs significantly from

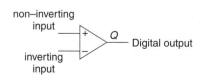

Figure 10.6 *Simple analogue comparator*

an operational amplifier because it is only able to output voltages at the two logic-compatible levels, usually 0 V and 5 V. Sometimes, the voltage at the comparator's inverting input is referred to as the 'threshold voltage', and the output is logic 0 or 1 depending upon whether the non-inverting input is at a voltage less than or greater than the threshold voltage. Usually, comparators are designed to have very low hysteresis values measured at their inputs. In many instances, one of the two inputs is kept at a constant or maybe a slowly varying voltage, and the comparator's job is then to indicate when the other input voltage rises above or falls below this 'reference'.

10.6 Flash converters

These types of A/D converters are conceptually the simplest of all. An A/D converter is required to produce one of a number of possible binary outputs, depending upon the input voltage; therefore, there is a certain input voltage range over which it will produce each unique possible output, and threshold voltages at which the output changes from one value to the next. A 'flash converter' consists of a number of analogue comparators, each set to trigger at a different one of these thresholds. An input voltage V will therefore trigger all the comparators that have threshold voltages less than V, and will *not* trigger the rest of the comparators that have thresholds greater than V. It follows that the outputs from the comparators indicate the value of the input voltage in a manner that can be interpreted by a logic system, but unfortunately not in a form that is particularly easy to use in subsequent circuitry; some further logic is needed to derive a conventional base 2 integer from all the trigger circuit outputs.

The basic principle is illustrated in Figure 10.7 and in the following table, showing a 2-bit conversion needing 3 comparators:

Input voltage/V	Comparator outputs			Output (base 2)
	C_2	C_1	C_0	
> 2.5	1	1	1	11
between 1.5 and 2.5	0	1	1	10
between 0.5 and 1.5	0	0	1	01
< 0.5	0	0	0	00

In this very simple example, C_0 is the output of the comparator set to the lowest threshold (0.5V), C_1 corresponds to a threshold of 1.5V, and C_2 corresponds to the highest threshold (2.5V). Clearly the MSB of the output word is equal to C_1 and, using the 4 unlisted 'don't care' terms in C_0, C_1 and C_2, a 3-variable K-map shows that the LSB of the output word is equal to $C_2 + \bar{C}_1 C_0$. The conversion logic required is similar to that used in a priority encoder (see section 5.13).

A flash converter having an n-bit binary integer output requires a total of $2^n - 1$ separate comparators as well as appropriate conversion logic, and a logic buffer for each output bit. Clearly, precision applications, for large values of n, require extremely large numbers of comparators, and so the flash conversion principle is only practical for modest numbers of bits output (typically up to $n = 10$, requiring 1023

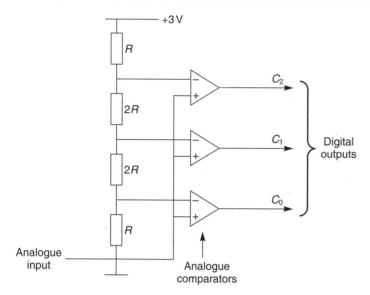

Figure 10.7 *Basic principle of a flash A/D converter, using three comparators*

comparators). However, flash converters operate considerably faster (hence their name) than any other type of A/D converter because of their simplicity, since all their components operate simultaneously, and also as there is no fundamental limitation upon the speed of their separate components.

General minimised logic for a flash converter

A flash converter producing an n-bit output requires logic to convert $2^n - 1$ comparator outputs to a conventional binary integer. Each of the binary output bits is set equal to 1 within a certain number of sub-ranges of the input voltage (and is cleared to 0 otherwise). Therefore, the minimised Boolean expression for each output bit consists of the Boolean OR of the conditions specifying each sub-range relevant to that particular output bit. The condition specifying most of the individual sub-ranges is of the form $S_{i,j} = C_{l,i,j} \cdot \bar{C}_{h+1,i,j}$, for sub-range number i relevant to output bit number j, where $C_{l,i,j}$ is the output of the lowest comparator (corresponding to the smallest analogue voltage value) within that sub-range, and $C_{h+1,i,j}$ is the output of the comparator that is one step higher than the top of that sub-range. The exception to this rule is that since the highest sub-range always includes the output binary integer with all bits set, there is therefore no next higher comparator; hence, the condition specifying the top sub-range for bit number j is simply $S_{\text{top},j} = C_{l,\text{top},j}$. Therefore, the general minimised logic for output bit b_j in a flash converter is of the form $b_j = C_{l,\text{top},j} + \sum_{i \neq \text{top}} C_{l,i,j} \cdot \bar{C}_{h+1,i,j}$. For the special case of the MSB of the binary output, there is only a single sub-range (i.e. the upper half of the table) and this formula simplifies to $b_{\text{MSB}} = C_{l,1,\text{MSB}}$. The output expressions for the simple case of a 2-bit/3-comparator flash converter shown in Figure 10.7 conform to these general rules.

10.7 Integrating A/D converter types

In this type of A/D converter, an operational amplifier is used as an integrator to integrate the input voltage over a specified time interval. In one well-known arrangement, known as the 'dual-slope integrating converter' and developed by Schlumberger, the input voltage is integrated for *one period, T,* of the AC mains supply. In Europe and other areas which have a 50 Hz supply, $T = (1/50)s = 20ms$; in North America and other areas which have a 60 Hz supply, $T = (1/60)s = 16.666ms$. The principle is illustrated in Figure 10.8. For clarity, a mechanical switch S is shown, but in practice, a mechanical switch would not operate fast enough and an active solid-state switch would be used instead. The input voltage is assumed for the moment to be constant. The integrator output starts from zero, and so after a time T its output has reached a voltage V_T proportional to the input voltage. The advantage of integrating over one mains time period is that any mains interference impressed upon the input voltage should, in principle, have *no effect* whatsoever upon the final output voltage. In effect, the input voltage is averaged over precisely one mains cycle.

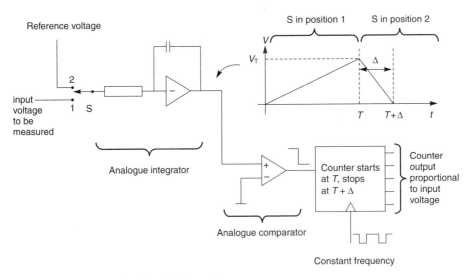

Figure 10.8   *The principle of the dual-slope A/D converter*

Any superimposed interference synchronous with the mains supply frequency may be regarded as a sine wave of a certain amplitude and phase shifted relative to the mains supply. This may be represented as a sum of sine and cosine voltage terms, of amplitudes V_s and V_c respectively, and having frequencies identical to the mains supply frequency, which will be integrated over one time period as follows:

$$\int_0^{(1/f)} [V_c \cos(2\pi ft) + V_s \sin(2\pi ft)]\mathrm{d}t$$
$$= (1/2\pi f)[V_c \sin(2\pi ft) - V_s \cos(2\pi ft)]|_0^{(1/f)}$$
$$= (1/2\pi f)[V_c \sin(2\pi) - V_s \cos(2\pi) - V_c \sin(0) + V_s \cos(0)]$$
$$= (1/2\pi f)[V_c.0 - V_s.1 - V_c.0 + V_s.1]$$
$$= 0,$$

giving no effect on the final output voltage, and interference at synchronous harmonics of the mains supply will similarly give zero contribution to the integrated output. The results of Fourier analysis may be used to show that any arbitrary interference waveform, *provided* it is mains-derived and therefore repetitive (but not necessarily sinusoidal) at the mains supply frequency, will give zero contribution to this integration.

At the end of this time period, typically timed with a crystal-controlled digital counter for accuracy and repeatability, a precise reference voltage of *opposite* polarity is immediately applied to the integrator without resetting it, and the time taken for the integrator output to return to zero is measured using the same digital counter. This time, Δ, is a measure of the input voltage required to be measured, and is transferred to the digital output. The standard reference voltage applied during the second integration will usually be equivalent to the maximum allowable analogue input voltage that can be applied during the first integration period, so that the second integration period is less than or equal to one mains time period T. Therefore, the complete conversion takes a maximum of $2T$ (i.e. 40 ms or 33.333 ms for 50 Hz or 60 Hz supplies respectively) and the maximum number of A/D conversions per second is half the mains supply frequency (i.e. 25Sa/s or 30Sa/s respectively). This type of A/D converter has a number of great advantages. It has high immunity to mains-borne interference; also, its accuracy depends only upon the stability of the timer and counter circuits, the accuracy of the standard voltage, and the stability of the integrator components over each measurement cycle. A highly stable timer is easy to achieve using a resonant crystal oscillator. Therefore, the dual-slope integrator is widely used in both general-purpose and high-precision digital voltmeters. Its disadvantage is that it is relatively slow, since the maximum number of conversions per second is seriously limited, and for faster conversions other types must be used. Sometimes a slow converter, such as an integrating type, is to be used even though it may be required that the A/D conversion is carried out at a precise time rather than by an integration over a substantial period, and in this case the converter must be preceded by a 'sample and hold' circuit that samples the incoming analogue voltage at a definite time and feeds this sampled analogue voltage to the converter.

10.8 A/D converter types using an embedded D/A converter

Another class of A/D converter is built around a D/A converter which undertakes the opposite conversion to that actually required (see Figure 10.9). Every time an A/D conversion is needed, a conventional binary counter is cleared and starts counting up from the starting value of $(0)_2$. The digital outputs of the binary counter are directly connected to the embedded D/A converter, so the output from this converter is an analogue voltage that rises steadily (in 'staircase' fashion) from 0 V. An analogue comparator continuously compares the output voltage from the embedded D/A converter with the analogue input voltage; at the *exact instant* that the output voltage from the embedded D/A converter has risen above the analogue input voltage, the counter's digital output value is stored in a set of D-type flip-flops. This D-type register must therefore now contain the digital equivalent output of the analogue input.

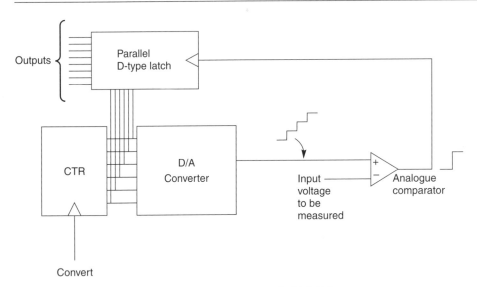

Figure 10.9   *The principle of the A/D converter using an embedded D/A converter*

This type of converter is quite slow as each conversion takes at least the time required for the counter to count from $(0)_2$ to the binary equivalent of the input analogue voltage (and potentially to the maximum binary output). However, the complexity of the flash converter is avoided. This type of A/D converter may operate faster than an integrating type, since the conversion period is not tied to the mains supply cycle, but does not have the advantage of rejecting mains-synchronous interference. In principle, the integrating type of A/D converter can also operate faster than is necessary to suppress mains-borne interference, but this would discard the great advantage of this type of converter.

There are two important variations of this type with an embedded D/A converter. In the first variation, the counter is *not* reset to $(0)_2$ at each conversion request, but instead the comparator is used to indicate whether the analogue input voltage is greater or less than the output from the embedded D/A converter retained from the previous conversion, and the counter then counts up or down, as appropriate, from the previous count value. If the input voltage has not changed very greatly from the time of the previous conversion, i.e. the input voltage is *slowly varying* and/or A/D conversions are required on a regular basis, then this modification offers the advantage of reduced conversion time. However, if the input voltage cannot be assumed to be slowly varying, or if conversions are only required on an irregular basis, then in principle the time taken for a new A/D conversion will still be equal to the time taken for the counter to count from $(0)_2$ to the maximum binary count value, i.e. the *worst case* conversion time. This disadvantage may be alleviated by arranging for internal conversions to be made continuously, but when an external conversion request is received, the conversion cycle restarts immediately with the counter starting from its current value.

In the second variation of the basic A/D converter with an embedded D/A converter, a binary counter is not used but instead is replaced by an *n*-bit digital storage register in which each of the *n* bits may be independently set or cleared under control of some extra logic. The basic principle is illustrated in the block circuit diagram shown in Figure 10.10, and a typical voltage waveform at the output of the embedded D/A

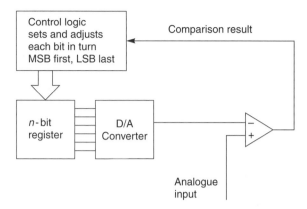

Figure 10.10 *The principle of the 'successive approximations' A/D converter*

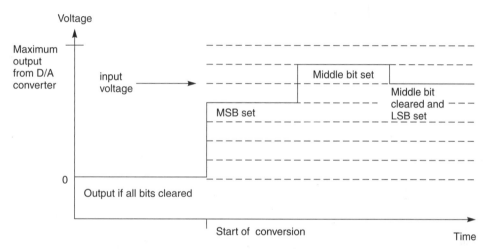

Figure 10.11 *Typical waveform produced by the embedded D/A converter in a 'successive approximations' A/D converter*

converter is shown in Figure 10.11 in the case of a simple 3-bit converter. In this simple example, the 3-bit D/A converter has $2^3 = 8$ possible output voltages ranging from 0 up to 7 times its basic output unit. Rather than having a counter starting to count from $(0)_2$ towards the final base 2 digital equivalent of the analogue input, the control logic first clears to 0 all the bits of the storage register except the MSB which is set to 1. Therefore, the output of the embedded D/A converter will be close to half its maximum value. The output is not exactly half because, as an example for the case of a 3-bit register, the maximum output from the embedded D/A converter is $(111)_2 = (7)_{10}$, whereas just setting the MSB to 1 as set up by the control logic gives the output $(100)_2 = (4)_{10}$. Then, the comparator is used to determine whether the embedded D/A converter is giving an output greater or less than the analogue input; if the embedded D/A converter is giving an output greater than the analogue input then the MSB is cleared to 0, but if not, the MSB is left at the value 1. At this stage, therefore, the control logic has adjusted the MSB to the correct value, and so moves on to the next most significant bit. The control logic sets this bit to 1 and again uses the comparator to determine whether the embedded D/A converter is now giving an output greater or less than the analogue input. If the embedded D/A converter output is greater than the analogue input, this bit is cleared to 0; if less than the analogue input, then this bit is left at 1. The control logic then moves on to the other bits in turn in order of their numerical significance, i.e. their position in the binary integer. This type of A/D converter is called a 'successive approximations' converter, as the converter is successively making better and better approximations to the final value of the digital output.

The advantage of the successive approximations A/D converter over the type using a counter is that it implements a *binary search* for the digital equivalent value of the analogue input, rather than a sequential search starting either from $(0)_2$ or from the previous output value. A binary search is a much more efficient method of finding an unknown value than a sequential search, and so in general the conversion times using a successive approximations converter will be much less than those achieved with a counter-type converter. This may be understood by examining the number of voltage comparisons needed for each A/D conversion. For an n-bit output, a successive approximations converter needs to undertake n comparisons per conversion, one for each bit. By contrast, the number of comparisons per conversion required by the counter type will range from 1 at minimum up to a maximum of $2^n - 1$. Since the number of required comparisons varies widely in the case of the counter type, occasionally the counter type will be faster than the successive approximations type. However, if *any* allowable value of the analogue input voltage is equally likely, and if the counter is reset for each conversion, then *on average* the number of comparisons required by the counter type is $(2^n)/2 = 2^{n-1}$, which will always be greater than the n comparisons required by a successive approximations converter, assuming that A/D conversions of more than $n = 2$ bits are needed.

The counter type where the counter is *not* reset for each conversion may operate much faster under favourable conditions, i.e. if conversions are required so frequently that the required output changes by no more than $(1)_{10}$ or $(2)_{10}$ or so at each conversion. However, if the input is varying so rapidly that there is no similarity between successive input voltages at the sampling times, or if *any* value of the analogue input voltage within the allowable conversion range is equally likely, then the *average* number of comparisons required per conversion will be approximately $(2^n)/4 = 2^{n-2}$, the same as for a counter type where the counter is reset to a value midway between minimum and maximum counts prior to either up or down counting under control of the comparator.

Note that in the case of the successive approximations A/D converter (unlike the type with a counter driving an embedded D/A converter), there is no advantage to be gained in not clearing the register before each conversion; each conversion still requires each of the n bits to be examined in turn. The equal conversion times from a successive approximations converter is a major advantage in certain applications where the variable conversion times from the counter types (and potentially very long conversion times when the counter must count over all or nearly all of its range) cannot be tolerated.

The familiar Compact Disc (CD) digital audio format stores analogue audio signals in the form of an optically-readable stream of digital values sampled regularly at 44.1kSa/s. It requires 16-bit conversions, A/D in recorders or D/A in CD players. However, professional-quality digital recorders will record at 20 or 24 bits (or greater resolution) and at greater conversion rates, partly so as to be compatible with improved digital formats, but mainly so that, on copying the master studio recording to the final CD, the recording level (i.e. the amplitude of the final analogue signal) may be increased to some extent if necessary while still maintaining at least 16-bits resolution in the digital information recorded on CD. The design of such high-resolution converters and their associated circuitry is not straightforward, because of the need for the analogue parts of the system to operate to the same precision or better. Although the standard CD format provides only 16-bit sampling precision, some CD players use various advanced techniques to increase the effective number of bits available for converting to the analogue signal, by reconstructing extra digital information according to some assumptions made about the nature of the audio signal, and the use of other digital signal-processing techniques. These players undertake D/A conversion at higher rates (known as *oversampling*), in attempts to increase the accuracy of the reconstructed audio signal. Similar principles are used in the design of digital audio tape (DAT) recorders and players. The required conversion rate prevents the use of integrating types of A/D converter in recorders, and the precision necessary prevents the use of flash converters. On the other hand, for digital video discs and tapes the bandwidth required is much greater (5.5MHz bandwidth for a typical conventional video signal, requiring a minimum sampling rate of 11MSa/s) but the necessary precision is poorer (8 bits per sample is usually ample for video information, because of the greater noise tolerance of the eye than the ear). Therefore, for digital video, if no data compression techniques are used, it would be necessary to record and reproduce of the order of 88 Mbit/s plus the bit rate needed for the accompanying audio and control signals, compared to the 1.4Mbit/s for 2-channel 16-bit audio at 44.1kSa/s. For this reason, the technical requirements for digital video are considerably more exacting than for digital audio, and commercial digital video systems have only been available since the mid-1990s. Even so, current digital video systems use some data compression techniques to reduce the necessary bit rate, whereas domestic digital audio has been a practical reality since the early 1980s. At the highest conversion rates, self-contained IC A/D converters capable of operating at 1GSa/s with 6- or 8-bit resolution are now readily available.

10.9 Shaft encoders and linear encoders

A shaft encoder is a sensor device that can be attached mechanically to a rotating shaft, and electrically connected to a logic system in order to feed information to the logic system regarding the rotation of the shaft. A linear encoder is a similar device that senses the linear motion of a slider, relative to a fixed body of the sensor. Shaft encoders are increasingly being used to sense the rotation of the manual controls in consumer items such as audio and video equipment, following their widespread employment in laboratory instruments for many years. The design concept is that the user turns the control knob fixed to a shaft encoder which sends signals to a microcomputer controlling the instrument, rather than the user actually turning the shaft of a variable resistor, capacitor, or other variable component. By this means, one control knob may be used to adjust several equipment functions. Linear encoders are widely used in the control of robots, x-y plotters, and other situations where linear motion must be sensed accurately for computer control. There are two main types of each encoder, 'absolute encoders' and 'incremental encoders'. The digital logic aspects of both shaft and linear encoders are similar, and these types only differ substantially in the mechanical arrangements used. The 'absolute' and 'incremental' types are, however, fundamentally different in their philosophy and in their digital logic aspects.

10.10 Sensing of motion

The precise manner of sensing the rotation or linear motion is of interest here insofar as there are various ways of achieving the same end. Usually the motion, rotating or sliding, is sensed by using an optical arrangement consisting of a light source, two or more optical detectors such as photodiodes or phototransistors, and an intervening 'screen' or 'reticle' with alternating opaque and transparent areas. When the opaque areas are aligned between source and detector, the detectors detect no light, and when the slider shifts or the shaft rotates slightly then a transparent area is substituted and the appropriate detector detects light transmission through the reticle. The basic arrangement for a linear incremental encoder, for example, is shown in Figure 10.12. In practice, a more sophisticated optical arrangement involving correct focusing of the optical beams would be used to optimise the optical performance of the unit. However, even when using precise optical imaging techniques, because of optical limitations the smallest distance between two points on the reticle that can be distinguished easily is of the order of one wavelength of the optical radiation employed, so the shortest wavelength radiation possible is usually used. The fabrication of the reticle, upon which the accuracy and resolution of

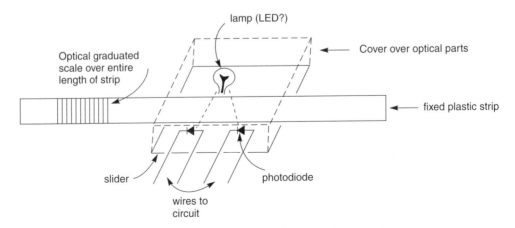

Figure 10.12 *Sketch of the basic mechanical arrangements for a linear incremental encoder*

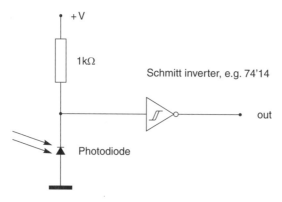

Figure 10.13 *Simple circuit for interfacing a photodiode to a logic system*

the encoder rests, may be undertaken using processes such as silk-screen printing for low-precision applications, or by high-precision manufacturing techniques such as microphotolithography.

Electrical interfacing between the optical detectors and the digital system is straightforward, remembering that optical detectors may not provide output voltages equal to the standard logic levels, so that Schmitt input gates are necessary. A typical circuit is shown in Figure 10.13, though the exact details will depend crucially upon the actual optoelectronic and optical components used; it may be necessary to include extra analogue amplification between the photodiode and logic gate, for example. Often in linear encoders the reticle is stationary and the optical source and detectors move with the slider, as shown in Figure 10.12; in shaft encoders the optical components are generally stationary and the reticle rotates with the shaft so that continuous shaft rotation by an unlimited number of revolutions is usually possible with such a system.

Other methods have also been used to detect the motion in encoders, such as primitive mechanical arrangements using toothed wheels operating mechanical switches, and more sophisticated magnetic encoders. The magnetic type is widely employed in electronic ignition systems in automotive engines where a magnetic toothed wheel is fixed to the camshaft adjacent to a stationary sensing coil. As the toothed wheel rotates, its magnetic field induces a varying voltage in the sensing coil. The shapes of the wheel and coil are such that the induced voltage changes rapidly at the instant when the camshaft rotates through the correct ignition point. This rapid voltage change is detected by the electronic ignition system and is used to trigger the ignition spark. These encoders are similar to motional feedback sensors, commonly attached to electric motors such as those used in washing machines, that produce an AC signal of frequency directly proportional to the motor speed. This signal is then fed to the motor speed controller. Magnetic encoders are often preferred over optical sensors in oily or dusty environments such as engines and motors where optical sensors would rapidly cease to function correctly as their optical paths become obstructed. Magnetic encoders have also been built using Hall effect sensors which sense magnetic fields directly, using the Hall effect in a semiconductor. In all these cases, the method of interfacing to the digital logic system is dependent upon the electrical characteristics of the technology used.

10.11 Absolute encoders

An absolute encoder is an encoder giving a digital output of a binary word indicating the current position of the shaft or slider. Usually there are as many optical detectors as there are bits in the final word required at the output, and the reticle contains opaque and transparent sections corresponding to the *Gray code* (see section 1.21) equivalent of the integer indicating the current position.

In this kind of encoder, even if power is temporarily removed from the digital system producing the output, the slider or shaft may be moved, and on restoring power the system will still give the correct indication of the current position because the reticle will have been moved mechanically and will still indicate the correct Gray code integer. This represents a major advantage of this kind of encoder over the 'incremental' type, but at the expense of requiring many more optical detectors.

Gray code must be used, rather than ordinary base 2, so that only one bit changes at a time as the motion continues. Otherwise, inaccurate readings could be obtained if ordinary base 2 were used. Consider the design of an encoder where the position of a shaft or slider is to be transmitted in binary form to a digital system. If conventional binary coding were used, then at certain positions of the moving shaft or slider several binary bits would have to change simultaneously. The problem in using conventional binary coding in an absolute position encoder is illustrated in the following table for a simple 3-bit system.

Step	Base 2 code			
	b_2	b_1	b_0	
$(0)_{10}$	0	0	0	
$(1)_{10}$	0	0	1	$\left.\begin{array}{l}\\ \\\end{array}\right\}$ b_1 and b_0 change, possible spurious readings 000 or 011
$(2)_{10}$	0	1	0	
$(3)_{10}$	0	1	1	$\left.\begin{array}{l}\\ \\\end{array}\right\}$ all bits change, several possible spurious readings
$(4)_{10}$	1	0	0	
$(5)_{10}$	1	0	1	$\left.\begin{array}{l}\\ \\\end{array}\right\}$ b_1 and b_0 change, possible spurious readings 100 or 111
$(6)_{10}$	1	1	0	
$(7)_{10}$	1	1	1	$\left.\begin{array}{l}\\ \\\end{array}\right\}$ all bits change, several possible spurious readings
$(0)_{10}$	0	0	0	

In this 3-bit system there are $2^3 = (8)_{10}$ possible encoded positions. Between the adjacent positions $(1)_{10}$ and $(2)_{10}$ two bits change. They must change exactly simultaneously, as otherwise two spurious readings shown may be given. The rest of the table is self-explanatory. For a shaft encoder, though not for a linear position encoder, all three bits also change as the shaft completes each full revolution between positions $(7)_{10}$ and $(0)_{10}$.

This required simultaneous changing of several bits is a serious problem because in practice the encoder cannot be made so precisely that all of the bits intended to change at each step do so absolutely simultaneously. Therefore, the bits required to change will do so over a small but non-zero range of position. The order that the bits change will appear to be random and in practice will be determined by the mechanical inaccuracies and continuing wear of the encoding system. Hence, motion between genuine positions may have a completely spurious reading(s) interposed between them, depending on the order in which the bits change. Note that these spurious readings are genuine inaccuracies because the encoder has *not* physically moved to the corresponding positions. Rather, these readings have come about because of shortcomings in the coding and mechanical systems used.

The solution to this problem is the adoption of *Gray code* rather than conventional base 2 coding in such applications. *Gray code* is a special binary coding where only *one bit* changes at each step of the count (see also section 1.21). Superficially, Gray code resembles conventional base 2 binary coding except that the ordering of the codes for successive steps is changed from the conventional base 2 order. Gray code is an *unweighted* code, because each bit position does *not* have an associated numerical value in the same manner as does base 2 binary coding. If values of a successful Gray code are plotted on a K-map, then the successive cells must trace out a locus on the K-map that moves only to *adjacent* cells (defined in the same manner as in section 3.8, i.e. for minimising Boolean functions) at each step, because then and only then can only one K-map variable change at a time. In addition, for a Gray code to be used for rotation encoder applications, the path must be *re-entrant*; that is, the cell for the final value must be adjacent on the K-map to the cell for the first value.

One of the most useful types of Gray code is formed by taking the Exclusive-OR of adjacent binary bits of the corresponding base 2 equivalent at each step. If the equivalent base 2 for any step of the code is composed of individual bits denoted b_k, where $k = 0$ corresponds to the least significant bit (LSB), and the Gray code is composed of individual bits denoted g_k, where $k = 0$ corresponds to the right hand bit, then the conversion from base 2 bits to Gray code bits is given by the equation

$$g_k = b_{k+1} \oplus b_k.$$

For an *n*-bit conversion, the binary bits are denoted b_0 (LSB) to b_{n-1} (the most significant bit, or MSB), and the value of b_n (one place more significant than the MSB) is taken as 0 (when required in the evaluation of g_{n-1}). The conversion table for a 3-bit Gray code, obtained by direct application of the above defining equation, is shown in the following table.

	Base 2 code			Gray code		
Step	b_2	b_1	b_0	g_2	g_1	g_0
$(0)_{10}$	0	0	0	0	0	0
$(1)_{10}$	0	0	1	0	0	1
$(2)_{10}$	0	1	0	0	1	1
$(3)_{10}$	0	1	1	0	1	0
$(4)_{10}$	1	0	0	1	1	0
$(5)_{10}$	1	0	1	1	1	1
$(6)_{10}$	1	1	0	1	0	1
$(7)_{10}$	1	1	1	1	0	0

Note that a characteristic of this particular Gray code is that the base 2 and Gray codes for $(0)_{10}$ and $(1)_{10}$ are identical, but that the base 2 and Gray codes corresponding to

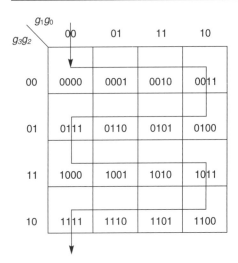

Figure 10.14   *Values of a 4-bit Gray code defined by $g_k = b_{k+1} \oplus b_k$ plotted on a K-map*

greater base 10 integers differ. Also, at each step of this particular Gray code, the next Gray code is formed by complementing the most right-hand bit possible that gives a code *not* used previously. So, for example, Gray code 000 is followed by 001, after complementing the most right-hand bit. Complementing the most right-hand bit again gives code 000 which is not the next Gray code as it has appeared previously in the sequence, but complementing the middle bit gives the correct next Gray code 011. In fact, there are many forms of Gray code, which can be seen easily from the table because a perfectly valid Gray code conversion would be obtained by a cyclic permutation of the given Gray code values, although the convenient bitwise conversion equation above would then not correspond to this new Gray code.

It is instructive to plot the Gray code values derived from the bitwise conversion equation on a K-map, as shown (for the 4-bit Gray code defined by the same bitwise defining equation $g_k = b_{k+1} \oplus b_k$) in Figure 10.14. The locus of cells is regular and covers the entire map. The locus moves only to adjacent cells at every step, and also is re-entrant (so that this code is suitable for rotational applications).

Proof of validity of the bitwise defining equation for Gray code

To prove that the bitwise defining equation $g_k = b_{k+1} \oplus b_k$ will always give a coding with the properties of Gray code, it is noted firstly that by inspection of the conversion table and Figure 10.14 the equation gives a successful Gray code in the cases of 2-, 3-, and 4-bit conversions (and $n = 1$ is a trivial case where the code given by the equation is a Gray code that is identical to the base 2 code).

Now, take an existing n-bit code, defined by $g_k = b_{k+1} \oplus b_k$, which it is *assumed* for the moment has the properties of a Gray code at each of its 2^n steps. To lengthen the code to give extra precision, a further additional bit g_n may be inserted to the left of the existing code, to form a new $(n+1)$-bit code having 2^{n+1} steps (not proven to be a Gray code yet). The values of the bits in the new code will still be produced using the same rule $g_k = b_{k+1} \oplus b_k$ from the corresponding $(n+1)$-bit base 2 code.

Imagine all the base 2 and the corresponding new codes written out in a table in numerical order of the base 2 codes, as in the 3-bit table in the text above. All the new code values calculated for every step having $b_n = 0$, i.e., the *top* half of the conversion table, are the same as those for the existing n-bit code but prefixed by $g_n = 0$ corresponding to the prefix $b_n = 0$ for all the base 2 codes, because the left-most bit of the new code is given by $g_n = b_{n+1} \oplus b_n = 0 \oplus b_n = b_n = 0$. Note that $b_{n+1} = 0$ for an $(n + 1)$-bit base 2 code, and the defining equation leaves the other bits unchanged. Therefore, the new $(n + 1)$-bit codes in the *top* half of the conversion table form a valid Gray code, provided the existing n-bit code was indeed itself a valid Gray code.

Next, all the new code bit values g_k corresponding to $b_n = 1$ (i.e. for the *bottom* half of the conversion table) have the new left-most bit $g_n = 1$, because $g_n = b_{n+1} \oplus b_n = 0 \oplus 1 = 1$. The next new bit on the right is given by $g_{n-1} = b_n \oplus b_{n-1} = 1 \oplus b_{n-1} = \bar{b}_{n-1}$, and so is the *logical complement* of bit b_{n-1} for the step in the conversion table written $(2^{n+1})/2 = 2^n$ steps previously, i.e. in the top half of the conversion table, for $b_n = 0$. The other bits g_k, for $k < (n - 1)$, are given by the usual formula $g_k = b_{k+1} \oplus b_k$ and so are unchanged from those of the step written 2^n steps previously in the conversion table. Hence, the new $(n + 1)$-bit codes in the *bottom* half of the conversion table also form a valid Gray code, provided the existing n-bit code was indeed itself a valid Gray code.

It is also necessary to establish that only one bit changes at the join between the top and bottom halves of the $(n + 1)$-bit conversion table, and also in 'rolling over' from the last code back to the first code. In moving one step from the top to the bottom half of the table, the base 2 code changes from 0 followed by a total of n binary 1s, to 1 followed by n binary 0s. Therefore, the left-most new bit g_n changes from 0 to 1, the next new bit to the right is $g_{n-1} = b_n \oplus b_{n-1} = (1 \oplus 0 \text{ or } 0 \oplus 1) = 1$ and so is unchanged, and all the other bits of the new code are $g_k = b_{k+1} \oplus b_k = (1 \oplus 1 \text{ or } 0 \oplus 0) = 0$ and so are also unchanged, and so this satisfies the Gray code condition (only one bit changes).

In 'rolling over' from the last code back to the first code, the $(n + 1)$-bit base 2 code changes from all binary 1s (a total of $(n + 1)$), to all binary 0s (a total of $(n + 1)$). Therefore, the left-most new bit g_n changes from 1 to 0, and all the other bits of the new code are $g_k = b_{k+1} \oplus b_k = (1 \oplus 1 \text{ or } 0 \oplus 0) = 0$ and so are unchanged, and so this also satisfies the Gray code condition (only one bit changes).

Finally, since it has been proven that if an n-bit code defined by $g_k = b_{k+1} \oplus b_k$ has the properties of a Gray code, then an $(n + 1)$-bit code defined by $g_k = b_{k+1} \oplus b_k$ will also have the properties of a Gray code, and also that the same defining equation correctly gives a Gray code in the case of $n = 1$. Therefore it has been established *by the method of induction* that $g_k = b_{k+1} \oplus b_k$ gives a Gray code correctly for all positive integer values of n.

10.12 Conversion from Gray code to base 2

The defining equation $g_k = b_{k+1} \oplus b_k$ is useful for generating a new code that is guaranteed to have the properties of a Gray code, but usually in designing encoders the raw output from the encoder head will consist of Gray code which, being an unweighted code, is unsuited for any numerical display, or other digital processing. Therefore, the designer must usually arrange that the first task of the associated digital system is to convert the raw Gray code provided by the encoder head to the corresponding base 2 code. This reverse conversion can be obtained from the truth-table of the defining equation for Gray code, $g_k = b_{k+1} \oplus b_k$, as shown in the table:

b_{k+1}	b_k	g_k
0	0	0
0	1	1
1	0	1
1	1	0

Re-ordering the columns of this truth-table gives the following table:

g_k	b_{k+1}	b_k
0	0	0
1	0	1
1	1	0
0	1	1

It is clear that the required reverse conversion is given by

$$b_k = g_k \oplus b_{k+1}$$

(and see also section 2.14) so that, to convert a Gray code defined by $g_k = b_{k+1} \oplus b_k$ back to the equivalent base 2 code, all that is necessary is to take each Gray code bit g_k and form the Exclusive-OR with the base 2 bit b_{k+1} (which is the base 2 bit that is one place *more significant than* bit b_k).

Of course, since the object of this exercise is to *find* the corresponding base 2 code in its entirety, this means that the conversion from Gray code to base 2 can only be performed for base 2 bit b_k *after* the bit b_{k+1} (the bit to its immediate left) has been found. This means, in turn, that the conversion must be performed in order from MSB to LSB. Therefore, the MSB (b_{n-1}) is found using $b_{n-1} = g_{n-1} \oplus b_n = g_{n-1} \oplus 0 = g_{n-1}$. This means that the base 2 MSB is always identical to the left-most bit of the Gray code. Having found the base 2 MSB, the equation $b_k = g_k \oplus b_{k+1}$ is used to find the next base 2 bit to the right, and then the next bit to the right of that, and so on, until the base 2 LSB has been reached. This conversion process *cannot* proceed from LSB to MSB, simply because using the equation $b_k = g_k \oplus b_{k+1}$ to find the LSB (i.e., b_0) requires knowledge of b_1 which would be unknown at that time. Based upon this process, designing a circuit to perform the conversion from Gray code to base 2 is straightforward (see Figure 10.15). For an n-bit conversion, a total of $(n-1)$ Exclusive-OR gates are required. For a large value of n and in a complex system controlled by a microcomputer, it may be more cost-effective to perform the conversion using appropriate microcomputer software.

10.13 Petherick code

Petherick code is a binary coded decimal equivalent of Gray code, and is normally used for encoding step values in the range 0 to 9 inclusive. Again, as in the case of Gray code, only one bit changes at each step, including the 'roll-over' step between maximum value (9) and minimum value (0), because the 'cycle length' is $(10)_{10}$. The conversion table is shown below:

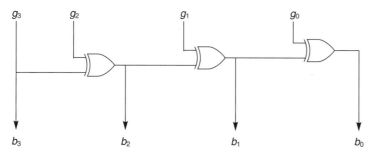

Figure 10.15 *Circuit diagram for conversion of Gray code to base 2*

Decimal	Base 2 code				Petherick code			
	b_3	b_2	b_1	b_0	P_3	P_2	P_1	P_0
0	0	0	0	0	0	1	0	1
1	0	0	0	1	0	0	0	1
2	0	0	1	0	0	0	1	1
3	0	0	1	1	0	0	1	0
4	0	1	0	0	0	1	1	0
5	0	1	0	1	1	1	1	0
6	0	1	1	0	1	0	1	0
7	0	1	1	1	1	0	1	1
8	1	0	0	0	1	0	0	1
9	1	0	0	1	1	1	0	1

The characteristics of this code are that, as in the case of a 4-bit Gray code defined by $g_k = b_{k+1} \oplus b_k$, the code for $(1)_{10}$ is $(0001)_{\text{Petherick}}$. Unlike Gray code, however, some 1s and some 0s are always present in Petherick code. This is a useful feature if an AC-coupled signal-recovery amplifier is used in conjunction with serial transmission of the code, as otherwise the DC levels of the 1s and 0s would be lost irretrievably. Using four K-maps, and treating the codes not defined in the Petherick code conversion table above as 'can't happen' terms, shows immediately that the conversion from the Petherick code defined above to four bits of NBCD is:

$$b_0 = (P_0 \oplus P_1) \oplus (P_2 \oplus P_3)$$
$$b_1 = P_1.\bar{P_2}$$
$$b_2 = P_1.(P_2 + P_3)$$
$$b_3 = \bar{P_1}.P_3.$$

Note that if more than one NBCD digit is required to be encoded, the use of several cascaded sets of Petherick code does *not* give a code where only one bit changes at each step, because whenever the least significant NBCD digit changes from 9 to 0 (or *vice versa*) then at least one of the more significant NBCD digits must also simultaneously change by 1, so that at least *two* bits in the entire cascaded sets would change simultaneously. In such circumstances, a better solution may be to use a genuine Gray code, convert that to base 2, and then to convert this to NBCD.

10.14 Incremental encoders

By contrast with absolute encoders, an incremental encoder of itself only indicates *relative* movement of the shaft or slider, and then only when the associated electronic system is powered. Other means must be used to find the absolute position; the encoder manufacturer may provide additional outputs such as extra data bits indicating when 'index' marks are passed, or the system designer may include external switches to indicate the end of the allowable travel. In some applications, for example a shaft encoder used to control the operation of an instrument, the lack of knowledge of the absolute position may be irrelevant in any case.

In a typical incremental encoder, *two* detectors, using optical or another technology, are used to provide digital signals X and Y from the reticle. Both these signals indicate the motion of the shaft or slider by toggling their logic state at regular intervals of shaft rotation or slider distance. However, the logic state changes in the signal X are shifted compared to those in signal Y. In terms of the phase of the underlying oscillatory signals that would be produced by constant velocity motion, the two are in *quadrature*, i.e. there is a phase difference between them of 90°. The key to the operation of this system is that when the motion of the slider or shaft changes direction, the 90° phase difference between X and Y reverses automatically as a result of the mechanical reversal. This phase reversal may be interpreted by a logic circuit such as that shown in Figure 10.16(a).

In Figure 10.16(a), the counter is such that rising edges applied to one input cause the output integer to count down, whereas pulses applied to the other input cause it to count up. Counting pulses are produced by the AND gate connected to X and Y, and the J-K flip-flop 'steers' these pulses to one or other of the counter clocking inputs, depending upon the direction of motion. This is shown in detail in the waveform diagrams of Figure 10.16(b). Note that it is essential in this circuit that the OR gate has a propagation delay greater than the data set-up time for the flip-flop, and that the shortest possible period of the X and Y waveforms is longer than this propagation delay. It follows that there is a maximum permissible mechanical speed that can be tracked by this circuit without error. Fortunately, most logic gates are so fast in operation that this is not usually a severe limitation.

Two signals, X and Y, are needed in this kind of encoder, in order to be able to detect the direction of the motion. In some applications, sensing of the direction of motion is not needed. For example, the motion direction is irrelevant when logging the total bearing wear in a rotating machine, and the motion direction may already be known in washing machine motors or in automotive electronic ignition systems. In this case a single output feeding a simpler subsequent circuit, such as a counter indicating 'total motion', may be used; such devices are often called 'tacho generators' or 'tachometers'.

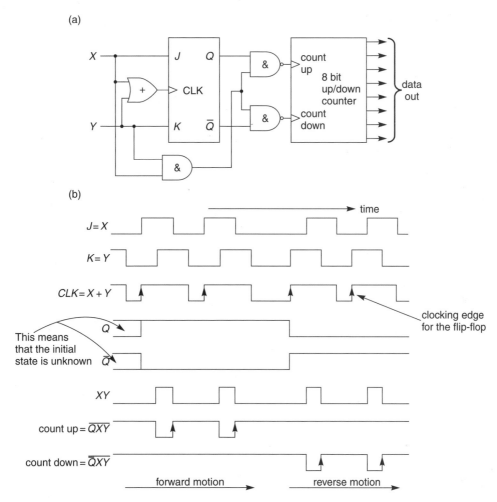

Figure 10.16 *(a) Circuit diagram of a simple incremental encoder interface (b) Typical waveforms when interpreting motion*

In the circuit shown in Figure 10.16, one count pulse is fed to the counter for every period of either the X or the Y waveforms, dependent upon the direction of motion. In fact there are four logic transitions within each of these periods, two from each of X and Y in quadrature, and so to obtain the greatest resolution from the encoder, by detecting and counting each and every logic transition, a much more complicated circuit is needed. Figure 10.17 shows a circuit designed to interpret all of the logic transitions at the encoder's output in this manner. In this circuit, RC networks, together with Schmitt inverters, are used to introduce well-defined signal delays of the order of $\tau = RC = 100\,\Omega \times 2\,\text{nF} = 200\,\text{ns} = 0.2\,\mu\text{s}$, and the associated XNOR gates produce logic high spikes lasting for this period. The full details of the operation of this circuit are left as an exercise. This circuit can be used with a typical commercial linear encoder of 1m length, and with $5\,\mu\text{m}$ resolution between each logic transition giving a complete encoder period of $20\,\mu\text{m}$ and a total possible count of $\pm 2 \times 10^5$ depending upon the zero starting position. To cope with this resolution, the binary counter chip would need to be expanded to 20 bits. A typical computer interface card would also give facilities for zeroing the counter

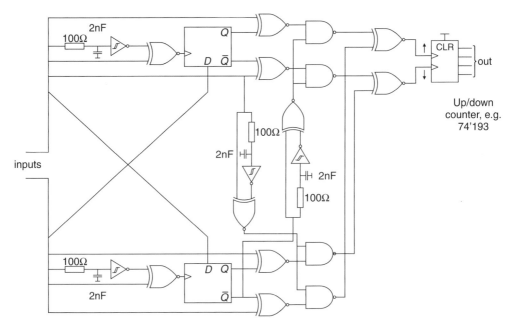

Figure 10.17   *Circuit diagram of an interface circuit interpreting all the logic transitions from an incremental encoder*

and for switching the counting direction that corresponds to forward motion. It may also be possible to synchronise with other similar cards the instant at which the counters are read. This can be important for a multi-axis system where it is required to know the (x,y) position of a driven component at certain instants specified by the computer.

10.15 Open collector and tri-state gates

'Open collector' and 'tri-state' gates are specialist types of digital logic gates that are frequently used in situations where a digital system must interface successfully to other components or systems that are generally regarded as operating in an analogue manner. To understand fully the operation of open collector and tri-state gates and how they can be connected together, it is necessary to examine the internal circuitry of the gates concerned. The important aspect to consider is the configuration of the output stage of the gates and the way the output terminal is connected to the internal circuitry. Most logic gates fabricated for use in TTL or CMOS technology contain two transistors connected in series across the supply rails, with their common connection taken as the gate output, as shown in Figure 10.18. This is often called a 'totem-pole' output stage, and it is reminiscent of an analogue power-amplifier 'Class B' output stage. These transistors act as switches, and can be either switched 'on', capable of passing current, or 'off', incapable of passing current. At any one time only one of these transistors can be 'on', and the other must be 'off'. If the upper transistor is 'on', the output voltage is at logic high level, and if the lower transistor is 'on' then the output is at logic low level. At no time, of course, may *both* transistors in the 'totem-pole' within any one gate be switched 'on', for this would result in rapid destruction of the gate by massive current flow through both transistors.

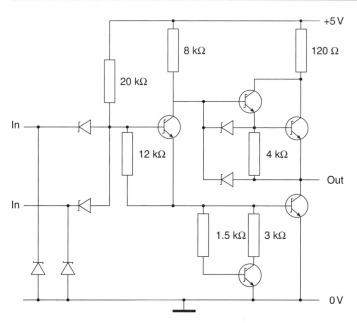

Figure 10.18 *Internal circuit of a typical integrated-circuit logic gate (similar to one gate in types 74LS00, 74LS04, 74LS10, and 74LS20) with a 'totem-pole' output stage*

Any output line connected to a logic gate will have some stray capacitance, and the charge needed to raise the voltage of this stray capacitance suddenly, from ground to the positive supply voltage, causes a large current to flow for a short time in the upper transistor. This can momentarily cause the supply voltage for the entire system to fall below acceptable levels unless precautions are taken. For this reason, logic IC manufacturers often advise connecting smoothing or 'decoupling' capacitors, usually of ceramic dielectric material and having a typical value $0.1\mu F$, between the positive supply line and ground for every five (or another number) IC packages.

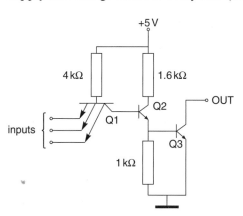

Figure 10.19 *Internal circuit of a typical integrated-circuit logic gate with an 'open-collector' output stage*

In a *tri-state gate*, there is an additional input, denoted the 'ENABLE' or 'CHIP SELECT' input, that must be connected by the user. Under control of this input it is possible for the output to be in an additional state, the high-impedance or 'Z state', where *both* transistors are switched 'off'. This may be thought of as equivalent to isolating the output terminal from the rest of the internal gate circuit. Gates with standard or 'two state' outputs do not have the extra input and cannot enter the Z state.

In an 'Open collector' gate, the internal circuitry of the gate, particularly the output stage, is rather different, as shown in Figure 10.19. Instead of a complete 'totem-pole', only the lower *npn* transistor switch is included, and its collector is connected directly to the output terminal of the gate. Such a gate can only be used with an external load resistor, often called a 'pull-up resistor', connected between the output terminal and the positive voltage supply rail. The transistor switch may be either 'on', when the output is at logic low level, or 'off', when the load resistor ensures that the output will be at logic high in the absence of any further current sinking. In circuit diagrams, an 'open collector' output stage is indicated by an asterisk (*) or the special symbol ◊ adjacent to the gate output concerned.

Logic gates intended for use with logic technologies other than TTL or CMOS may have different internal construction for which the terms 'open collector' or 'tri-state' may not have any meaning.

10.16 Use of open collector gates

Use of an occasional isolated open collector gate within a system otherwise composed entirely of conventional gates is often not wise design. This is because at transitions from low to high logic level at the output of the open collector gate, the single output transistor is turned 'off' and current is supplied through the load resistor only. The inevitable stray capacitance associated with any logic output line means that these transitions are relatively slow, with a decaying exponential rise of voltage towards the maximum value governed by the time-constant $\tau = R_L C_s$ (where R_L is the load resistance value and C_s is the stray capacitance value). Even a small value of C_s and a moderate value of R_L can lead to substantial slowing of the rising edges in practice, compared to the performance expected from conventional gates. The use of two output transistors in the 'totem pole' circuit, where the upper transistor can supply a large current when in the high logic level, allows them to achieve faster rising edges. However, by connecting together several open collector gate outputs together with a load to the positive supply, the output voltage will be at logic low level when *any* of their output transistors is switched 'on', and will only be at high logic level when *all* the transistors are switched 'off'. Therefore, the output voltage level corresponds to the Boolean AND of the individual gate outputs. This arrangement is often called 'wired AND' or 'collector dotting' and enables the AND of many signals to be obtained easily provided fast operation is *not* needed. Serious reservations remain concerning the operating speed of the open collector gates unless C_s can be reduced to a negligible value.

For example, in Figure 10.20(a) the outputs of a pair of two-input open collector NAND gates are connected to the same external pull-up resistor R_L. If the output of one of the NAND gates is low, the final circuit output will be pulled low, irrespective of the other NAND gate. The AND function is performed by a 'gate' that has no physical reality apart from the connection between the gates and as a consequence is called a *phantom AND gate*. The output of the network is

$$Z = PQ = \overline{AB} \cdot \overline{CD}$$

If inverted variables had been applied instead at the inputs, then the output would have been

$$Z = PQ = \overline{\overline{A}\,\overline{B}} \cdot \overline{\overline{C}\,\overline{D}} = (A + B) \cdot (C + D)$$

Similarly, if two open collector NOR gates are connected, as shown in Figure 10.20(b), the output Y will be the AND of the output of two NOR gates. Hence, the output is given by

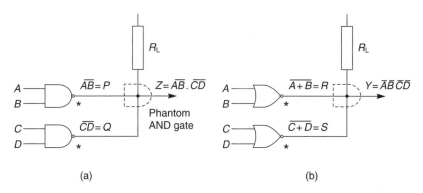

Figure 10.20 *Combination of (a) two open collector NAND gates and (b) two open collector NOR gates. Since the phantom AND gate is not a real component, its symbol is often omitted from circuit diagrams*

$$Y = RS = \overline{(A+B)} \cdot \overline{(C+D)} = \overline{A+B+C+D}$$

and if inverted variables had been applied instead at the input then

$$Y = \overline{\bar{A} + \bar{B} + \bar{C} + \bar{D}} = ABCD.$$

The logic designer has the responsibility of determining a suitable value for the pull-up resistor R_L. Since the values of R_L and the stray capacitance C_s determine the rise-time of the logic transition from low to high voltage, the designer must decide what maximum transition time is required, and estimate the stray capacitance C_s in the circuit layout to be used, in order to determine R_L. Clearly, for the fastest performance, the smallest possible value of R_L should be used, but in practice this leads to large currents flowing through the open collector gates when switched on. This in turn leads to large power dissipation in R_L itself, leading to problems with overall power consumption and, in extreme cases, the power handling capacity of R_L might be exceeded. Thus, the designer must take account of the maximum current allowable through the gates and pull-up resistor.

As an example, suppose that a logic system uses a 5 V supply (V_s) and a time constant less than 5 ns is required on rising edges, with a stray capacitance of 18 pF. Therefore, the required value of pull-up resistor is given by

$$R_L < \frac{\tau}{C_s} = \frac{5 \times 10^{-9} \text{s}}{18 \times 10^{-12} \text{F}} = 278 \,\Omega.$$

In practice, the nearest preferred value resistor, 270 Ω, would probably be used. Then, the maximum current through the load resistor when in logic 0 state will be

$$I = \frac{V_s}{R_L} = \frac{5\text{V}}{270\,\Omega} = 18.5 \,\text{mA}$$

and the gates must be capable of passing this current without damage. Hence, the peak power dissipated in R_L is given by

$$P = \frac{V_s^2}{R_L} = \frac{(5\text{V})^2}{270\,\Omega} = 92.6 \,\text{mW}.$$

The actual average power dissipated in R_L will be a proportion of this value, depending on the proportion of time the gates spend in the low logic state, but in any case it is good practice to assume a worst-case power dissipation of 92.6 mW in order to cater for possible fault conditions. In practice, component tolerances will cause some further variations in these calculated figures, but whether this power dissipation is too great either for the overall consumption of the circuit or for the resistor R_L to handle depends upon other details of the design.

Another use of open collector gates is in driving logic lines to other equipment where in principle there may be more than one device capable of placing data on the lines or logic 'bus'. Usually, for correct operation, only one gate should be in command of each logic line at a time, as otherwise the operation of the bus would become immensely confused, but sometimes during testing, or due to a fault, it happens that more than one gate is indeed placed in charge of a logic line. No harm will come to open collector drivers used in this way.

A third very common use of open collector gates is to drive external components that are not specifically designed for direct connection to conventional logic gates in the same way that other logic gates of the same family may be inter-connected to build up a complex logic system. As an example, it may be required to connect a Light Emitting Diode (LED) to a logic system in order to indicate the logic state of one of the gates. This is a common requirement in logic systems that show information to the user on alpha-numeric LED displays which will need to be driven electrically from the logic system. LEDs are *pn* junction diodes that are especially designed to emit light when they are forward biased. Although they will also withstand reverse bias of a few volts without damage, and without emitting light, they are not usually intended to be reverse biased in normal operation. Electrically they behave in a similar manner to ordinary small-signal rectifying diodes except that they are usually fabricated from a III-V semiconductor (such as GaAlAs or a related material such as InP) rather than Si. As a result, the voltage drop at moderate currents is rather higher, usually around 1.6 V at typically 10 or 20 mA for a small LED intended for use as an indicator on small equipment, rather than around 0.6 V for a typical Si diode.

Conventional logic gates with totem-pole output stages are not primarily intended as current sources or sinks, and so it is not usual to drive LEDs from totem-pole output gates. The solution generally adopted is to place the LED in series with a current-limiting (or 'dropper') resistor between the output connection of an open collector gate and the positive supply rail, as shown in Figure 10.21. The respective manufacturers' data sheets must be consulted to ascertain, firstly, the current requirements of the LED at the required brightness and the voltage drop at this current value, and secondly, the current sinking capability of the open collector output transistor in the logic gate to be used. Assuming that its current sinking capability is adequate for the intended application, the value of the series resistor may be found after calculating the voltage drop across the resistor, equal to the supply voltage less the voltage drop across the LED. This assumes that the voltage drop in the output transistor when switched 'on' is of the order of 0.1 V and may be neglected. As an example, consider driving a typical LED requiring 15 mA at 1.6 V drop using an open-collector gate

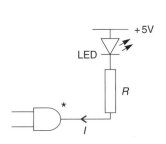

Figure 10.21 *Connection of an LED to an open collector gate*

connected to a 5 V supply. Assuming that the gate is able to supply 15 mA without damage, the value of the series resistor needed is equal to

$$\frac{5.0\,\text{V} - 1.6\,\text{V}}{15\,\text{mA}} = \frac{3.4\,\text{V}}{15\,\text{mA}} = 227\,\Omega$$

and the nearest preferred value used might be 220 Ω.

A similar application is the use of an open collector output gate to drive a small electromechanical relay, in order to control a component requiring substantially more current than can be supplied by the output transistor itself. In certain cases a small relay may have a sufficiently small current driving requirement that it can be driven directly from an open collector gate, in a manner similar to an LED. Sometimes the relay is connected between the open collector terminal and a positive supply voltage rather greater than the normal logic circuit supply. A very small relay will typically require 5 V or 12 V or more, at a current of 10 or 20 mA, to operate correctly, and a series dropper resistor is not needed with typical relays (but see Figure 10.22). Note that the open collector output stage allows the use of a component, the relay, that requires a larger driving voltage than the normal logic system supply rails can provide. This is a direct result of the collector of the output transistor being connected only to the external relay, and not to any other internal components within the logic gate. Open collector output gates are available that are able to switch typically up to 40 mA at up to around 30 V. If a relay is to be used that needs a greater driving voltage or current, then further external active components must be added to provide the increased driving capacity.

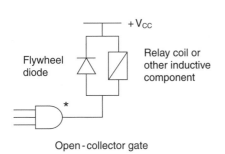

Figure labels: +V_CC, Relay coil or other inductive component, Flywheel diode, Open-collector gate

Figure 10.22 *The 'Flywheel diode' connected across an inductive component and necessary to protect the driving gate*

Use of relays and other inductive components

When the transistor driving a relay (or another component with a large self-inductance) is turned 'off', the *rate of change* of current can be very large because the time taken for the current to reduce from its normal 'on' value to zero can be very short. Therefore, the self-induced back-EMF, $V = -L(\text{d}I/\text{d}t)$, is potentially very large if the product of L and $(\text{d}I/\text{d}t)$ is large. In some cases this back-EMF can be so large that the driving transistor within the logic gate, if direct drive from an open collector gate is being used, is in danger of being damaged by the repeated application of excessive voltages. Fortunately, the remedy is simple, since by Lenz's law the back-EMF spikes have the opposite polarity to the driving voltage, and all that is needed is to insert a diode in *parallel* with the relay or inductance, with polarity such that it is *reverse biased* by the driving voltage in normal operation. Therefore, this 'flywheel diode' has no effect upon the driving current to the relay since a reverse-biased diode draws no current away from the relay, but it prevents large back-EMF spikes reaching the driving transistor. When forward biased, the 'flywheel diode' must be capable of carrying the normal 'on' driving current of the relay, because at the instant of turning the driving transistor 'off' the relay will attempt to maintain the energy stored in its magnetic field by inducing the same magnitude of current through the 'flywheel diode' for an instant, before the relay and diode currents collapse.

10.17 Use of tri-state buffers and gates

A tri-state buffer is a logic inverter or a non-inverting buffer with a tri-state output stage. The four possible configurations are shown in Figure 10.23 and the truth table for the type in Figure 10.23(a) is also shown.

The input denoted E can be regarded as an *enable* line, which may require either an active low or active high input signal, and when activated it will allow the gate to output either the true or inverted data. When the enable line is not activated the buffer output stage has a high output impedance (i.e., the Z *state*, as described above in section 10.15) and transmission of data is prevented. An active high enable input is also sometimes referred to as the *Chip-Select* input, or *CS* (mainly in the case of VLSI chips having this input). In the case of gates or chips where the Enable input is active low, it is sometimes referred to as an *Inhibit* input, *I*, as, when taken high, it inhibits the gate operation.

The main use of tri-state gates is in driving logic lines or the connectors in a *data bus* (a contraction of the older term 'bus-bar' meaning a conductor providing a voltage or current, often from a power source, to many other devices). For example, in the connection of a microprocessor to RAM chips, it is necessary at some time that the microprocessor sends binary data to the RAM; and at other times, the microprocessor must read that data back from storage in the RAM chips. It is conventional to use the same connecting pins for routing this data to and from the microprocessor, which is hence known as a *bi-directional* bus. It is therefore necessary that in the first instance, the microprocessor must be in control of the logic state of the bus lines in placing the data to be stored upon the bus, and in the second case control is allowed to a RAM chip which then places its data upon the bus. This is not possible using conventional two-state outputs, which if used in both the microprocessor and RAM would soon lead to the destruction of both whenever there was *bus contention*, i.e. one component's output stage driving a bus line high, while at the same time the other component's

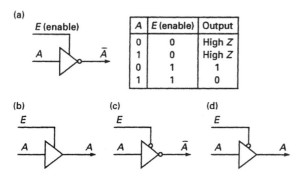

Figure 10.23 *Tri-state buffers (a) Inverting, active high enable with truth table (b) Non-inverting, active high enable (c) Inverting, active low enable (d) Non-inverting, active low enable*

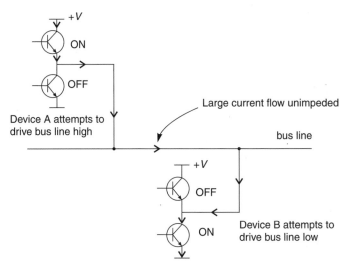

Figure 10.24 *Bus contention when using 2-state drivers (or tri-state drivers both enabled in error)*

output drives it low, leading to a large and destructive current flow through both of the 'on' transistors as well as along the bus line, as shown in Figure 10.24. The solution to this problem is to use tri-state outputs on both microprocessor and RAM, and to arrange some control circuitry such that at any one time only *one* of these components can have its outputs *enabled*, i.e. placed in either the 1 or 0 conventional output states. At this time all the other component outputs connected to the same bus lines must be *disabled* by being placed in the Z state. Reliable operation of this control circuitry is essential if destruction of the components is to be avoided. The 'bus' may therefore be regarded as a highway for data, and the high impedance property of the tri-state gate allows the data lines leaving a particular device to be effectively isolated at will from the bus system.

Figure 10.25 shows a diagram illustrating the use of tri-state buffers. Data from Device *A* can only be transferred to the *system interconnecting bus* when an active low signal is applied to the tri-state buffers via E_A and can be transferred from the bus to

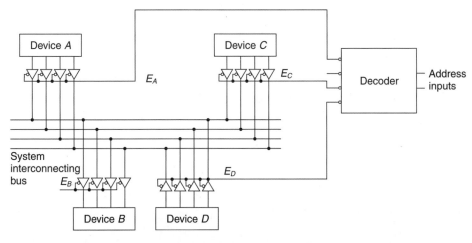

Figure 10.25 *Device connection to system data bus via tri-state buffers*

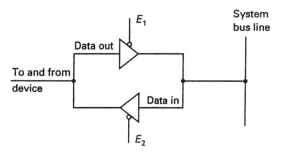

Figure 10.26 *Bi-directional tri-state connection of a device to a system bus*

Device B when an active low signal is applied via E_B. To ensure that none of the other devices connected to the bus system is simultaneously transferring data to the bus, their enable signals must not be activated. This could be achieved by connecting all of the enable lines controlling the output devices A, C, and D in Figure 10.25 to successive active low outputs of a decoder chip (see Chapter 5) only one of whose outputs may be low at any instant. Nevertheless, if (due to a fault) there is more than one enabled gate connected to any one logic line then it is likely that all the gates so enabled in error will be damaged. Use of open collector drivers avoids this problem at the expense of slower bus operation. This theme is developed in more detail in the discussion of ROM devices in Chapter 11.

In some cases, data has to be transmitted both from the device to the bus as well as from the bus to the same device. This will require a *bi-directional* capability, as illustrated in Figure 10.26. The transmission and the receipt of data is controlled by the two enable signals E_1 and E_2. When $E_1 = 0$ and $E_2 = 1$, data is transmitted from the device to the system bus; when $E_1 = 1$ and $E_2 = 0$, data can be received by the device from the system bus. Clearly $E_1 = E_2 = 0$ is not allowed in this circuit. If $E_1 = E_2 = 1$, both the device buffer and the bus buffer are in the Z state and data cannot pass in either direction. In certain applications, this last case may not be required, in which case the data transmission direction can be controlled by a single line E_1, and E_2 is connected to E_1 through a logic inverter.

10.18 Other interfacing components

The variety of other interfacing components that may also be used between digital systems and non-digital systems is almost inexhaustible. They include the following:

1. Electromagnetic actuators of many designs, which may be regarded here as inductive coils that may be driven in a manner similar to electromechanical relays.
2. Limit switches that sense the position of mechanical components, often operating at the allowable extremes of motion, either for signalling an absolute reference position to an incremental encoder, or for emergency signalling to a digital system of an undesirable mechanical condition which is about to occur.
3. Stepper motors, which are used for driving mechanical components under digital control. The most common types have a shaft that rotates a small angle (often $1/200$ or $1/400$ part of one revolution) on receipt of one set of driving pulses, although linear steppers are also available. Mostly these are magnetically driven,

but the piezoelectric effect has also been used in precision stepper motors. The design of driving systems for stepper motors is a specialist topic, and a non-specialist will normally use one of the proprietary systems available.

4. Electro-rheological components, employing a special fluid whose mechanical properties change, e.g. reversibly from a liquid to a solid, under the application of an electric field typically of the order of $5\,\mathrm{kV}\,\mathrm{mm}^{-1}$. These components often have special driving requirements (several kV, in order to generate the required electric fields). These are usually catered for by custom-designed systems.

5. Displays of various kinds, including cathode ray tubes (CRTs), alpha-numeric LED displays, LCD panels of varying sizes and performance (colour or mono-chrome), and plasma display panels. Mostly the systems involved in driving these displays will include D/A converters for controlling the individual beam brightnesses in CRTs, and open collector drivers for the other types of displays.

6. Magnetic disc drives and CD-ROM/RAM drives, for data storage in computing equipment. Mostly these will be controlled by dedicated IC driver systems, incorporating several specialised sub-systems. These include mechanical actuators interfaced as electromechanical relays, and stepper motors for positioning the read/write heads. For rotating the disc, conventional motors are used, also interfaced in a manner similar to electromechanical relays. Signal recovery amplifiers are interfaced to the digital system using Schmitt input gates. Sub-systems for data recording usually require TTL-compatible voltage level inputs from conventional gates.

Specialist knowledge of the characteristics and intended use of all these components is needed for the design of the digital systems which need to use them.

Problems

10.1 In Figure P10.1, one input of the NAND gate is taken to logic level 1 and a slow square wave logic signal is connected to the other input. Sketch the waveforms at points A and B. The frequency of the square wave is then increased gradually. Explain, with sketches, what you expect to happen to the waveforms at points A and B. What happens if one input of the NAND gate is now taken to low logic level and the other is taken to the same square wave as before?

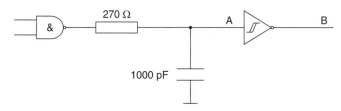

Figure P10.1

10.2 A certain video monitor can display video signals with a maximum frequency (i.e. its *video bandwidth*) of $5\,\mathrm{MHz}$ and it operates at 25 frames (i.e. complete pictures) per second. Suppose that one frame is stored in binary form by measuring the displayed brightness twice per cycle of the maximum displayable video frequency (i.e. 10×10^6 times per second), and each of these measurements is converted into a one-byte binary integer. How many bytes of storage (approximately) would be needed to store one frame?

10.3 A standard CD plays for a maximum of 74 minutes, and the two-channel audio signal is sampled at 44.1 kHz with 16 bit precision. Estimate the maximum storage capacity of a CD, assuming that no data compression is used. (This crude estimate will not account for any space taken up by the file directory structure.)

10.4 A digital audio tape system stores all samples to 20 bit precision. Estimate the signal-to-noise ratio of replayed sounds, assuming that all the noise generated in the system is 'quantisation noise', originating from the approximation of the required analogue signal as a series of voltage levels, equally-spaced by a difference corresponding to the significance of the least significant bit.

10.5 In general, if n bits are needed to represent a particular range of values in binary, what is the minimum number of bits needed to represent the same range in Gray code?

10.6 Using the conversion equation $g_k = b_{k+1} \oplus b_k$ (where b_k is bit number k (numbering from right to left, i.e. LSB to MSB) of the binary code, and g_k is bit number k of the Gray code), design a *sequential* circuit which will convert any 3-bit binary code to Gray code, using only *one* XOR gate, *one* shift register, and some other logic. The input binary code is to be held initially in the shift register, to be replaced eventually by the final Gray code. (Assume that a suitable clock line is available.)

10.7 The following table shows a conversion from binary $DCBA$ to a different coding $D'C'B'A'$ (actually a modified form of Petherick Code). Assuming that unlisted states are 'don't care' states, use four Karnaugh maps to find expressions for D', C', B', and A' in terms of A, B, C, D.

D	C	B	A	D'	C'	B'	A'
0	0	0	0	0	0	1	0
0	0	0	1	0	1	1	0
0	0	1	0	0	1	1	1
0	0	1	1	0	1	0	1
0	1	0	0	0	1	0	0
0	1	0	1	1	1	0	0
0	1	1	0	1	1	0	1
0	1	1	1	1	1	1	1
1	0	0	0	1	1	1	0
1	0	0	1	1	0	1	0

10.8 The circuit diagram of an incremental shaft encoder is shown in Figure P10.8, together with the two waveforms X and Y (in quadrature) produced as the shaft rotates. If the rotation changes direction, the phase difference between X and Y reverses. For X and Y waveforms as shown, draw the waveforms at CLK and Q, and explain how the circuit operates. What is the function of the *RC* network?

10.9 For the circuit in Figure P10.9, find the Boolean Algebra function for the output f and write down the truth table.

10.10 The 'totem-pole' TTL output in Figure P10.10 is at a *high* logic level. Deduce whether Q3 and Q4 are ON or OFF.

10.11 Logic gates made using a certain process have the following characteristics:

	Minimum	Maximum
Current available at logic low from gate output:	4 mA	10 mA
Current available at logic high from gate output:	−15 mA	−30 mA

It is required to use one of these gates to drive an LED indicating the logic state of the gate. The LED requires 10 mA and drops 1.4 V. The supply lines, and the logic levels, are at 0 V and +5 V. Show how you would connect the LED and the gate, using other passive components but no other active components.

10.12 A CMOS logic gate drives a long transmission line which has a stray capacitance of 150 pF between the signal conductor and ground. The signal to be transmitted is a logic signal of average frequency 20 MHz, and the logic voltages are 0 V and 5 V respectively. The current drawn by the logic gate is dominated by that needed to drive the transmission line. Estimate the average current drawn from the supply.

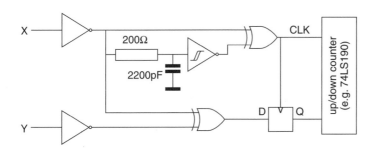

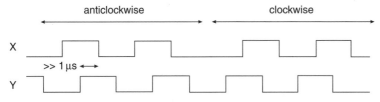

Figure P10.8

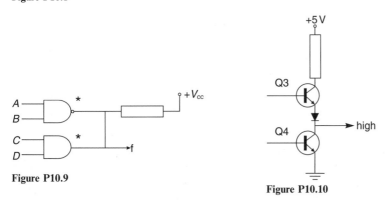

Figure P10.9

Figure P10.10

11 Programmable logic devices

11.1 Introduction

Traditional methods of logic design employing SSI and MSI circuits have been strongly challenged in recent years as a result of advances in technology which have led to the development of LSI *programmable logic devices* (or PLDs). The programmable logic devices currently available to the logic designer may broadly be classified as Read Only Memories (ROMs), a variety of different types of programmable logic arrays (PLAs), and several different brands of field programmable gate arrays (FPGAs) which may roughly be considered as developments of the PLA concept.

The basis of these devices is two-dimensional arrays of identical cells, usually laid out on the silicon chip in regular rows and columns. These arrays can be *programmed* or loaded with digital logic information specific to the particular application intended in various ways. Programming may be done by the manufacturer in accordance with programming information supplied by the designer (usually necessary when storage of the information using the particular technology concerned requires expensive microelectronics equipment and expertise) or, alternatively, field programmable devices may be programmed by the designer or user with relatively cheap equipment. Programmable logic devices may be faster than their MSI and SSI equivalents and their use simplifies the task of the logic designer for medium- and large-scale logic circuit designs.

11.2 Read only memory

A read only memory (ROM) chip in its most basic form stores a large number of binary integers, one at each unique value of the ROM *address* which acts in the same way as a 'house number' and identifies each stored integer or *binary word* by its *memory location*. When the external logic system presents an address or memory location to the ROM, the ROM returns the data stored in the *register* or memory storage at that address. Each register is capable of storing one binary integer, originally placed there either by the chip manufacturer working from data supplied by the logic system designer, or by the system designer taking the ROM chip through a special programming process. A typical ROM consists of an array of addressable registers of identical length (number of bits); each register or 'memory location' has a unique address (a binary integer in the range 0 to one fewer than the total number of locations) and can be selected by circuitry included in the ROM designed to read and interpret the address number required (similar to an address decoder as described in Chapter 5). A block diagram showing the basic components of a typical ROM is

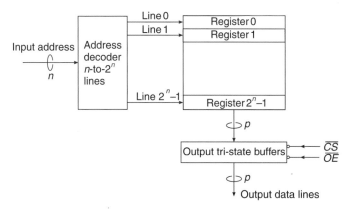

Figure 11.1 *The internal block structure of a ROM*

shown in Figure 11.1. The ROM has n address lines and, since there are 2^n possible combinations of n binary digits, the chip will house 2^n registers. Each register is identified or addressed by one of the 2^n output lines of the internal address decoder contained within the ROM chip.

In the ROM shown in Figure 11.1, each register contains p bits, and so the total storage capacity of the ROM is $p \times 2^n$ bits. For a typical word length $p = 8$ and a typical number of address lines $n = 12$, the total storage capacity is $8 \times 2^{12} = 32768$ bits. A group of eight binary digits is often referred to as a *byte*, so that the storage capacity of this particular ROM is $2^{12} = 4096$ bytes, or 4Kbyte, where K means 1024 and is pronounced 'kilo' by analogy with the usual measurement unit prefix. This memory chip may also be described as a $4K \times 8$ ROM, or as a 4K byte-organised ROM.

When a ROM is incorporated into a digital system where communication between devices is via an interconnecting bus system, two control signals are normally required. In many applications, for example a microprocessor system, where a number of ROMs may be used to store a program, only one ROM must be connected to the bus system at any given instant. The ROM to be connected to the bus will be identified by activating its *chip select* (*CS*) signal. Additionally, the ROMs may be connected to the bus system via tri-state gates which are in the high impedance state until they are enabled by an *output enable* (*OE*) signal. Once enabled, the data at the input to the tri-state buffers will be transferred to the bus.

Computer systems also use large numbers of *random access memory* (RAM) chips to store temporary results of computations and processing. There are two main types of RAM: static RAM, in which each bit of data is stored on the equivalent of a single D-type flip-flop, and dynamic RAM, in which each bit of data is stored as an electrical charge on the gate capacitor of a MOSFET. Since the capacitors are not perfect and the charge leaks away after 1ms or so, the charge must be '*refreshed*' regularly. The advantage of static RAM is that refreshing is not needed, whereas the advantage of dynamic RAM is that the '*packing density*' (number of stored bits per chip) of available devices is much greater than on available static RAM devices. RAM chips have an internal structure similar to ROM chips except that data can be stored an unlimited number of times in *any* or all of the memory locations. This data is generally lost when power is removed from the RAM chip, that is, the data, is said to be 'volatile', although special 'non-volatile' RAM chips are also available. Therefore, a RAM needs a third control signal, the *write* (*WR*) or $\overline{read(RD)}$ signal. If *WR* is activated simultaneously with *CS*, data is transferred from the RAM data lines to the internal data register selected. However, if *WR* is *not* activated then the RAM behaves similarly to a ROM chip. Apart from this extra signal, RAM circuitry is in principle similar to ROM circuitry, except that to be useful RAM must first have data stored in it and this limits its use almost exclusively to computer and microprocessor systems which are outside the scope of this text.

ROMs are, by definition, non-volatile memories because the program written into the memory, when it is initially programmed, remains stored when the power is removed. Because of its non-volatility, ROM is typically used for basic program storage and also for the storage of unchanging data patterns.

There are several main categories of ROMs currently available:

1. *Mask programmed by manufacturer*. The data stored in the ROM, the 'contents', are programmed by the manufacturer during fabrication according to a specification supplied by the customer. This type of ROM is only suitable when the designer's required data or program has been extensively tested and verified to avoid errors, as it is not possible to change the stored data after fabrication and packaging. Programming these devices during manufacture requires expensive equipment and is economic only for very high volume applications and, in addition, there may be some delays before the final devices are produced.

2. *PROMs* (*Programmable ROMs*). The PROM contents are written into the PROM by the user with the aid of a piece of equipment known as a 'PROM programmer'. Programming this type of ROM is essentially an irreversible process, so this type is sometimes referred to as 'One-time programmable' (OTP). Since PROMs are relatively cheap, they are often used in the early stages of product development when considerable changes may have to be made to the stored program, as the changes can be made by simply programming another PROM by the user. When the design has been finalised, the data may be sent to a ROM manufacturer for mass production of a high-volume mask-programmed ROM dedicated to the proven design. Alternatively, low-volume applications can continue to use individually programmed PROMs.

3. *EPROMs* (*Erasable PROMs*). The contents are programmed electrically by the user but can be subsequently erased, followed by loading new programming information. This is achieved by shining Ultra-Violet (UV) light, from a special UV source designed for EPROM erasure, for a period of 10 to 20 minutes through a transparent window on top of the ROM package. This type of ROM may therefore be recognised by the presence of this window, usually around 10 mm × 10 mm, through which the actual ROM chip may be seen. Like PROMs, EPROMs can be used for system development as well as for low-volume production, in which case it is normal to cover the window with opaque tape to prevent inadvertent erasure of the EPROM contents. Often the manufacturers state a limit of perhaps 100 UV erasures that can be undertaken with any one EPROM before the erasure and storage become unreliable.

4. *EEPROMs* (*Electrically Erasable Programmable ROMs*). This type of user-programmable ROM can have its program completely erased electrically. However, there is a limit to the number of times that the stored data can be erased and the device reliably reprogrammed, so EEPROMs are not a substitute for genuine RAM.

A typical example of an EPROM is the TMS27128 containing 131072 bits (16Kbyte). Before programming, the chip is erased by UV radiation (so that all bits are set to 1), and after erasure, 0s are programmed in those locations specified by the designer. The TMS27128 EPROM is packaged as a 28-pin IC; further increase in storage capacity (with the same control facilities) requires an IC having more than 28 pins. The TMS47256 ROM has a storage capacity of 262144 bits (32Kbyte) but with simpler control facilities fabricated as a 28-pin IC. The Appendix on Functional Logic Symbols describes in detail the symbols for these devices.

11.3 ROM timing

The time between the arrival of a valid address at the address pins and the appearance of valid data at the data output pins is termed the memory access time. For typical EPROMs, the chip select and output enable lines also determine the time taken to read data from the chip.

To specify this more precisely, manufacturers publish *timing diagrams* showing diagrammatically the typical logic waveforms to be expected at the various pins of the ROM while it is undertaking certain tasks. The usual conventions used on timing diagrams relevant to the read timing cycle are shown in Figure 11.2. Transitions between 0 and 1 are shown as sloping lines; this indicates that logic transitions take a non-zero time. For example, the transition time for a change of address is t_a. The ROM data output lines will often start in the high impedance Z state (indicated schematically by a level halfway between 0 and 1)

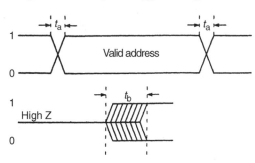

Figure 11.2 *Timing diagram conventions*

before changing to the correct data (either a 1 or 0 output, indicated by *two* levels simultaneously on the diagram as either value could be selected) during a time less than t_b.

For example, the *read cycle* of the TMS2764 (an 8 Kbyte EPROM), which uses these conventions, is illustrated in Figure 11.3, and the following timing parameters are specified on that diagram:

1. $t_{a(A)}$: *memory access time* – the maximum time taken from the arrival of the address at the address pins to the appearance of valid data at the output pins.
2. $t_{a(E)}$: *chip select access time* – the maximum time taken from $\bar{E}$ $(= \overline{CS})$ becoming active (low logic state) to the appearance of valid data at the output pins.
3. $t_{en(G)}$: *output enable time* – the maximum time taken from the enable signal $\bar{G}$ $(= \overline{OE})$ becoming active (low logic state) to the appearance of valid data at the output pins.
4. $t_{dis(G)}$: *output disable time* – the maximum time taken to disable the tri-state gates at the output and return them to the high-impedance state.

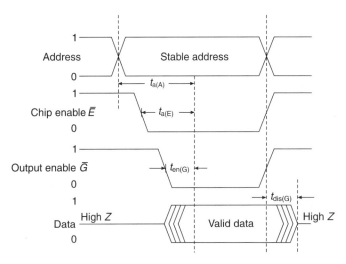

Figure 11.3   *ROM read cycle timing for TMS2764*

11.4 Internal ROM structure

Both bipolar and MOS technology are used in the fabrication of ROMs. The significant differences between the two technologies are *speed* and *packing density*. Bipolar ROMs generally have shorter access times while MOS ROMs have a higher packing density. This means that they can accommodate a larger number of memory cells in a given space.

The array of registers shown in Figure 11.1 is frequently called the memory matrix. A simple ROM matrix is shown in more detail in Figure 11.4, and consists of two sets of intersecting and orthogonal bus-bars. The vertical lines, connected to the output of the address decoder, are called the word lines, and in this simple model there is

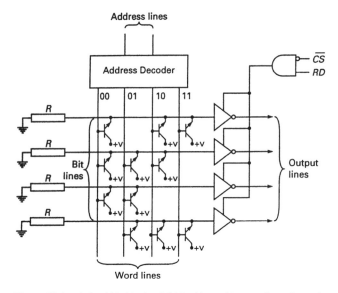

Figure 11.4   *A 4 × 4-bit bipolar ROM addressed in one dimension only*

a separate word line corresponding to each addressed ROM register. The horizontal lines, called the bit lines, are connected to the inputs of the tri-state buffers whose outputs are the data outputs of the ROM. Figure 11.4 shows a trivial ROM with capacity 4×4 bits.

In the case of a ROM manufactured using bipolar transistor technology, words are programmed into the ROM at each register address by making a connection to a bipolar *npn* transistor at each bit location required to be at logic 1. (This is shown in Figure 11.4 for the top three bits of location $(00)_2$, for example; the full ROM contents in Figure 11.4, given in hexadecimal for ascending addresses, are E, 7, D, and 9). The output on a bit line depends on whether it is electrically connected to the addressed word line *via* an *npn* transistor. If the connection exists when a word line is addressed, the bit line is raised to 1; if not, it remains at 0. In this way the programmed word at the selected address is transferred to the inputs of the tri-state buffers. The tri-state gates are enabled when the chip select signal $\overline{CS}$ is low and the read signal *RD* is high. When this condition is satisfied, the selected word is transferred to the data lines.

PROMs fabricated using bipolar transistors have an overall internal structure almost identical to that of the ROM shown in Figure 11.4. However, a 'fusible link' is connected in series with the emitter of a transistor at every position of the array (see Figure 11.5). A fusible link is a tiny fuse that can be either shorted, as manufactured, or open, following vaporisation by sending a large current through it. As all the fusible links are originally shorted, all the corresponding data bits are initially set to 1; the PROM programming equipment changes the bits required to be 0 by vaporising the corresponding fusible links so that they are open. Once programmed, bits already changed to 0 clearly cannot be changed back to 1, although if necessary it is usually possible to change to 0 a bit that is still equal to 1, thus giving a very limited amount of reprogrammability. If the word line associated with the transistor is selected, it is turned on by the potential applied to the base so that the voltage between collector and emitter is approximately zero and the voltage *V* is transferred to the bit line.

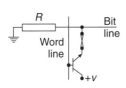

Figure 11.5 *Fusible links used in a bipolar PROM*

EPROMs and EEPROMs also use the same basic structure but have specially developed technology using floating gate MOSFETs (i.e. with no electrical connection to the gate) as a basic charge storage element, rather than fusible links. This is similar to the technology of dynamic RAMs but in this case the charge does not leak away significantly over a long period of time. The charge stored on the floating gate can be released, thus erasing the data, by UV radiation in the case of EPROMs, or electrically in the case of EEPROMs.

11.5 Implementation of Boolean functions using ROMs

A schematic way of representing a programmed 64-bit ROM is shown in Figure 11.6. Each word line corresponds to one of the eight minterms possible from three address input lines *ABC*. Eight different Boolean output functions are shown as examples, each one corresponding to one of the available eight bit lines. The intersections of the

(a)

Address			Output functions							
A	*B*	*C*	Z_1	Z_2	Z_3	Z_4	Z_5	Z_6	Z_7	Z_8
0	0	0	1	1	0	1	1	0	0	1
0	0	1	0	0	0	1	0	0	0	0
0	1	0	1	0	1	0	1	0	1	0
0	1	1	1	0	1	0	1	0	0	0
1	0	0	1	0	1	0	1	0	0	0
1	0	1	0	0	0	1	0	1	1	1
1	1	0	0	0	1	0	0	0	0	0
1	1	1	0	1	1	1	1	0	0	1

(b)

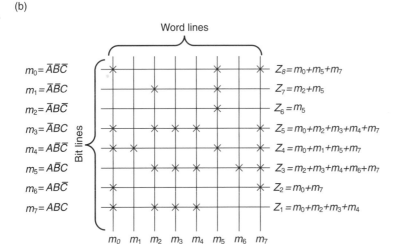

Figure 11.6 *Implementation of eight Boolean functions using a 64-bit ROM (a) Truth table (b) Connection matrix*

bit lines and word lines marked with a cross are those bits in the ROM that are set at logic 1; the unmarked intersections are those set at logic 0. Any particular bit line takes on the Boolean value of the bit stored at its intersection with the currently selected word line. Therefore, a cross indicates that its corresponding minterm is one of the terms in a Boolean canonical sum-of-products expression for its corresponding output function; in other words, each output function is the Boolean OR of the minterms marked with a cross on that bit line. When implementing Boolean functions with a ROM, simplification of the functions is neither necessary nor relevant, as they are implemented directly as a sum of minterms. For example, the output Z_8 in Figure 11.6 is given by

$$Z_8 = m_0 + m_5 + m_7.$$

Many of the functions available on MSI chips, such as those described in Chapter 5, can be readily implemented by a ROM. For example, a BCD to 7-segment decoder can be implemented with a 16-byte (128 bit) ROM as shown in Figure 11.7. Since there are only seven segments, one column in the truth table, corresponding to the eighth output

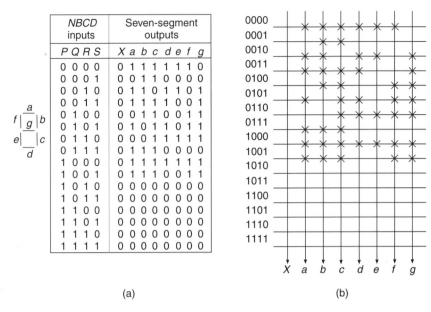

NBCD inputs	Seven-segment outputs
P Q R S	X a b c d e f g
0 0 0 0	0 1 1 1 1 1 1 0
0 0 0 1	0 0 1 1 0 0 0 0
0 0 1 0	0 1 1 0 1 1 0 1
0 0 1 1	0 1 1 1 1 0 0 1
0 1 0 0	0 0 1 1 0 0 1 1
0 1 0 1	0 1 0 1 1 0 1 1
0 1 1 0	0 0 0 1 1 1 1 1
0 1 1 1	0 1 1 1 0 0 0 0
1 0 0 0	0 1 1 1 1 1 1 1
1 0 0 1	0 1 1 1 0 0 1 1
1 0 1 0	0 0 0 0 0 0 0 0
1 0 1 1	0 0 0 0 0 0 0 0
1 1 0 0	0 0 0 0 0 0 0 0
1 1 0 1	0 0 0 0 0 0 0 0
1 1 1 0	0 0 0 0 0 0 0 0
1 1 1 1	0 0 0 0 0 0 0 0

(a) (b)

Figure 11.7 *(a) Truth table for an NBCD to 7-segment decoder (b) Decoder implementation using a 128-bit ROM*

line on the memory matrix, can be regarded as a 'don't care' column whose entries can be either 1s or 0s at will. In this example 0s have been selected for all the entries in the column labelled X. Similarly, for those addresses selecting the invalid codes, the entries in the truth table are 0. Alternative designs might either use output X to indicate out-of-range input values, or else decode the full range of input minterms to give the appropriate hexadecimal displays required of a 4-bit binary to 7-segment decoder. The completed truth table can now be used as programming instructions for the ROM to be used as a decoder.

Static 1-hazards, similar to those generated in combinational circuits implemented with conventional gates, can also be generated in ROM implementations of Boolean functions for exactly the same reason – i.e., because of the time delay that exists between a true and complemented variable when the complemented variable has been generated by inverting the true variable. In fact they are more likely in the ROM-based version, precisely because there is no possibility of using redundant groupings in the ROM to eliminate them. For example, suppose a 4-variable function f implemented by a ROM contains the two minterms $\bar{A}B\bar{C}\bar{D}$ and $\bar{A}BC\bar{D}$. When either of the two combinations of the variables A, B, C and D representing these minterms appears at the input terminals of the address decoder the function has the value $f = 1$. If, initially, the input is $\bar{A}BC\bar{D}$ and it then changes to $\bar{A}B\bar{C}\bar{D}$, the output of the ROM internal decoder for word line 6 will change from 1 to 0 when C changes from 1 to 0, and the ROM internal decoder for word line 4 will change from 0 to 1 when $\bar{C}$ changes from 0 to 1. Since $\bar{C}$ is generated internally by inverting input C, the change in level of word line 4 occurs a short time after the change in level of word line 6. For this reason, f will fall to 0 for a very short period of time before returning to the value $f = 1$, thus generating a static-1 hazard.

11.6 Internal addressing techniques in ROMs

The connection matrix shown in Figure 11.8(a) is for a ROM addressed in one dimension only. The total capacity of this ROM is $2 \times 8 = 16$ bits, and the Boolean functions generated by it are

$$Z_1 = m_2 + m_5 + m_7$$

and

$$Z_2 = m_0 + m_4 + m_5 + m_7.$$

For this method of addressing, a total of ten connections to the ROM matrix is required, made up of eight word lines and two bit lines.

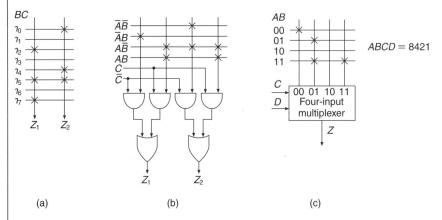

Figure 11.8 *(a) A 2 × 8 bit word ROM addressed in one dimension, and (b) in two dimensions (c) Generation of one four-variable Boolean function using a two-dimensional addressing scheme*

An alternative, two-dimensional, method of addressing a ROM is illustrated in Figure 11.8(b). Examination of the connection matrix shows that the same Boolean functions are generated as in the previous example. In effect, each bit line has now been split into two sections, and selection of the appropriate section is done by a 2-to-1 multiplexer which is controlled by the Boolean variable C. Using this two-dimensional addressing technique, there is a reduction in the number of connections to the ROM matrix. Now, a total of only eight connections to the ROM matrix is required, made up of four word lines and four bit lines, a reduction of two when compared with the one-dimensional addressing technique.

The larger the number of input variables, the more significant this reduction in matrix connections becomes. For example, the same ROM matrix in conjunction with a 4-to-1 multiplexer can be used to generate the 4-variable function $Z = \sum 0, 5, 13, 15$ as illustrated in Figure 11.8(c). Again, a total of just eight matrix connections is required in this implementation. However, if the same function had been generated by a ROM addressed in one dimension, 17 matrix connections would have been required, consisting of 16 word lines and one bit line.

There are significant advantages in addressing large-capacity ROMs internally in two dimensions. For example, a 1024×8-bit ROM, using one-dimensional addressing, would require a 10-to-1024 line internal address decoder and eight bit lines, giving a total of 1032 connections to the ROM matrix. By comparison, the two-dimensional addressing scheme shown in Figure 11.9 uses six of the input variables, *A, B, C, D, E* and *F*, to drive a 6-to-64 line address decoder, while the other four variables, *G, H, I* and *J*, are used to provide the control signals to eight 16-to-1 multiplexers. For this scheme, a total of only 192 connections needs to be made to the ROM matrix, consisting of 64 input (word) lines and $8 \times 16 = 128$ output (bit) lines. Since connections take up valuable space on the ROM chip, this represents a considerable saving when compared with the one-dimensional addressing scheme.

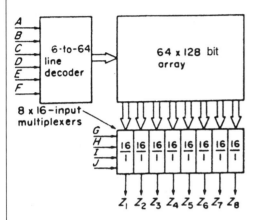

Figure 11.9   *Structure of a 2K-byte ROM addressed in two dimensions*

11.7 Memory addressing

One typical application for memory chips is to provide storage for programs and data in a microprocessor system. It is common practice for a number of memory chips to provide this function, each of them having their output lines connected to the system data bus via tri-state gates. At any given instant only one address location can be accessed, so that only one memory chip can be connected to the system data bus at that time. Consequently a common task facing the system designer is to arrange the selection of one out of a number of memory chips by the microprocessor.

For example, consider the microprocessor system shown in Figure 11.10. The microprocessor itself has 8 data lines and a total of 20 address lines (i.e., a possible total addressing capacity of $8 \times 2^{20} = 1$Mbyte). The total storage capacity required by this system, however, is only 128Kbyte which is provided by eight 16Kbyte memory chips. Address lines $A1$ to $A14$ inclusive are required to address the 16384 locations on each memory chip. A 3-to-8 line decoder is used to select a single memory chip (actually almost always providing active-low $\overline{CE}$ signals, one for each memory chip). Selection of one of the eight chips is provided by the three address lines A_{15}, A_{16} and A_{17} which are used as the input select signals to the decoder.

The remaining three address line outputs from the microprocessor could be left unconnected. However, in this case, to ensure that a particular location in the memory can only be accessed by a unique address, a system of *absolute decoding* is employed. Address lines A_{18}, A_{19} and A_{20} are connected to an OR gate whose output is connected to the active low enable inputs of the decoder. The decoder is enabled only when

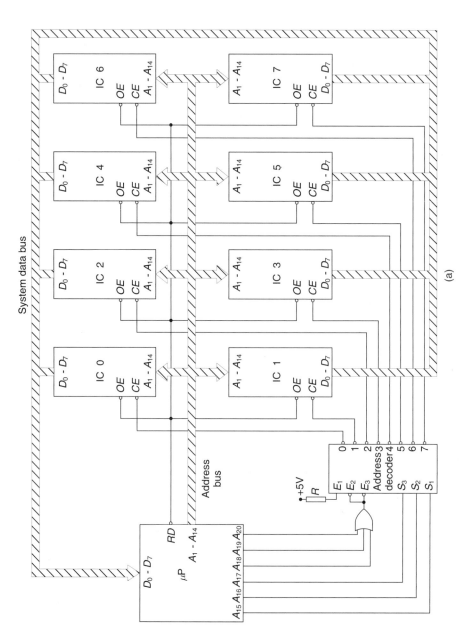

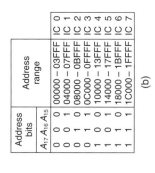

Address bits			Address range	
A_{17}	A_{16}	A_{15}		
0	0	0	00000 – 03FFF	IC 0
0	0	1	04000 – 07FFF	IC 1
0	1	0	08000 – 0BFFF	IC 2
0	1	1	0C000 – 0FFFF	IC 3
1	0	0	10000 – 13FFF	IC 4
1	0	1	14000 – 17FFF	IC 5
1	1	0	18000 – 1BFFF	IC 6
1	1	1	1C000 – 1FFFF	IC 7

(b)

Figure 11.10 *(a) Memory address decoding for eight 16Kbyte memory chips in a microprocessor system (b) Address ranges*

$A_{18} = A_{19} = A_{20} = 0$. Any other combination of these three address signals will disable the decoder. If this were not done, and the address decoder were always enabled, then each memory location could be accessed at any of eight possible addresses sent out by the microprocessor, i.e. at its *base address* plus an integer multiple (up to 7 times) of 2^{17}. Sometimes this can be advantageous to the programmer, and is termed 'memory fold-back'. The various combinations of the decoder select signals A_{15}, A_{16}, and A_{17} are tabulated in Figure 11.10(b) with the corresponding address range for each memory chip using the assumption that $A_{18} = A_{19} = A_{20} = 0$.

A common output enable (OE) signal is supplied to each of the eight memory chips. This signal, in conjunction with the individual chip enable signals, enables the output tri-state gates of the selected memory chip.

11.8 Design of sequential circuits using ROMs

ROMs can also be used for the implementation of clock-driven sequential circuits and, as an example, the NBCD invalid code detector, designed in Chapter 8, using JK flip-flops and NAND gates will be implemented here using a ROM.

In this problem, serial NBCD data arrives on line X, with the most significant digit first. Each data bit is synchronised with a clock pulse. It is required to design a circuit using a ROM that generates a fault signal $Z = 1$ each time an invalid code is received.

The block diagram and the internal state diagram are shown in Figures 11.11(a) and (b). The state table (Figure 11.11(c)) is shown in a suitable form for programming a ROM. For example, in the first row of the table, the current input to the ROM is $A = 0$, $B = 0$, $C = 0$, and $X = 0$, and the ROM output word is $A = 1$, $B = 0$, $C = 0$ and $Z = 0$. Using this state table, the ROM design can be developed, as illustrated in Figure 11.11(d). This implementation (using two-dimensional addressing) requires an $8 \times 8 = 64$-bit ROM.

Besides the ROM, additional logic is required to produce the output signal $Z = ABC\sqcap$, and also three D-type flip-flops are required, one in each address line to buffer the ROM outputs from the inputs and to synchronise the operation of the circuit to the clock. These additional components with their connections to the ROM are shown in the circuit diagram in Figure 11.11(e). The outputs from the ROM on the lines A, B and C are transferred back to the input of the ROM on the trailing edge of the clock pulse.

11.9 Programmable logic devices (PLDs)

Clearly, each additional address line or input variable *doubles* the size of ROM required. A ROM having 12 address inputs requires $2^{12} = 4096$ internal word lines and the storage capacity is $4096 \times 8 = 32768$ bits. In a ROM of this size, all of the 4096 possible minterms are represented internally, and any eight of the 2^{4096} possible Boolean functions of 12 variables can be generated. In practice, the designer may require to use only a small percentage of these and this would mean uneconomic use of the ROM. Because of this disadvantage, programmable logic devices (PLDs) have been developed that provide the advantages of ROM-based design, tailored to typical system requirements, and which can provide a more economic implementation of Boolean functions in those situations where a ROM would not be economic.

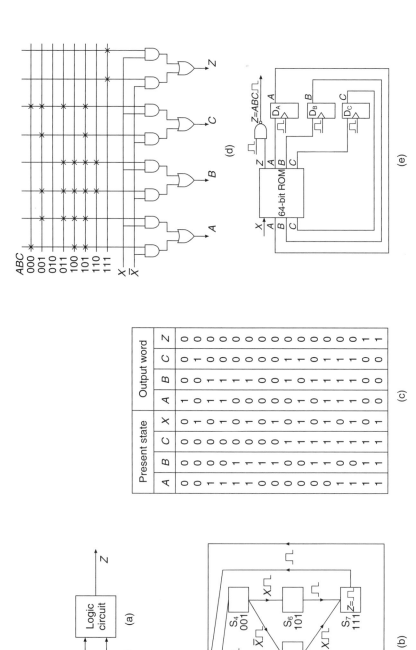

Figure 11.11 The invalid code detector (a) Overall system diagram (b) Internal state diagram (c) State table (d) Connection matrix (e) Circuit diagram for ROM implementation

Several main categories of programmable logic devices are available. The Programmable Gate Array (PGA), the first PLD to be developed, provides a single level of logic such as an array of multi-input AND gates. Developed from these was the Programmable Logic Array (PLA), which, in essence, actually consists of two logic arrays, a programmable AND array and a programmable OR array. A PLA is capable of implementing any logic function in two level sum-of-products form. A special case of the PLA, and one of the most widely used programmable logic devices, is the Programmable Array Logic (PAL) device, consisting of a programmable AND array with a *fixed* OR array and bi-directional input/output pins. The Programmable Logic Sequencer (PLS) provides two levels of logic, usually an AND/OR array together with a number of on-chip single-bit memory elements, some of whose outputs are fed back to the inputs of the programmable array while others function as output latches.

Uncommitted Logic Arrays (ULAs) are to PGAs and PLAs what mask programmed ROMs are to PROMs: i.e., the logic designer must send to the ULA manufacturer a complete and tested gate design to be implemented using that manufacturer's own general-purpose chip consisting of basic logic 'cells' that can be connected during manufacture in various different ways to produce different logic gates and therefore ultimately a *semi-custom* logic chip, economic only for complex systems in large quantities. For very complex systems in very large quantities, or if the system required has some feature(s) that cannot easily be implemented on a general-purpose programmable device, it may be worthwhile using *fully-custom* logic design, where a dedicated IC is designed from scratch, specifically for the logic system required.

The prototyping tool most often used now for medium-sized logic circuit development is the Field Programmable Gate Array (FPGA). These currently represent the most complex form of development of PLDs. The distinctions between the various types of PLDs, PGAs, PALs, and FPGAs are largely ones of scale and the complexity and sophistication of the tools, both hardware and software, available from the manufacturers to aid design and prototyping. A PAL implies a design approach positioned, roughly speaking, halfway between that of a PGA and a FPGA.

11.10 Programmable gate arrays (PGAs)

As in the case of a PROM, programming of these devices is carried out by blowing fusible links at points in the logic array specified by the designer. A typical arrangement is shown in Figure 11.12(a) where the input X and its complement $\bar{X}$ are both connected to the next gate in the array *via* separate fusible links. X and $\bar{X}$ are produced internally by buffer circuits having properties similar to open collector gates, and the vertical connecting line behaves as a pull-up resistor except that, as a result of special circuitry included in the PLD, there is no cross-interaction (*via* the vertical connecting line) from any variable to any other variable. Therefore,

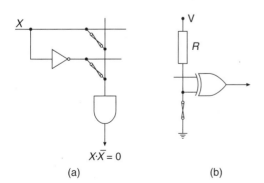

Figure 11.12   *Fusible link arrangements for a PGA (a) The programmable AND array (b) The output XOR gates*

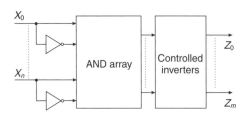

(a)

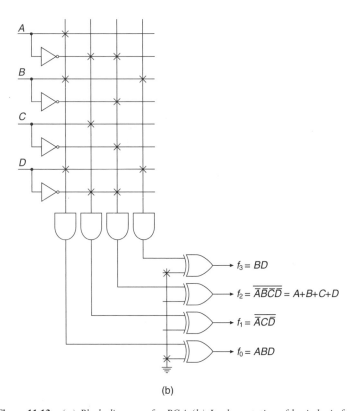

(b)

Figure 11.13   *(a) Block diagram of a PGA (b) Implementation of basic logic functions using a PGA*

the wired AND of all the connected variables is produced. In the context of PLDs, this is usually indicated by drawing a single input into an AND gate symbol, and as many variables or their complements as required can be connected to the input of the AND gate symbol. When neither of these links has been blown, the gate output is $X \cdot \bar{X} = 0$; by blowing one or other of the two fusible links, either X or $\bar{X}$ are left connected to the gate input, and this will be ANDed with any other variables or complements also left connected to the same line. Alternatively, if both fusible links are blown, the variable X is disconnected and does not appear in the expression for the output from this AND gate symbol. These devices are 'one-time programmable' (OTP).

A PGA may also have XOR gates on each of its output lines, connected as controlled inverters (see section 4.15). Typical connections to the XOR gate are shown in

Figure 11.12(b). When the fusible link is intact, the lower input to the XOR gate is grounded and it operates in transmission mode. If the fusible link is blown, that input rises to logic level 1, courtesy of the pull-up resistor, and the gate inverts the signal present at its other input.

A block diagram of a typical PGA is shown in Figure 11.13(a). It consists of a number of input lines, one for each Boolean variable. Within the device, the complement of each of these variables is generated. The input variables and their complements are then fed to a programmable array of AND gates whose outputs are fed to the inputs of an array of XOR gates, operating as controlled inverters.

A circuit diagram for a 4-input, 4-output PGA is shown in Figure 11.13(b). The main part of the AND array consists of intersecting vertical and horizontal lines. A cross at an intersection indicates the presence of an intact fusible link (i.e. a fusible link that the designer has not blown), and means that the variable identified by the intersecting horizontal line is one of the inputs to the wired AND gate identified by the intersecting vertical line. Since PGAs provide the designer with only a single level of logic, they have only very limited application.

11.11 Programmable logic arrays (PLAs)

A PLA consists of a programmable AND array, similar to that in a PGA, and which can be regarded as a Boolean product generator, together with a programmable OR array similar to the AND array except that each single-input OR gate symbol produces the OR of all the connected variables. The PLA can be regarded as a logical sum generator. Usually there are also programmable XOR gates, acting as controlled inverters, on each of the output lines. Some PLAs have tri-state buffers, having a common enable line, between the programmable XOR gates and the chip outputs. The connection matrices are again conventionally drawn as a group of intersecting lines, as shown in Figure 11.14, and the presence of a fusible link connection in the AND and OR gates is indicated by crosses at those intersections specified by the logic designer. Both mask and field programmable logic arrays are available; mask programmable devices (like mask programmed ROMs) are programmed by the manufacturer acting on the instructions of the customer, while field programmable devices (FPLAs) are programmed locally by the purchaser. A typical FPLA has 16 inputs, 48×32-input AND gate equivalents, 8×48-input OR gate equivalents, and 8 XOR output gates, or, alternatively, 8 tri-state output gates. Clearly, implementation on this FPLA is limited to functions that have no more than 48 product terms.

A typical application for a PLA arises in code conversion. For example, the conversion from XS3 code to NBCD is shown in Figure 11.15(a). From this table the K-maps for the NBCD output signals P, Q, R and S are plotted and minimized in Figure 11.15(b). Some codes cannot occur in the XS3 code, giving rise to some 'don't care' cells (marked 'X') which aid the minimisation (see section 3.10). The equations of the output signals are:

$P = AB + ACD$
$Q = \bar{B}\bar{C} + \bar{B}\bar{D} + BCD$
$R = \bar{C}D + C\bar{D} = C \oplus D$
$S = \bar{D}$

and the implementation of these functions on a PLA is illustrated in Figure 11.15(c).

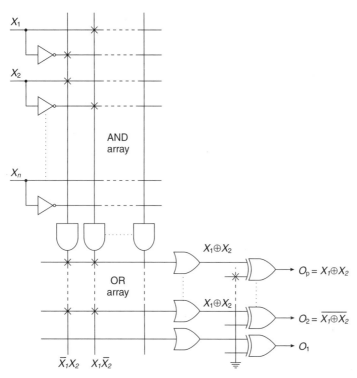

Figure 11.14 *Simplified diagram of a programmable logic array (PLA). Using the connections shown, this PLA produces the XOR and XNOR of two of the input variables*

It often happens that when an FPLA has been programmed, the AND gates may not all have been utilised. However, the programmer need not blow the corresponding fusible links since when they are intact the output of an unused AND gate is $A \cdot \bar{A} \cdot B \cdot \bar{B} \cdot \ldots \cdot N \cdot \bar{N} = 0$ and so it will not affect the output of the OR array to which it is connected. Leaving unused fusible links intact will also provide a limited amount of flexibility to the programmer who, at some later stage, may wish to introduce an additional product term or, alternatively, modify one or more of the functions already programmed.

PLAs can also be used for implementing more complex Boolean functions, such as (for example) the following four 6-variable functions:

$$
\begin{array}{ccccc}
& 1 & 2 & 3 & 4 \\
f_1 = & AC\bar{D}E\bar{F} + & \bar{A}E\bar{F} + & A\bar{C}D + & B\bar{D} \\
& 5 & 3 & 4 & 6 \\
f_2 = & AB\bar{E}F + & A\bar{C}D + & B\bar{D} + & EF \\
& 7 & 8 & 9 & 10 \\
f_3 = & ACDE\bar{F} + & AB\bar{C}\bar{D} + & \bar{A}\bar{C}D + & ABD \\
& 11 & 2 & 4 & 12 & 6 \\
f_4 = & A\bar{C}\bar{D}\bar{E} + & \bar{A}E\bar{F} + & B\bar{D} + & CD + & EF
\end{array}
$$

There are 12 separate product terms in these four equations, each of them numbered, and none of them minterms. Since there are 12 product terms, the PLA must have 12

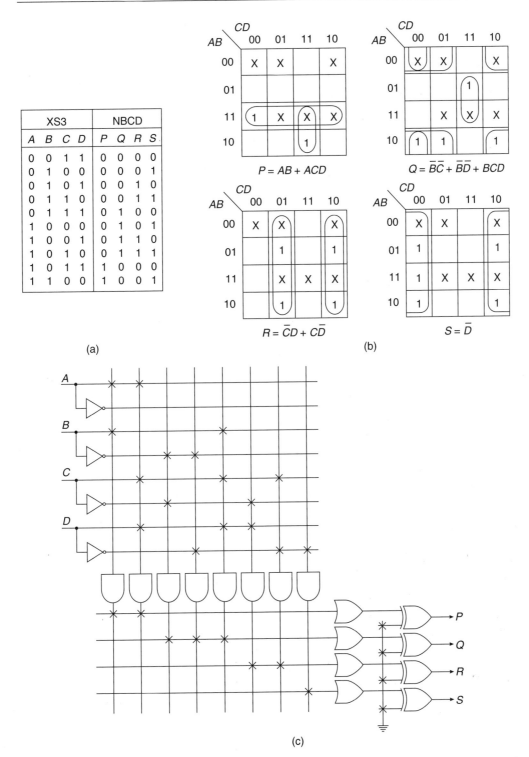

Figure 11.15 *(a) The XS3 and NBCD codes (b) K-maps for the XS3 to NBCD conversion (c) Implementation of XS3 to NBCD code converter using a PLA*

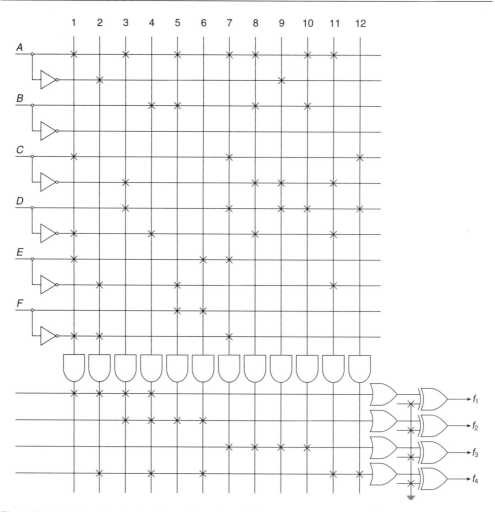

Figure 11.16   *Boolean function implementation using a PLA*

AND gate equivalents for generating these terms. The AND gate outputs are fed to an OR array which needs the equivalent of four gates. There is a total of $(12 \times 12) + (4 \times 12) + 4 = 196$ locations for fusible links (each of which must be programmed as 0 or 1) on the PLA in the design shown in Figure 11.16. To implement the same four functions using a ROM would require a storage capacity of 2^6 bits for each function, or a total of $2^6 \times 4 = 256$ bits (i.e., 256 fusible links), and the functions would need to be expressed in canonical form. However, programming PLAs is often more complex than programming ROMs because PLAs are frequently used to implement Boolean functions with a larger number of variables.

It will be recalled that the Boolean function $f = \bar{A}B + AC$ generates a static 1-hazard (if implemented as a minimised sum of products) when $B = C = 1$. In this case the equation reduces to $f = \bar{A} + A = 1$, but because of the propagation delay in the inverter producing $\bar{A}$, $f = 0$ is generated for a short period after A has changed from 1 to 0, giving a typical static 1-hazard. In general, this hazard is eliminated by adding the consensus term BC to the original expression which then becomes $f = \bar{A}B + AC + BC$. For the condition $B = C = 1$, the equation now reduces to

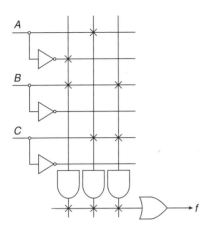

Figure 11.17 *Hazard elimination when implementing Boolean functions with a PLA*

$f = \bar{A} + A + 1$ and (irrespective of the timing of the changes in A and $\bar{A}$) the output remains at $f = 1$ throughout, and the hazard is eliminated. In a discrete gate (SSI) design this requires an extra gate; when using a PLA implementation, elimination of the hazard requires programming the additional product term BC, as illustrated in Figure 11.17.

When implementing several Boolean functions on a PLA, minimisation of these functions does not necessarily offer the optimum implementation. For example, the two 4-variable functions

$$f_1 = \sum 1, 2, 3, 5, 6, 7, 10, 11, 12, 13$$
$$f_2 = \sum 2, 4, 6, 9, 12, 13$$

have been plotted in Figure 11.18 and minimised in the normal way. The minimal equations are

$$f_1 = \overset{1}{\bar{A}D} + \overset{2}{\bar{A}C} + \overset{3}{\bar{B}C} + \overset{4}{AB\bar{C}} + \overset{5}{B\bar{C}D}$$

$$\text{and } f_2 = \overset{6}{\bar{A}C\bar{D}} + \overset{7}{B\bar{C}\bar{D}} + \overset{4}{AB\bar{C}} + \overset{8}{A\bar{C}D}$$

Each of the prime implicants has been numbered; eight product terms cover both functions. Therefore, for their implementation on a PLA, eight AND gates are needed to generate the eight product terms.

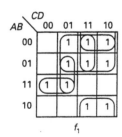

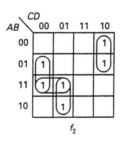

Figure 11.18 *K-map plots for two functions to be implemented on a PLA*

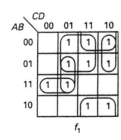

Figure 11.19 *Modified loopings for f_1 giving optimum PLA implementation*

An economy of space on the PLA can, however, be achieved by covering the function f_1 with the alternative set of loopings shown in Figure 11.19. The function f_1, now in non-minimal form, may be written

$$f_1 = \bar{A}D + \bar{B}C + \bar{A}C\bar{D} + AB\bar{C} + B\bar{C}D$$

The prime implicant $\bar{A}C$ no longer appears in this equation and has been replaced by the product term $\bar{A}C\bar{D}$ which is now common to the equations for f_1 and f_2. Because of the modification of the map loopings, the number of product terms to be implemented has been reduced from eight to seven, thus providing a more economic utilisation of space on the PLA. This theme has already been developed in more detail in sections 3.18 and 3.19.

11.12 Programmable array logic (PAL)

The versatility of a PLA stems from the fact that the output of any of the programmable AND gates can be shared among a number of OR gates. There are many applications where this provision is not required, and so PALs have been developed in which gates in a programmable AND array are dedicated to a particular output OR gate, making the OR matrix fixed instead of programmable. The number of AND gates associated with each OR gate is typically 2, 4, 8 or 16. This gives a programmable device that is, in principle, less flexible than a PLA but which is often easier to use. PALs often also include a '*security fuse*' which is maintained intact while programming, then after programming the fuse is blown and as a result the PAL is protected from copying or further programming. This can be a useful feature for a designer wishing to prevent unauthorised reading and decoding of the PAL matrix contents, known as 'reverse engineering' of the design.

When implementing Boolean functions with a PAL, the logic equations must be simplified as for a PLA, and since the AND gates are dedicated to a particular OR gate, term sharing is not possible. This means that there is little advantage in searching for optimum solutions as there would be for PLA implementations.

A simple PAL structure is illustrated in Figure 11.20(a). This device has two input lines, two AND gate equivalents, each having programmable input connections to four vertical lines, and one non-programmable (fixed circuit) 2-input OR gate. With the fusible links intact, the output of both AND gates is $A \cdot \bar{A} \cdot B \cdot \bar{B} = 0$ and so the OR gate output is $f = 0$. In Figure 11.20(b) this elementary PAL has been programmed by selectively blowing the fusible links so that the XOR function $f = \bar{A}B + A\bar{B}$ is realised.

Additional flexibility can be provided if the manufacturer places tri-state buffers between the outputs of the OR gates and the external connecting pins. In the elementary PAL structure shown in Figure 11.21, a programmable AND gate provides the enable signal X for the tri-state output buffer. There is also a buffer with its *input* connected to

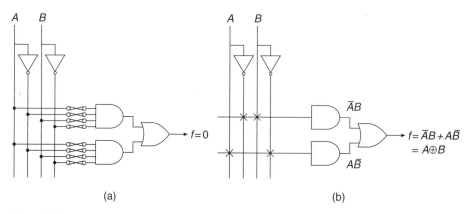

(a)   (b)

Figure 11.20   *A simple PAL (a) before programming, (b) after programming*

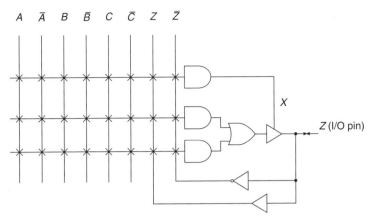

Figure 11.21 *Programmable I/O connection pin in a simple unprogrammed PAL*

Z, sending Z and $\bar{Z}$ signals to the AND array. This arrangement is reminiscent of the bi-directional tri-state connection of a device to a system bus, and gives bi-directional capability to the connection pins of the PAL. The external connection pin Z can now be used in four different operating modes, depending upon the programming of the enable signal X:

1. *Dedicated input pin*: the AND gate generating X is programmed so that $X = 0$ always, and the tri-state buffer is permanently disabled (i.e. is always in the high impedance state). Pin Z is now always used as an input pin and has direct access to the programmable AND array, and in this case another pin(s) would need to be used for output.
2. *Dedicated output pin*: the AND gate generating X is programmed so that $X = 1$ always, and the tri-state buffer is permanently enabled. Pin Z is now always used as an output pin.
3. *Controlled output pin*: the AND gate generating X is programmed so that X can be either 0 or 1 depending upon the present state of its input signals. Pin Z can now be either an output pin ($X = 1$) or is a disabled (or input) pin ($X = 0$).
4. *Output pin with feedback*: pin Z provides controlled feedback to the AND array. When the output of the controlling AND gate is $X = 0$, the tri-state gate is disabled and there is no feedback; when $X = 1$, the tri-state gate is enabled and signal Z is fed back to the AND array.

Unlike PLAs, PALs have manufacturers' type numbers that indicate directly the basic PAL internal structure. For a PAL having type number pXq, p is the number of inputs to the AND matrix, q is the number of outputs, and X indicates whether the outputs are active high (X replaced by H), active low (X replaced by L), programmable (X replaced by P), or other possibilities. The other important information required by the designer is the number of inputs per OR gate, which must be obtained from the manufacturer's catalogue; many PALs are manufactured having two inputs per OR gate.

As an example of the use of a PAL, a circuit for converting from NBCD to the 5421 code will be designed. The truth table for the conversion is shown in Figure 11.22(a) and the K-maps for the 5421 code outputs, P, Q, R, and S are plotted in Figure 11.22(b).

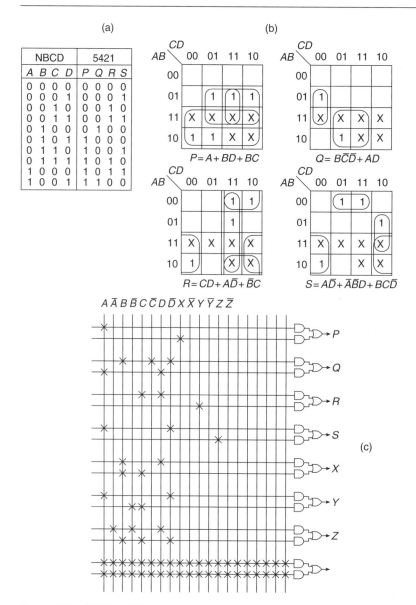

Figure 11.22   *NBCD to 5421 code converter (a) Truth table (b) K-map plots (c) Implementation using a type 10H8 PAL*

(The 'don't care' terms in the NBCD code have been used to simplify the final equations.) Each of the outputs has been simplified and their minimal functions are written under the relevant K-map plot.

The designer must now choose a suitable PAL for implementing the four output functions. Since there are only four input signals to the code converter, a PAL having the smallest number of available inputs would be selected. For example, PAL type 10H8 has 10 inputs and 8 outputs. A further examination of the manufacturer's data sheet might reveal that the selected PAL only provides two inputs per OR gate, whereas the output signals *P*, *R*, and *S* are each obtained from a sum of *three* terms.

This problem can be overcome by splitting the equations for P, R, and S into two sections. For example,

$$P = A + (BD + BC) = A + X$$

where $X = BD + BC$. X can be generated at the output of one of the eight 2-input OR gates and may be fed back to one of the six unused input pins of the PAL. It can then be combined with A to form the term $A + X$ at the output of one of the remaining unused 2-input OR gates. Similarly, the functions R and S can also be sectionalised so that

$$R = CD + (A\bar{D} + \bar{B}C) = CD + Y$$
$$S = A\bar{D} + (\bar{A}\bar{B}D + BC\bar{D}) = A\bar{D} + Z$$

where clearly $Y = A\bar{D} + \bar{B}C$ and $Z = \bar{A}\bar{B}D + BC\bar{D}$. Signals Y and Z are fed back to the input of the PAL and combined with CD and $A\bar{D}$ respectively to form terms R and S at the output.

Implementation of the encoder is shown in Figure 11.22(c). Because of the feedback of X, Y, and Z, seven out of the ten input pins are utilised. Also, because of the sectionalisation of the output signals P, R, and S, seven out of the eight output pins are used. However, in this implementation, the signals P, R, and S are generated after two passes through the PAL giving increased propagation delay in these signals.

11.13 Programmable logic sequencers (PLSs)

The essential features of a PLS are illustrated in the block diagram shown in Figure 11.23. In addition to the *programmable* AND and OR arrays provided on a PLA, the PLS

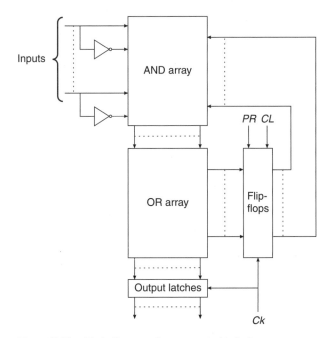

Figure 11.23 *Block diagram of a programmable logic sequencer*

has a number of on-chip single-bit memory elements which may be SR, JK or D-type flip-flops. In some cases, control of the flip-flops is available so that JK flip-flops can be converted to D-type flip-flops. Additionally, facilities are provided for latching the outputs.

PLSs are primarily intended for implementing synchronous sequential state machines of either the Mealy or the Moore type (see chapter 8). The flip-flop outputs represent the state variables of the state machine. Some of the flip-flop outputs are fed back to the programmable AND/OR arrays where they can be combined with the machine inputs to generate the flip-flop input signals. Other flip-flop outputs can be combined with machine inputs to generate the machine outputs. A clock signal is provided by an external source, and asynchronous preset and clear facilities for the flip-flops may also be available.

As an example of the use of a PLS, a hexadecimal counter will be designed. It will be assumed that the on-chip single-bit memory elements on the selected PLS are D-type flip-flops and the output of the counter will be decoded ready for directly driving a conventional 7-segment display.

The state table for the hexadecimal counter is shown in Figure 11.24(a). The inputs to the four flip-flops required for each state change are shown on the right of the state table, and have been obtained with the aid of the steering table for the D-type flip-flop shown in Figure 11.24(b). K-maps have been plotted and simplified for each flip-flop input in Figure 11.24(c), and the minimum form of the input equations obtained from them are

$$D_D = \bar{C}D + \bar{B}D + \bar{A}D + ABC\bar{D} \qquad D_B = A\bar{B} + \bar{A}B$$
$$D_C = AB\bar{C} + \bar{A}C + \bar{B}C \qquad D_A = \bar{A}.$$

The segment allocation for the seven-segment display is defined in Figure 11.25(a), as well as the segmental representation of each of the 16 hexadecimal digits. A truth table for the seven-segment decoder is shown in Figure 11.25(b), and the implementation of counter and display decode logic is shown in Figure 11.25(c).

Sixteen product lines are required for decoding the hexadecimal digits. As an example, the hexadecimal digit A (corresponding to the binary code 1010) requires that segments P, Q, R, T, U and V should be illuminated. Hence, for this binary combination, the signal for driving each of these segments must be set to 1, and the signal for segment S must be set to 0. The functions for all of the segments are easily read from the truth table. For example, the function for segment P is

$$P = \sum 0, 2, 3, 5, 6, 7, 8, 9, A, C, E, F$$

and the corresponding expressions for the other segments are obtained similarly.

As a further example of implementing a synchronous sequential machine using a PLS, consider the design of an invalid code detector for XS3 codes using a PLS with on-chip D-type flip-flops. Four-bit XS3 codes are fed to the detector, most significant digit first, and the machine is to be designed to give an active high output when an invalid code is received.

The ASM chart for the detector is shown in Figure 11.26(a). Since there are eleven states on the chart, four flip-flops ($2^4 = 16, 2^3 = 8$) are required to implement the machine. The column headed 'next state' in the state table (see Figure 11.26(b)) is a tabulation of the flip-flop input functions. These functions are relatively sparse

(a)

Present state				Next state				Flip-flop inputs			
D	C	B	A	D	C	B	A	D_D	D_C	D_B	D_A
0	0	0	0	0	0	0	1	0	0	0	1
0	0	0	1	0	0	1	0	0	0	1	0
0	0	1	0	0	0	1	1	0	0	1	1
0	0	1	1	0	1	0	0	0	1	0	0
0	1	0	0	0	1	0	1	0	1	0	1
0	1	0	1	0	1	1	0	0	1	1	0
0	1	1	0	0	1	1	1	0	1	1	1
0	1	1	1	1	0	0	0	1	0	0	0
1	0	0	0	1	0	0	1	1	0	0	1
1	0	0	1	1	0	1	0	1	0	1	0
1	0	1	0	1	0	1	1	1	0	1	1
1	0	1	1	1	1	0	0	1	1	0	0
1	1	0	0	1	1	0	1	1	1	0	1
1	1	0	1	1	1	1	0	1	1	1	0
1	1	1	0	1	1	1	1	1	1	1	1
1	1	1	1	0	0	0	0	0	0	0	0

(b)

Q^t	$Q^{t+\delta t}$	D^t
0	0	0
0	1	1
1	0	0
1	1	1

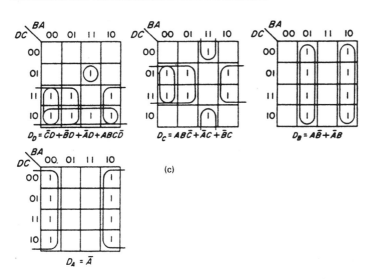

$$D_D = \bar{C}D + \bar{B}D + \bar{A}D + ABC\bar{D}$$

$$D_C = AB\bar{C} + \bar{A}C + \bar{B}C$$

$$D_B = A\bar{B} + \bar{A}B$$

$$D_A = \bar{A}$$

(c)

Figure 11.24 *(a) State table for the hexadecimal counter (b) Steering table for the D-type flip-flop (c) K-maps for the hexadecimal counter*

(few 1s and 'don't care' terms compared to the total number of minterms), so they have been plotted on the reduced dimension maps (see sections 3.20 to 3.23) shown in Figure 11.27(a). After simplification, minimised excitation functions have been read directly from these maps and are written below the maps.

The output function Z has been plotted on a 5-variable K-map in Figure 11.27(b), and the following equations are obtained from the $X = 0$ and $X = 1$ maps respectively:

$$Z_{X=0} = \bar{X}PS + \bar{X}PR + \bar{X}QRS$$
$$Z_{X=1} = XPS + XQR + XQRS$$

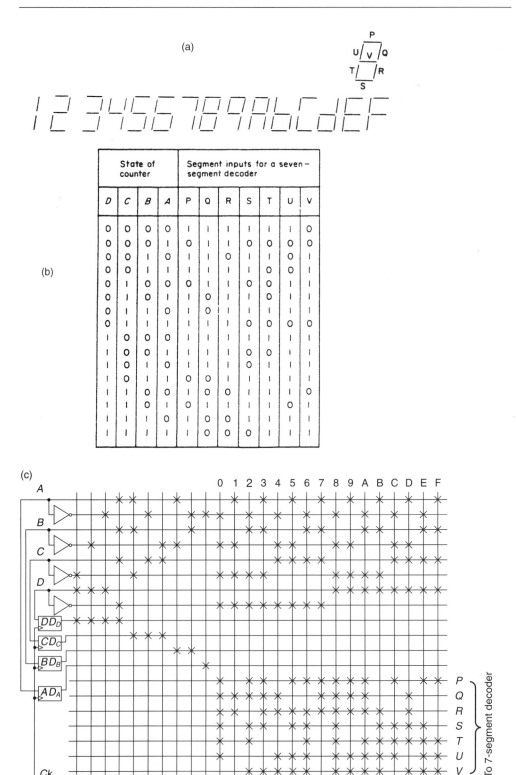

State of counter				Segment inputs for a seven-segment decoder						
D	C	B	A	P	Q	R	S	T	U	V
0	0	0	0	1	1	1	1	1	1	0
0	0	0	1	0	1	1	0	0	0	0
0	0	1	0	1	1	0	1	1	0	1
0	0	1	1	1	1	1	1	0	0	1
0	1	0	0	0	1	1	0	0	1	1
0	1	0	1	1	0	1	1	0	1	1
0	1	1	0	1	0	1	1	1	1	1
0	1	1	1	1	1	1	0	0	0	0
1	0	0	0	1	1	1	1	1	1	1
1	0	0	1	1	1	1	0	0	1	1
1	0	1	0	1	1	1	0	1	1	1
1	0	1	1	0	0	1	1	1	1	1
1	1	0	0	1	0	0	1	1	1	0
1	1	0	1	0	1	1	1	1	0	1
1	1	1	0	1	0	0	1	1	1	1
1	1	1	1	1	0	0	0	1	1	1

Figure 11.25 *(a) Seven-segment display representation of hexadecimal digits (b) Truth table for seven-segment decoder (c) Implementation of the hexadecimal counter using a PLS*

(a)

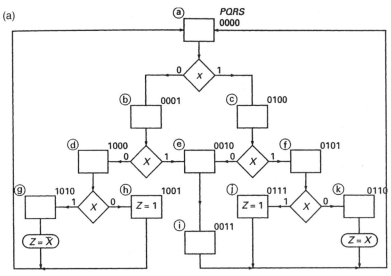

(b)

Input X	State symbol	Present state P Q R S	Next state Flip-flop inputs D_P D_Q D_R D_S
0	ⓐ	0 0 0 0	0 0 0 1
0	ⓑ	0 0 0 1	1 0 0 0
0	ⓔ	0 0 1 0	0 0 1 1
0	ⓘ	0 0 1 1	0 0 0 0
0	ⓒ	0 1 0 0	0 0 1 0
0	ⓕ	0 1 0 1	0 1 1 0
0	ⓚ	0 1 1 0	0 0 0 0
0	ⓙ	0 1 1 1	0 0 0 0
0	ⓓ	1 0 0 0	1 0 0 1
0	ⓗ	1 0 0 1	0 0 0 0
0	ⓖ	1 0 1 0	0 0 0 0
0	–	1 0 1 1	X X X X
0	–	1 1 0 0	X X X X
0	–	1 1 0 1	X X X X
0	–	1 1 1 0	X X X X
0	–	1 1 1 1	X X X X
1	ⓐ	0 0 0 0	0 1 0 0
1	ⓑ	0 0 0 1	0 0 1 0
1	ⓔ	0 0 1 0	0 0 1 1
1	ⓘ	0 0 1 1	0 0 0 0
1	ⓒ	0 1 0 0	0 1 0 1
1	ⓕ	0 1 0 1	0 1 1 1
1	ⓚ	0 1 1 0	0 0 0 0
1	ⓙ	0 1 1 1	0 0 0 0
1	ⓓ	1 0 0 0	1 0 1 0
1	ⓗ	1 0 0 1	0 0 0 0
1	ⓖ	1 0 1 0	0 0 0 0
1	–	1 0 1 1	X X X X
1	–	1 1 0 0	X X X X
1	–	1 1 0 1	X X X X
1	–	1 1 1 0	X X X X
1	–	1 1 1 1	X X X X

Figure 11.26 *Design of an XS3 invalid code detector (a) ASM chart (b) State table*

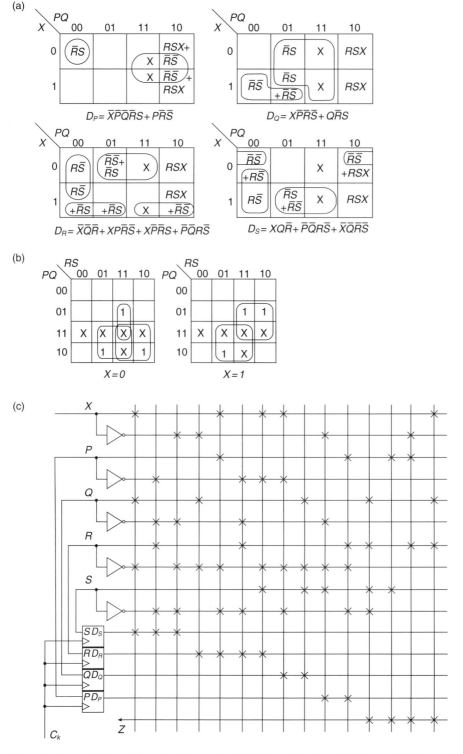

Figure 11.27   *(a) Reduced-dimension K-maps for flip-flop input functions (b) Five-variable K-map for output function Z (c) PLS implementation*

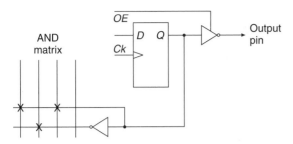

Figure 11.28 *Output connections on a PAL-based PLS*

Combining these two equations gives

$$Z = PS + \bar{X}PR + XQR + QRS.$$

(Alternatively, the same result is obtainable from the equivalent reduced-dimension map for the output function, although with greater difficulty, because of the complex functions involved.) The PLS implementation of the invalid code detector is shown in Figure 11.27(c).

Manufacturers have also modified PALs to behave as PLSs by incorporating D-type flip-flops on the PAL chip. A typical example is the type 16R6 which provides 16 AND gate inputs and six rising edge-triggered D-type flip-flops, having a common external clock connection. The flip-flop outputs are taken to the external pins via tri-state inverting buffers having a common output enable signal (OE). When this device is used in the design of a synchronous sequential machine, the flip-flops (providing feedback to the AND matrix) can change state when the tri-state output buffers are disabled, as shown in Figure 11.28.

11.14 Field programmable gate arrays (FPGAs)

FPGAs are types of VLSI chip that have most of the superficial characteristics of PALs, i.e. many types of complex circuits can be designed using one general-purpose IC as the basis. However, the design is initially produced using specially-designed CAD software. The necessary software is specifically intended for, and dedicated to, the particular brand of FPGA used, as there are several very different and mutually exclusive FPGA 'architectures', or organisations of the FPGA chip, available commercially. When the design in the CAD system is complete, the information relating to the design may then be sent electrically to the FPGA device which is capable of being configured internally in order to produce a single IC solution to the design. Thus, the design philosophy is moving away from the designer having to understand the detailed internal workings of the PAL chip in order to make best use of the resources available, and instead the designer now specifies the final result required and expects the dedicated software to be able to make these decisions for the FPGA to be used.

Usually also, custom FPGA software is able to implement multi-level or hierarchical logic designs, where a sub-circuit consisting of basic logic gates may be defined and used as a single block, perhaps many times in the design. Further combinations of several blocks may be defined as a further block which may itself be used many times, with usually little or ideally no limitations on the internal size of the blocks or how many block definitions may be 'nested' in this manner. The latest versions of the design software from the leading companies in this field may incorporate the possibility of using option packages to help designers working in particular specialist fields, such as digital audio. Such sophistication of design is beyond the capability of simple PAL architectures which are designed to implement only Boolean canonical forms and relatively simple Boolean expressions.

The FPGA solution for logic circuits is now commonly used for medium-sized logic designs where the complexity is too great for a cost-effective solution using individual component ICs and where the final circuit must be produced on only a single IC. For the large-scale manufacture of a design where very large numbers of identical logic circuits are required, typically one or more for each system unit manufactured, the most cost-effective route remains an uncommitted logic array (ULA) or a fully-custom- or semi-custom-designed IC. An FPGA is necessarily more complex than would be a custom-designed IC or a mask-programmed ULA.

The detailed design of custom and semi-custom ICs is complex and depends fundamentally upon the particular methods of, and IC architecture provided by, the manufacturer, and are necessarily beyond the scope of this general text. However, various types of FPGAs are now available and their characteristics may be summarised as follows. Some are programmable only once (OTP) whilst others are programmable several times, depending upon the particular technology used to store the user's logic circuit design on-chip.

Clearly, the most flexible approach is to store the design information on-chip in a section of the IC that bears many resemblances to conventional RAM which, in principle, is capable of being written to, and read from, an unlimited number of times. However, just as with conventional RAM, the programming information stored on-chip is completely and immediately lost when power is removed, i.e. the RAM is 'volatile'. These types of FPGAs are usually used with external 'non-volatile' memory, for long-term storage of the chip configuration patterns, together with a hardwired subsystem for automatically loading the FPGA with its intended programming information. This approach usually gives the slowest logic and the greatest propagation times in practice. This type of FPGA is particularly suited for 'proof of concept' prototyping, where the problems associated with the volatile configuration memory are minimal.

Other approaches include the use of field-programmable fusible links or 'anti-fuses' of varying types, which operate faster than RAM and are useful where increased speed is required. These types are 'non-volatile' and retain their programmed information after the removal of power, thus eliminating the need for outboard storage of the programming information. However, just as with fusible link PROMs, once a fusible link is open-circuited, or an 'anti-fuse' is short-circuited, the process cannot be reversed, so that apart from minor changes where an extra fuse is opened or an additional 'anti-fuse' is shorted the programming information cannot be revised once it has been committed to the chip. This means that, as with programming information stored in mask-programmed ROM, the logic design must be essentially complete at the time of programming the chip, with little or no practical possibility of modifying the circuit corresponding to this stored information, other than by starting again with a new and unprogrammed chip. These characteristics mean that these types of FPGA may be considered suitable for small-volume manufacturing runs once the design has been finalised.

A typical FPGA chip is organised around a large two-dimensional array of programmable logic block elements together with a number of input/output blocks at the chip periphery. These handle the interfacing from the internal chip architecture to the circuitry external to the chip, including the logic signals to and from the programmable logic blocks as well as the signals required for programming, or setting up, the appropriate logic configurations inside the programmable logic blocks. There will also be internal wiring between the input/output blocks and the array of programmable logic blocks, and interconnection switches for connection of the logic blocks to the input/output blocks.

The differences between the various available brands of FPGAs arise mainly from differences in the types and complexities of the different manufacturers' programmable logic blocks, the details of the programming signals required, and the overall size and complexity of the chips that are manufactured. Unfortunately, at the time of writing, there is no 'industry-standard' FPGA chip architecture, in contrast to, for example, the standard 74TTL series of basic logic gates that are produced to substantially the same specifications by many companies world-wide. This means that FPGA chips from one manufacturer are usually incompatible with those from others, and without a complete system redesign are often not replaceable by another manufacturer's FPGA. However, just as SSI and MSI logic chip pinout design eventually converged upon one common industry-standard (i.e., the 74TTL series of chips, now extended to CMOS technologies), in the future it may be that FPGA design will also converge upon one type of industry-standard architecture or at least, perhaps, evolve in such a way that the available software packages may program a variety of differing FPGA chips without the user needing to know which particular FPGA will be used. There are already signs of a certain amount of convergence from some companies which offer devices pre-programmed according to the output from other companies' dedicated FPGA CAD systems. However, at the time of writing, it is not clear what that standard will be as the technology is too immature. In all cases, intending users must refer to the full data available from the respective manufacturers, as only a broad outline of the principles involved can be given here.

The 'Xilinx' family of FPGAs has been chosen here as the most typical exemplar of this class of device, and a brief outline of the principal features of some other types is also given.

11.15 Xilinx field programmable gate arrays

The Xilinx family of FPGAs is amongst the most popular types of FPGA, and is frequently encountered in prototyping contexts. This type of FPGA is configured by on-chip CMOS static RAM. Each programmable element or 'configurable logic block', CLB, is controlled by a corresponding memory cell, in which binary values are stored during programming in order to define the function and connectivity of that programmable element (see Figure 11.29). Multiplexers (MUXs) figure prominently in the organisation and operation of this type of FPGA, as they offer a flexible approach to designing reprogrammable logic (see Figure 11.30). Amongst the logic functions implemented in the Xilinx repertoire are:

1. Multiplexers taking previous signals as data inputs and whose address lines are connected to binary values downloaded, i.e. set up, by the controlling software or

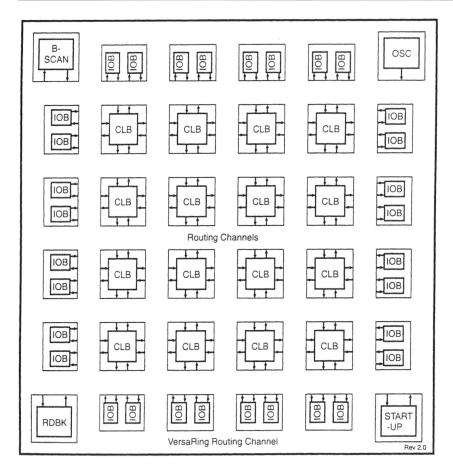

Figure 11.29   *Basic block diagram of Xilinx 'Spartan' FPGAs*

configuration program. So, for example, a 2-to-1 multiplexer may be used to determine which one of any two possible signals should be used to clock a flip-flop. This would be chosen at the time of programming the FPGA, and then during its normal operation would not be changed unless an error were found in the logic design. This simple programmable component is intended to be used as a means of introducing flexibility into the logic circuit, without the necessity of making and breaking physical connections to route the correct clocking signal to the flip-flop.

2. 'Look-up tables' (LUTs), which are analogous to logic functions implemented using a 2^n-to-1 multiplexer. They may be regarded as multiplexers having n address lines, and the 2^n data input lines are connected to binary values (0 or 1) downloaded by the controlling software. Thus, in a simple example using a 4-to-1 multiplexer controlled by address inputs A and B (and numbering the data inputs respectively as $(00)_2$, $(01)_2$, $(10)_2$, $(11)_2$), downloading the binary values $\{0, 0, 0, 1\}$ respectively would lead to the logic function A AND B, which could easily be changed to A XOR B by downloading the binary values $\{0, 1, 1, 0\}$. Clearly, as there is one data input on the multiplexer corresponding to each possible minterm, any Boolean function of several binary variables may be programmed in this manner, and may be reprogrammed to any other Boolean function simply by downloading a different set of binary values.

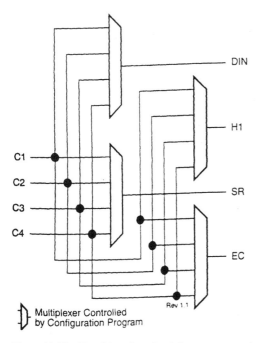

Multiplexer Controlled
by Configuration Program

Figure 11.30 *Signal interface circuit between external signals and configurable logic blocks (CLBs) in Xilinx 'Spartan' FPGAs. The externally applied signals are C1, C2, C3, and C4, and the CLBs are controlled by the signals DIN, H1, SR, and EC*

3. Programmable elements (CLBs) also contain 'general purpose' logic components sometimes known colloquially as 'glue logic', since they are needed to 'glue' the various other logic components together electrically and to perform other mundane logic functions that may be required. These may include D-type flip-flops with direct Set and Reset inputs connected to programmable MUXs and LUTs in a manner that hopefully offers the user most flexibility in programming required logic circuits (see Figure 11.31).

4. There are also programmable input/output blocks (IOBs) around the periphery of the chip for interfacing the external signals of various types to the internal chip workings (see Figure 11.32). These incorporate clocked flip-flops, so that (as well as PLAs and PALs) the functions of PLSs may be implemented by these devices.

Also included in the Xilinx architecture are signal routing arrangements for directing signals to and from the various programmable CLBs. There are direct horizontal and vertical interconnects between adjacent CLBs giving the minimum signal delay, and fast interconnection lines running vertically and horizontally across the entire chip. These are intended primarily for use by critical global signals such as clocking signals that must be synchronised as accurately as possible over the entire design. There are also general-purpose interconnects running vertically and horizontally, together with programmable routing switches that can route horizontal signals to a vertical interconnect and *vice-versa*, and that route non-critical signals between CLBs whose delays depend upon the details of the chip positions of the sending and receiving CLBs (see Figure 11.33). The details of the structure of the CLBs and the interconnects will often be irrelevant to a typical user who employs the dedicated software to lay out the required design, as the software will attempt to ensure that the available CLBs are utilised in the optimal manner.

Obviously, the performance of a system designed using these chips depends critically upon the exact programming implementation used for any particular circuit, and so the design software will typically undertake an analysis and simulation to predict the various timing delays in the final circuit. One typical member of the Xilinx FPGA family, the XC3090, contains 144 IOBs and 320 CLBs, claimed to be the equivalent of 9000 conventional two-input logic gates.

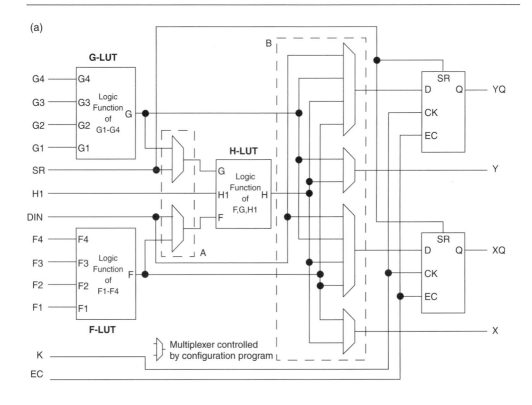

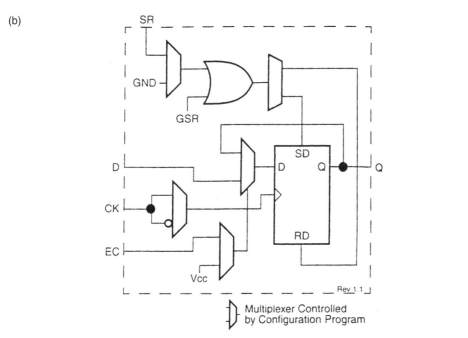

Figure 11.31 *(a) Simplified logic diagram of a configurable logic block (CLB) in Xilinx 'Spartan' FPGAs (b) The enhanced functionality of the basic flip-flops used in this family of FPGAs, showing how their operation is controlled by the configuration program*

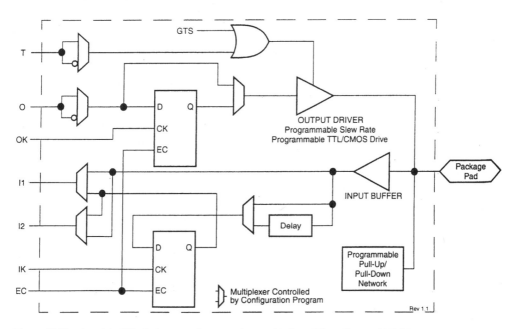

Figure 11.32 *Simplified block diagram of an input/output block in Xilinx 'Spartan' FPGAs*

11.16 Actel programmable gate arrays

The Actel Programmable Gate Array architecture uses a unique 'anti-fuse' technology, where small semiconductor regions defined in the chip structure initially have a high resistance when manufactured, but the application of a high current will modify the semiconductor material so that it has a low resistance. This action is similar to the well-known and unwanted failure mechanism often observed in poor designs using discrete semiconductor devices! Clearly, by arranging for the members of a large array of such 'anti-fuses' to be individually addressable externally by a suitable current source, it is possible to store programming information that can be later used to route internal logic signals to appropriate destinations, thus giving significant flexibility in designing specific logic circuits starting from an original general-purpose uncommitted array. The 'anti-fuses' have a chip area much smaller than more conventional fusible links or the individual RAM cells in FPGAs, using onboard RAM to store programming information, so that relatively complex general-purpose arrays may be fabricated using this technology.

Actel gate arrays use input/output modules for communicating with the external circuitry, logic modules consisting basically of multiplexers with programmable inputs, inter-module interconnections, special programming logic for managing the 'anti-fuses', and logic for testing purposes. The logic modules are arranged in rows separated by interconnects, which run the width of the chip but can also be subdivided to make several adjacent interconnects. There are also some additional interconnects running perpendicular to the rows of logic modules. 'Anti-fuses' are placed at the intersection of horizontal and vertical interconnects, so that logic signals may be routed along rectilinear paths between modules and may change direction from horizontal to vertical and *vice-versa* several times between modules.

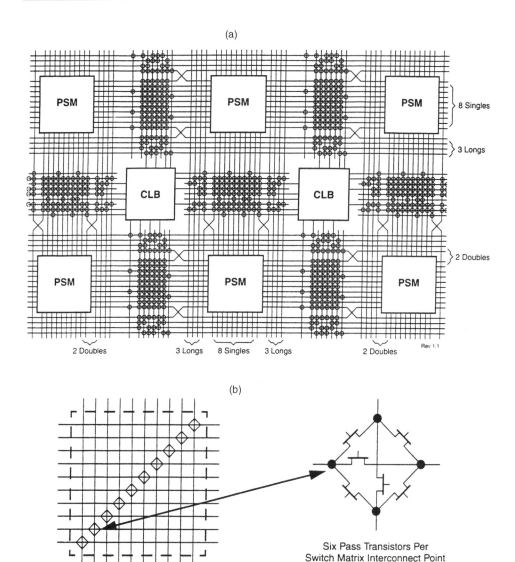

Figure 11.33 *(a) CLB routing channels and interface block diagram for Xilinx 'Spartan' FPGAs. Blocks labelled 'PSM' are programmable switch matrices (b) Programmable switch matrix in Xilinx 'Spartan' FPGAs. This circuit allows logic signals on the horizontally-running bus to be selectively routed to and from logic signals on the vertically-running bus*

Additionally, sufficient 'blank' interconnections are provided so that most signal connections can, in practice, be implemented using at most four 'anti-fuse' crossings. There is a speed penalty introduced by each 'anti-fuse' traversed by a logic signal, because the electrical resistance R associated with an 'anti-fuse' is not zero, as it would be if its performance were ideal, and so in conjunction with the stray capacitance C of the relatively long interconnect lines there is a time delay of the order of RC per 'anti-fuse' crossed. However, the most critical signals, such as clocks, must be delayed least and can usually be routed using just two 'anti-fuses'. The less critical signals can be accommodated by using a greater number of 'anti-fuse' crossings.

The basic Actel logic module consists of three 2-to-1 multiplexers with separate programmable single-bit address lines, one having its address line driven from a 2-input OR gate. With this structure, it is possible to construct a wide variety of basic logic elements, such as look-up tables of Boolean functions, in a manner similar to that used in Xilinx FPGAs. Circuits behaving as simple latches and flip-flops may be constructed by connecting a logic module output to one of its own inputs, giving the requisite logic signal feedback needed.

The first generation of the Actel FPGAs contained up to 2000 logic gates. The manufacturer claimed a flexibility in practice equivalent to around three times as many gates in a more conventional FPGA, because so much of the programming is done using 'anti-fuses' instead of RAM cells, and also because the chip architecture used is particularly versatile. More recent devices, such as the eX256, contain up to 12000 logic gates including 256 registers and 512 logic cells, have 130 input/output lines, and the internal clock operates at up to 350 MHz with propagation delays as low as 4.1ns. This type of technology, because of its advantage of non-volatility, gives particular advantages in high-volume production.

Other companies, such as Integrated Logic Systems, make broadly similar programmable gate arrays using a technology called 'Metal Bridge Architecture' which involves customising a metal layer at manufacture. Their devices can currently accommodate designs using 40000 gate equivalents, 40 kbits of RAM, and provide 324 pins for connecting to external circuitry.

11.17 Altera erasable programmable logic devices

Early types of PLDs available from Altera were based upon UV EPROM technology. Using this technology in conjunction with PLD technology on the same chip produces EPLDs (UV Erasable Programmable Logic Devices). Clearly, such devices have the advantage of re-usability in case incorrect programming is discovered later, or if revised programming becomes necessary, and so are best suited to prototyping or low-volume production runs.

More recent devices are based around EEPROMs. These devices have two advantages over EPLDs. Firstly, less expensive packages can be used since a UV window is not needed, and secondly, if a small amount of reprogramming becomes necessary then it is possible to reprogram a small part of the design without erasing the entire contents of the device and starting again. Whether this is feasible in any particular situation depends upon the extent of the reprogramming necessary, of course.

As in the case of the other FPGAs discussed in previous sections, the basis of the logic functions implementable using Altera PLDs is a general-purpose logic block, here called a 'macrocell' or 'logic array block' (LAB). Macrocells and LABs can be configured in a variety of ways to perform a number of different elementary logic functions as flexibly as possible.

Altera macrocells contain an AND/OR lattice for producing a Boolean function in the form of a sum-of-products of the logic signals input to that macrocell, an XOR gate to invert the sum-of-products according to whether a further 'invert control' input is high or low, a D-type flip-flop for 'sample-and-hold' under control of a clock signal, some programmable logic for taking the output signal back to the AND/OR lattice to produce logic functions with feedback, and a tri-state buffer for driving the external pins of the chip. The main programmability in these devices arises from the AND/OR lattice that is programmed in a manner broadly similar to those in PALs except that when the device is UV-erased then all the possible lattice 'connections' are made, and those not wanted must be 'destroyed' by changing the respective bits in the associated onboard EPROM controlling the chip's programming. Multiplexers within the macrocell allow feedback, clock, and output sections to be programmed independently. The clock section can be programmed to be under the control of global clock signals (*synchronous* mode) or a clock signal generated within the same macrocell from the AND/OR lattice (*asynchronous* mode). JK flip-flops and SR latches are not directly provided, but can easily be constructed by the programmer using the basic D-type flip-flop available, together with some appropriate extra logic implemented using the AND/OR lattice.

LABs, as used on the later Altera FLEX devices, consist of a number, typically eight or 16, of individual macrocells or Logic Elements (LEs) containing basic logic functions such as a four-input look-up-table (LUT), a programmable register, and a capability for producing additional Boolean products. The main benefit of this organisation is that Boolean product terms needed in several parts of the logic design may be shared between macrocells, without any need to repeat their generation within each macrocell. This can increase the flexibility of the chips in producing large designs that require some product terms to be used in several places within the design. The FLEX chips also include a smaller number of 'Embedded Array Blocks' (EABs) that are intended for use as memories of various types. Each EAB provides the equivalent of up to 4kbits of fast-access memory. A more recent family of Altera devices, the 'APEX' series, contain 'Embedded System Blocks' (ESBs) to implement products of Boolean variables, LUTs, RAM, ROM and other memory types including Content Addressable Memory or CAM. CAM is designed so that a complete data record may be retrieved from memory by specifying a part only of the record required, rather than by specifying its address within the memory. All these features once again are intended to maximise the flexibility of using these devices in practical circuits, since the manufacturer has no idea of the details of the design that will eventually be implemented using one of these PLDs but assumes that most PLDs will be used to implement one of the more widely-used approaches to logic design.

The early EP300 series chips contained eight independently programmable macrocells, and the later EP1800 series contains 48 independently programmable macrocells, roughly equivalent to around 100 to 1000 conventional gates depending upon their usage within the circuit design. The even more recent MAX7000 series chips contain the equivalent of up to 10000 usable gates structured as up to 512 macrocells, with propagation delays of 3.5 ns. The MAX9000 series uses slightly different technology and is slightly slower (10 ns propagation delays) but contains the equivalent of up to 12000 gates in up to 560 macrocells. This series of chips may be programmed after they are connected to the final system, eliminating faults due to accumulated static charges arising from manual handling, and making upgrading particularly easy as the chip does not have to be removed from the circuit to perform a system upgrade. This is achieved using an industry-standard 4-pin ISP (In-System Programmability) interface as specified by JTAG (the Joint Test Action Group). The Altera FLEX series contains up to 250000 gate equivalents and the APEX series up to 1 million gates, running at clock speeds as fast as 622 MHz. Special measures, such as the use of phase-locked-loop techniques borrowed from communications engineering, must be used to minimise 'clock skew', or phase differences in the clock signals applied to different parts of the chip caused by transmission delay over the relatively long distances involved in such large chips. Once again, the complexity of such large chips requires dedicated design software in order to ease the burden of taking full advantage of the chip structure available.

This market is now becoming so mature that other companies, such as Clear Logic, offer to manufacture devices that are *pin-compatible*, by working from the designer's Altera CAD programming output file. As far as the external circuit is concerned, these devices operate in an identical manner to, and are functionally indistinguishable from, an Altera MAX device.

Problems

11.1 A scale-of-10 counter is controlled by a signal X. When $X = 1$ the circuit counts in the normal binary sequence. When $X = 0$ it counts in a Gray code sequence. Select a suitable Gray code sequence, construct an ASM chart, and hence determine the state table. Implement the design with

(a) a ROM and JK flip-flops, and
(b) a PLA having on-chip D-type flip-flops.

11.2 Four 4-variable functions are defined by the following equations:

$$f_1(A, B, C, D) = \sum 2, 3, 6, 7, 11, 15$$
$$f_2(A, B, C, D) = \sum 0, 4, 8, 9, 11, 15$$
$$f_3(A, B, C, D) = \sum 1, 3, 5, 7, 10, 11$$
$$f_4(A, B, C, D) = \sum 0, 2, 4, 6, 8, 9, 11, 12, 13, 15$$

Show how these functions can be implemented on a PLA having an 8×8 AND array and a 4×8 OR array.

11.3 The 2-out-of-5 code for decimal digits given below is to be converted to the seven-segment code which is then used to give a decimal display. Implement the code converter using a ROM having a capacity of 256 bits. The eighth output line should give a logic high output when an invalid code is received (and a low output otherwise).

Decimal digit	2-out-of-5 code (*EDCBA*)
0	00011
1	00101
2	00110
3	01001
4	01010
5	01100
6	10001
7	10010
8	10100
9	11000

11.4 The 6-variable function

$$f(A,B,C,D,E,F) = \sum 0,3,4,7,11,16,18,19,20,31,36,41,42,50,51,52,55,57,63$$

is to be implemented by cascading two ROMs, as shown in Figure P11.4. Construct the connection matrix for ROM1 and ROM2.

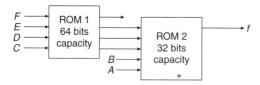

Figure P11.4

11.5 Implement the ASM chart shown in Figure P11.5 with a PLA having on-chip JK flip-flops.

11.6 Five 1 Kbyte ROMs are to be used to provide a system with 5 Kbyte of ROM. The system has 16 address lines and 8 data lines. Each ROM has an active low chip enable pin ($\overline{CE}$) and an active low output enable pin ($\overline{OE}$). A 3-to-8 line decoder having two active low and one active high chip enable pins as well as a selection of SSI gates are available. Using an absolute addressing scheme design a circuit diagram for the system. Also give the address ranges for each of the five ROMs using hexadecimal notation.

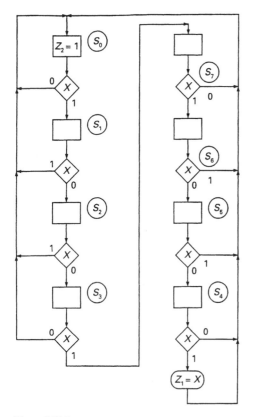

Figure P11.5

12 Arithmetic circuits

12.1 Introduction

One important aspect of digital design with MSI circuits not dealt with in earlier chapters is the design and implementation of arithmetic circuits. Originally, the basic arithmetic circuits were designed using discrete components, but this method has long been superseded by the introduction of MSI circuits. Multi-bit adders, arithmetic logic units and other circuits are now readily available as medium scale integrated circuits.

In some cases, a required arithmetic function is not available in a standard MSI package and modifying logic may be required. A typical example of this is the implementation of a binary adder/subtractor or a circuit used for the implementation of BCD arithmetic. The modifying logic can be provided by discrete gates or by another MSI circuit, so that some arithmetic circuits may be implemented by a combination of MSI and SSI chips.

Progammable logic devices may also be used in arithmetic applications. For example, ROMs programmed as look-up tables can implement the multiplication process, while a combination of multi-bit adders and ROMs, in some cases, can extend the range of multiplication that can be provided.

12.2 The half adder

A half adder is used for adding together the two least significant digits in a binary sum such as the one shown in Figure 12.1(a). The four possible combinations of two binary digits A and B are shown in Figure 12.1(b). The sum of the two digits is given for each of these combinations, and it will be noticed for the case $A = 1$ and $B = 1$ that the sum is $(10)_2$ where the 1 generated is the carry to the next stage of the addition. In the sum shown in Figure 12.1(a), a carry is generated in the least significant column and is then added in at the second stage where a further carry is generated. The carry has rippled through two stages of the addition. Carry ripple, through many stages, in adder circuits generates unacceptable delays, and methods are now available to eliminate this problem.

The additions shown in Figure 12.1(b) are tabulated in the truth table (see Figure 12.1(c)). The columns headed A and B display every combination of the two binary digits to be added, while the third and fourth columns are the corresponding tabulations of the sum S and carry C, respectively. The Boolean equations for the sum and carry read directly from the truth table are:

$S = \bar{A}B + A\bar{B} = A \oplus B$

$C = AB$

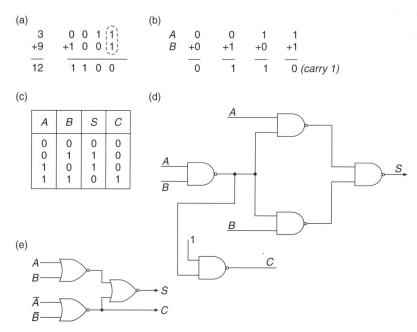

(a)

3	0 0 1 ⌐1¬
+9	+1 0 0 ⌊1⌋
12	1 1 0 0

(b)

A	0	0	1	1
B	+0	+1	+0	+1
	0	1	1	0 *(carry 1)*

(c)

A	B	S	C
0	0	0	0
0	1	1	0
1	0	1	0
1	1	0	1

Figure 12.1 *(a) Binary addition. The half adder is used for adding together the two least significant bits (dotted) (b) The addition of the four possible combinations of two binary digits A and B (with a carry to the next most significant stage of addition) (c) Truth table for the half adder (d) NAND implementation of the half adder (e) NOR implementation of the half adder*

The implementation of the sum and carry functions using NAND and NOR logic is illustrated in Figure 12.1(d) and 12.1(e).

12.3 The full adder

When adding any pair of digits other than the least two significant digits a *full adder* is required. The full adder circuit has three inputs and two outputs which are shown in the block diagram (see Figure 12.2(a)). These are the two binary digits A and B and the input carry C_{in} from the stage on the immediate right, the sum output S and the carry-out to the next most significant stage of the addition, C_{out}.

The truth table for the full adder is shown in Figure 12.2(b) and the Boolean equations for the sum and carry-out read from the truth table are:

$$S = \bar{A}\bar{B}C_{in} + \bar{A}B\bar{C}_{in} + A\bar{B}\bar{C}_{in} + ABC_{in}$$
$$C_{out} = \bar{A}BC_{in} + A\bar{B}C_{in} + AB\bar{C}_{in} + ABC_{in}$$

The equation for the sum may be rewritten as:

$$S = \bar{A}(\bar{B}C_{in} + B\bar{C}_{in}) + A(\bar{B}\bar{C}_{in} + BC_{in})$$
$$= \bar{A}(B \oplus C_{in}) + A(\overline{B \oplus C_{in}})$$
$$= A \oplus B \oplus C_{in}$$

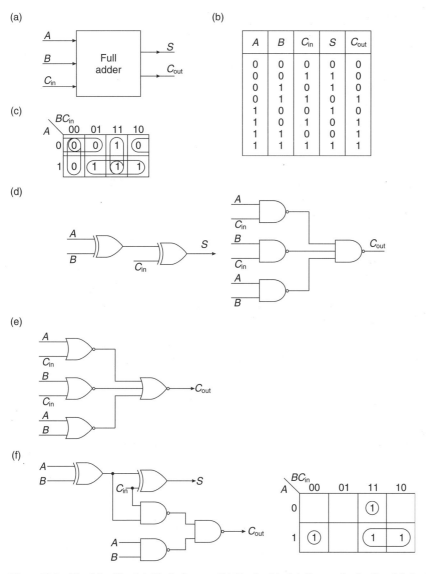

Figure 12.2 *The full adder (a) Block diagram (b) Truth table (c) K-map plot for C_{out} (d) Implementation of full adder (e) NOR implementation of C_{out} (f) Alternative implementation of full adder with K-map showing presence of static hazards*

The carry-out equation is plotted on the K-map shown in Figure 12.2(c). After simplification, the carry-out equation may be written as:

$$C_{out} = AC_{in} + BC_{in} + AB$$

An implementation of the full adder is shown in Figure 12.2(d).

Simplifying the 0's plotted on the K-map gives the minimum inverse function:

$$\bar{C}_{out} = \bar{A}\bar{B} + \bar{A}\bar{C}_{in} + \bar{B}\bar{C}_{in}$$

and inverting:

$$C_{out} = (A + B)(A + C_{in})(B + C_{in})$$

This is the minimum P-of-S form of the equation for C_{out} which can be implemented by the 2-level NOR circuit shown in Figure 12.2(e).

An alternative implementation of the full adder can be obtained by factorising the C_{out} equation taken directly from the truth table:

$$C_{out} = (\bar{A}B + A\bar{B})C_{in} + AB(\bar{C}_{in} + C_{in})$$
$$= (A \oplus B)C_{in} + AB$$

Implementation of this equation, along with the equation for the sum, is shown in Figure 12.2(f). Although the implementation of C_{out} requires less hardware, the time delay for the carry-out has been significantly increased.

There is also an additional difficulty with the implementation of this form of the C_{out} equation. Expanding the above equation for C_{out} gives

$$C_{out} = (\bar{A}B + A\bar{B})C_{in} + AB$$
$$= \bar{A}BC_{in} + A\bar{B}C_{in} + AB$$

A K-map of this function is also shown in Figure 12.2(f) and it can be seen that there are 1's in adjacent cells not covered by the same prime implicant and this indicates the presence of static 1 hazards. To eliminate the static hazards, two extra gates would be required. The lesson for the designer is that the simplest function implementation does not necessarily provide a hazard-free solution.

12.4 Binary subtraction

The binary subtraction of the four possible combinations of two binary digits, $X - Y$, is shown below:

$$
\begin{array}{ccccc}
X & 0 & 0 & 1 & 1 \\
-Y & -0 & -1 & -0 & -1 \\
\hline
\Delta & 0 & 1 & 1 & 0
\end{array}
$$

The only result that requires an explanation here is the second from the left, in which the difference $0 - 1$ has to be found. In order to perform this subtraction a digit has to be borrowed from the next highest column of the subtraction and the operation then becomes $(10)_2 - (1)_2 = (1)_2$. Having borrowed a digit B_{in} from the next most significant stage it is clear that the borrow has to be replaced.

It is now possible, using the above rule, to develop the truth table for the *full subtractor* as shown in Figure 12.3(a). The columns headed B_{in} and B_{out} represent the borrow needed to enable the subtraction to take place and the replacement borrow respectively. The equations for the difference Δ and the borrow in B_{in} can be read from the truth table, and after algebraic manipulation the following two equations are obtained:

$$\Delta = X \oplus Y \oplus B_{in}$$
$$B_{out} = \bar{X}B_{in} + \bar{X}Y + YB_{in}$$

In practice it is simpler to invert the subtrahend, using a controlled XOR gate, and perform an addition, using a full adder, after connecting $\bar{B}_{in}$ to the C_{in} input (see Figure 12.3(b)). The inverse of the borrow-out will appear at the C_{out} terminal. The full adder has been converted to a full subtractor using the method of 2's complement arithmetic.

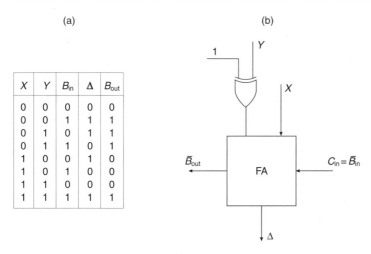

(a)

X	Y	B_{in}	Δ	B_{out}
0	0	0	0	0
0	0	1	1	1
0	1	0	1	1
0	1	1	0	1
1	0	0	1	0
1	0	1	0	0
1	1	0	0	0
1	1	1	1	1

(b)

Figure 12.3 *(a) Truth table for a full subtractor (b) A single bit binary subtractor*

12.5 The 4-bit binary full adder

It is now a simple matter to build a 4-bit adder from four single-bit adders. The block schematic for such an adder is shown in Figure 12.4. For the least significant full adder, the carry-in input C_{in} is grounded and consequently this stage operates as a half-adder.

This type of circuit is referred to as a *ripple-through adder* because a carry from one stage of the adder may ripple through a number of the succeeding stages. In the worst case it is possible for a carry generated in FA_0 to ripple through the carry circuits of all the four full adders before it appears as the carry-out from the final stage of addition. For example, if the following addition has to be performed

```
  1111        15
+ 1001       +09
 11000       +24
 ↗↗↗↗
  1111             Carries
```

a carry is generated in the least significant stage of the addition and it ripples through each successive stage of the addition until it appears at the carry-output terminal of the most significant stage, where it becomes the sum digit for what is, in effect, the fifth bit of the sum. Under these circumstances C_3 ripples through four 2-level logic circuits and the sum is finally completed after eight gate delays. For this kind of adder the maximum delay is directly proportional to the number of stages, n.

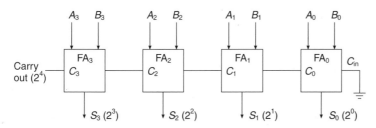

Figure 12.4 *A four-bit parallel adder*

The four full adders shown in Figure 12.4 can all be implemented on a single 16-pin chip to provide a 4-bit MSI adder. Eight inputs are required for the operands, four for the sum outputs, one each for the carry-in and carry-out, and two pins for the supply voltage. A typical example of a 4-bit adder in the TTL family is the 74283 and the new functional logic symbol for this chip is shown in Figure A.24 (see appendix).

12.6 Carry look-ahead addition

The performance of the 4-bit parallel adder described in the previous section can be improved by increasing the speed of operation. This can be achieved by using gates having a reduced propagation delay or by designing a circuit that minimises the delay generated by the carry circuit. In practice such a circuit requires more hardware and the improvement gained is a trade-off between cost and increased speed. Several methods have been developed for reducing the addition time and one of these, the *carry look-ahead* technique, will be described here.

The carry-output equation for a full adder may be written:

$$C_{\text{out}} = (A \oplus B)C_{\text{in}} + AB$$

or as:

$$C_{\text{out}} = PC_{\text{in}} + G$$

where $P = A \oplus B$ is referred to as the *propagation* term, and $G = AB$ is called the *generation* term. If $G = 1$, then $A = B = 1$, and a carry is generated in the stage defined by the C_{out} equation. Additionally, if the carry into the stage $C_{\text{in}} = 1$, and either A or B is 1, then the input carry will be propagated to the next stage. For a 4-bit adder the generation and propagation terms for each stage are:

$$G_0 = A_0 B_0 \qquad P_0 = A_0 \oplus B_0$$
$$G_1 = A_1 B_1 \qquad P_1 = A_1 \oplus B_1$$
$$G_2 = A_2 B_2 \qquad P_2 = A_2 \oplus B_2$$
$$G_3 = A_3 B_3 \qquad P_3 = A_3 \oplus B_3$$

while the carries for the various stages are:

$$C_0 = P_0 C_{\text{in}} + G_0$$
$$C_1 = P_1 C_0 + G_1$$
$$C_2 = P_2 C_1 + G_2$$
$$C_3 = P_3 C_2 + G_3$$

Substituting for C_0 in the C_1 equation and similarly in successive equations, leads to the following equation for the carry out C_3 from the most significant stage of a 4-bit adder:

$$C_3 = P_3 P_2 P_1 P_0 C_{\text{in}} + P_3 P_2 P_1 G_0 + P_3 P_2 G_1 + P_3 G_2 + G_3$$

This carry-out equation can be implemented by the 2-level AND/OR circuit, shown in Figure 12.5, but fan-in problems will occur as the number of bits to be added is greater than four. Only two levels of logic are required to generate the carry-out in the 4-bit CLA scheme, compared with eight levels of logic needed for the 4-bit ripple-through adder. However, the number of gates required by the CLA scheme is significantly

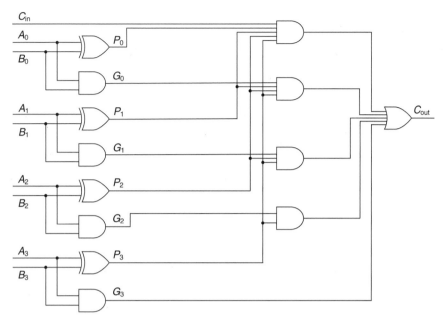

Figure 12.5 *4-bit carry look-ahead logic*

greater than the gate requirement for the ripple-through adder. This is an example of the trade-off between speed and cost.

The carry-out equation may be written in the following form:

$$C_3 = PC_{in} + G$$

where $G = P_3P_2P_1G_0 + P_3P_2G_1 + P_3G_2 + G_3$

and $P = P_3P_2P_1P_0$

The 74181 Arithmetic/Logic Unit provides 4-bit addition without the carry look-ahead facility. To overcome this difficulty, the arithmetic section of the package can be operated in conjunction with the 74182, a carry look-ahead generator. Connecting four 74181s in cascade will provide 16-bit addition, and the four carry look-ahead units in the 74182 will provide carry look-ahead for each of the 4-bit adders. An arrangement for 16-bit addition with carry look-ahead facilities is shown in Figure 12.6.

12.7 The 74283 4-bit carry look-ahead adder

The 74283 performs the addition of two 4-bit words and full internal carry look-ahead facilities are provided in the package. The sum bits and the carry-out from the fourth stage are available at the output pins. Typically, the carry-out delay is of the order of 10ns.

The carry-out equation of the full adder developed earlier may be written in the following form:

$$C_{out} = (A + B)C_{in} + AB$$

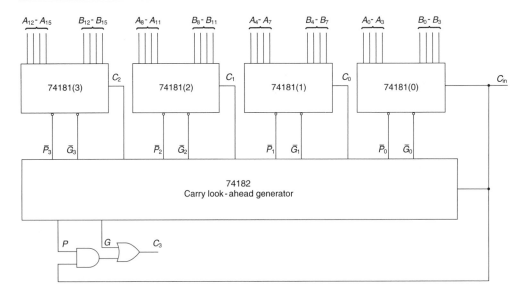

Figure 12.6   *16-bit addition using carry look-ahead generator. Each 74181 also produces four sum output bits.*

where for this circuit design, the propagation term is defined as $P = A + B$, while the generation term retains its previous definition $G = AB$. The implication of this definition of the propagation term is that it contains a generated carry.

A further modification can be made to the carry-out equation since:

$$PG = (A + B)AB = AB = G$$

Hence the equation for the carry-out may be written:

$$C_{out} = P(G + C_{in})$$

With the aid of this equation, the carries for the four stages of the 74283 can be developed as shown below:

$$C_0 = P_0(G_0 + C_{in})$$

and $C_1 = P_1(G_1 + C_0)$

and substituting in this equation for C_0 gives:

$$C_1 = P_1G_1 + P_1P_0G_0 + P_1P_0C_{in}$$

Making two further substitutions leads to the following equation for the carry out C_3 of the adder:

$$C_3 = P_3G_3 + P_3P_2G_2 + P_3P_2P_1G_1 + P_3P_2P_1P_0G_0 + P_3P_2P_1P_0C_{in}$$

This expression is in the sum-of-products form and it is left to the reader to show that the product-of-sums form is

$$C_3 = P_3(G_3 + P_2)(G_3 + G_2 + P_1)(G_3 + G_2 + G_1 + P_0)(G_3 + G_2 + G_1 + G_0 + C_{in})$$

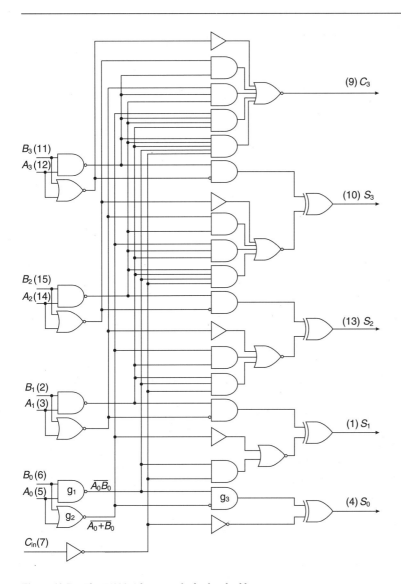

Figure 12.7 *The 74283 4-bit carry look-ahead adder*

The logic diagram is shown in Figure 12.7 and it is left to the reader to verify that the output carry of the circuit is given by either of the above two equations.

An inspection of the logic diagram shows that gates g_1 and g_2 produce the inverse of the generation and propagation terms respectively, and that the output of gate g_3 is $\bar{G}_0 P_0$. The equation for S_0 is:

$$S_0 = \bar{G}_0 P_0 \oplus C_{in}$$
$$= (\overline{A_0 B_0})(A_0 + B_0) \oplus C_{in}$$
$$= (\bar{A}_0 + \bar{B}_0)(A_0 + B_0) \oplus C_{in}$$
$$S_0 = A_0 \oplus B_0 \oplus C_{in}$$

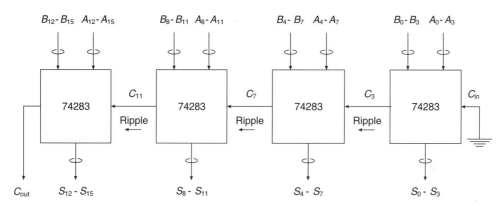

Figure 12.8 *16-bit addition*

A 16-bit adder can be formed from a cascade of four 74283s, as illustrated in Figure 12.8. The cascade of adders provides full carry look-ahead for each adder module with carry ripple from module to module. The carry out delay time for each of the modules is of the order of 10 ns and the total carry propagation delay for the 16-bit adder will be of the order of 40 ns.

12.8 Addition/subtraction circuits using complement arithmetic

Addition is carried out in all cases, irrespective of whether the operands are positive or negative. The sign bits are included in the addition, and any carry-out from the sign bit position is ignored. If the resulting answer is positive the sign bit is 0 and the numerical part of the answer is expressed in magnitude form. If the resulting answer is negative, the sign bit is 1, and the numerical part of the answer is expressed in 2's complement form.

An adder/subtractor using 2's complement arithmetic is illustrated in Figure 12.9. The number $A_2A_1A_0$ is the augend in the addition mode and the minuend in the subtraction mode, while the number $B_2B_1B_0$ is the addend in the addition mode and the subtrahend in the subtraction mode. The sign bits for the two numbers are A_3 and B_3 respectively. The circuit can be implemented with a 74283 4-bit adder and the most significant bits on the chip are used as sign digits. A 7486 quad XOR package is used as a controlled inverter, and inverts the B digits in the subtraction mode. The mode signal is used to select the addition and subtraction modes. When $M = 0$ and $C_{in} = 0$, the least significant stage of the adder acts as a half adder and the 74283 is in the addition mode. When $M = 1$ and $C_{in} = 1$ the adder is in the subtraction mode. In this mode the 7486 is acting as an inverter and additionally a 1 is added in at the least significant stage of the 74283 to form the 2's complement of the subtrahend.

An additional complication arises if 1's complement arithmetic is used and this is illustrated in the two examples shown below:

$$
\begin{array}{rl}
+4 & 0,100 \\
-3 & 1,100 \\
\hline
+1 & 10,000 \\
\end{array}
\qquad
\begin{array}{rl}
-4 & 1,011 \\
-3 & 1,100 \\
\hline
-7 & 10,111 \\
\end{array}
$$

$$
\begin{array}{ll}
\xrightarrow{\quad}1 \quad \text{EAC} & \xrightarrow{\quad}1 \quad \text{EAC} \\
0,001 & 1,000
\end{array}
$$

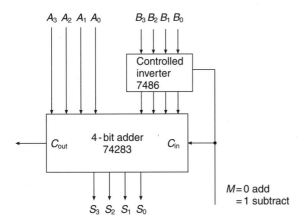

Figure 12.9 *An adder/subtractor using 2's complement arithmetic*

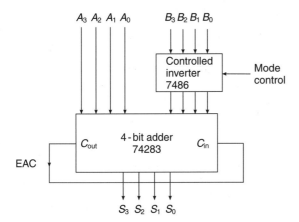

Figure 12.10 *1's complement adder/subtractor*

This carry, called the end-about carry EAC, is returned to the least significant place of the adder where it is added in. This requires a modification to the 2's complement adder shown in Figure 12.9. The C_{out} terminal is now connected directly to the C_{in} terminal, as shown in Figure 12.10.

12.9 Overflow

In certain circumstances, when an adder/subtractor circuit is employing signed arithmetic, there is arithmetic overflow from the most significant magnitude bit into the sign bit. This will occur for example, if a 4-bit arithmetic result is required when two 3-bit numbers are added together and where the fourth bit in the circuit has been assigned the task of indicating the sign of the answer. The consequences of overflow when it occurs are:

1. The addition of two positive numbers gives a negative answer
2. The addition of two negative numbers gives a positive answer

An example of four possible situations that may arise is given below for a 4-bit word ($n = 4$) and for each case the carries from the $(n - 1)$th and nth stages have been displayed.

$C_n = 0, C_{n-1} = 0$		$C_n = 0, C_{n-1} = 1$	
+1	0,001	+5	0,101
+3	0,011	+6	0,110
+4	0,100	+11	1,011	interpreted as −5

$C_n = 1, C_{n-1} = 1$		$C_n = 1, C_{n-1} = 0$	
+5	0,101	−5	1,011
−3	1,101	−6	1,010
+2	0,010	−11	0,101	interpreted as +5

The reader will observe that when either a positive or negative sum of $(11)_{10}$ is required, the magnitude of this number, either in its positive or negative form, cannot be expressed in terms of three binary digits, and the resulting answer is both incorrect and has the wrong sign.

A truth table for C_{n-1} and C_n is shown in Figure 12.11 and it is clear from this table that the overflow function is the XOR of C_n and C_{n-1}. Hence the equation for the overflow flag is:

$$O = C_n \oplus C_{n-1}$$

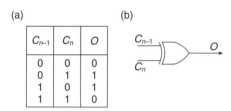

(a)

C_{n-1}	C_n	O
0	0	0
0	1	1
1	0	1
1	1	0

(b)

Figure 12.11 *(a) Truth table for the overflow function (b) Implementation of overflow function*

When using an MSI package such as 74283, C_{n-1} is not available as an output, and overflow has to be expressed differently. It is left to the reader to confirm that an alternative Boolean expression that can be used is:

$$O = \bar{A}_3\bar{B}_3S + A_3B_3\bar{S}$$

where S is the sign of the result and A_3 and B_3 are the sign digits of the two 4-bit numbers.

12.10 Serial addition and subtraction

For parallel addition a full adder is required for each stage of the addition and carry ripple can be eliminated if carry look-ahead facilities are available. An alternative approach is to use a serial addition technique which requires a single full adder circuit and a small amount of additional logic for saving the carry. Serial addition takes longer, but a smaller quantity of hardware is required and the selection of serial or parallel addition depends upon the trade-off between speed and cost.

A serial adder uses a sequential technique and may be regarded as a very simple finite state machine. The basic element of the circuit is a full adder which is

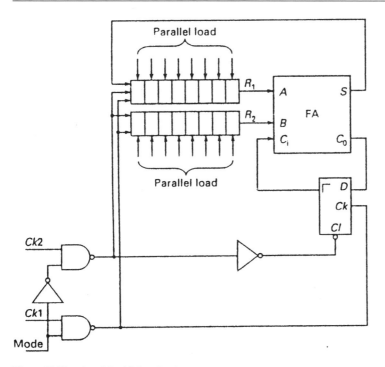

Figure 12.12 *A serial addition circuit*

operated in conjunction with a DFF and a pair of shift registers which have parallel loading and shift right facilities controlled by $Ck1$ and $Ck2$. The selection of either of the two clock pulses is a function of the mode control M (see Figure 12.12). With $M = 0$, $Ck2$ is enabled, the flip-flop is cleared, and the registers are loaded with the two numbers to be added so that the two least significant bits are available at terminals A and B. The corresponding sum and carry-out appear at the output terminals of the full adder. With $M = 1$, $Ck2$ is disabled and $Ck1$ is enabled. $Ck1$ is now used to shift right the digits in registers R_1 and R_2, thus presenting the next most significant pair of digits at terminals A and B. Additionally C_o is clocked to the output of the flip-flop and becomes the next C_{in}, while the sum of the two least significant digits is clocked into the left-hand end of R_1. This process is repeated on receipt of each clock pulse ($Ck1$) until the two numbers stored initially in R_1 and R_2 have been added and the resulting sum has been clocked back into the register R_1. If at the termination of the addition $C_o = 1$, this will represent the most significant digit of the sum.

The serial adder can also be used in the subtraction mode, as shown in Figure 12.13. The B digits are inverted when the mode signal $M = 1$ but an initialisation pulse I of short time duration is required at the input of g_1 at the same time that the least significant pair of digits appear at the full adder inputs. The initialisation pulse is used to preset the DFF to 1, thus forming the 2's complement of the number entering sequentially at the B input. A similar arrangement is made when the adder is in the addition mode. The mode signal $M = 0$ and a short initialisation pulse is needed at the input of g_2 to clear the DFF so that $C_{in} = 0$.

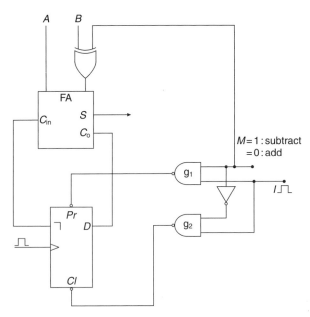

Figure 12.13 *Serial adder/subtractor*

12.11 Accumulating adder

A list of numbers can be summed by operating a group of single-bit adders in parallel. The numbers to be added are stored in memory and can be accessed by a counter. Memory having eight address lines and four outputs can store 256 four bit numbers. The maximum numerical value of a 4-bit number is $(1111)_2 = (15)_{10}$, and the maximum total that can be achieved by the multi-bit adder is $256 \times 15 = (3840)_{10} = (111100000000)_2$. Three single-bit full adders and eight half adders are required along with a 12-bit register for holding the total. The numbers are fetched from memory and are presented to the multi-bit adder along with the previous sum held in the 12-bit register. A block diagram of the multi-bit accumulating adder is shown in Figure 12.14.

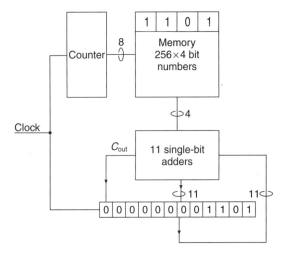

Figure 12.14 *Multi-bit accumulating adder*

12.12 Decimal arithmetic with MSI adders

It is sometimes desirable to perform arithmetic operations using binary coded decimal numbers. Such a requirement occurs where the result of the operation is to be displayed directly in decimal form using seven-segment indicators. Decimal numbers are commonly represented by the 4-bit NBCD code tabulated in Figure 12.15.

When two unsigned NBCD numbers are added together, incorrect answers are obtained in some cases. There are three cases to consider:

Case 1. $0 \leq S \leq 9$

$$
\begin{array}{r}
4 \\
+5 \\
\hline
9
\end{array}
\qquad
\begin{array}{r}
0100 \\
+0101 \\
\hline
1001
\end{array}
$$

In this range the sum is correct and no correction is required.

Case 2. $9 < S \leq 15$

$$
\begin{array}{r}
7 \\
+6 \\
\hline
13
\end{array}
\qquad
\begin{array}{r}
0111 \\
+0110 \\
\hline
1101 \\
0110 \\
\hline
1,0011
\end{array}
\qquad \text{Add 2's complement of } (10)_{10}
$$

Addition generates an invalid code. Correction is made by subtracting $(10)_{10}$, that is, by adding $(6)_{10}$, the 2's complement of $(10)_{10}$. This process also generates the required carry.

Case 3. $15 < S \leq 19$

$$
\begin{array}{r}
9 \\
+8 \\
\hline
17
\end{array}
\qquad
\begin{array}{r}
1001 \\
+1000 \\
\hline
1,0001 \\
0110 \\
\hline
1,0111
\end{array}
\qquad \text{Add } (6)_{10}
$$

Addition generates a valid but incorrect code and a carry-out. Correction is made by adding $(6)_{10}$.

Summarising the algorithm for adding two decimal digits represented by the NBCD code:

1. If $0 \leq S \leq 9$ no correction is required on addition
2. If $9 < S \leq 19$ the required correction is to add $(6)_{10}$.

When $9 < S \leq 15$, a carry is required for the next most significant stage of the addition and a logic function must be developed which will detect the six invalid codes: 1010, 1011, 1100, 1101, 1110 and 1111. The invalid combinations are shown plotted on the K-map in Figure 12.15. After simplification, the carry function for the given range is:

$$f_c = AB + AC$$

For the range $15 < S \leq 19$ a carry C_{out} is generated when the two decimal digits are added, and the equation for the carry-out C'_{o} which becomes the carry in of the next stage of addition is given by:

$$C'_{\text{o}} = C_{\text{out}} + AB + AC$$

(a)

(b)

Dec. digit	NBCD 8 4 2 1
0	0 0 0 0
1	0 0 0 1
2	0 0 1 0
3	0 0 1 1
4	0 1 0 0
5	0 1 0 1
6	0 1 1 0
7	0 1 1 1
8	1 0 0 0
9	1 0 0 1

(c)

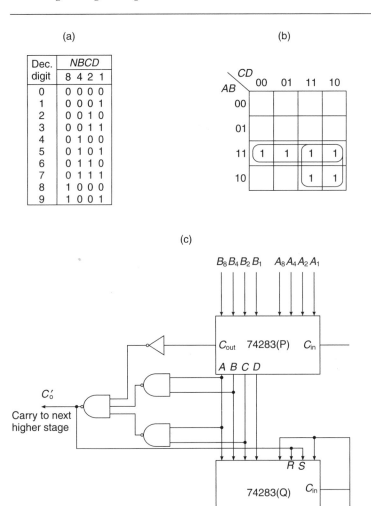

Figure 12.15 *(a) The NBCD code (b) The forbidden code combinations plotted on a K-map (c) A single-stage NBCD adder*

The implementation of a single stage NBCD adder is shown in Figure 12.15. It requires two 74283 4-bit adders, three NAND gates and one inverter. The adder marked P is used for the addition of two 4-bit NBCD codes $A_8A_4A_2A_1$ and $B_8B_4B_2B_1$. Outputs from this adder, in conjunction with its carry-out, are fed to a NAND gate circuit which generates C'_0. They are also fed to the adder Q whose only function is to add in $(0110)_2$ to the total when the sum of two decimal digits is in the range $9 < S \leq 19$. When this condition exists, a carry to the next stage C'_0 is generated and is also fed to the inputs marked R and S on adder Q, thus generating an input of $(0110)_2$ at its four right-hand terminals. For $S \leq 9$, $C'_0 = 0$ and the input to the four right-hand terminals is $(0000)_2$.

In the case of an adder which operates with a word length of eight bits, two NBCD digits can be allocated to one word. The decimal number range is then 0 to 99 inclusive. An example of the addition of a pair of two decimal digit NBCD numbers is shown below:

```
49      0100    1001        Augend
33      0011    0011        Addend
        0111    1100
        0000    0110
        0111   ╱0010        NBCD correction

         1       0          Carry
82      1000    0010        Sum
```

The reader should observe that an NBCD correction in the least significant nybble (LSN) may itself produce a requirement for an NBCD correction in the next most significant nybble (MSN).

12.13 Adder/subtractor for decimal arithmetic

In order to transform the NBCD adder shown in Figure 12.15 to an adder/subtractor, additional logic circuitry is required. When subtracting binary numbers 2's complement, arithmetic is used, but when dealing with NBCD, subtraction is carried out using either 10's or 9's complement arithmetic. As 10 is the radix in the decimal system, the 10's complement is defined as:

$$[X]_{10} = 10^n - X$$

where n is the number of decimal digits contained in the decimal number X and $[X]_{10}$ represents its 10's complement. For $X = 823$, and $n = 3$:

$$[X]_{10} = 10^3 - 823 = 177$$

Subtraction can now be carried out as an addition, using the 10's complement form for negative numbers. Two cases are considered:

Case 1: Addition of positive and negative numbers where the positive number has the greatest magnitude. The subtraction 62–55 can be performed by taking the 10's complement of 55 and adding it to 62:

```
                        62      0110    0010
10's complement of 55 =  45     0100    0101
       Discard (1) ──▶ 07       1010    0111
                                0110    0000    NBCD correction
              Discard (1) ──▶   0000    0111
```

Case 2: The subtraction to be carried out is 55–62, that is the subtrahend is greater than the minuend.

```
                        55      0101    0101
10's complement of 62 =  38     0011    1000
                        93      1000    1101
                                0000   ╱0110    NBCD correction

                                 1      ___     Carry
                                1001    0011
```

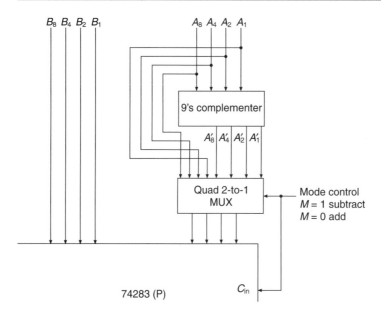

Figure 12.16   *Conversion of NBCD adder to adder/subtractor*

Since the answer is negative it is expressed in 10's complement form. To find the magnitude, the 10's complement of 93 is taken, which gives the required magnitude 07.

When the difference is positive as in *Case 1*, a carry is generated in the most significant place. The generation of the carry in a 10's complement adder/subtractor circuit can be used to distinguish between positive and negative results.

The diminished radix in the decimal system is 9 and the 9's complement is defined as:

$$[X]_9 = 10^n - (X + 1)$$

For $X = 823$, $[X]_9$ has a value of 176.

Additional logic circuitry is needed for the 10's complement adder/subtractor. These extra requirements are shown in Figure 12.16. The 9's complement of the NBCD code and the code itself are both connected to a 2-to-1 MUX. Selection of one out of these two forms of the code is provided by the mode control M. If $M = 1$ the 9's complement of the code is fed to adder P and the 10's complement is formed by adding $M = 1$ at the carry input terminal of adder P. If $M = 0$ the NBCD code is selected, $C_{in} = 0$ and addition takes place. If the answer is negative, the output of the adder/subtractor will be in 10's complement form, and a further controlled 10's complementer is required at the output of each stage if the answer is required in magnitude form.

The input 9's complementer circuit is designed using the normal combinational logic techniques. The four left-hand columns of the truth table shown in Figure 12.17 give a listing of the NBCD code $A_8 A_4 A_2 A_1$ while the four right-hand columns give the corresponding 9's complement of each code $A_8' A_4' A_2' A_1'$. The K-maps for $A_8' A_4' A_2' A_1'$ are given in Figure 12.17, and the Boolean equations obtained from these maps are

$$A_8' = \bar{A}_8 \bar{A}_4 \bar{A}_2 \qquad A_2' = A_2$$
$$A_4' = A_4 \oplus A_2 \qquad A_1' = \bar{A}_1$$

Implementation of the 9's complementer is shown in Figure 12.17.

(a)

(c)

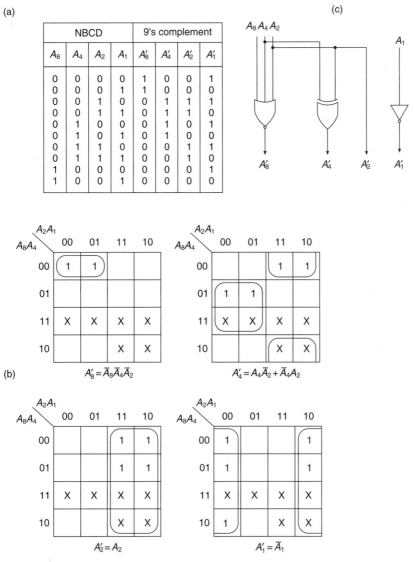

	NBCD			9's complement			
A_8	A_4	A_2	A_1	A_8'	A_4'	A_2'	A_1'
0	0	0	0	1	0	0	1
0	0	0	1	1	0	0	0
0	0	1	0	0	1	1	1
0	0	1	1	0	1	1	0
0	1	0	0	0	1	0	1
0	1	0	1	0	1	0	0
0	1	1	0	0	0	1	1
0	1	1	1	0	0	1	0
1	0	0	0	0	0	0	1
1	0	0	1	0	0	0	0

(b)

$$A_8' = \bar{A}_8\bar{A}_4\bar{A}_2$$

$$A_4' = A_4\bar{A}_2 + \bar{A}_4A_2$$

$$A_2' = A_2$$

$$A_1' = \bar{A}_1$$

Figure 12.17 *Design of 9's complementer (a) Truth table (b) K-maps (c) Implementation*

A 10's complementer could have been designed using the same techniques, but the Boolean equations obtained are more complicated and require a greater amount of hardware for their implementation. In this case it is much simpler to add $M = 1$ at the carry-in of adder P to form the 10's complement. An adder/subtractor for decimal arithmetic which uses the XS3 code provides a much simpler design, and the reader is referred to the design example at the end of the chapter.

12.14 The 7487 true/complement unit

A number of different arithmetic operations can be performed by controlling one set of inputs to a 4-bit adder. This control can be achieved by inserting the true/complement

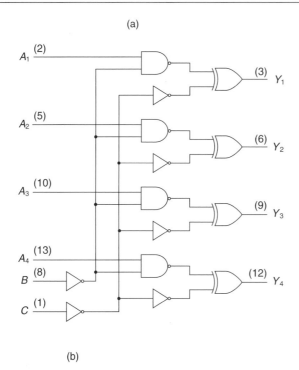

(a)

(b)

Figure 12.18   *The 7487 True/Complement Unit (a) Logic diagram (b) Truth table*

Select Signals		Output Word			
B	C	Y_4	Y_3	Y_2	Y_1
0	0	$\overline{A}_4$	$\overline{A}_3$	$\overline{A}_2$	$\overline{A}_1$
0	1	A_4	A_3	A_2	A_1
1	0	1	1	1	1
1	1	0	0	0	0

unit between one set of input lines and the adder. Such a unit is the 7487 which, besides the true/complement facility, also provides all 0's or all 1's. The logic diagram and the truth table for this device are shown in Figure 12.18.

A 4-bit adder such as the 74283 operating in conjunction with the 7487 is shown in Figure 12.19. The functional behaviour depends upon the logical value of the two select signals S_0 and S_1 and the presence, or absence, of the carry input C_{in}. There are eight possible combinations of the three signals, and the behaviour for the first two of these combinations is illustrated in the block diagrams shown in Figure 12.19. The functional outputs of the controlled adder for each of the eight combinations are also tabulated in the function table shown in Figure 12.19.

12.15 Arithmetic/logic unit design

A block diagram for an *n*-bit ALU is shown in Figure 12.20. The output function $f = (f_{n-1} \ldots f_0)$ is generated by performing a logic or alternatively an arithmetic

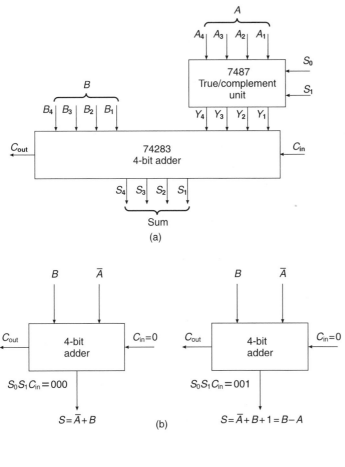

Figure 12.19 *(a) Block diagram of controlled adder (b) Function implementation of controlled adder (c) Function table*

operation on the two *n*-bit inputs $A = (A_{n-1} \ldots A_0)$ and $B = (B_{n-1} \ldots B_0)$ to be determined by $S = (S_{k-1} \ldots S_0)$ selection bits.

The functions to be generated are tabulated in Figure 12.21. Since there are eight functions in the table, a total of three selection variables S_2, S_1 and S_0 is required.

The design will be based on 4-bit modules such as the 74283 4-bit adder, and when it is completed it will consist of an interconnection of a number of MSI packages. The desired range of numbers to be operated on by the ALU will determine the number

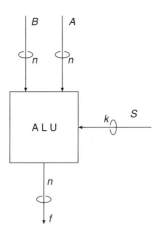

Figure 12.20   *Block diagram of n-bit ALU*

S_2	S_1	S_0	ALU Function	
0	0	0	$-A+B$	Subtract
0	0	1	$A+B$	Add
0	1	0	$B-1$	Decrement
0	1	1	$B+1$	Increment
1	0	0	$A \cdot B$	AND
1	0	1	$A+B$	OR
1	1	0	$\bar{A}$	NOT
1	1	1	$A \oplus B$	XOR

Figure 12.21   *ALU function table*

of bits n. Because of the availability of 4-bit packages it would be desirable that n should be a multiple of 4 so that the ALU will consist of a cascade of 4-bit slices, as shown in Figure 12.22.

The block diagram representing the basic 4-bit slice of the n-bit ALU is shown in Figure 12.23. It consists of a logic unit LU and an arithmetic unit AU working in conjunction with a quadruple 2-to-1 MUX. The output of the MUX can be either a logic or an arithmetic function. If the selection variable $S_2 = 0$ the arithmetic unit is selected, and for $S_2 = 1$ the logic unit is selected.

First, consider the design of the least significant 4-bit arithmetic unit. The block diagram for this section of the 4-bit slice of the ALU is shown in Figure 12.24(a) along with the truth table of the adder functions in Figure 12.24(b). It consists of the 7487 True/Complement Unit interposed between the A input lines and the A' inputs to the 74283 carry look-ahead 4-bit adder.

In order to decrement a number such as $B_3 B_2 B_1 B_0 = 1001$ it is only necessary to add the number 1111, as illustrated in the example shown below:

$$B_3 B_2 B_1 B_0 = 1001$$
$$\underline{1111}$$
$$1000$$

and hence in order to decrement the B input, the selection signals $S_1 S_0 = 10$ and the output of the True/Complement Unit in Figure 12.24(a) is then $A'_3 A'_2 A'_1 A'_0 = 1111$. The logic equation for the carry-in can be read directly from the truth table shown in Figure 12.24(b) and is:

$$C_{in} = \bar{S}_1 \bar{S}_0 + S_1 S_0$$
$$= \overline{S_1 \oplus S_0}$$

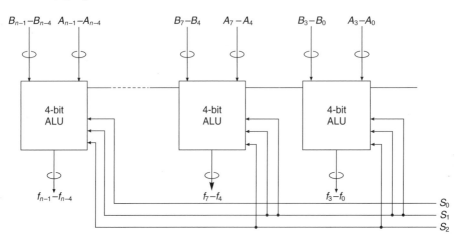

Figure 12.22   *Structure of an n-bit ALU made up from 4-bit slices*

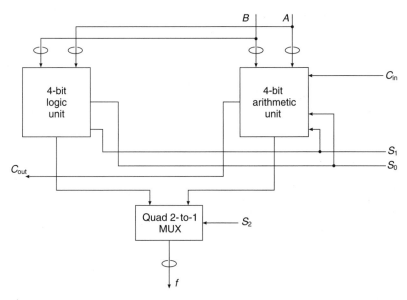

Figure 12.23 *Basic 4-bit slice interconnections*

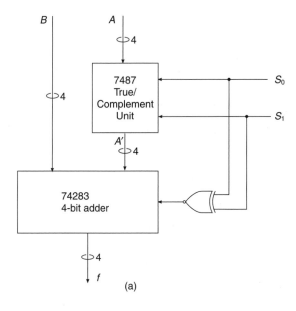

(a)

S_1	S_0	f_{AU}	
0	0	$B-A$	Subtract
0	1	$A+B$	Add
1	0	$B-1$	Decrement
1	1	$B+1$	Increment

(b)

Figure 12.24 *(a) 4-bit ALU Arithmetic section (b) Arithmetic section truth table*

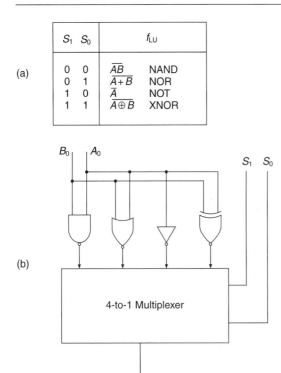

S_1	S_0	f_{LU}	
0	0	$\overline{AB}$	NAND
0	1	$\overline{A+B}$	NOR
1	0	$\overline{A}$	NOT
1	1	$\overline{A \oplus B}$	XNOR

(a)

(b)

Figure 12.25   *(a) Function table for logic slice of ALU (b) Implementation for one pair of input bits for the slice*

The function table for designing the 4-bit logic slice of the ALU is shown in Figure 12.25. A simple approach to the design is to use the basic logic gates in conjunction with a pair of dual 4-to-1 MUXs such as the 74353, as shown in Figure 12.25 for a single bit pair only. An alternative approach can be developed with the aid of the truth table shown in Figure 12.26 for the pair of single bits A_0 and B_0. The logic function f_{LU} is plotted on the K-map and simplified. The hazard free function read from the map is:

$$f_{LU} = \bar{A}_0 \bar{B}_0 + \bar{S}_0 \bar{A}_0 + \bar{B}_0 S_1 \bar{S}_0 + A_0 B_0 S_1 S_0$$

and its implementation is shown along with the K-map in Figure 12.26.

12.16 Available MSI arithmetic/logic units

Two examples of ALUs available in the TTL family are the 74381 and the 74382. These two 4-bit devices perform eight arithmetic/logic operations defined by the function table shown in Figure 12.27 and selected by the three function select lines, $S2$, $S1$ and $S0$. The difference between these two devices is that the 74381 has two outputs, one a carry generation output G and the other a carry propagation output P, allowing a group of ALUs to be cascaded. A group carry look-ahead facility can then be provided as illustrated earlier in this chapter in Figure 12.6 by the 74182 carry look-ahead generator. The 74382, on the other hand, provides a ripple carry output C_{n+4} to the succeeding ALU input and a 2's complement overflow output OVR. Traditional logic symbols for the two ALUs are also shown in Figure 12.27.

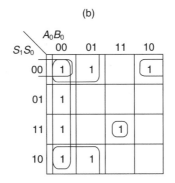

(a)

S_1	S_0	A_0	B_0	f_{LU}	Function
0	0	0	0	1	
0	0	0	1	1	
0	0	1	0	1	NAND
0	0	1	1	0	
0	1	0	0	1	
0	1	0	1	0	NOR
0	1	1	0	0	
0	1	1	1	0	
1	0	0	0	1	
1	0	0	1	1	NOT
1	0	1	0	0	
1	0	1	1	0	
1	1	0	0	1	
1	1	0	1	0	XNOR
1	1	1	0	0	
1	1	1	1	1	

(b)

(c)

S_1 S_0 A_0 B_0

f_{LU}

Figure 12.26 *Alternative design for 4-bit logic slice of ALU (a) truth table (b) K-map and (c) implementation of design*

The 74181, a 4-bit ALU/Function generator, is also available in the Type 74 family. This MSI circuit provides a much more comprehensive range of arithmetic and logic functions than the 74381/382. It can be operated with either active high or active low input data and the arithmetic functions generated depend upon the absence or presence of a carry-in. Like the 74381, a cascade of 74181s can be operated in conjunction with the 74182 carry look-ahead generator to provide group carry look-ahead. The function table for active low input data, assuming $C_{in} = 0$, is shown in Figure 12.28 along with the traditional logic symbol.

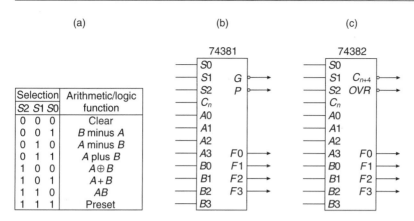

Figure 12.27   *(a) Function table for 74381/382 ALUs (b) Logic symbol for 74381 and (c) logic symbol for 74382*

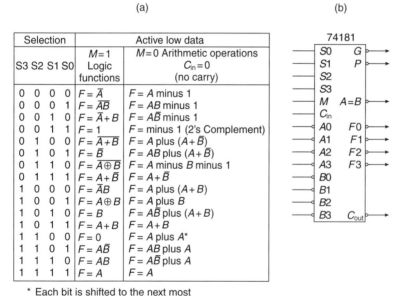

* Each bit is shifted to the next most
significant position

Figure 12.28   *(a) Function table and (b) logic symbol for the 74181 4-bit ALU*

12.17 Multiplication

Multiplication of binary numbers can be achieved either combinationally or sequentially. The simplest form of binary multiplication is multiplication by the base 2. When multiplying by the base 10 in the decimal number system a shift to the left of one place occurs; for example, $9 \times 10 = 90$ and the 9 has moved one place to the left. Similarly, if a binary number such as 1101 is multiplied by the base 2 it becomes 11010, and if the number had been stored in a register this would have represented a shift left of one place in the register. In this example $(1101)_2 = (13)_{10}$ and $(11010)_2 = (26)_{10}$.

12.18 Combinational multipliers

The 2×2 multiplier is a simple example of a combinational multiplication circuit which multiplies two binary numbers A_2A_1 and B_2B_1. The truth table for the multiplication is shown in Figure 12.29 and the most significant term of the product is represented by the single minterm $P_3 = A_2A_1B_2B_1$. The remaining three product terms can be obtained by plotting and simplifying their respective K-maps. For example, the K-map for P_2 is shown in Figure 12.29 and the simplified function obtained from this map is:

$$P_2 = A_2\bar{A}_1B_2 + A_2B_2\bar{B}_1$$

It is left to the reader to determine the Boolean equations for the remaining two product outputs.

One of the simplest and fastest methods of multiplying employs a combinational logic circuit which is composed of AND gates and full adders. The method depends upon the fact that the rules of Boolean multiplication are based upon those of binary multiplication and consequently a series of AND gates can be used for forming the products that occur in the multiplication process.

If the 4-bit binary number $A_3A_2A_1A_0$ is to be multiplied by $B_3B_2B_1B_0$, the 'pencil and paper' method that would normally be employed is illustrated in Figure 12.30.

(a)

A_2	A_1	B_2	B_1	P_3	P_2	P_1	P_0
0	0	0	0	0	0	0	0
0	0	0	1	0	0	0	0
0	0	1	0	0	0	0	0
0	0	1	1	0	0	0	0
0	1	0	0	0	0	0	0
0	1	0	1	0	0	0	1
0	1	1	0	0	0	1	0
0	1	1	1	0	0	1	1
1	0	0	0	0	0	0	0
1	0	0	1	0	0	1	0
1	0	1	0	0	1	0	0
1	0	1	1	0	1	1	0
1	1	0	0	0	0	0	0
1	1	0	1	0	0	1	1
1	1	1	0	0	1	1	0
1	1	1	1	1	0	0	1

(b)

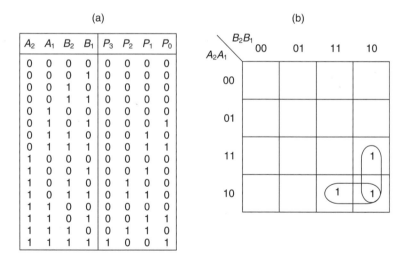

Figure 12.29 *2×2 multiplier (a) Truth table (b) K-map for P_2*

$$
\begin{array}{ccccccccc}
 & & & & A_3 & A_2 & A_1 & A_0 & \text{multiplicand} \\
 & & & & B_3 & B_2 & B_1 & B_0 & \text{multiplier} \\
\hline
 & & & & A_3B_0 & A_2B_0 & A_1B_0 & A_0B_0 & \\
 & & & A_3B_1 & A_2B_1 & A_1B_1 & A_0B_1 & & \\
 & & A_3B_2 & A_2B_2 & A_1B_2 & A_0B_2 & & & \\
 & A_3B_3 & A_2B_3 & A_1B_3 & A_0B_3 & & & & \\
\hline
P_7 & P_6 & P_5 & P_4 & P_3 & P_2 & P_1 & P_0 & \text{double length product}
\end{array}
$$

Figure 12.30 *'Pencil and paper' multiplication*

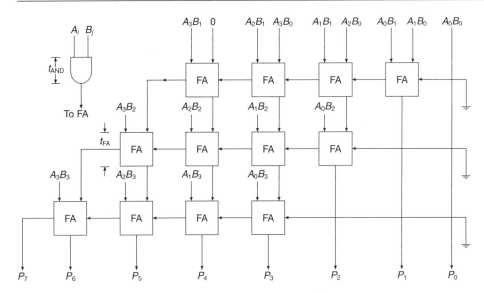

Figure 12.31   *Array multiplier*

For the first row of the multiplication, the least significant bit of the multiplier multiplies each bit of the multiplicand in turn, forming four partial product terms, A_3B_0, A_2B_0, A_1B_0 and A_0B_0. The second row of the multiplication is obtained by shifting one place to the left and multiplying each term in the multiplicand by the B_1 term in the multiplier to form four more partial products. This procedure is continued to obtain the last two rows of partial products. The columns of partial products now have to be added, and it is clear that from the second column onwards, carries can be generated which have to be carried forward to the next column of partial products and added into the sum for that column.

This multiplication process can be represented by the array multiplier shown in Figure 12.31 which consists of a number of AND gates and full adders. The overall delay of this array is given by the largest value of:

$$t_{AND} + 6t_c; \; t_{AND} + t_{FA} + 5t_c; \; t_{AND} + 2t_{FA} + 4t_c; \; \text{or} \; t_{AND} + 3t_{FA}$$

where t_{AND} is the propagation delay of the 2-input AND gates, t_c is the carry delay of a full adder, and t_{FA} is the sum output delay of a full adder, where it is assumed that each of the AND gates, and each of the full adders, have identical time delays.

For this kind of parallel multiplier the amount of combinational logic required increases with the number of bits in the multiplier and multiplicand, and a register is also needed to store the double length product. Before the advent of LSI chips the amount of combinational logic required was a deterrent to using this technique, but now that LSI circuits are readily available, fast multiplier chips using this method of multiplication are available.

12.19 ROM implemented multiplier

Binary multiplication can be achieved by using a ROM as a 'look-up' table. For example, multiplication of two 4-bit numbers requires a ROM having eight address lines, four of them, $X_4X_3X_2X_1$, being allocated to the multiplier, and the remaining four, $Y_4Y_3Y_2Y_1$ to the multiplicand. Since the multiplication of two 4-bit numbers can result in a double-length product, the ROM should have eight output lines, and a ROM with a capacity of 256 bytes is required. A block diagram of the multiplier is shown in Figure 12.32.

The stored product method described above clearly has its limitations. If, for example, the product of two 8-bit numbers is to be stored, then $2^{16} = 65536$ memory locations and 16 output lines for the double-length product are needed. This requires a ROM capacity of $65536 \times 16 \approx 10^6$ bits or 128 Kbytes. For 16 bit multiplication the ROM capacity required is quite formidable. The number of address lines is 32 and the number of output lines is 32, so that the ROM capacity required is $2^{32} \times 2^5 = 2^{37}$ bits $= 2^{34}$ bytes $= 2^{24}$ Kbytes $= 2^{14}$ Mbytes $= 16$ Gbytes.

In the case of the 8-bit multiplier, it is possible to partition the problem by splitting both the multiplier and the multiplicand into two 4-bit words. For example, the 8-bit multiplier $N_1 = 10010010$ can be regarded as two separate 4-bit words, $H_1 = 1001$ and $L_1 = 0010$. Then:

$$N_1 = (2^4 H_1 + L_1)$$

where H_1 is shifted four places to the left relative to L_1 by the shift operator 2^4. Similarly, the multiplicand $N_2 = 01111001$ can also be regarded as two separate 4-bit words, $H_2 = 0111$ and $L_2 = 1001$. Then:

$$N_2 = (2^4 H_2 + L_2)$$

Multiplying out $N_1 N_2 = (2^4 H_1 + L_1)(2^4 H_2 + L_2)$
$$= 2^8 H_1 H_2 + 2^4 H_1 L_2 + 2^4 H_2 L_1 + L_1 L_2$$

The four products in the above equation, $L_1 L_2$, $H_2 L_1$, $H_1 L_2$ and $H_1 H_2$ can each be generated by a 256-byte ROM as described previously in Figure 12.32. The individual 8-bit products generated in this way then have to be summed, with proper regard being paid to their position in the final double-length 16-bit product.

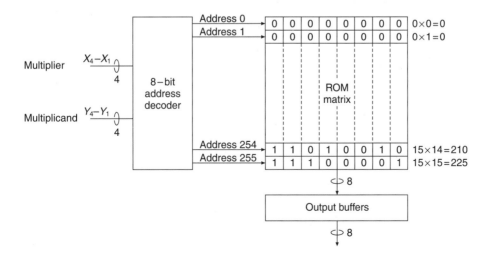

Figure 12.32 *Binary multiplication of two 4-bit words using a ROM*

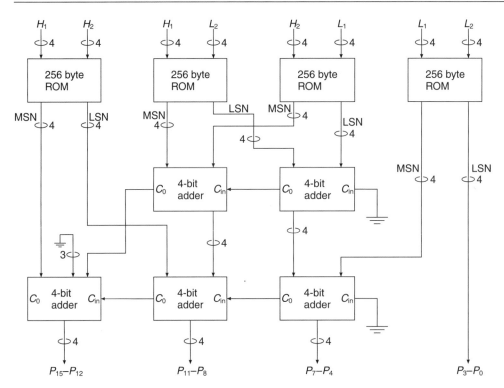

Figure 12.33　*Binary multiplication scheme for two 8-bit words*

If the final product is represented by bits P_0 to P_{15}, the ROM generating L_1L_2 provides bits P_0 to P_3 of the final product and a component of bits P_4 to P_7. The two ROMs generating the 8-bit products H_2L_1 and H_1L_2 have their outputs shifted four places to the left by the shift operator 2^4 and each provide 8-bit components of the product bits P_4 to P_{11}. Finally, the ROM generating the 8-bit product H_1H_2 has its output shifted eight places to the left by the shift operator 2^8. This ROM contributes bits P_{12} to P_{15} of the final product and a component of bits P_8 to P_{11}. The various components of P_4 to P_7 and P_8 to P_{11} are summed in a number of 4-bit adders, as illustrated in Figure 12.33.

12.20 The shift and add multiplier

The 'pencil and paper' method for multiplying together two binary integers is again illustrated in the example shown below:

Multiplicand	1110	14
Multiplier	×1010	×10
Partial Product 1	0000	
Partial Product 2	1110	
Partial Product 3	0000	
Partial Product 4	1110	
Result	10001100	140

There are three main features to the process:

1. If the multiplier bit is 1, then a partial product is formed by writing down the multiplicand. Alternatively, if the multiplier bit is 0, then the partial product is formed by writing down a row of 0's.
2. Four partial products are formed, one for each bit of the multiplier, and they are all added together to form the final product.
3. As the multiplication progresses from the least to the most significant bit of the multiplier, each succeeding partial product is shifted one place to the left.

To implement binary multiplication using a digital machine, two processes introduced earlier have to be performed, namely addition and shifting. Addition can be carried out using a 4-bit adder. The result of the addition is loaded into the product register when the adder output is enabled by AE, the adder enable signal, while the shifting process is achieved by generating a shift pulse for the product register. A multiplier designed on the basis of these two processes is called a shift and add multiplier.

There would, in practice, be one change to the 'paper and pencil' method in the machine implementation. The above example shows that all the partial products are formed before the addition takes place to generate the product of the two numbers. In a machine this would require four registers, one for each partial product, and clearly this would increase as the number of multiplier bits increases. From the hardware point of view this would be extremely uneconomic, and in practice, addition takes place each time the multiplicand appears as a partial product, that is, every time the multiplier bit is 1.

An examination of the multiplication of two 4-bit numbers indicates the following preliminary list of hardware requirements:

1. A 4-bit register for the multiplicand,
2. A 4-bit register for the multiplier,
3. A double-length 8-bit register for the product,
4. A 4-bit adder,
5. Control logic for controlling the add and shift operations.

In practice, this preliminary list is more than is required. It is clear that initially, the double-length register for the product contains no data at all, and it would seem reasonable to use a portion of this register for holding the multiplier on a temporary basis. As the multiplication progresses and successive bits of the multiplier are used, they are moved out of the product register one bit at a time, thus leaving space available for the accumulation of the partial products. This portion of the product register is referred to as the accumulator.

A basic diagram for the machine is shown in Figure 12.34(a). It consists of the 4-bit multiplicand register, a 4-bit adder, an 8-bit product register which consists of accumulator and multiplier sections, and a box labelled 'control logic'. The control logic box has three functions. It must examine the multiplier bit to determine whether it is 0 or 1, and it must generate the shift and adder enable signals. The control logic is also supplied with clock and 'start' signals for synchronising and starting the multiplier operation. A scale-of-4 counter is used to count the number of shift and add operations.

If the multiplier bit is 1, an adder enable signal AE is generated by the control logic. The contents of the multiplicand register and the least four significant bits of the

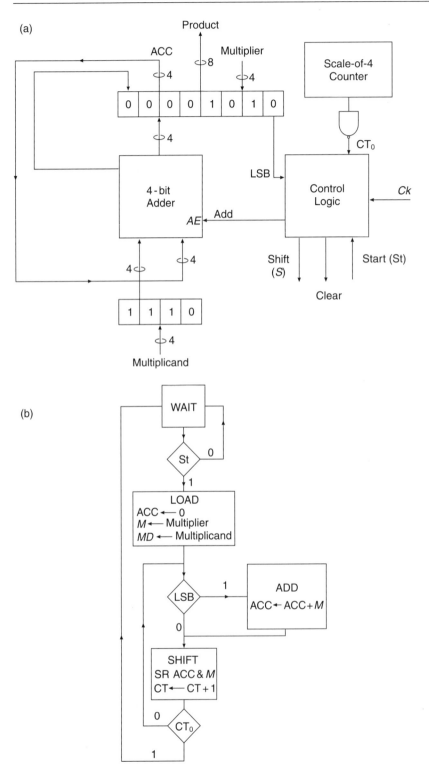

Figure 12.34 *(a) Block diagram of Shift and Add Multiplier (b) ASM Chart (c) State diagram (d) implementation of control logic using a PLS (e) idealised timing diagram for* 1011 × 1010

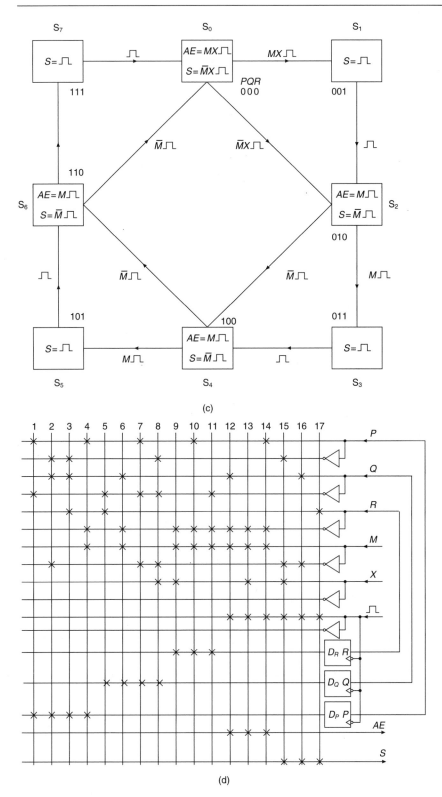

(c)

(d)

Figure 12.34 *(Continued)*

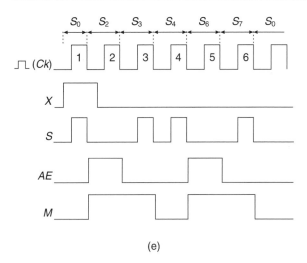

(e)

Figure 12.34   *(Continued)*

accumulator, having been connected to the inputs of the 4-bit adder where addition takes place, are now returned to the accumulator. A shift pulse is generated, and the data stored in the product register is shifted one place to the right, thus moving the least significant bit of the multiplier out of the product register and replacing it with the next most significant bit. In the event of the multiplier bit being zero, a shift pulse S is generated, and no addition takes place. Comparing the machine operation with the 'paper and pencil' multiplication it will be noticed that, on paper, the multiplicand shifts left and the multiplier remains in a fixed position relative to the multiplicand, which remains in a fixed position.

An ASM chart for the multiplier is shown in Figure 12.34(b) and it consists of four states, WAIT, LOAD, ADD and SHIFT:

WAIT: The multiplier is in the quiescent state waiting for the START signal.

LOAD: The accumulator is cleared and the multiplier and multiplicand registers are loaded.

ADD: The multiplier enters this state if the LSB $= 1$. The mutiplicand is added to the contents of the accumulator. This state will be bypassed if LSB $= 0$ and the multiplier will go directly to the SHIFT state.

SHIFT: The accumulator will shift one place right moving the least significant bit of the multiplier out of the multiplier register. The count will advance by 1. If the counter output CT_0 is 0 the machine will return to the LSB decision box. If the counter output is 1, the end of the count cycle has been reached and the machine returns to the WAIT state.

A suitable state diagram for the machine is shown in Figure 12.34(c). Once start signal X is received, an adder enable or shift pulse is generated in state S_0, depending upon the value of the multiplier bit M. Assuming that $M = 1$, an adder enable pulse is generated on the trailing edge of the next clock pulse to be received, and a transition will be made to S_1, the sum having been returned to the accumulator. Alternatively, if $M = 0$ a shift pulse is generated, and a transition is made to S_2 on the trailing edge of the next clock pulse. The outer square will be traversed if all four multiplier bits are 1's.

Alternatively, if all four multiplier bits are 0's then a transition path will be traced round the inner diamond of the state diagram.

Since the control logic state diagram has eight states, three flip-flops, P, Q and R are required for its implementation. A programmable logic sequencer (PLS) which has on-board D type flip-flops can be used for the implementation. The input equations for the DFFs are derived using the methods described in Chapter 8. It is left to the reader to construct a state table, plot the flip-flop input equations and the shift and adder enable functions on 5-variable K-maps and then simplify them. The following results with the state allocation given on the state diagram should be obtained:

$$D_P = P\bar{Q} + \bar{M}\bar{P}Q + \bar{P}QR + MP\bar{R}$$
$$D_Q = \bar{Q}R + MQ\bar{R} + \bar{M}P\bar{Q} + \bar{M}X\bar{P}\bar{Q}$$
$$D_R = MX\bar{R} + MP\bar{R} + M\bar{Q}\bar{R}$$
$$AE = MX\bar{R} + MP\bar{R} + MQ\bar{R}$$
$$S = R + \bar{M}Q + \bar{M}PX$$

The implementation of the control logic is shown in Figure 12.34(d) and an idealised timing diagram, assuming leading-edge triggered DFFs, is shown in 12.34(e).

12.21 Available multiplier packages

The 74284 and the 74285 in the Type 74 series can be used in combination to provide 4-bit-by-4-bit parallel multiplication. When the two chips are connected as shown in Figure 12.35 an 8-bit product is generated. The individual chips have a pair of enable pins G_A and G_B and open collector outputs which should be connected to a pull-up resistor.

In practice, it is now unlikely that these two chips would be used for multiplication. They were originally manufactured in the early days of the Type 74 series when pin limitation was a crucial design factor. More recently, single chips have been designed which are capable of much higher orders of multiplication. However, higher orders of multiplication can be achieved by an array of these two multiplier packages. For example, 8×8 multiplication can be achieved using an array of four pairs of these two chips using the following equation which was developed in section 12.19:

$$P = 2^8 H_1 H_2 + 2^4 H_1 L_2 + 2^4 H_2 L_1 + L_1 L_2$$

The outputs of the multipliers are summed using an array of 74283 4-bit adders, bearing in mind that proper regard must be paid to the shift operators in the above equation. Implementation of the multiplier is identical to the ROM/Adder array in Figure 12.33, the ROMs being replaced by 4×4 multipliers. It is left to the reader to calculate the worst case delay of the multiplier array from input to output.

12.22 Signed arithmetic

Consider the multiplication of two 4-bit numbers where the multiplicand is negative and the multiplier is positive. As an example, the product of the two numbers is calculated in 4-bit 2's complement arithmetic and must not be more negative

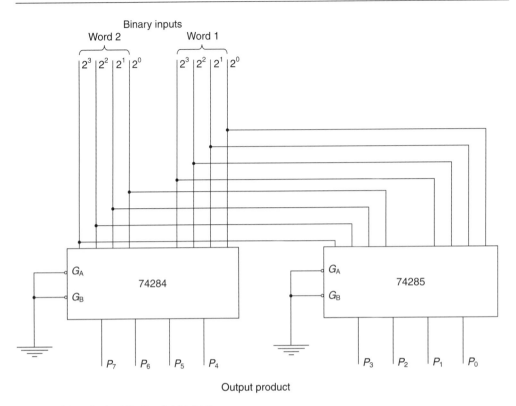

Figure 12.35   *4-bit-by-4-bit Parallel Multiplier*

than $(-15)_{10}$, the lowest number allowed when a 5-bit register is used to hold the product.

Multiplicand	$(-7)_{10}$	1001	2's complement of $(7)_{10}$
Multiplier	$(+2)_{10}$	0010	
		0000	
		1001	
		0000	
		0000	
		0010010	

The four least significant bits of this multiplication are the 2's complement of the required answer, $(-14)_{10}$. It will be observed that all that was required to complete the multiplication was a single left shift of the multiplicand. It follows that if the multiplier consists of a series of 0's followed by a single 1, reading from the LSB towards the MSB, the multiplication process would only require a series of left shifts of the multiplicand. For example, in the case of the 8-bit multiplier $(00010000)_2$, four left shifts of the multiplicand are required as shown in the example below:

Multiplicand	$(-7)_{10}$	11111001	2's complement of $(7)_{10}$
Multiplier	$(+16)_{10}$	00010000	
Product	$(-112)_{10}$	10010000	4 left shifts of the multiplicand

The product obtained in this case is the 2's complement of $(112)_{10}$.

Consider now the multiplication of the multiplicand P by n 1's. This may be written as:

$$P(2^n - 1) = P \times 2^n - P$$

The first term in this equation represents a left shift of the multiplicand P by n places, while the second term represents the addition of the 2's complement of the multiplicand to the result of the left shift. The calculation of the product $(-5)_{10} \times (15)_{10} = (-75)_{10}$ using the above equation is shown below:

Multiplicand	$(-5)_{10}$	11111011	2's complement of $(5)_{10}$
Multiplier	$(+15)_{10}$	00001111	
Product	$(-75)_{10}$	10110000	4 left shifts
		00000101	Subtract $(-5)_{10}$, i.e. add $(5)_{10}$
		10110101	2's complement of $(+75)_{10}$

12.23 Booth's algorithm

It is clear from the examples in the previous section that arithmetic operations arising from the multiplication process only take place when the multiplier bits change from 0 to 1 or from 1 to 0. Based on these observations, A D Booth developed the following multiplication algorithm:

1. If two adjacent multiplier bits are the same (00 or 11) do nothing, and shift the partial product left one place
2. If a bit of the multiplier is 1 and the next least significant bit is 0, subtract the multiplicand from the accumulated partial product and shift left one place
3. If a bit of the multiplier is 0 and the next least significant bit is 1, add the multiplicand to the accumulated partial product and shift left one place.

Two examples of the application of Booth's algorithm follow, firstly for a pair of positive numbers and secondly for a positive multiplicand and a negative multiplier. The multiplier is defined by the equation $M = M_7 M_6 M_5 M_4 M_3 M_2 M_1 M_0$ and in order to start the multiplication process a digit $M_{-1} = 0$ is placed behind the LSB of the multiplier.

Multiplicand	$(+7)_{10}$	00000111		
Multiplier	$(+9)_{10}$	00001001(0)		
		11111001	$M_0 M_{-1} = 10$	Subtract 7 and shift left
		00000111	$M_1 M_0 = 01$	Add 7 and shift left
		00000000	$M_2 M_1 = 00$	Shift left
		11111001	$M_3 M_2 = 10$	Subtract 7 and shift left
		00000111	$M_4 M_3 = 01$	Add 7 and shift left
	$(63)_{10}$	100100111111		Product

The eight least significant digits represent the product $(63)_{10}$. In practice, the multiplier would accumulate the partial products as the calculation proceeds whereas in this example the individual partial products have been summed after they have all been formed.

For the second example the multiplier is negative and is expressed in 2's complement form:

Multiplicand	$(+7)_{10}$	00000111
Multiplier	$(-9)_{10}$	11110111(0)

$$
\begin{array}{llll}
& 11111001 & M_0M_{-1} = 10 & \text{Subtract 7 and shift left} \\
& 00000000 & M_1M_0 = 11 & \text{Shift left} \\
& 00000000 & M_2M_1 = 11 & \text{Shift left} \\
& 00000111 & M_3M_2 = 01 & \text{Add 7 and shift left} \\
& 11111001 & M_4M_3 = 10 & \text{Subtract 7 and shift left}
\end{array}
$$

Product $(-63)_{10}$ $\overline{00011000001}$

The least significant eight bits are the 2's complement of $(63)_{10}$.

12.24 Implementation of Booth's algorithm

The basic building blocks for a multiplier utilising Booth's algorithm are shown in Figure 12.36. It consists of an 8-bit register which holds the multiplicand. The output

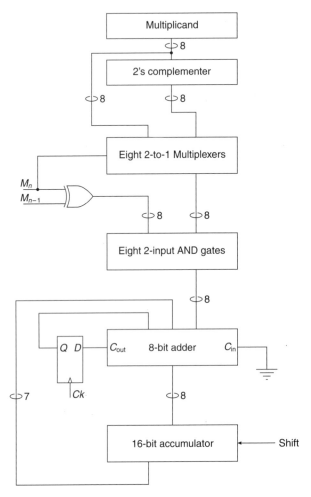

Figure 12.36   *Block diagram for a Booth's multiplier*

of the register is fed directly to one of the inputs of each one of an array of 2-to-1 MUXs, while the output of the 2's complementer is fed to the second input of each of the MUXs. If the multiplier bit $M_n = 1$ the output of the MUXs is the 2's complement of the multiplicand. Alternatively, if $M_n = 0$, the output of the MUXs is the multiplicand. Each output of the MUXs is fed to one of the inputs of a bank of eight 2-input AND gates. The AND gates are enabled/disabled by the output of an XOR gate whose inputs are M_n and M_{n-1}. Assuming that each of these multiplier bits are either 00 or 11 the AND gates are disabled. For the other two possible XOR inputs, i.e. 01 or 10, the XOR output is 1 and the AND gates are enabled. The outputs of these gates are fed to an 8-bit adder in conjunction with the seven least significant bits of an accumulator. For each bit of the multiplier arrangements should be made to shift the partial product in the accumulator. It is left to the reader to design a complete hardware implementation using Booth's algorithm.

The 74384 8-bit by 1-bit 2's complement multiplier package is shown in Figure 12.37 along with the function table. This package employs Booth's algorithm to implement the multiplication of a pair of numbers, both expressed in 2's complement form. The 8-bit multiplicand is stored in an array of eight latches which are controlled by the clear input $\overline{\text{CLR}}$. When the $\overline{\text{CLR}}$ input is low all the latches are cleared and they are now able to receive an 8-bit multiplicand. When the clear input is high any further input to the array of latches is inhibited.

The multiplier is fed to the package via the Y input in a serial bit stream, least significant bit first, and the product is clocked out of the chip on the line labelled PROD. Multiplication of an x-bit multiplicand by a y-bit multiplier generates a product of length of $(x + y)$ bits and the clock must provide $(x + y)$ shift pulses to produce the 2's complement product. Facilities are also available for extending the range of multiplication. This can be achieved by connecting the PROD output to the K input of the next multiplier in the array.

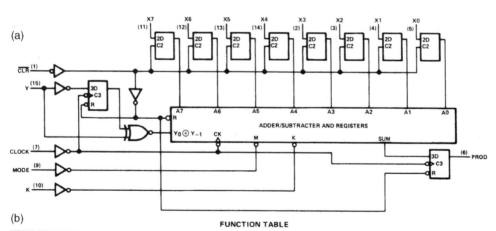

(a)

(b)

FUNCTION TABLE

CLR	CLK	Xᵢ	Y	INTERNAL Y₋₁	OUTPUT PROD	FUNCTION
L	X	Data	X	L	L	Load new multiplicand and clear internal sum and carry registers
H	↑	X	L	L	Output	Shift sum register
H	↑	X	L	H	per	Add multiplicand to sum register and shift
H	↑	X	H	L	Booth's	Subtract multiplicand from sum register and shift
H	↑	X	H	H	algorithm	Shift sum register

H = high-level, L = low-level, X = irrevelant, ↑ = low-to-high-level transition

Figure 12.37 *(a) The 74384 block diagram of the 2's complement Booth multiplier (b) the function table*

Problems

12.1 Develop a combinational logic circuit that will convert 4-bit binary numbers into their corresponding 2's complement form.

12.2 Develop a combinational logic circuit that will generate the 10's complement of the decimal digits where these digits are expressed in the NBCD code.

12.3 The code tabulated below is the XS3 code representation for the decimal digits.

Develop a set of rules for adding together two decimal digits expressed in XS3 form, and hence perform the operation $(34)_{10} - (19)_{10}$ using 9's complement arithmetic.

dd	XS3 code	dd	XS3 code
0	0011	5	1000
1	0100	6	1001
2	0101	7	1010
3	0110	8	1011
4	0111	9	1100

12.4 The decimal digits 0 to 9 are represented by an 8421 NBCD code (normal binary coded decimal). A logic circuit is required which will convert the decimal numbers expressed in 8421 NBCD code into the decimal numbers expressed in the corresponding XS3 code. Design such a code converter using (two-input) NAND gates only.

12.5 Develop the circuit for a 3-decade, XS3 decimal adder/subtractor using 7483 4-bit adders and any additional discrete logic that may be required. 9's complement arithmetic is to be used for the subtraction process.

12.6 Develop an algorithm for the addition of two positive binary coded duodecimal (base 12) numbers.

With the aid of the algorithm, design one stage of an n-stage binary coded duodecimal adder/subtractor circuit which uses 11's complement arithmetic. A 4-bit binary adder is to be used as the basic building block in conjunction with any other logic gates required.

12.7 The decimal digits are to be represented by the 2421 self-complementing code. Develop an algorithm for adding together any two decimal digits using this code. With the aid of this algorithm, design a single stage 2421 adder/subtractor circuit. A 4-bit binary adder is to be used as the basic building block in conjunction with any other necessary logic gates and/or MSI chips.

12.8 An arithmetic circuit has two selection signals, S_0 and S_1. The circuit is required to perform the operations listed below.

$$F = A + B \qquad F = A + B + 1$$
$$F = \bar{A} \qquad F = \bar{A} + 1$$
$$F = A + \bar{B} \qquad F = A + \bar{B} + 1$$
$$F = \bar{B} \qquad F = \bar{B} + 1$$

Using a 4-bit adder as the basic building block, design the circuit that will implement the above operations.

12.9 Design a binary multiplier that multiplies a 4-bit number, $B_3 B_2 B_1 B_0$, by a 3-bit number $A = A_2 A_1 A_0$. The circuit is to be implemented using AND gates and full adders.

12.10 A parallel binary multiplier for 4-bit positive numbers, using the shift and add technique, is illustrated in Figure P12.10. The multiplier is controlled from the box labelled 'Control logic' in the diagram. The inputs to the controller are (a) the clock signal X, (b) the start signal N, and (c) the multiplier bit M. The outputs from the controller are (a) the shift pulse S, (b) the add pulse A, and (c) the reset pulse r.

The control logic is to be designed such that if the multiplier bit, M, is 1 at a given clock time, addition takes place. The multiplier bit should then be reset to 0 and at the next clock time a shift takes place.

Using synchronous sequential design techniques, develop the control logic for the multiplier.

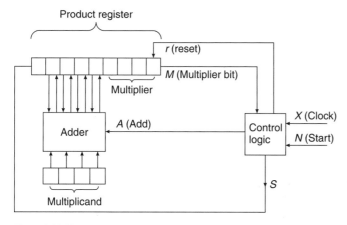

Figure P12.10

12.11 Design a serial binary multiplier using Booth's method.

12.12 Calculate the product $(-9_{10}) \times (-13)_{10}$ using Booth's algorithm.

13 Fault diagnosis and testing

13.1 Introduction

Successful and efficient fault-finding is an art as well as a science, for experienced practitioners learn which faults are likely to occur and which are not, and use tests that isolate typical faults most quickly. It is also technically extremely exacting, as it requires an understanding not only of how the circuit is designed to operate, but also of how the circuit will behave under a multitude of fault conditions.

In fact, broadly similar approaches are needed to establish why a new design does not operate as expected, or why a newly constructed circuit made to a known good design does not operate the first time it is tested, or why a circuit that previously operated correctly has suddenly started showing fault symptoms. For some commercial designs, component-level fault finding in the field is discouraged and faulty modules must be replaced as a whole, but in many of these cases fault-finding is thereby merely transferred to a central depot having the necessary specialist facilities. In any case, design engineers still need the skills that will enable them to find faults in prototype designs.

Before undertaking any electrical tests, it is always a good idea to carry out a physical inspection of the circuit, as many faults are caused by mechanical problems, such as smashed components or faulty soldering, that are clearly visible. Only when a faulty circuit has no obvious visual defects are electrical fault-finding methods then needed. At this point it is highly desirable to have available the technical details of the design. If these details are not immediately to hand, finding an elusive fault is made very much more difficult and may well be uneconomic. It is usually an early task, therefore, to try to obtain these technical details.

A simple way of detecting faults in a combinational logic circuit is to apply every possible input combination and compare the circuit response with the known truth table of the circuit, or with the response of a known faultless version of the circuit. There are clearly limitations to this method because, as the number of inputs (n) increases, the number of tests required increases exponentially and is equal to 2^n in the case of a combinational circuit. Potentially a much greater number of tests than this would be needed for a sequential circuit where the order of the applied inputs is significant.

In a limited number of situations, an examination of the circuit in conjunction with a tabulation of the expected circuit response can reveal the nature of the fault. For example, the circuit of Figure 13.1(a) gives the column headed f in the truth table of Figure 13.1(b) when operating correctly, while the columns headed f_1 and f_2 are given if there is a fault present in the circuit. Response f_1 is obtained when input line q is held permanently and erroneously at 1, and the output of g_1 is then 1 whenever

(a)
                       (b)

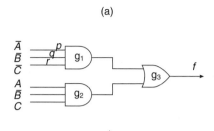

Inputs			Fault free response	Faulty response	
A	B	C	f	f_1	f_2
0	0	0	1	1	0
0	0	1	0	0	0
0	1	0	0	1	0
0	1	1	0	0	0
1	0	0	0	0	0
1	0	1	1	1	1
1	1	0	0	0	0
1	1	1	0	0	0

Figure 13.1   *An informal approach to fault diagnosis (a) Circuit implementation of $f = \bar{A}\bar{B}\bar{C} + A\bar{B}C$ (b) Circuit response for fault-free (f) and faulty (f_1 and f_2) conditions*

$A = C = 0$. Response f_2 is obtained when any one or more of the input lines p, q or r is (are) erroneously held at 0.

The fault f_1 is revealed by applying $A = 0, B = 1, C = 0$, and examining the output, while the fault f_2 is revealed by the test $A = B = C = 0$. The first fault is distinguishable from all other possible faults, while the second test reveals *seven* indistinguishable faults, three corresponding to one input line being erroneously low, three where any two of these three input lines are erroneously low, and one where all three input lines are erroneously low.

This limited and informal fault analysis of the circuit shown in Figure 13.1(*a*) has been carried out by inspection. However, it is desirable that more formal techniques for testing should be developed and it is the purpose of this chapter to examine fault analysis and methods for generating an adequate test set for a specified circuit. This process is frequently termed *test pattern generation (TPG)*.

13.2 Fault detection and location

Inexperienced engineers, faced with a circuit containing a fault or faults unknown, often resort to removing semiconductor components at random and attempting to test them or to replace them with new components without testing the old. This is a *completely incorrect* approach, for the following reasons:

1. In a typical system with many components, it is unlikely that the faulty component will be chosen at random for removal.
2. Often, either perfectly good components, or the circuit board, or both, will be damaged by inexpert removal and replacement.
3. In many systems, the most common faults are mechanical failure of switches and similar components, followed by connection (soldering) faults, open circuit resistors, and open circuit capacitors. Failure of ICs and other semiconductor devices, although possible, is relatively unlikely.
4. Sometimes a fault will cause an associated semiconductor device to fail. If this associated device is replaced at random, the replacement will be destroyed as well with no further progress to show as a result.

In all cases, a good working rule is that *no component (or sub-system, or minimal replaceable part) should be removed or replaced until it has been proven faulty (beyond any doubt) and it has also been established that there is no associated fault that will cause the replacement to fail.* To *prove* that a component in the circuit is faulty requires that voltage and perhaps other electrical checks should be performed upon the circuit while it is powered, and the component should be subjected to typical signals whilst the output is examined for the correct output signals.

In order to locate a fault in a typical system consisting of a chain of *n* components connected in sequence, often *signal tracing* will be employed in situations where the intermediate signals are accessible at each point in the chain. Using this technique, the signals at the outputs of components are examined and compared with what would be expected, assuming no fault were present. If correct signals are observed at the input to a certain component but not at its output, then it is fairly clear that the fault must be associated closely with that component. This still does not mean that this particular component is faulty, for it is also possible that the power supply to this component has failed or that the function of this component has been affected by some other fault. Therefore, exhaustive voltage checks centred upon this component must still be undertaken to investigate these other possibilities before condemning this component.

How can this faulty component be located? Suppose that the fault is equally likely to occur in *any* of the *n* links of the chain. The correct procedure is *not* to trace the signal *in turn* through each component in the signal chain, for this method requires, on average, $n/2$ tests before the faulty stage is located (of course, one could be lucky and locate the fault immediately with one or two tests, but equally one may be unlucky and need as many as *n* tests to locate the fault). Much more efficient is the procedure of *binary division* where firstly the presence or absence of the correct signal is established *halfway through* the chain of *n* stages. Depending upon the result of this single test, either the entire first half or the entire last half of the circuit can therefore be eliminated from further attention, and subsequent tests confined to the faulty half. The same principle can then be applied to this half of the circuit, and so on, by successive binary divisions the fault is isolated to one single stage. With a large number of stages, this procedure requires only $\log_2(n)$ tests to locate the faulty stage, considerably less than the average of $n/2$ tests required for tracing the signal *in turn* through each stage. For example, for $n = 16$, sequential signal tracing requires 8 tests on average, whereas binary division signal tracing requires 4 tests.

There are two types of test which can be carried out on digital circuits:

1. *Fault detection tests*, which are used to reveal faults, and
2. *Fault location tests*, which are designed to reveal, locate and identify faults.

The testing method used for the circuit of Figure 13.1(a) in section 13.1 above was exhaustive, in that all possible tests were applied. In practice, it is possible to devise shorter test procedures which will detect and locate faults. However, before developing realistic test methods, a number of simplifying assumptions will be made relating to the circuits to be tested and the types of faults occurring in those circuits:

1. The procedures to be dealt with in this chapter are concerned with the detection, location, and diagnosis of *single* faults. This is not to exclude the possibility of the occurrence of multiple unrelated faults, but the probability of such faults occurring is small in comparison with the probability of a single fault occurring. In a complex

system where multiple faults exist, it is likely that each fault will affect a different part of the system, so that in practice the system may be divided into sub-systems in each of which there is, at most, only one fault. Of course, it is also possible that one fault may cause failure of a closely related component, so that when a faulty component is located, the question must always be asked whether that component failed as a direct consequence of another failure. If so, replacing the first component will merely cause the replacement to fail again.

2. It will be assumed that the faults being detected are permanent rather than intermittent. Intermittent faults are, in general, extremely difficult to locate, as, after replacing a suspect component, it is impossible to say without doubt whether subsequent correct operation is a result of replacing the faulty component or is a result of inadvertently *disturbing* a different actual faulty component so that it is now (temporarily) operating correctly. It is a truism that it is impossible to find a fault in a system that is, even if temporarily, operating according to its specification. In many cases, when faced with an intermittent fault whose location is not obvious, the best course of action is either to replace the entire faulty system or sub-system, or else to wait until the intermittent fault becomes permanent. Replacement may be the best option if this system or sub-system has a critical function, whereas waiting may be a cheaper option if the system or sub-system does *not* have a critical function as intermittent faults usually become permanent eventually. After waiting, the fault diagnosis will be made much easier by having a permanent fault to locate. Some intermittent faults are thermally activated, produced by a faulty connection that is good at certain temperatures but at other temperatures the differing thermal expansion coefficients of the various materials involved in the connection cause electrical contact to be lost. In this case, it may be worthwhile using an aerosol freezing spray to cool isolated parts of the circuit, hopefully inducing the fault to be present for longer periods, in order to locate the fault.

To start with, it will be assumed that all faults are such that a certain logic line is either *stuck-at-0* (*s-a*-0) or *stuck-at-1* (*s-a*-1), i.e., the line is permanently at logic level 0 or 1 respectively, regardless of what logic level is actually supposed to be present on that line. This widely used fault model does not cover all possible faults, but its use is justified on the grounds that most circuit failures exhibit symptoms corresponding to this model. For example, a short circuit of any line to ground can be represented by a *s-a*-0 fault, while an open circuit on an input line to a TTL gate will cause that input to 'float' at a voltage corresponding to an unreliable and noisy logic 1, causing an *s-a*-1 fault. Other common types of fault occur when one of the transistors in the usual 'totem-pole' output stage of a gate becomes short circuited and the other becomes open circuited, which can again be represented by an *s-a*-0 or an *s-a*-1 fault as appropriate. Equivalent faults can also occur in 'open collector' gate output stages.

Later (section 13.8), *bridging faults* will be examined, where two logic lines are inadvertently connected together. However, other faults are also possible in practice. For example, an open circuit on an input line to a CMOS gate results in that input being connected neither to logic 0 nor to logic 1, and, assuming the usual 'totem-pole' gate output configuration, it is then possible that neither output transistor is turned fully off. Therefore, the gate output stage will draw substantially more current than it is designed to carry, which can lead to serious overheating of that gate, as well as excessive current consumption. Overheating can lead to gate failure which will itself usually be well represented by the *s-a*-0 or *s-a*-1 fault models, but the excessive current consumption can also lead to reduced power supply voltage or even power supply failure, causing erratic operation elsewhere in the circuit. This type of fault is best located by careful voltage checks on the IC pins. Note that in this case, merely replacing the overheating IC will clearly not solve the problem, unless the cause of the fault is an internal open circuit between the IC connection pin and the actual gate input transistor.

Another possibility with CMOS gates is that if a gate input is a poorly connected though not quite open circuit, there may be a large effective resistance R_s in series with the gate. In conjunction with the effective capacitance C_{eff} of the gate input this can give correct logic operation after a significant delay governed by the time constant $\tau = R_s C_{eff}$ which in some cases can be as much as several seconds. This is an example of a *delay fault*, characterised by excessive propagation time through a particular part of the circuit. In the case of a slow logic system, the presence of such a fault may not matter or may not even be detectable, but in the case of a fast system where such a delay is significant such symptoms can be baffling to the inexperienced engineer, causing unexpected racing hazards for example.

A further type of fault worthy of mention here, encountered mainly, though not necessarily exclusively, with VLSI chips such as memories, is the *pattern-sensitive* fault, where a repeatable and demonstrable logic error occurs whenever some particular logic pattern is set up elsewhere in the circuit. As an example of this kind of fault, it is possible that if one particular data word is output by a memory chip, then and only then is incorrect data read from the data bus. This can be caused by poor supply line decoupling, so that a data word that happens to require a large current drive to the memory output data buffers momentarily takes the supply line below its specified voltage, thus affecting the memory or data bus operation. This type of fault can be extremely difficult to find as for most of the time the circuit operates correctly, without any 'stuck-at' or bridging-type behaviour at all. It is distinguishable from an intermittent fault only in that the fault is electrically repeatable, if only the correct conditions for repeating it can be found, and also it usually does not change in severity over time.

13.3 Gate sensitivity

One concept that is central to the testing of all circuits is the provision of a sensitive path between input and output so that any signal transitions appearing at the input will be transmitted to the output. At the gate level this concept depends upon the transmission properties of each gate. For example, consider the 2-input AND gate shown in Figure 13.2(a):

1. If $B = 1$, signal transmission through the gate is *enabled*; $Z = 0$ for $A = 0$, and $Z = 1$ for $A = 1$. Any logic transitions appearing at input A are directly transmitted to the gate output, Z.
2. If $B = 0$, signal transmission through the gate is *disabled*; $Z = 0$ irrespective of the value of A. Logic transitions appearing at input A are not transmitted to the gate output Z.

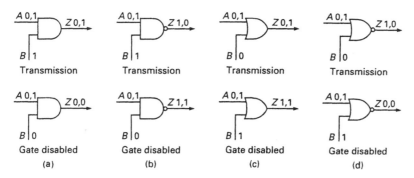

Figure 13.2 *The transmission characteristics of (a) AND, (b) NAND, (c) OR, and (d) NOR gates*

Similarly, a 2-input NAND gate is enabled when one of its inputs is held at 1, while 2-input OR and NOR gates are enabled when one of their inputs is held at 0. The transmission properties of these gates are also illustrated in Figure 13.2.

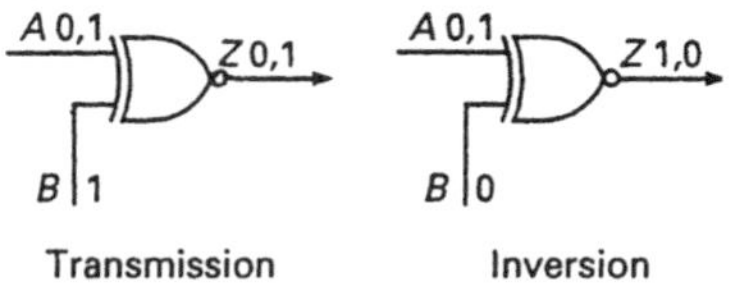

It is interesting to note that the XOR and XNOR gates do not exhibit the same gate sensitivity as the other four gates, as neither of them can be disabled. The behaviour of the XNOR gate is illustrated in Figure 13.3 and, like the XOR gate, it behaves as a controlled inverter. Both of these gate types are therefore permanently enabled.

Figure 13.3 *The transmission characteristics of the XNOR gate*

13.4 A fault test for a 2-input AND gate

The AND gate shown in Figure 13.4(a) has two input lines, labelled p and q respectively, and one output line labelled r. In all, there are six possible single faults for which tests can be made. They are: any one of p, q, and r s-a-1 or, alternatively, any one of p, q and r s-a-0.

Suppose that line p has a s-a-0 fault; it is required to determine the test that will detect this fault. Circuit input B must be maintained at logic 1 level to enable the gate, otherwise the gate's output will be permanently held at 0, while the other circuit input (A) must be held at the complement of the 'stuck-at' fault value of p – i.e., in this case, 1. Hence, the required test is $A = B = 1$, and the results of this test are summarised in Figure 13.4(a). It is also clear that the same test conditions will also detect line q s-a-0 and line r s-a-0.

To determine the test for line p s-a-1, the input B must be maintained at logic 1 level to enable the gate, while input A is held at 0 (the inverse of the s-a-1 value of p). Hence, the required test is $A = 0$ and $B = 1$, and the results of this test are summarised in Figure 13.4(b). This same test will also detect whether r is s-a-1.

Finally, to test for q s-a-1, input $A = 1$ is the gate enabling signal, while the complement of the 'stuck-at' fault value must be applied at input B, giving $B = 0$. Therefore, the required test is $A = 1$ and $B = 0$, and the possible test results are shown in Figure 13.4(c).

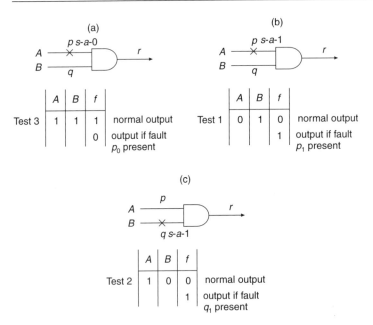

Figure 13.4   *Fault test for a 2-input AND gate (a) Test for p s-a-0 (b) Test for p s-a-1 (c) Test for q s-a-1*

The results deduced above can be tabulated as follows, using the shorthand notation that p_1 signifies *p s-a*-1 and other faults are indicated similarly:

Test number	A	B	f	Faults detected by test
0	0	0	–	[Not required]
1	0	1	1	p_1, r_1
2	1	0	1	q_1
3	1	1	0	p_0, q_0, r_0

The table shows that a test using $A = B = 0$ is not needed, and that the remaining three tests will detect all possible faults. Hence, the minimal test set for an AND gate is $T = \{1, 2, 3\}$, and therefore this analysis has shown that there is a 25% reduction in the tests needed, compared to the test set using all entries in the truth table.

13.5 Path sensitisation

The determination of a test set for a single gate, where there is direct access to the input and the output, is achieved by enabling or *sensitising* the gate. However, it often happens that when testing a combinational circuit such as the one shown in Figure 13.5, the outputs D and E of gates g_1 and g_2, respectively, are *not* directly accessible so that a fault at p, for example, must be detected at the output F. For example, points D and E may be within an IC which must be tested by applying input signals whilst examining the output signals. This requires the sensitisation of two gates, g_1 and g_3, which then provide a sensitised path from input A to output F.

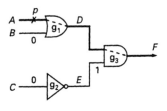

Figure 13.5   *Path sensitisation for testing for a s-a-0 fault on line p*

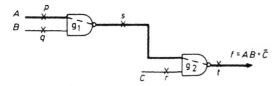

Figure 13.6 *Circuit to be tested by path sensitisation method*

To determine whether line p shows an s-a-0 fault, gate g_1 is sensitised by setting $B = 0$. In order to sensitise gate g_3, point E must be set to 1 which requires input $C = 0$. The path through g_1 and g_2 is now sensitised, and to test for p_0 then $A = 1$, the inverse of the fault for which a test is required. Hence the test for p_0 is $(A,B,C) = (1,0,0)$.

As an example of path sensitisation, a test set covering all the possible faults will be obtained for the NAND gate combinational network shown in Figure 13.6. There are three possible paths through this circuit: *pst, qst* and *rt*. On the diagram, as an example, path *pst* has been marked in bold. The analysis proceeds by sensitising each path in turn, and the tests for the various s-a-0 and s-a-1 faults associated with each path determined. The results are summarised in the following table:

Path (Figure 13.6) →	*pst*		*qst*		*rt*	
Gate sensitisation input signals:	$B = C = 1$		$A = C = 1$		$(A,B) = (0,0)$ or $(0,1)$ or $(1,0)$	
Assumed fault:	p_0	p_1	q_0	q_1	r_0	r_1
Test (complement of assumed fault):	$A = 1$	$A = 0$	$B = 1$	$B = 0$	$C = 1$	$C = 0$
Full test conditions (A,B,C):	$(1,1,1)$	$(0,1,1)$	$(1,1,1)$	$(1,0,1)$	$(0,0,1),$ $(0,1,1),$ or $(1,0,1)$	$(0,0,0),$ $(0,1,0),$ or $(1,0,0)$
Normal output f:	1	0	1	0	0	1
Faulty output (inverse of normal f):	0	1	0	1	1	0
Other faults detected by this test:	s_1 & t_0	s_0 & t_1	s_1 & t_0	s_0 & t_1	t_1	t_0

In these tests, for paths *pst* and *qst* the sensitising signals sensitise the selected path in the forward transmission direction and the sensitising process is referred to as the *forward trace* step. However, for path *rt* the sensitising signal for g_2 is $s = 1$, and values therefore have to be assigned to A and B which will set up this sensitising signal. This is referred to as the *backward trace* step, and for the circuit of Figure 13.6 there are three possible combinations of the input signals A and B that will set up the required sensitising signal, as shown in the table.

Additionally, it should be noted that when a path has been sensitised, 'stuck-at' faults can be detected at *each* point along the sensitised path. The faults that are automatically detected in this way correspond to the logical complement of the normal logic level expected.

A similar table summarising the results for other circuits can be prepared by adopting the following procedure:

1. Select the fault for which a test or tests are to be determined, and identify a path from the site of the fault to the circuit output.
2. Sensitise the path using the forward trace step.
3. Establish the network inputs, if necessary, by the backward trace step.

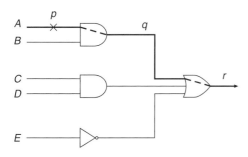

Figure 13.7   *Circuit with independent inputs and gate fan-out of 1*

The path-sensitising technique is always satisfactory for circuits in which all the inputs are independent of each other and where the fan-out of each gate in the circuit is unity. In such cases, this method will generate tests for all possible faults in the circuit because consistent input combinations always exist, irrespective of which path is sensitised, and consequently all paths in the circuit can be sensitised. Another example of such a circuit is shown in Figure 13.7, where the tests for the faults p_0, q_0 and r_0 are the input combinations $(A,B,C,D,E) = (1,1,0,X,1)$ or $(1,1,X,0,1)$,   where   X   indicates   a   'don't   care'. The remaining four sensitisable paths can be used to determine tests for all other possible faults in this circuit.

13.6 Path sensitisation in networks with fan-out

A circuit incorporating fan-out is illustrated in Figure 13.8(a). The signal path for input variable B branches into the two lines s and t, one branch being connected to the lower input of g_1 and the second branch being connected to the upper input of g_2. After passing through gates g_1 and g_2, these two paths reconverge at g_3. This is an example of *reconvergent fan-out*. Further inspection of this circuit shows that there are two single paths that can be sensitised independently, i.e. *qsuw* (Figure 13.8(a)) and *qtvw* (Figure 13.8(b)). Additionally, there is a *multiple path* (Figure 13.8(c)) consisting of these same two paths together, which can also be sensitised. The following table shows the tests possible for detecting the fault q_0:

Path (Figure 13.8) →	*qsuw*	*qtvw*	*qsuw* & *qtvw*
Gate sensitisation input signals:	$A = 1$	$C = 1$	$A = C = 1$
	$v = 1, \Rightarrow C = 0$	$u = 1, \Rightarrow A = 0$	
Assumed fault:	q_0	q_0	q_0
Test (complement of assumed fault):	$B = 1$	$B = 1$	$B = 1$
Full test conditions (A,B,C):	$(1,1,0)$	$(0,1,1)$	$(1,1,1)$
Normal output f:	1	1	1
Faulty output (inverse of normal f):	0	0	0
Other faults detected by this test:	$s_0, t_0, u_1,$ & w_0	$s_0, t_0, v_1,$ & w_0	$s_0, t_0,$ & w_0

In this example, the fault q_0 can be transmitted along either of the two independently sensitised paths and can be detected at the output. For the case of dual path sensitisation, if the input at B changes from $1 \rightarrow 0$ the outputs of g_1 and g_2 both change from $0 \rightarrow 1$; the fault is transmitted along both paths simultaneously. This is an example of *positive reconvergence*.

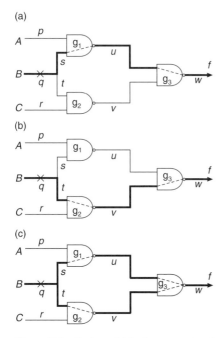

Figure 13.8 *Path sensitising in a network with fan-out (a) Single-path qsuw (b) Single-path qtvw (c) Dual-path qsuw and qtvw sensitised simultaneously*

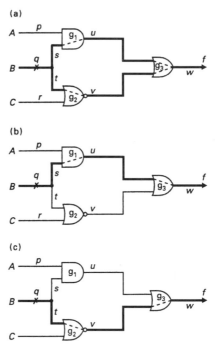

Figure 13.9 *Path sensitisation for circuit having unequal inversion parity along its reconvergent branches (a) Dual-path sensitisation test fails (b) and (c) Single-path sensitisation tests succeed*

It is tempting at this point to conclude that dual path sensitisation will always prove satisfactory in circuits with reconvergent fan-out. However, this is not so, as shown in the following example. A second circuit incorporating reconvergent fan-out is shown in Figure 13.9(a). The signal path for the B input again branches into the two lines s and t, one path going *via* g_1 and g_3, the second path going *via* g_2 and g_3. Assuming that q is s-a-0, the sensitising signals for gates g_1 and g_2 (thus sensitising both possible paths) are $A = 1$ and $C = 0$. To test for q_0, input B is set to 1, the logical complement of this fault. The outputs of g_1 and g_2 are then expected to be 1 and 0 respectively, and the circuit output f at g_3 is 1. However, if the fault q_0 is present, the outputs of g_1 and g_2 will be 0 and 1 respectively, giving an *unchanged* circuit output $f = 1$. Clearly, for the circuit in Figure 13.9(a), $(A,B,C) = (1,1,0)$ is *not* a valid test for q_0.

Why has the path sensitisation method failed in this case? The fault is transmitted along the two sensitised paths $qsuw$ and $qtvw$, but the changes generated by the fault along the two paths are $0 \rightarrow 1$ and $1 \rightarrow 0$, respectively. Thus, fault q_0 leads to a change of the signals at the input of g_3 from $(0,1)$ to $(1,0)$. Such a change at the inputs of an OR gate does not generate a change at its output, and consequently the effect of the fault is not transmitted through g_3. The failure of the path sensitisation test in this example is due to the unequal *inversion parity* of the two paths. In contrast with the positive reconvergence illustrated in Figure 13.8(c), this is an example of *negative reconvergence*.

Although dual sensitisation does not generate a test for q_0, independent sensitisation of path *qsuw* (Figure 13.9(b)) or of path *qtvw* (Figure 13.9(c)) will do so, as shown in the following table:

Path (Figure 13.9) $\rightarrow$	*qsuw*	*qtvw*
Gate sensitisation input signals:	$A = 1$	$C = 0$
	$v = 0, \Rightarrow C = 0$	$u = 0, \Rightarrow A = 0$
Assumed fault:	q_0	q_0
Test (complement of assumed fault):	$B = 1$	$B = 1$
Full test conditions (A,B,C):	$(1,1,1)$	$(0,1,0)$
Normal output f:	1	0
Faulty output (inverse of normal f):	0	1
Other faults detected by this test:	$s_0, t_0, u_0,$	$s_0, t_0, v_1,$
	& w_0	& w_0

This last example has shown that multiple path sensitisation does not necessarily produce a valid test for a given fault. At this point it may appear that multiple path sensitisation should be avoided and that, in general, it would be better to concentrate on single path sensitisation. Unfortunately, this conclusion is also not sustainable since the following example shows that single path sensitisation may itself fail to produce a valid test in certain circumstances.

In the circuit shown in Figure 13.10, it will be assumed that fault p_1 is present, and the path selected for sensitisation is *via* gates g_2, g_5 and g_8. Gate g_8 is sensitised by holding all its inputs at 1 with the exception of the sensitised input, and gate g_5 is sensitised by holding its upper input at logic level 1. To ensure that the output of gate g_6 is 1, irrespective of whether fault p_1 is present, its lower input must be held at 0, hence input $D = 0$. With $D = 0$, the lower input of g_3 is also held at 0, the output of g_3 is 1 and so the upper input of g_7 is 1. To ensure that the output of g_7 is held at 1, its lower input must be held at 0, hence $C = 0$. Since $C = 0$, the lower input of g_1 is held at 0 and the output of gate g_1 is therefore 1. This output is connected to the lower input of gate g_4 and so to ensure that the output of gate g_4 is 1, its upper input must be held at 0,

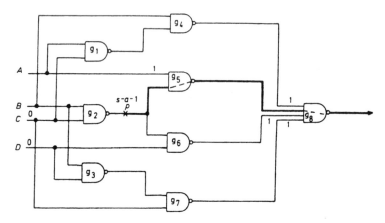

Figure 13.10   *Failure of single-path sensitisation*

hence input $B = 0$. For the chosen sensitised path, $B = C = 0$, but these inputs are inconsistent with the path sensitisation procedure where, if fault p_1 is present, the output of gate g_2 should be equal to the complement of the fault, i.e. 0. (To achieve this condition it would be necessary to have $B = C = 1$.) Hence, in this example, single path sensitisation fails to produce a valid test for fault p_1. It can be shown that, in this case, if the two paths g_2-g_5-g_8 and g_2-g_6-g_8 are simultaneously sensitised, a valid test for fault p_1 can be found.

It will be clear from all the preceding examples that it is not possible to specify one prescribed series of steps which will be successful for all circuits when using the path sensitisation technique. In some cases, the desirability of providing a simple method of testing may provide additional design philosophy in determining the best implementation of a given logic circuit specification. However, the following general guidelines can be applied to all circuits when trying to set up a fault detection strategy:

1. Attempt to derive fault tests using single paths only. In circuits without fan-out, single path sensitisation is the only possibility.
2. Assign logical value(s) at the input(s) which will produce the logical complement of the fault value at the point of the fault.
3. Sensitise all the gates along the path.
4. Use the backward trace technique where necessary. Provided that a self-consistent input combination can be found, then a valid test exists.
5. If, for a selected path, a valid fault test does not exist, repeat the procedure specified in rules (1) to (4) for other single paths in the circuit which may be sensitised.
6. In the event of failure of single path sensitisation, attempt to derive a valid test by sensitising two or more paths using the procedure described above. If necessary, try every possible combination of the single paths in the circuit.

13.7 Undetectable faults

The fault-free response of the combinational circuit shown in Figure 13.11(a) is tabulated in the column headed f in Figure 13.11(b). If fault p_0 is present, the output of gate g_2 is held permanently at logic 0 and for this condition the response of the circuit appears in Figure 13.11(b) under the heading f_1. Comparison of the two columns shows that the responses are identical, and consequently the fault p_0 is *undetectable*.

The reason for this is explained by examining the Boolean equation of the circuit, which is:

$$f = C + \bar{A}BC + A\bar{B}$$

$$= C(1 + \bar{A}B) + A\bar{B}$$

$$= C + A\bar{B}$$

and so the circuit is *not* the minimal implementation of f. Gate g_2 generating the term $\bar{A}BC$ is redundant and could be omitted, using a two-input OR gate for g_3 instead, without any effect on the circuit output. The K-map for this function is shown in Figure 13.11(c). Therefore, in this example, the output of the redundant gate g_2 can be *s-a-0* without affecting the circuit operation.

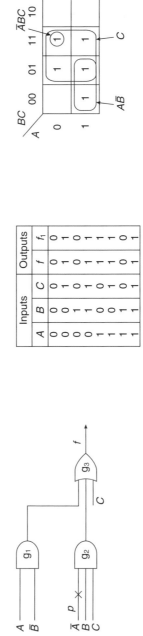

Figure 13.11 (a) Implementation of a function, $f = \bar{A}BC + A\bar{B} + C$, containing redundancy (b) Fault-free response f, and response f_1 with fault p_0 present (c) K-map of function f

(a)

(b)

Inputs			Outputs	
A	B	C	f	f_1
0	0	0	0	0
0	0	1	1	1
0	1	0	0	0
0	1	1	1	1
1	0	0	1	1
1	0	1	1	1
1	1	0	0	0
1	1	1	1	1

(c)

In practice, there may be similar undetectable faults in any circuit containing redundancy. Sometimes, redundancy is deliberately introduced into a combinational network in order to ensure that it is *hazard-free* (see Chapter 9). The result of including such redundancy does not show clearly in a static truth table analysis. Nevertheless, if there is a fault associated with the redundancy and a redundant gate is effectively disabled, then the hazard-free property of the design will not be fulfilled and hazards may be produced by the circuit. As an example, an undetectable, but nevertheless important, fault may arise in the hazard-free implementation of the function $f = AB + \bar{A}C$. The K-map of this function is shown in Figure 13.12(a), and the function generates a static 1-hazard when $B = C = 1$ and A makes a $1 \rightarrow 0$ transition. Elimination of the hazard is achieved by adding the consensus term BC to the original equation which then becomes $f = AB + \bar{A}C + BC$. The minimal NAND implementation of this hazard-free function is shown in Figure 13.12(b) and a test for fault p_1 is sought according to the following table:

Path (Figure 13.12(b)) $\rightarrow$	*pu*
Gate sensitisation input signals:	$q = 1, \Rightarrow C = 0$ or $A = 1$
	$r = 1, \Rightarrow A = 0$ or $B = 0$
Assumed fault:	p_1
Test (complement of assumed fault):	$p = 0, \Rightarrow B = C = 1$

Since the test requires $B = C = 1$, the input signals for the correct path sensitisation must be $A = 1$ (giving $q = 1$) and simultaneously $A = 0$ (giving $r = 1$). This is clearly impossible, and so fault p_1 is *undetectable* using the path sensitisation technique if the

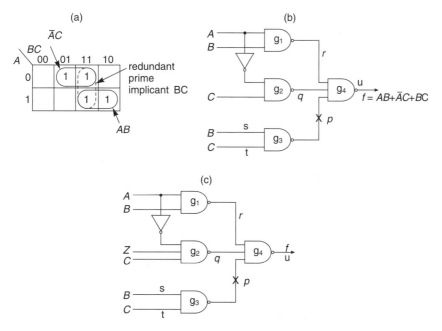

Figure 13.12 *(a) K-map of the function $f = AB + \bar{A}C$ (b) Hazard-free circuit with undetectable fault (c) Circuit with additional test input*

redundant gate g_3 is included in the circuit. Therefore, the circuit is not completely testable using this technique. Alternatively, if the redundant gate g_3 is omitted, thus using a two-input NAND gate for g_4 instead, the hazard will be generated. Faced with this dilemma, the designer may prefer to make output g_3 directly available to the tester at a *test point* (thus bypassing gate g_4), whereupon applying inputs $B = C = 1$ will establish the presence or absence of fault p_1. However, another possibility is to add an additional *test input* (Z) to gate g_2, which will now need three inputs, as shown in Figure 13.12(c). In normal operation Z is permanently held at 1, while Z is taken to 0 specifically for testing the circuit. If this is done, the test developed in the following table is now possible:

Path (Figure 13.12(c)) $\rightarrow$	*pu*
Gate sensitisation input signals:	$q = 1, \Rightarrow Z = 0$
	$r = 1, \Rightarrow A = 0$ or $B = 0$
Assumed fault:	p_1
Test (complement of assumed fault):	$p = 0, \Rightarrow B = C = 1$
Full test conditions (A,B,C,Z):	(0,1,1,0)
Normal output f:	1
Faulty output (inverse of normal f):	0
Other faults detected by this test:	$s_0, t_0, \& u_0$

Alternatively and equivalently, a similar test input could be added to gate g_1. A further alternative (without the test input Z) is to attempt to test for the presence of the static hazard directly in the circuit of Figure 13.12(b) by applying the test conditions $(A,B,C) = (\downarrow,1,1)$, where $\downarrow$ indicates the hazard-producing falling edge $1 \rightarrow 0$ on input A. Presence of the hazard at the output implies the failure of the redundant part of the circuit. Of course, this requires the use of a high-speed oscilloscope (and may also require repeated application of the falling edge), and so may not be practical or economic.

13.8 Bridging faults

Another possible fault which may occur in a combinational circuit is a bridge or a short between two lines, as shown in Figure 13.13(a) and (b). This type of fault often occurs through careless soldering that leaves a solder bridge between two adjacent lines that are supposed not to be connected. If inputs B and C happen to have equal logic values (i.e., both either 0 or 1), the interconnection pq causes no

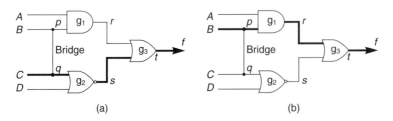

Figure 13.13   *Tests for bridging faults (a) Using sensitised path qst (b) Using sensitised path prt*

detectable fault since both gate inputs are held at their correct values. If, however, the inputs are driven to complementary logic levels so that $(B,C) = (0,1)$ or $(1,0)$, then several possibilities may occur:

1. whichever input signal is supposed to be set at 1 may be pulled down to 0, which often happens in practice with TTL gates, or
2. whichever input signal is supposed to be set at 0 may be pulled up to 1, which often happens in practice with ECL gates, or
3. both inputs B and C may be pulled to an indeterminate voltage which cannot be interpreted reliably as either logic 1 or logic 0, or
4. either or both of the driving gates may fail, possibly causing an additional 'stuck-at' fault.

Notwithstanding possibility no. 4 above, note that a bridging fault is fundamentally different from a 'stuck-at' fault, as the faulty lines may take on either logic value according to how they are driven. Another complication is that a bridging fault might occur between two logic lines in such a way as to form an effective feedback path around some combinational logic, so that a purely combinational circuit can be transformed into a faulty circuit having some of the characteristics of a flip-flop, i.e. a sequential circuit, which is considerably more difficult to analyse. The circuit outputs will then depend upon the previous outputs of the circuit.

However, in practice, it is usually unnecessary to develop tests for bridging between all possible pairs of lines, as it would seem highly unlikely that any other than physically adjacent lines would suffer bridging faults. Which lines are affected therefore depends upon the physical layout of the circuit. The practical tests for revealing the presence of the bridge pq for the two cases, $(B,C) = (0,1)$ and $(B,C) = (1,0)$, depend upon which of the four possibilities listed above actually occurs in practice. In the following development of the tests, it is assumed that whichever input signal is supposed to be set at 1 will always be pulled down to 0 (possibility number 1):

Path (Figure 13.13) $\rightarrow$	*qst*	*prt*
Gate sensitisation input signals:	$D = 0$	$A = 1$
	$r = 0, \Rightarrow B = 0$	$s = 0, \Rightarrow D = 1$
Assumed fault:	pq bridge	pq bridge
Test (complement of assumed fault):	$(B,C) = (0,1)$	$(B,C) = (1,0)$
Full test conditions (A,B,C,D):	$(X,0,1,0)$	$(1,1,0,1)$
Normal output f:	0	1
Faulty output (inverse of normal f):	1	0
Other faults detected by this test:	$q_0, s_1, \& t_1$	$p_0, r_0, \& t_0$

In this table, a valid test using path qst is obtained regardless of the value of input A, which is therefore a 'don't care'.

13.9 The fault detection table

Clearly, it is desirable to determine a minimal set of tests that can be applied to a given circuit, which can be guaranteed to find all possible faults. The 'fault table' (or 'fault matrix') method can be used to achieve this. The method will be explained by working through a typical example. The circuit shown in Figure 13.14(a) has seven

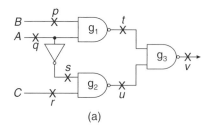
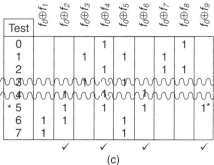

(a)

Fault:					p₀ t₁	q₀	r₀ s₀ u₁	t₀ u₀ v₁	v₀	p₁	q₁	r₁	s₁
Test	A	B	C	f_0	f_1	f_2	f_3	f_4	f_5	f_6	f_7	f_8	f_9
0	0	0	0	0	0	0	0	1	0	0	0	1	0
1	0	0	1	1	1	1	0	1	0	1	0	1	1
2	0	1	0	0	0	0	0	1	0	0	1	1	0
3	0	1	1	1	1	1	0	1	0	1	1	1	1
4	1	0	0	0	0	0	0	1	0	1	0	0	0
5	1	0	1	0	0	1	0	1	0	1	0	0	1
6	1	1	0	1	0	0	1	1	0	1	1	1	1
7	1	1	1	1	0	1	1	1	0	1	1	1	1

(b)

Test	$f_0 \oplus f_1$	$f_0 \oplus f_2$	$f_0 \oplus f_3$	$f_0 \oplus f_4$	$f_0 \oplus f_5$	$f_0 \oplus f_6$	$f_0 \oplus f_7$	$f_0 \oplus f_8$	$f_0 \oplus f_9$
0				1				1	
1			1		1		1		
2				1			1	1	
3	*(deleted — wavy line)*								
4	*(deleted — wavy line)*								
* 5		1		1		1			1*
6	1	1			1				
7	1				1				
		✓			✓		✓		✓

(c)

Figure 13.14 *(a) NAND logic implementation of $f = AB + \bar{A}C$ (b) Fault-free (f_0) and faulty (f_1 to f_9) responses for the circuit in (a) (c) Fault detection table*

lines, i.e. p, q, r, s, t, u and v. The function implemented by this circuit is $f = AB + \bar{A}C$. However, if there is a s-a-0 or a s-a-1 fault on any one of these seven lines, the function implemented by the circuit will be modified to be one of 14 possible faulty responses. (The presence of other faults, such as bridging faults, or multiple faults, would produce further possible faulty responses which could additionally be considered if desired.) If the symbol f_{x_i} denotes the faulty response when the single 'stuck-at' fault x_i is present, then the possible faulty responses are:

'Stuck-at-0'	'Stuck-at-1'
$f_{p_0} = \bar{A}C;$	$f_{p_1} = A + C;$
$f_{q_0} = C;$	$f_{q_1} = B;$
$f_{r_0} = AB;$	$f_{r_1} = \bar{A} + B;$
$f_{s_0} = AB;$	$f_{s_1} = AB + C;$
$f_{t_0} = 1;$	$f_{t_1} = \bar{A}C;$
$f_{u_0} = 1;$	$f_{u_1} = AB;$
$f_{v_0} = 0;$	$f_{v_1} = 1.$

Since $f_{p_0} = \bar{A}C$ and $f_{t_1} = \bar{A}C$ are identical functions, the circuit response will be identical for these two faults, and they are *indistinguishable*. In fact, there are nine *distinguishable* functions contained in the total of 14 faulty functions, as follows:

$$f_1 = f_{p_0} = f_{t_1} = \bar{A}C$$
$$f_2 = f_{q_0} = C$$
$$f_3 = f_{r_0} = f_{s_0} = f_{u_1} = AB$$
$$f_4 = f_{t_0} = f_{u_0} = f_{v_1} = 1$$
$$f_5 = f_{v_0} = 0$$
$$f_6 = f_{p_1} = A + C$$
$$f_7 = f_{q_1} = B$$
$$f_8 = f_{r_1} = \bar{A} + B$$
$$f_9 = f_{s_1} = AB + C$$

The table in Figure 13.14(b) lists all the possible input combinations to the circuit, and the column headed f_0 lists the fault-free response. Each input combination represents a distinct test, and the *test number* allocated to each row is the decimal equivalent of the binary representation of the input combinations for that row. The remaining columns in the table list the circuit response for each of the distinguishable fault conditions.

Examination of the table shows that if test 0 is applied to the circuit, the response to that combination of input variables when the circuit is fault-free differs from the response when, for example, r is *s-a*-1, as indicated in the column headed f_8. Clearly, $(A,B,C) = (0,0,0)$ is a test for r *s-a*-1. A further examination of this column reveals that the combination $(A,B,C) = (0,1,0)$ is also a test for the same fault, r_1. For both combinations, the fault-free response is 0, and the response with fault r_1 present is 1.

Formalising this result, input combination $(X_1, X_2, X_3, \ldots, X_n)$ will only be a test for the fault f_m provided that

$$f_0(X_1, X_2, X_3, \ldots, X_n) \oplus f_m(X_1, X_2, X_3, \ldots, X_n) = 1.$$

That is, to determine all the tests that will detect f_m, it is only necessary to take the XOR (i.e., the modulo 2 sum) of those columns in Figure 13.14(b) headed f_0 and f_m. The valid tests are indicated by those rows where the result of this operation is 1. It follows that if

$$f_0 \oplus f_m = 0$$

for every input combination, the fault(s) corresponding to f_m is (are) undetectable, usually a result of the circuit containing redundancy. Also, if

$$f_0 \oplus f_m = f_0 \oplus f_l$$

for every possible input combination $(X_1, X_2, X_3, \ldots, X_n)$, then fault number m and fault number l are indistinguishable. If, however,

$$f_0 \oplus f_m \neq f_0 \oplus f_l$$

for some or all of the input tests $(X_1, X_2, X_3, \ldots, X_n)$, then fault number m and fault number l are distinguishable.

A fault detection table can now be constructed from the information tabulated in Figure 13.14(b). In the table, shown in Figure 13.14(c), there is a column for the XOR of the fault-free response (f_0) with each of the fault conditions f_1 to f_9. The result of this XOR operation is shown as either a 1 or a blank (indicating 0).

Finding a minimum set of tests from this table is identical to the method used for reducing prime implicant tables. Any line containing all 1s and no blank entries corresponds to a valid test for all possible faults; there is, however, no such line in this example. Any column having a single 1 entry identifies an *essential* test, for this test is the only test that can detect this fault. An examination of Figure 13.14(c) shows that test 5 is the only one that will detect the fault associated with f_9, i.e. s_1, and consequently it is an *essential* test (and has been marked with asterisks). The same test detects the faults associated with f_2, f_4 and f_6, and so the columns corresponding to f_2, f_4, f_6 and f_9 have been ticked ($\checkmark$) to indicate that these faults have been covered by choosing test 5. Test 5 dominates test 4 (i.e., the line for test 5 includes all the 1s in the line for test 4), so test 4 has been deleted from the table. Test 1 dominates test 3 (i.e., the line for test 1 includes all the 1s in the line for test 3), so test 3 has been deleted from the table and test 1 chosen. The selection of test 1 allows the detection of faults associated with f_3, f_5 and f_7. There remain two 1s in the column corresponding to f_1, and two 1s in the column corresponding to f_8. The faults associated with f_1 can be detected by either tests 6 or 7, and the fault associated with f_8 can be detected by either tests 0 or 2. Hence, a minimal test set required for detecting all possible faults is

$$T = \{[0 \text{ or } 2], 1, 5, [6 \text{ or } 7]\}.$$

Which of the four possible alternative test sets implied here is actually chosen makes little difference, but as test 2 dominates test 0, and test 6 dominates test 7, one possible choice (to minimise ambiguity if a fault were actually detected) is

$$T = \{0, 1, 5, 7\}.$$

It is now a simple matter to devise a practical fault detection scheme to test the circuit of Figure 13.14(a). The order of the tests is largely immaterial, but to make the scheme as rapid as possible the tests detecting the largest number of faults might be applied first. Therefore, a suitable test routine would be:

Apply test 5: If $f = 1$ a fault exists; terminate experiment
If $f = 0$ proceed to next test ⤸
Apply test 1: If $f = 0$ a fault exists; terminate experiment
If $f = 1$ proceed to next test ⤸
Apply test 0: If $f = 1$ a fault exists; terminate experiment
If $f = 0$ proceed to next test ⤸
Apply test 7: If $f = 0$ a fault exists; end of experiment
If $f = 1$ circuit is fault free.

Once a fault has been detected using this test scheme, then the fault detection table in Figure 13.14(c) can be used to indicate the nature of the fault. For example, if tests 5, 1, and 0 give the correct fault-free output f, but then finally applying test 7 gives the output $f = 0$, this indicates in principle either faulty response f_1 or faulty response f_5. However, response f_5 has already been eliminated by test 1 conducted previously (which gave a fault-free output), so the circuit must therefore have a fault corresponding to response f_1. The possible faults, therefore, are p_0 in which case the fault must either reside in gate g_1 or in the circuitry providing input B, or t_1 in which case the fault must reside in gates g_1 or g_3 or their interconnection. These faults are indistinguishable, unless further tests can be made at the connections of the gates concerned.

It is possible with the data provided by the fault detection table to derive a *fault location table* or a *fault dictionary* which will identify the fault responsible for any given circuit responses to any set of tests. However, for a circuit with n inputs and x circuit nodes, there are at least $2x$ possible faults, counting the two possible 'stuck-at' faults but ignoring bridging and other faults, and there are 2^n possible tests that can be applied at the inputs. If all the faults are distinguishable, there will be $2x$ columns in the fault detection table, and even if some of these faults are indistinguishable the number of columns will still be of the order of $2x$, so the total number of entries in the table will be of the order of $x2^{n+1}$. As the complexity of the circuit increases rapidly with the number of interconnections and variables, the construction and reduction of the fault location table, either manually or with a computer, is extremely time consuming and is therefore impractical. Even the development of the fault detection table for any but the simplest combinational circuit is also impractical.

To extend the fault table method to address the problem of general fault detection and identification, the concept of 'adaptive testing' can be used. This method is used to determine *whether* a fault exists, and if so *what* it is. In this method, after a test has been conducted, the result of this test determines which further tests are to be carried out. In the example of Figure 13.14, the same series of tests as determined previously (test 5, then test 1, then test 0, then test 7) can be used to determine the absence of faults, but if faulty responses are given at any stage then a different series of tests can be used to identify precisely which fault is present.

As an example, suppose that for a certain circuit there are four distinguishable output responses. One method of identifying the particular fault present is illustrated in Figure 13.15(a). The tester starts at the top of the 'tree' and applies a certain test which results in either one of the possible faulty responses (F) or a correct response (C). In the case of a faulty response, no further tests are necessary as the fault has been identified, but in the case of a correct response further tests must be undertaken to isolate the fault. Each subsequent test then produces either one faulty response or a further correct response. However, an alternative method is illustrated in Figure 13.15(b). In this case, a different initial test is applied, which produces either two faulty responses (F) or a correct response. The next test to be applied depends upon whether a faulty or a correct response was obtained; in either case, the test applied distinguishes between the two alternatives possible at that point. The diagram of Figure 13.15(b) is known as a 'binary tree' because each test divides the possible faulty responses into two sets, as

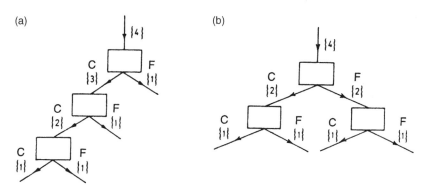

Figure 13.15 *(a) Adaptive tree identifying one fault only at each test node (b) Binary adaptive tree dividing the faults equally at each test node*

equal as possible in size. Clearly, for a total of y distinguishable faulty output responses, performing a total of y non-adaptive separate tests will allow complete identification of the distinguishable faults. However, by arranging the tests in a 'binary tree' the number of required tests may, in principle, be reduced to a minimum value of $\log_2(y+1)$. In this expression the value 1 is added to y because at least one test is required for determining whether a fault exists at all. In practice, perfect division of the responses into equal sized sets is usually impossible, so that $\log_2(y+1)$ is the lower bound and y the upper bound on the number of tests necessary.

Unfortunately, there is no known formal method that can be used for minimising the size of this tree, and so it is usually produced on the basis of trial and error using the designer's experience. Again, except for the simplest of circuits, designing such an adaptive tree for all possible faults is likely to represent a considerable investment of time and effort, and computing help must be sought.

13.10 Two-level circuit fault detection in AND/OR circuits

This section describes some specific techniques for finding the minimal test set that can be used to test the common AND/OR circuit architecture implementing functions expressed as Boolean sums of product terms. The methods described in this section are applicable in principle to fault-finding in PLAs and PALs (see sections 11.12 and 11.13) which are based upon the same fundamental AND/OR structure. These methods will be explained by examining a simple example.

The function generated by the two-level AND/OR circuit shown in Figure 13.16(a) is

$$f = A\bar{C} + BD + \bar{A}CD$$
$$\qquad \uparrow \qquad \uparrow \qquad \uparrow$$
$$\qquad p_1 \qquad p_2 \qquad p_3$$

This function is plotted on the K-map shown in Figure 13.16(b). Each term in this equation is a prime implicant of the function, and is plotted as an enclosed group on the map. Because the function contains no redundancy, each prime implicant encloses at least one unique cell not covered by any other group. For example, the cell unique to prime implicant $\bar{A}CD$ is cell $\bar{A}\bar{B}CD$.

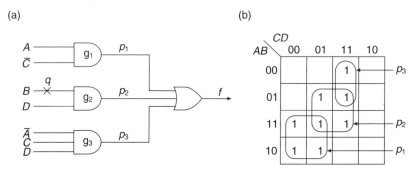

Figure 13.16   *Determination of s-a-0 test for a two-level AND/OR circuit (a) Circuit implementation (b) K-map for* $f = A\bar{C} + BD + \bar{A}CD$

If one of the input lines q of g_2 is *s-a-0* (Figure 13.16(a)), the output of g_2 is 0 and the group $BD = p_2$ no longer appears on the K-map of the function. To test for the absence or presence of group p_2, it is necessary to check for the absence or presence of one of the unique cells associated with it. Inspection of the map shows that cell $\bar{A}B\bar{C}D$ is one of the two cells unique to group p_2. Hence, a test for the presence or absence of p_2 is $(A,B,C,D) = (0,1,0,1)$. An alternative test would be $(A,B,C,D) = (1,1,1,1)$, corresponding to the other unique cell associated with p_2. If one of these combinations of variables is applied to the circuit inputs, then an output of $f = 0$ indicates that term p_2 is missing and there is a *s-a-0* fault associated with gate g_2. However, if $f = 1$, then it is clear that there is no *s-a-0* fault associated with gate g_2 in this circuit.

A complete *s-a-0* test for this circuit will therefore consist of three tests, one for each of the prime implicants p_1, p_2 and p_3. Since p_1 has three unique cells associated with it (numbers 8, 9 and 12), and any one of these cells can be used as a test for a *s-a-0* fault associated with g_1, the complete test set is

$$T_0 = \{3, [5 \text{ or } 15], [8 \text{ or } 9 \text{ or } 12]\}.$$

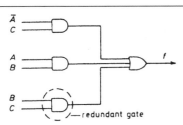

Figure 13.17 *Two-level AND/OR circuit for function $f = \bar{A}C + AB + BC$ having redundancy*

The effects of redundancy on circuit testing using the technique described above are shown in the circuit of Figure 13.17 which implements the function

$$f = AB + \bar{A}C + BC.$$

This is the same Boolean function as that implemented using NAND gates in Figure 13.12 and studied earlier in section 13.7. As discussed in that section, the prime implicant BC is redundant, since the cells associated with this prime implicant are already covered by the prime implicants AB and $\bar{A}C$. Consequently, there are no unique cells associated with term BC, and so (in the AND/OR implementation) a test of the circuit output value f cannot detect an *s-a-0* fault associated with term BC.

Having found a test for all possible *s-a-0* faults for the circuit in Figure 13.16(a), a method will now be developed for finding a series of tests that will detect all possible *s-a-1* faults in the same circuit. If the input line of g_2, labelled q in the diagram, is assumed to be *s-a-1*, then the output of g_2 is

$$D \cdot 1$$

$$= D(B + \bar{B})$$

$$= BD + \bar{B}D.$$

The first term in this expression is the required prime implicant (BD), while the second term ($\bar{B}D$) represents an unwanted product term generated by the fault. The unwanted term $\bar{B}D$ differs from the wanted prime implicant BD by *one* variable only. Two product terms differing in one digit place only are *adjacent* product terms on a K-map. The adjacent product terms for each of the prime implicants generated in the circuit of Figure 13.16(a) are tabulated below and plotted in Figures 13.18(a), (b) and (c).

Prime implicants	Adjacent product terms
$p_1 = A\bar{C}$	$\bar{A}\bar{C}, AC$
$p_2 = BD$	$\bar{B}D, B\bar{D}$
$p_3 = \bar{A}CD$	$ACD, \bar{A}\bar{C}D, \bar{A}C\bar{D}$

If a *s-a-1* fault exists at one of the circuit inputs, one of the adjacent product terms will be present at the output, so to test for *s-a-1* faults it is necessary to test for the presence of any of the seven adjacent product terms tabulated previously. However, in selecting the valid tests for the adjacent product terms, cells must be selected on the K-map that are *not* included in the original function $f = A\bar{C} + BD + \bar{A}CD$. Therefore, the 1s defining this original function (omitting the prime implicant groupings for clarity) are plotted again in Figure 13.18(d), along with the adjacent product terms. To test for the presence of an adjacent product term, a cell must be selected that is enclosed by that adjacent product term but which does *not* contain a 1. In this example, the number of tests is *minimised* by selecting cells that are common to as *many* adjacent product terms as possible. (Alternatively, cells could be selected that are enclosed by only *one* or as *few* adjacent product terms as possible, in which case the number of tests would not be minimised, but instead the tests would indicate the precise fault(s) present more clearly.) Suitable cells have been marked on the map by

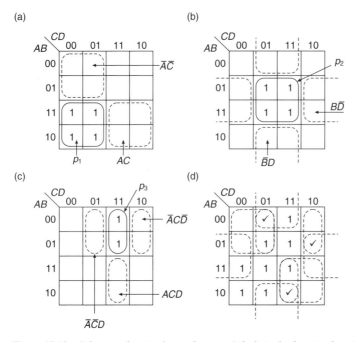

Figure 13.18   *Selection of minimal tests for a s-a-1 fault in the function $f = A\bar{C} + BD + \bar{A}CD$ (a), (b) and (c) Plots of the adjacent product terms corresponding to the terms p_1, p_2, and p_3 in this function (d) Ticked (✓) cells specify tests for s-a-1 faults*

a tick ($\checkmark$) and the adjacent product terms tested by each of the ticked cells are tabulated below in decimal minterm form:

Test	Terms tested
$(1)_{10}$	$\bar{A}\bar{C}, \bar{A}\bar{C}D, \bar{B}D$
$(6)_{10}$	$B\bar{D}, \bar{A}C\bar{D}$
$(11)_{10}$	$AC, ACD, \bar{B}D$

Note that each adjacent product term must enclose at least one ticked ($\checkmark$) cell. (If any adjacent product term contains only 1s with no space for a tick ($\checkmark$), this indicates that the original function has not been minimised correctly.) Hence a suitable *s-a*-1 test for the circuit in Figure 13.16(a) is:

$$T_1 = \{1, 6, 11\}$$

and the presence of an *s-a*-1 fault is indicated if the faulty output $f = 1$ is obtained when any one of these input combinations is applied to the circuit. The table above indicates the nature of the fault, as nearly as can be determined using this particular test set. If the output $f = 1$ is obtained from all three of these applied tests, then either there are three (or more) separate *s-a*-1 faults or else there is an *s-a*-1 fault at one (or more) of the AND gate outputs, the interconnections between the OR and AND gates, or directly associated with the OR gate. The full test of this circuit consists of the six input combinations contained in T_0 and T_1.

In some circuits having multiple outputs, the optimal implementation is achieved by sharing terms between functions (as described in sections 3.18, 3.19, and 11.12). For example, in the optimal implementation of the two functions $f_1 = \bar{A}\bar{B} + \bar{B}C$ and $f_2 = AC + BC$ shown in Figure 13.19(a) (in fact, the same as the circuit of Figure 3.28 that was designed in section 3.18), the common term is the product function $f_1 \cdot f_2$ as shown in Figure 13.19(b). In a PLA implementation, by sharing terms, the number of lines of the AND array used in a PLA may be reduced even though the Boolean expressions are not completely minimised. The consequence of this optimisation, as far as testing is concerned, is that the adjacent product terms corresponding to the common term(s) rather than to the expected minimal terms must be used, and the proposed tests need to be examined carefully. Since the optimal circuit is *not* minimised, it is possible for the optimal circuit to have *s-a*-1 faults where adjacent product terms coincide with one or more of the output functions. So, in the example of Figure 13.19(a), the common term is $A\bar{B}C$ which gives adjacent product terms $\bar{A}\bar{B}C$, ABC, and $A\bar{B}\bar{C}$. The adjacent product term $\bar{A}\bar{B}C$ coincides entirely with prime implicant $\bar{A}\bar{B}$ of f_1 and so a *s-a*-1 fault on input line A is untestable through output f_1, although this fault is testable through f_2. Similarly, the adjacent product term ABC coincides entirely with prime implicant BC of f_2 and so a *s-a*-1 fault on input line $\bar{B}$ is untestable through output f_2, although this fault is testable through f_1. If this situation occurs, the designer must make a decision on whether to use the optimal implementation and to accept its reduced testability, or whether to prefer the minimised but more complicated yet more testable circuit.

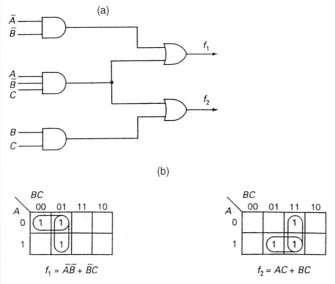

Figure 13.19   *(a) Optimal implementation of the two functions $f_1 = \bar{A}\bar{B} + \bar{B}C$ and $f_2 = AC + BC$ (b) K-map plots of the functions f_1 and f_2*

13.11 Two-level circuit fault detection in OR/AND circuits

The OR gate shown in Figure 13.20(a) implements the function $f = A + B + C$, which has been plotted on the K-map in Figure 13.20(b). The 0 inserted in only one cell indicates that if the inputs $(A,B,C) = (0,0,0)$ are applied to the gate, its output $f = 0$. If, however, one or more inputs to the gate have *s-a*-1 faults, then the output of the gate will be $f = 1$. Clearly, if $(A,B,C) = (0,0,0)$ is used as a test input, then the output $f = 1$ indicates the presence of one or more *s-a*-1 faults associated with this gate.

In general, valid tests for *s-a*-1 faults in an OR/AND circuit correspond to cells that should give a fault-free output of 0. To find these cells, the function can be inverted, and the 0 s of the original function are the same as the 1s of the inverted function.

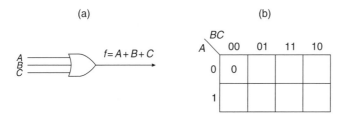

Figure 13.20   *(a) Implementation of function $f = A + B + C$ (b) K-map plot of the function f*

As an example, consider the OR/AND circuit shown in Figure 13.21(a). The function implemented by this circuit is

$$f = (A + C)(\bar{B} + \bar{D})(\bar{A} + \bar{B} + \bar{C}),$$

and the Boolean complement of this function is:

$$\bar{f} = \bar{A}\bar{C} + BD + ABC$$
$$\quad\uparrow\quad\quad\uparrow\quad\quad\uparrow$$
$$\quad p_1\quad\quad p_2\quad\quad p_3$$

using De Morgan's theorem. The zeros of f, derived directly from the prime implicants of $\bar{f}$, are shown plotted on a K-map in Figure 13.21(b). The cells marked with 0's on this map define those combinations of the variables for which $f = 0$.

Each term in the equation for $\bar{f}$ is a prime implicant of the inverse function and appears as an enclosed group on the K-map. In this example, each of the three prime

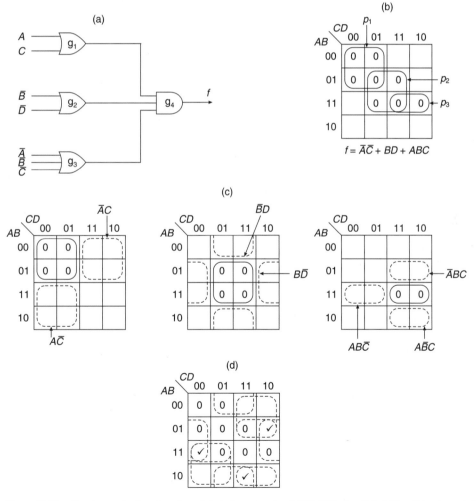

Figure 13.21 *Development of s-a-1 and s-a-0 tests for a two-level OR/AND circuit (a) Circuit implementation of $f = (A + C)(\bar{B} + \bar{D})(\bar{A} + \bar{B} + \bar{C})$ (b) K-map of the zeros of function f, obtained from $\bar{f}$ (c) Adjacent product terms of the inverse function (d) Cells marked with ticks ($\checkmark$) identify the tests for s-a-0 faults in the original circuit*

implicants encloses at least one unique cell, indicating that the function has no redundancy. For example, the cell unique to p_3 is defined by $(A,B,C,D) = (1,1,1,0)$.

If one of the input lines of g_3 is s-a-1, the output of g_3 will be 1 and p_3 will be missing from the map of 0's of the function f. To test for the absence or presence of p_3, it is necessary to check for the absence or presence of 0 in the unique cell associated with it. A cell common or overlapping with another prime implicant cannot be used as the other prime implicant may still be present. Hence, a valid test for the presence of p_3 is $(A,B,C,D) = (1,1,1,0)$. If this combination of the variables is applied to the circuit and it is found that the circuit output is $f = 1$, an s-a-1 fault exists associated with gate g_3. On the other hand, if the output is $f = 0$, then p_3 is present and there is no s-a-1 fault associated with g_3.

Hence, the complete s-a-1 test for the circuit will consist of three tests, one for each of the three prime implicants, and so is:

$$T_1 = \{[0 \text{ or } 1 \text{ or } 4], [7 \text{ or } 13], 14\}$$

as there are three possible testable cells for p_1 and two alternatives for p_2.

The s-a-0 test for the OR/AND circuit of Figure 13.21(a) is found by analogy with the method that was used to find the s-a-1 test in the AND/OR circuit. However, for the OR/AND circuit the adjacent product terms of the *inverse* plot are used to define the set of tests for s-a-0 faults.

The output of g_1 is $(A + C)$, and if there is an s-a-0 fault on line A, this expression becomes

$$0 + C$$
$$= A\bar{A} + C$$
$$= A\bar{A} + C(1 + A + \bar{A})$$
$$= (A + C)(\bar{A} + C),$$

where $(A + C)$ is the sum term required from g_1, while $(\bar{A} + C)$ is an unwanted sum term generated by the fault. The complement of this expression is $\bar{A}\bar{C} + A\bar{C}$, where $A\bar{C}$ is the additional product term generated by the s-a-0 fault on line A. As before, this additional product term is adjacent to the complement's prime implicant $p_1 = \bar{A}\bar{C}$; similarly, a s-a-0 fault on line C will produce the different adjacent product term $\bar{A}C$. The product terms adjacent to all the prime implicants of the inverse function are tabulated below:

Prime implicants	Adjacent product terms
$p_1 = \bar{A}\bar{C}$	$\bar{A}C, A\bar{C}$
$p_2 = BD$	$\bar{B}D, B\bar{D}$
$p_3 = ABC$	$\bar{A}BC, A\bar{B}C, AB\bar{C}$

and are plotted on the three K-maps shown in Figure 13.21(c).

If an s-a-0 fault exists at the circuit inputs, then one of the adjacent product terms must be present; it follows that the test for s-a-0 faults is to test for the presence of one of the seven possible adjacent product terms using cells not included in the prime implicants. In order to choose the correct cells, the zeros of the function $f = (A + C)(\bar{B} + \bar{D})(\bar{A} + \bar{B} + \bar{C})$ have been replotted in Figure 13.21(d). Again, as before, the lines enclosing the three prime implicant groups have been omitted, but the seven adjacent product terms tabulated above are shown enclosed by dashed lines on the map. Therefore if, for example, there is an *s-a*-0 fault present at one or both of the inputs of gate g_1, then adjacent product term $\bar{A}C$ or adjacent product term $A\bar{C}$, or both, are present. To test for the presence of $A\bar{C}$, one of the input combinations $(A,B,C,D) = (1,0,0,0)$ or $(1,0,0,1)$ or $(1,1,0,0)$ can be applied to the circuit. All these cells are included in term $A\bar{C}$, and a fault-free circuit will give an output $f = 1$ in each case; with an *s-a*-0 fault present on line A, then the output $f = 0$. Any of these input combinations will suffice to test for term $A\bar{C}$ but to keep testing to a minimum, cells are selected that are common to as many as possible adjacent product terms. (An alternative testing philosophy might be to choose tests that enable the quickest determination of which fault is present, in which case, as far as possible, cells *not* common to any other adjacent product terms should be chosen.) The chosen cells have been ticked (✓) in Figure 13.21(d), and the terms tested by each of the selected cells are tabulated below in decimal minterm form:

Test	Terms tested
$(6)_{10}$	$\bar{A}C,\ \bar{A}BC,\ B\bar{D}$
$(11)_{10}$	$\bar{B}D,\ A\bar{B}C$
$(12)_{10}$	$B\bar{D},\ A\bar{C},\ AB\bar{C}$

Hence, the complete *s-a*-0 test for the circuit of Figure 13.21(a) is

$$T_0 = \{6, 11, 12\}$$

and when the input combinations specified by test set T_0 are applied to the circuit in turn, output $f = 0$ indicates that an *s-a*-0 fault exists. The table above indicates the nature of the fault, as nearly as can be determined using this particular test set. If the output $f = 0$ is obtained from all three of these applied tests, then either there are three (or more) separate *s-a*-0 faults or else there is an *s-a*-0 fault at one (or more) of the OR gate outputs, the interconnections between the AND and OR gates, or directly associated with the AND gate. The full test of this circuit consists of the six input combinations contained in T_0 and T_1.

13.12 Boolean difference

This section examines the method of *Boolean differences*. This is an algebraic procedure for determining test sequences for combinational circuits using the Boolean equation which represents the circuit to be tested.

The *Boolean difference* is the Boolean analogy of the *partial differential* or *derivative* of a function of continuous variables. For a function $f(x, y)$ of continuous variables x and y, by definition

$$\frac{\partial f(x, y)}{\partial x} = \lim_{(\Delta x \to 0)} \left(\frac{f(x + \Delta x, y) - f(x, y)}{\Delta x} \right) = \lim_{(\Delta x \to 0)} \left(\frac{\Delta f}{\Delta x} \bigg|_{y\ \text{constant}} \right).$$

If the Boolean function of the logic circuit to be tested is $f(X_1, X_2, \ldots, X_n)$, then by analogy with the partial derivative of a function of continuous variables

$$\frac{\partial f(X_1, X_2, \ldots, X_n)}{\partial X_i}$$

$$= \frac{\Delta f}{\Delta X_i}\bigg|_{\substack{\text{all other variables}\\\text{constant}}}$$

$$= \frac{\text{change in value of } f}{\text{change in value of } X_i}\bigg|_{\substack{\text{all other variables}\\\text{constant}}}$$

However, X_i can only change from 1 to 0, and hence ΔX_i in this equation can only take a value of 1. Also, in the Boolean context it is usual to take the *modulus* of the corresponding change in the value of f, and so Δf in this equation only takes the values of either 0 or 1. Noting that the Boolean XOR function produces the modulus of the numerical difference between its two arguments, the *Boolean difference* of function f with respect to the variable X_i is therefore *defined* as

$$\frac{\partial f}{\partial X_i} = f(X_1, \ldots, X_{i-1}, 1, X_{i+1}, \ldots, X_n) \oplus f(X_1, \ldots, X_{i-1}, 0, X_{i+1}, \ldots, X_n)$$

$$= f_i(1) \oplus f_i(0)$$

where

$$f_i(x) = f(X_1, \ldots, X_{i-1}, x, X_{i+1}, \ldots, X_n)$$

is the modified function obtained by putting $X_i = x$ (itself taking the value 0 or 1 only) in the original function expression. In other words, the Boolean difference with respect to variable X_i is defined as the XOR of the function $f|_{X_i=1}$ and the function $f|_{X_i=0}$; like Δf, it can only take the values 0 or 1.

Since $f_i(1)$ and $f_i(0)$ differ only in a change of the binary value of X_i from 1 to 0, there are four possibilities for the numerical values of $f_i(x)$. These are tabulated below:

$f_i(1) =$ value of f when $X_i = 1$	$f_i(0) =$ value of f when $X_i = 0$	$\frac{\partial f}{\partial X_i} = f_i(1) \oplus f_i(0)$
0	0	0
0	1	1
1	0	1
1	1	0

This table shows that if there is *no change* in the value of f as a consequence of the change of X_i between 0 and 1, then $\partial f/\partial X_i = 0$ (i.e., the value of f is *independent* of the value of X_i). However, if there is a *change* in f as a consequence of the change in X_i between 0 and 1, then $\partial f/\partial X_i = 1$ (i.e., the value of f is *dependent* on the value of X_i). Thus, the Boolean difference can be regarded as a flag indicating dependence or not of a function upon a specified Boolean variable. Note that dependence on, or independence of, a variable X_i may itself depend upon the values of the other variables X_j defining f.

Given a function $F(X, Y)$, to find the Boolean difference $\partial F(X, Y)/\partial X$ it is often easiest to calculate the value of $F(1, Y) \oplus F(0, Y)$ directly. However, as with differentiation of functions of a continuous variable, it is sometimes useful to have available a 'library' of standard analytical results for the Boolean difference. The following results (where X and Y are independent logic inputs, and F and G are Boolean functions of X and Y) may be confirmed by examining the relevant truth tables of the variables and the values of their derivatives. As in conventional differentiation, here 'd' is used instead of '∂' where only one variable is involved:

$$\frac{\mathrm{d}(\bar{X})}{\mathrm{d}X} = \frac{\mathrm{d}(X)}{\mathrm{d}X} = 1 \qquad \frac{\partial(X \oplus Y)}{\partial X} = 1$$

$$\frac{\partial(X \oplus F)}{\partial X} = \overline{\left(\frac{\partial F}{\partial X}\right)} \qquad \frac{\partial(F \oplus G)}{\partial X} = \frac{\partial F}{\partial X} \oplus \frac{\partial G}{\partial X}$$

$$\frac{\partial(XY)}{\partial X} = Y \qquad \frac{\partial(X + Y)}{\partial X} = \bar{Y}.$$

If $E(F)$ is a Boolean function of the previous output F, then the definition of $\partial E/\partial X$ gives

$$\frac{\partial E(F)}{\partial X} = \frac{\mathrm{d}E}{\mathrm{d}F} \cdot \frac{\partial F}{\partial X}$$

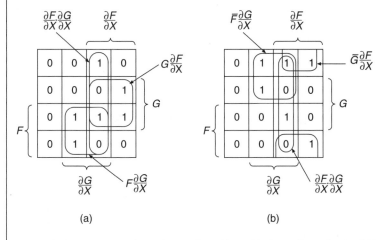

(a)   (b)

Figure 13.22 *(a) K-map for evaluating $\partial(FG)/\partial X$; (b) K-map for evaluating $\partial(F + G)/\partial X$.*

K-maps for the Boolean difference of the AND and OR functions of F and G are shown in Figure 13.22, constructed by inspection; the results may be expressed as:

$$\frac{\partial(FG)}{\partial X} = \left(F\frac{\partial G}{\partial X}\right) \oplus \left(G\frac{\partial F}{\partial X}\right) \oplus \left(\frac{\partial F}{\partial X} \cdot \frac{\partial G}{\partial X}\right);$$

$$\frac{\partial(F+G)}{\partial X} = \left(\bar{F}\frac{\partial G}{\partial X}\right) \oplus \left(\bar{G}\frac{\partial F}{\partial X}\right) \oplus \left(\frac{\partial F}{\partial X} \cdot \frac{\partial G}{\partial X}\right)$$

Finally, using the fundamental definition of Boolean difference ($\partial f/\partial X_i = f_i(1) \oplus f_i(0)$), and also the fact that the XOR operation is both Associative and Commutative, it is easy to show that

$$\frac{\partial^2 F(X, Y)}{\partial X \partial Y}$$

$$= \left.\frac{\partial F}{\partial Y}\right|_{X=1} \oplus \left.\frac{\partial F}{\partial Y}\right|_{X=0}$$

$$= F(1,1) \oplus F(1,0) \oplus F(0,1) \oplus F(0,0)$$

$$= [F(1,1) \oplus F(0,1)] \oplus [F(1,0) \oplus F(0,0)]$$

$$= \frac{\partial^2 F(X, Y)}{\partial Y \partial X}.$$

or, in other words, the order of taking the Boolean difference with respect to two independent input variables is immaterial, just as the order of differentiation with respect to two continuous independent variables is immaterial.

Using these standard results in appropriate combinations, the Boolean difference of other functions may be calculated directly without the necessity of evaluating the function f for the two possible values of X. Sometimes, however, this is an error-prone procedure, simply because of its complexity.

To find tests for either a *s-a-0* or a *s-a-1* fault on the X_i line, the first objective is to find those combinations of the input variables for which $\partial f/\partial X_i = 1$, where f is an accessible circuit output. Then, in order to test for a *s-a-0* fault on the X_i line, the procedure is to apply the inverse of the fault at the X_i input, i.e. $X_i = 1$. Hence, the valid test(s) that will detect a *s-a-0* fault on the X_i line are the solution(s) to the equation $\frac{\partial f}{\partial X_i} = 1$ while $X_i = 1$ simultaneously, i.e.

$$X_i \frac{\partial f}{\partial X_i} = 1.$$

Similarly, to test for an *s-a-1* fault on line X_i, the required input to the line is $X_i = 0$, so the valid test(s) that will detect this fault are the solution(s) to the equation $\frac{\partial f}{\partial X_i} = 1$ while $\bar{X}_i = 1$ simultaneously, i.e.

$$\bar{X}_i \frac{\partial f}{\partial X_i} = 1.$$

As an example of the application of Boolean difference, consider the function

$$f = (X_1 X_2 + \bar{X}_3) \cdot (X_3 + X_4).$$

A direct gate implementation of this equation is shown in Figure 13.23. To find tests for faults on the X_3 input, it is necessary to calculate $\partial f/\partial X_3$. To do this, here

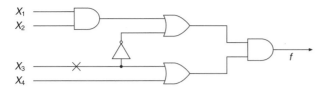

Figure 13.23   *Implementation of $f = (X_1 X_2 + \bar{X}_3) \cdot (X_3 + X_4)$*

it is simplest to use the fundamental definition of Boolean difference. Firstly, calculate

$$f_3(1) = (X_1 X_2 + 0) \cdot (1 + X_4) = X_1 X_2$$

and

$$f_3(0) = (X_1 X_2 + 1) \cdot (0 + X_4) = X_4.$$

Then, therefore

$$\frac{\partial f}{\partial X_3} = f_3(1) \oplus f_3(0)$$

$$= X_1 X_2 \oplus X_4$$

$$= X_1 X_2 \bar{X}_4 + \overline{(X_1 X_2)} X_4$$

$$= X_1 X_2 \bar{X}_4 + \bar{X}_1 X_4 + \bar{X}_2 X_4.$$

The condition for an *s-a*-0 fault on the X_3 line is

$$X_3 \frac{\partial f}{\partial X_3} = 1,$$

that is,

$$1 = X_1 X_2 X_3 \bar{X}_4 + \bar{X}_1 X_3 X_4 + \bar{X}_2 X_3 X_4$$

$$= X_1 X_2 X_3 \bar{X}_4 + \bar{X}_1 (X_2 + \bar{X}_2) X_3 X_4 + (X_1 + \bar{X}_1) \bar{X}_2 X_3 X_4$$

$$= X_1 X_2 X_3 \bar{X}_4 + \bar{X}_1 X_2 X_3 X_4 + \bar{X}_1 \bar{X}_2 X_3 X_4 + X_1 \bar{X}_2 X_3 X_4 + \bar{X}_1 \bar{X}_2 X_3 X_4$$

$$= X_1 X_2 X_3 \bar{X}_4 + \bar{X}_1 X_2 X_3 X_4 + \bar{X}_1 \bar{X}_2 X_3 X_4 + X_1 \bar{X}_2 X_3 X_4.$$

To satisfy this equation, any one of the terms in the last line can equal 1. Hence, any of the input combinations $(X_1, X_2, X_3, X_4) = (1,1,1,0)$ or $(0,1,1,1)$ or $(0,0,1,1)$ or $(1,0,1,1)$ is a valid test for an *s-a*-0 fault on the X_3 line, and the fault is detected if the output is the complement of that given by the original equation $f = (X_1 X_2 + \bar{X}_3) \cdot (X_3 + X_4)$.

For an *s-a*-1 fault on the X_3 line, the condition to be satisfied is

$$\bar{X}_3 \frac{\partial f}{\partial X_3} = 1,$$

that is,

$$1 = X_1 X_2 \bar{X}_3 \bar{X}_4 + \bar{X}_1 \bar{X}_3 X_4 + \bar{X}_2 \bar{X}_3 X_4$$

$$= X_1 X_2 \bar{X}_3 \bar{X}_4 + \bar{X}_1 (X_2 + \bar{X}_2) \bar{X}_3 X_4 + (X_1 + \bar{X}_1) \bar{X}_2 \bar{X}_3 X_4$$

$$= X_1 X_2 \bar{X}_3 \bar{X}_4 + \bar{X}_1 X_2 \bar{X}_3 X_4 + \bar{X}_1 \bar{X}_2 \bar{X}_3 X_4 + X_1 \bar{X}_2 \bar{X}_3 X_4 + \bar{X}_1 \bar{X}_2 \bar{X}_3 X_4$$

$$= X_1 X_2 \bar{X}_3 \bar{X}_4 + \bar{X}_1 X_2 \bar{X}_3 X_4 + \bar{X}_1 \bar{X}_2 \bar{X}_3 X_4 + X_1 \bar{X}_2 \bar{X}_3 X_4.$$

To satisfy this equation, any one of the terms in the last line can equal 1. Hence, any of the input combinations $(X_1, X_2, X_3, X_4) = (1,1,0,0)$ or $(0,1,0,1)$ or $(0,0,0,1)$ or $(1,0,0,1)$ is a valid test for an *s-a*-1 fault on the X_3 line, and the fault is detected if the output is the complement of that given by the original equation $f = (X_1 X_2 + \bar{X}_3) \cdot (X_3 + X_4)$.

This technique of finding valid tests is best suited to logic functions defined in algebraic terms that do not readily lend themselves to direct implementation in one of the standard OR/AND or AND/OR forms, for which the methods discussed in sections 13.10 and 13.11 are inapplicable. Note that these tests have been derived by a *purely algebraic procedure*, without reference to the circuit diagram. However, there is an alternative method of finding the Boolean difference by manipulation of the K-maps of the function concerned. Note firstly that from its definition, the Boolean difference may also be written as

$$\frac{\partial f}{\partial X_i} = f_i(X_i) \oplus f_i(\bar{X}_i) = f \oplus f_i(\bar{X}_i).$$

This is because X_i may take only the values 0 or 1; so, regardless of which of these values X_i actually takes, the XOR is evaluated of the function f (with 1 replacing X_i) and of the function f (with 0 relacing X_i). Since the XOR operation is commutative, this gives the same result as the previous definition of the Boolean difference.

To explain the K-map method of evaluating the Boolean difference, this method will be used to confirm the previous calculation of Boolean difference for the function

$$f = (X_1 X_2 + \bar{X}_3) \cdot (X_3 + X_4) = X_1 X_2 X_3 + X_1 X_2 X_4 + \bar{X}_3 X_4$$

This function is plotted on the K-map shown in Figure 13.24(a). Next, the function

$$f_3(\bar{X}_3) = (X_1 X_2 + X_3) \cdot (\bar{X}_3 + X_4) = X_1 X_2 \bar{X}_3 + X_1 X_2 X_4 + X_3 X_4$$

is plotted on the K-map shown in Figure 13.24(b). Therefore, to find the K-map of the Boolean difference, all that is needed is to find the XOR of these two K-maps. This is done by comparing cell by cell the two maps of Figure 13.24(a) and (b), and transferring to the corresponding cell of a new map, either the result 0 if both starting cells contain the same value, or the result 1 if the two starting cells contain different values. The result is shown in Figure 13.24(c), and agrees with the previous algebraic result

$$\frac{\partial f}{\partial X_3} = X_1 X_2 \bar{X}_4 + \bar{X}_1 X_4 + \bar{X}_2 X_4.$$

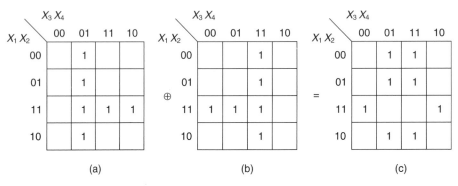

Figure 13.24   *K-map determination of the Boolean difference (a) K-map of $f = (X_1 X_2 + \bar{X}_3) \cdot (X_3 + X_4)$ (b) K-map of $f_3(\bar{X}_3)$ (c) K-map of $\partial f / \partial X_3$*

13.13 Compact testing techniques

In practice, the methods described earlier in this chapter for fault diagnosis and the generation of test sequences often have limited use. The fault table method can, in principle, always be used to generate a minimal test sequence, but as the number of variables increases, the computational time required increases significantly and computer assistance becomes necessary for large circuits. Similarly, although the method of Boolean differences is a useful and revealing technique for fault diagnosis and the generation of test sequences, its use is limited to small circuits. Because of the restrictions of these techniques, other methods of fault diagnosis have been developed.

Clearly, fault-finding is simplified if test points are available and are distributed throughout the circuit and the signal to be expected at each test point under fault-free conditions is documented by the designer. In order to undertake such a test, it is usual to initialise the circuit in some prescribed manner, so that the circuit is starting from known conditions. A specified test sequence is then applied at the circuit input and the resulting sequence at the selected test point is checked against the fault-free sequence for that point which has previously been stored. A block diagram of a typical testing system employing this technique is shown in Figure 13.25. An XOR gate is shown symbolically as the logic comparator for the two data streams; it will give an output 0 when the sequence to be checked agrees with the correct sequence, and it will give an output 1 in the case of any difference. Therefore, a fault is indicated by the appearance of a logic 1 at the output of the XOR comparator gate.

In choosing the test sequence applied at the input to the circuit under test, clearly it will be most useful if as many as possible of the circuit properties are tested systematically, but in practice this may be difficult to arrange. However, the compact testing technique is particularly versatile because almost *any* input test sequence can represent a valid test, provided that it is repeatable and that the expected output can be deduced or recorded from a known fault-free circuit. One test sequence often used is the output from a pseudo-random binary sequence generator (see Section 7.25).

Ideally, the test points will be approximately equally spaced throughout the circuit, and the sequence can be tested at test points chosen according to the 'binary division' method (Section 13.2). Using this procedure, a section of the circuit will be found where the fault first manifests itself by providing an incorrect sequence. More detailed checks applied to this section will then isolate the fault. The data streams to be compared may typically be thousands of bits long or even more for each defined test

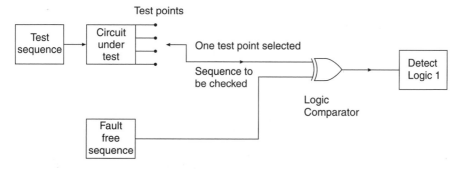

Figure 13.25 *Block diagram for testing using known sequences*

point. The problem with this technique is that the correct sequence for each test point has to be stored, and in the case of a large digital system (with a large number of test points as well as long test sequences) this may require an inordinately large amount of storage.

Because of these difficulties, methods have been developed where the fault-free sequences have been stored in a *compacted* form. The output of a digital circuit is a bit stream of 0's and 1's. Instead of storing directly the fault-free output sequence generated by the input test pattern, data compaction can be achieved by either (1) detecting and counting the number of $0 \rightarrow 1$ or $1 \rightarrow 0$ transitions, or (2) detecting and counting the number of 1's or the number of 0's. In each of these cases, the amount of storage required to represent a long fault-free test sequence is considerably reduced (e.g., a sequence of 1000 bits may contain around 500 logic 1's, the exact number of which may be represented by a 9-bit binary integer; or, it may contain around 250 logic $0 \rightarrow 1$ transitions, the exact number of which may be represented by an 8-bit integer) so that considerable data compaction takes place and the storage requirements are greatly reduced. Unfortunately, there are inherent difficulties with both of these compaction techniques since a faulty output may well have the same number of transitions or the same number of logic 1's or 0's as the fault-free output. This problem has led to the development of *signature analysis*, a method where the probability of mistaking faults for correct operation is very low.

13.14 Signature analysis

The technique of signature analysis was developed by Hewlett Packard for testing large digital systems. Overall, the method is akin to a compact test; the bit sequence from a specified test point in the circuit under test is passed through another circuit, called the *compacter* or *signature analyser*, which generates at its output a shorter bit sequence than that applied to its input. The output of the compacter is termed the *signature*. As before, the generated signature is then checked against the fault-free signature obtained either from a known fault-free circuit or, alternatively, by simulation. As shown below, the signature always contains the same number of bits, regardless of the length of the test sequence, and this gives a considerable reduction in the amount of storage required for examining the results of tests. A block diagram illustrating the method is shown in Figure 13.26.

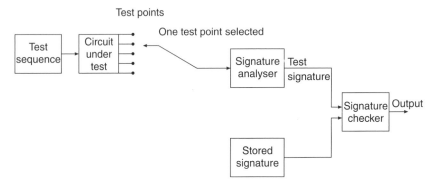

Figure 13.26 *Block diagram illustrating signature analysis*

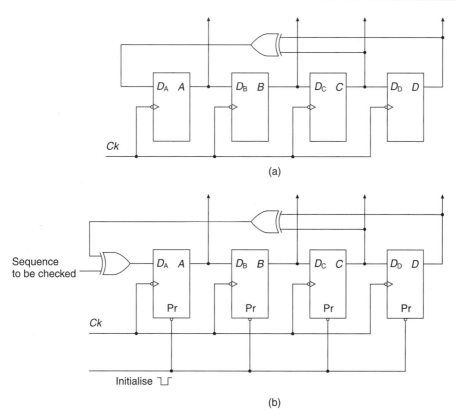

Figure 13.27   *(a) Pseudo-random binary sequence generator (b) Pseudo-random binary sequence generator modified for serial signature analysis*

The basic compacter circuit for signature analysis is a pseudo-random binary sequence generator similar to that described in Section 7.25, and shown in Figure 13.27(a). This circuit is a standard FIFO shift register with the feedback from the last and penultimate stages taken through an XOR gate. For use in signature analysis, this circuit must be modified by adding a second XOR gate as shown in Figure 13.27(b). This circuit was originally called a *cyclic code checker* and is also known as a *serial signature analyser*.

Conducting the signature analysis is similar to undertaking a compact test, except that as well as initialising the circuit under test the serial signature analyser must also be initialised simultaneously to a known state. In the example shown in Figure 13.27(b), initialising the serial signature analyser sets all its flip-flop outputs to logic 1 level. Following initialisation, as in compact testing, the test sequence is applied to the circuit under test. The sequence of states of the flip-flop outputs in the circuit of Figure 13.27(b) will depend upon the bit sequence generated by the circuit under test combined with the feedback signal; this combination is undertaken by the additional XOR gate at the input to the shift register. If the sequence being checked just happens to consist entirely of 0's, then the shift register input is identical to the feedback signal. In this case, the compacter circuit behaves exactly as a pseudo-random binary sequence generator and after being initialised produces a repeatable sequence of pseudo-random numbers. If the sequence being checked is not entirely 0's, then the feedback signal is modified and clearly the regular cycle of pseudo-random numbers is upset, but it is still predictable and should be the same each time that same sequence is applied. The expectation is that if there is an error, caused by a fault, in the bit sequence applied, then the compacter output number sequence is changed, and at the end of the sequence the shift register holds an incorrect signature.

It is nevertheless possible for a fault in the circuit under test to produce a correct signature by chance. Note that although the standard pseudo-random binary sequence generator of Figure 13.27(a) can never enter the state with all its flip-flop outputs at logic level 0, the feedback path in the compacter of Figure 13.27(b) is modified by the input bit sequence and so all possible output states may in principle be entered. Therefore, if n is the number of stages in the shift register, there are 2^n different possible signatures. So, the probability of a fault producing the correct signature by chance is roughly 2^{-n}, assuming of course that the test sequence is not strongly *correlated* with the feedback signal, and so the different possible signatures are all approximately equally likely. As n increases, the probability of mistaking an error for the correct signature should be reduced exponentially to a very small value. A typical implementation uses a 16-stage shift register (i.e., $n = 16$), so that the probability of confusing a faulty sequence with the correct sequence is approximately $2^{-16} \approx 1.5 \times 10^{-5}$.

In principle, there is no limit on the length of test sequences that can be used with signature analysis. The advantage of the method is that instead of storing the full length of the fault-free sequences, only the n bits of the correct signature at each test point need to be stored and compared, giving a massive reduction in required storage in the case of long test sequences. However, a major disadvantage of the method is that if a faulty signature is detected, then there is little likelihood of being able to deduce the nature of the fault from the faulty signature value. As the faulty sequence is combined with a pseudo-random bit sequence in the compacter, it is difficult to work backwards from the faulty signature to find which bits in the test sequence were incorrect. As the number of possible faults is usually very large indeed, it is likely that several different faults could produce identical faulty signatures. In principle, the most common faulty signatures could be stored, and the faulty signature compared with those, but in practice it is likely that so many faults are possible that it would defeat the object of the compacter to store all the possible faulty signatures together with an indication of their corresponding fault location.

13.15 The scan path testing technique

This technique is used for designing a synchronous state machine that is more easily testable than the basic form of the machine described in Chapter 8. As an example, it will be assumed that D-type flip-flops have been used in the machine design. When constructing the state machine, a 2-to-1 multiplexer is used at the input of each of the D-type flip-flops, as illustrated in Figure 13.28. (In some cases, flip-flops containing an internal multiplexer are available to simplify the construction of such circuits.)

A block diagram showing the necessary arrangements for scan path testing appears in Figure 13.29. When the multiplexer selection signal $G = 0$, the state machine is in its normal operational mode;

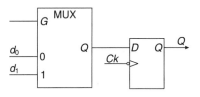

Figure 13.28   *D-type flip-flop with additional multiplexer, for scan path testing of state machine circuits*

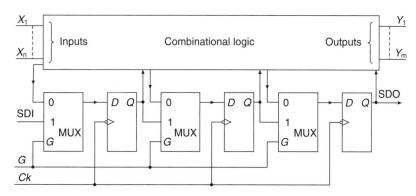

Figure 13.29 *Block diagram for a state machine with scan path testing facility*

the outputs of the combinational logic are selected and appear at the inputs to the D-type flip-flops. However, when $G = 1$, the combinational logic is disconnected and the machine flip-flops are now connected in cascade to form a shift register. The testing procedure for the synchronous state machine can therefore be carried out as follows:

1. Set $G = 1$ and move a string of 1s and 0s through the shift register using the SDI (serial data input) line. To verify the four possible transitions of each flip-flop $(0 \rightarrow 0, 0 \rightarrow 1, 1 \rightarrow 1, 1 \rightarrow 0)$, the input string should take the form 00110011 ...
2. If the flip-flops are functioning correctly and with G still held at 1, set the flip-flops in the machine to the state pattern required for starting the machine test, by moving the appropriate bit pattern through the flip-flops using the SDI line and with repeated clock pulses applied to the Ck line.
3. Set the input values $X_1 \ldots X_n$ required for starting the machine test.
4. Place the machine in the operational mode by setting $G = 0$, and after allowing the combinational logic to settle to its final condition, check the output values $Y_1 \ldots Y_m$.
5. Clock the machine once using the Ck line.
6. Set $G = 1$ and examine the state of the machine by applying further clock pulses to the Ck line and examining the SDO (serial data out) line to confirm that the machine has entered the correct next state.
7. Return to step (2) above to repeat for as many further state transitions as must be tested.

The scan path testing technique allows all the internal states and transitions to be thoroughly examined. Any internal state of the machine can be set when operating in the shift register mode, either for testing or initialisation. However, scan path testing requires serial setting of the starting state for each test, and serial readout of the final state of each test, and so will be slower than parallel setting and reading of the flip-flops in the machine if this is feasible.

IBM developed the *level sensitive scan design* (*LSSD*) method, a scan path technique which is designed to overcome such problems as hazards, races and sensitivity to timing constraints. It depends upon the use of a specially designed *shift register latch* (*SRL*) that effectively combines into a single device a storage facility and a selection process equivalent to multiplexing. The SRL is a master/slave latch circuit which has two separate input ports, one for the normal machine operating mode and the second for the shift register mode.

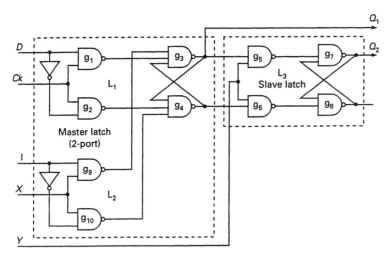

Figure 13.30   *The shift register latch*

The gate connections for an SRL are shown in Figure 13.30. The main master latch L_1 consists of gates g_1, g_2, g_3, and g_4, with data input line D and clock Ck. When $Ck = 0$, L_1 is disabled, but after Ck has made a $0 \rightarrow 1$ transition the data is transferred to the output Q_1 of the latch (normally made available as a test point). To ensure that the data has settled before the latch is enabled it is arranged that changes in D only take place while $Ck = 0$. (The nomenclature 'level sensitive' refers to the connection arrangement of the clock line to the input gates.)

The outputs of L_1 are the inputs to the slave latch L_3 which consists of gates g_5, g_6, g_7 and g_8. Latch L_3 is enabled when its clock signal Y is at logic 1 level, and the data at Q_1 now appears at output Q_2. Output Q_2 is taken directly to the combinational section of the machine so that the machine operates normally when L_1 and L_3 are clocked appropriately.

The subsidiary master latch L_2, consisting of gates g_9, g_{10}, g_3, and g_4, provides the second input port of the two-port master latch. The data input line of L_2 is I and its clock line is X. Normally, $X = 0$ and L_2 is disabled; when $X = 1$, L_2 is enabled and data on line I (normally connected to the output Q_2 of the preceding SRL) is transferred to the output of L_2. The slave latch L_3 is again enabled when clock signal Y is at logic 1 level, and the data at the output of L_2 now appears at output Q_2 which is also connected to input I of the next SRL. In this mode, all the SRLs are now connected in cascade and form a shift register.

It is important to ensure that no two of the clock signals X, Y, and Ck are ever simultaneously at logic high level. This is because the slave must be enabled only when both of the masters are disabled. This can be arranged by external logic where $Y = \bar{X}$ in shift register mode, and $Y = \overline{Ck}$ in the normal operating mode.

A skeleton circuit diagram of a synchronous state machine designed using the double latch LSSD technique is shown in Figure 13.31. The connections for the normal operating mode of the state machine have been highlighted by the solid black lines, and the thinner lines represent the additional connections for shift register mode. SRL latches must be used instead of conventional flip-flops for every stage of the machine.

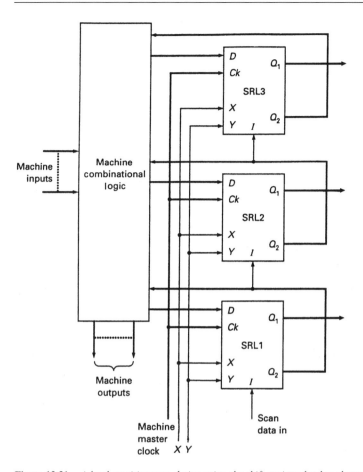

Figure 13.31 *A level sensitive scan design using the shift register latches shown in Figure 13.30*

13.16 Designing for testability

It will have been obvious from the foregoing that testing a digital circuit is usually far from being trivial, except in the simplest of cases. The largest VLSI chips now manufactured cannot be fully tested, for the entire range of possible inputs is so vast that at the normal maximum operating speed of the chip it would take many years to cover all possible input combinations. Therefore, there is considerable interest in designing modern circuits and systems *specifically* with ease of testing in mind. Designing for testability (DFT) is an enormous subject and is becoming an issue of great importance, and in this section it is possible only to outline the steps that a prudent designer will take in order to increase the ease of testing a new design.

Some methods of improving testability have already been mentioned. In circuits employing redundant gates, it is good practice to include accessible test points, or test inputs, in order to allow undetectable faults to be isolated and detected. Sometimes there are unused connection pins on a module connector, and these can usefully be employed as test inputs or test points. If there are no spare connections, then it is sometimes useful to employ 2-to-1 multiplexers to allow shared use of the pins that are available. In normal operation, all the multiplexers route the correct internal signals to the next logic stages, but on receipt of a special test signal at their select inputs, all the multiplexers connect important internal gate inputs directly to accessible external connections or test points so that they may be driven by known signals. Multiplexers may also be used to route intermediate outputs to external connections when being tested. These techniques can considerably simplify the problems of identifying a fault buried deep inside a logic system and otherwise needing complicated path sensitisation to unmask the fault, but it should be borne in mind that these multiplexers can themselves be a source of faults, and they may also increase problems caused by propagation delays.

In some cases a module of logic is so large that fault-finding access is improved by subdividing the module, usually in such a way that both sub-modules are approximately the same size, and using the principle of binary division explained in Section 13.2. This can be done physically, or if this is not possible then the technique of 'degating' can be used, in which two extra gates are introduced into the signal path of the logic circuit (see Figure 13.32). When $(A,B) = (1,1)$, then the operation of the circuit is normal; but if $A = 0$, then the value of $\bar{B}$ is fed into the second half-module, which can then be directly tested with a known input. Again, added propagation delay and additional possible faults are the penalties for using this technique.

Often, logic components such as gates and flip-flops will have unused inputs that must be tied to high or low logic level for correct operation. For maximum testability, these inputs should be tied to their respective logic levels through individual resistors, one for each input, so that, should the need arise, they can be driven individually by an external source. This would be impossible if these inputs were connected directly to a supply rail.

Many logic systems, particularly sequential designs, must be initialised on first being powered, and this power-on initialisation can also be used to drive all the flip-flops to a known and predictable state. If this is done, then the initial state of the circuit is known and the subsequent response of the circuit is well characterised, enabling fault conditions to be recognised rapidly. When highly developed, such as in a computer, it is possible for the system to undertake its own basic tests of the important sub-systems, and to deduce, for example, the major parameters of the functional parts of the circuit. These may then be reported to the user for further consideration and comparison with the known fault-free parameters.

Other techniques for enhancing testability depend upon constructional methods. It is still usual to mount the most complex chips using sockets, preferably of the 'zero insertion force' type which make removing and refitting the chip virtually free of the risk of physical damage, despite the disadvantages associated with possible contact resistance at each pin connection and slightly degraded high-frequency performance. It is wise to ensure that physical access to test points and component terminals on printed circuit boards is as easy as possible. Additionally, it is frequent practice to provide one or more guided probes at selected points in

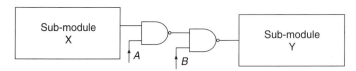

Figure 13.32   *The principle of "degating"*

the system which can be attached to each of the test points in a properly ordered sequence according to a test routine devised according to the principles outlined in this chapter.

Problems

13.1 Determine a complete test set for the 3-input NAND gate shown in Figure P13.1.

13.2 Using the path sensitisation technique, obtain a complete test set for the circuit shown in Figure P13.2.

13.3 Using the path sensitisation technique, determine a test set for the circuit shown in Figure P13.3.

13.4 Using the path sensitisation technique, determine the input test set for the fault p_1 in the circuit shown in Figure P13.4, and check the answer using the Boolean difference method.

13.5 The NAND implementation of an Exclusive-OR gate is shown in Figure P13.5. Determine a complete fault detection test set for each of the following faults: p_0; q_1; r_0.

13.6 For each of the circuits shown in Figure P13.6 determine the minimal fault detection test set.

13.7 For the circuit shown in Figure P13.7, develop the fault table and find a minimal fault detection test set.

13.8 Determine the complete test set for all *s-a-0* and *s-a-1* faults in the four 2-level circuits shown in Figure P13.8.

13.9 Determine the Boolean difference $\partial F/\partial X_2$ for the following functions:

(a) $F = X_1 X_2 + X_1 X_3 + X_2 X_3$

(b) $F = (X_1 + X_2) \cdot (X_1 + X_3) \cdot (X_2 + X_3)$.

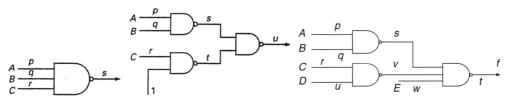

Figure P13.1   **Figure P13.2**   **Figure P13.3**

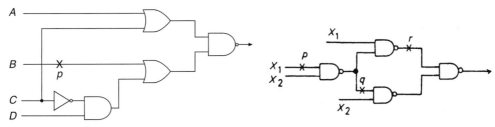

Figure P13.4   **Figure P13.5**

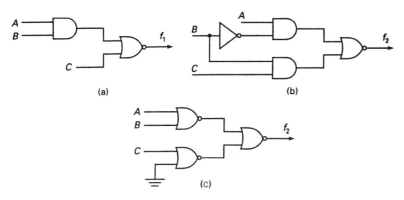

Figure P13.6

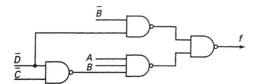

Figure P13.7

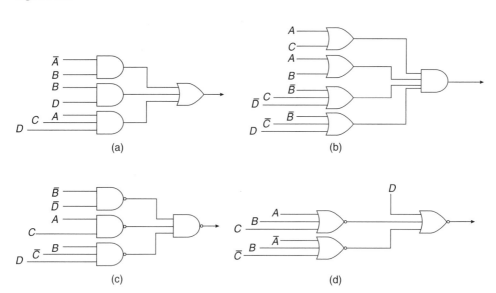

Figure P13.8

Appendix
Functional logic symbols

A.1 Introduction

Over a period of more than twenty-five years, during which time the technology of digital systems has been developing at an unprecedented rate, methods of depicting digital systems in circuit diagrams using standardised logic symbols have similarly been developing. This has resulted in two completely separate sets of logic gate symbols that can be used in drawing circuit diagrams, the 'old' or 'conventional' system and the 'new' or 'functional' logic symbols. In the UK the 'new' standards for drawing circuit diagrams appear in BS 3939, Section 21, and in the USA they appear in IEEE Std. 91/ANSI Y32.14 and IEC publication 617–12. Perhaps the most significant advantage of the standardized system is that it enables 'functional' circuit diagram symbols to be defined, corresponding to most or all SSI and MSI components that describe the logic function of the component in a consistent and logical manner. It is not practical to indicate the detailed logic functions of all VLSI components but the symbolic methodology can still be used to indicate the functions of major parts of these components. This is an international standard so that circuit diagrams drawn using the 'new' system will, in principle, be understood in many different countries.

One frequently voiced criticism of the functional symbol system is that it was originally devised when CAD software was capable of drawing only rectangles, and so all the various types of simple logic gates have functional symbols that are rectangular. The older 'conventional' symbols are completely different for the various kinds of simple logic gates and so are much less prone to confusion. Furthermore, a sizeable number of professional logic designers, having used the distinctive shape symbols for many years, are reluctant to abandon the symbols and conventions with which they are familiar and which are still in common use, and for these reasons the older symbols are used elsewhere throughout this text. However, a professional designer should at least be familiar with the new standards since they are now widely used in many manufacturers' data sheets and published logic circuit diagrams. This Appendix is intended to give sufficient information to allow interpretation of the new symbols for typical devices described in this text, and to use them in simple cases, but it is beyond the scope of this summary to give all the detailed rules of the system to enable symbols for newly available devices to be derived.

A.2 Basic principles of the functional symbol system

In the functional symbol system, all logic gates and other fundamental logic units or *elements* are indicated by basic rectangular outlines, each labelled with a *general*

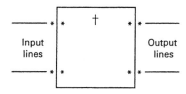

Figure A.1   *General format of a functional symbol for a logic element*

qualifying symbol defining its function. For example, the symbol '&' signifies an AND gate. Additionally, each element has both input and output lines with the possibility of additional qualifying symbols relating specifically to these inputs and outputs. The general principle is shown in Figure A.1, where the dagger ('†') indicates the position of the general qualifying symbol and the asterisks ('*') indicate possible positions of qualifying symbols for the input or output lines. The *inversion circle*, used at the inputs and outputs of conventional symbols and introduced in Chapter 2, is a simple example of an input or output qualifying symbol in the functional system.

The *internal state* means the logic state existing inside the rectangular symbol outline at an input or output, and the *external state* means the logic state existing external to the complete symbol, i.e. outside of any qualifying symbols applying to that input or output.

Although a logic element is usually described by a rectangular outline in this system, the distinctive conventional shape symbols for gates as used elsewhere in this text are still allowed. However, using the conventional distinctive shape symbols loses one major advantage of the functional system, which is that symbols for complex components may be built as combinations of simple gate outline symbols. Three examples of abutting elements are illustrated in Figure A.2. Symbols joined in a vertical direction, and joined by a horizontal boundary, are part of the same IC package but, apart from power supply lines are not electrically connected unless indicated by other means such as a general qualifying symbol. However, symbols joined in a horizontal direction, and joined by a vertical boundary, have a single internal connection, usually from left to right. Multiple internal connections can be indicated by the appropriate number of short perpendicular strokes across the mutual boundary, and a single stroke would confirm a single connection. Therefore, in Figure A.2(a), there are no connections between the adjacent elements, while in Figure A.2(b) a vertical line separates the two elements and indicates that there is a single internal connection between them. In Figure A.2(c) there are three interconnections between the abutting elements.

A *common control block* associated with a number of logic elements is indicated by an element with a special outline, as shown in Figure A.3. This is the only special outline used in the functional system.

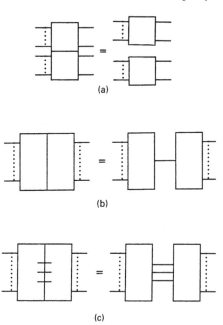

Figure A.2   *Combination of logic elements (a) No internal connection (b) A single internal connection (c) Multiple internal connection*

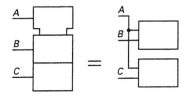

Figure A.3   *Common control block (top) for two logic elements*

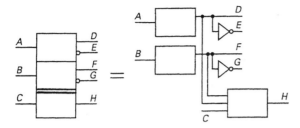

Figure A.4 *Logic array with common output element (bottom)*

&	AND gate
≥1	OR gate (at least one input must be active to activate the output)
=1	2-input XOR gate (one input only must be active to activate the output)
2k+1	Multi-input XOR gate (an odd number of inputs must be active to activate the output)
1	Buffer gate (the single input must be activated to activate the output)
MUX	Multiplexer
X/Y	Decoder or code converter
DMUX or DX	Demultiplexer
CTR *m*	Counter with *m* bits
CTR DIV *m*	Counter with cycle length *m* clock cycles
SRG *m*	Shift register with *m* bits
EPROM, ROM	[Erasable programmable] read only memory
RAM	Random access read/write memory
Σ	Adder
P – Q	Subtractor

Figure A.5 *Some general qualifying symbols Some of the internal qualifying symbols shown in figure A.6 may also be used as general qualifying symbols if they apply to all inputs or outputs*

In the example shown in Figure A.3 there are two logic elements, not connected to each other but both controlled by input *A*.

When an array of abutting elements has a common output element, this is indicated by drawing a *double line* at the boundary between the output element and the rest of the array. In the example shown in Figure A.4 the common output element also has an external input (*C*) connected to it.

Some of the possible *general qualifying symbols* are shown in Figure A.5, and some of the possible *qualifying symbols* for inputs and outputs are shown in Figures A.6 and A.7. The general qualifying symbols for simple OR and XOR gates and buffers indicate the number of inputs that must be active in order to activate the output. The functional system includes an alternative to the inversion circle, the *polarity indicator*. In a *positive logic* circuit diagram this means the same as, and is completely interchangeable with, the inversion circle, with the proviso that the direction of the arrow indicates the direction of signal flow. None of these tables is intended to be complete, however, and only those symbols necessary for understanding the components described in this text are shown. The simplest Boolean logic gates have functional symbols that are simply rectangles labelled with the appropriate general qualifying symbol, as shown in Figure A.8.

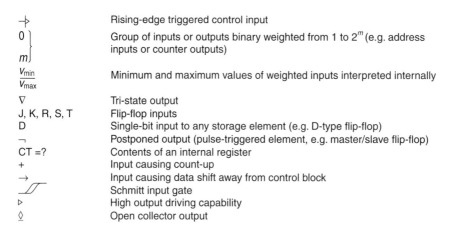

$\rightarrow$	Rising-edge triggered control input
$0\\rceil$ $m\\rfloor$	Group of inputs or outputs binary weighted from 1 to 2^m (e.g. address inputs or counter outputs)
$\dfrac{V_{min}}{V_{max}}$	Minimum and maximum values of weighted inputs interpreted internally
∇	Tri-state output
J, K, R, S, T	Flip-flop inputs
D	Single-bit input to any storage element (e.g. D-type flip-flop)
$\neg$	Postponed output (pulse-triggered element, e.g. master/slave flip-flop)
CT =?	Contents of an internal register
+	Input causing count-up
$\rightarrow$	Input causing data shift away from control block
	Schmitt input gate
$\triangleright$	High output driving capability
$\Diamond$	Open collector output

Figure A.6 *Some qualifying symbols for use inside the element outline with logic inputs and outputs*

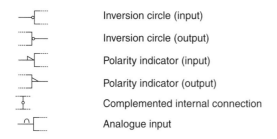

Inversion circle (input)

Inversion circle (output)

Polarity indicator (input)

Polarity indicator (output)

Complemented internal connection

Analogue input

Figure A.7 *Some qualifying symbols for use outside the element outline with logic inputs and outputs*

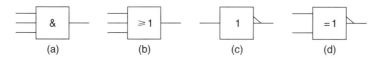

| (a) | (b) | (c) | (d) |

Figure A.8 *Some examples of functional logic symbols for simple Boolean logic gates (a) AND gate (b) OR gate (c) Inverting buffer (d) 2-input XNOR gate*

A.3 Dependency notation

The systematic 'dependency notation' is a completely new feature of the functional logic symbols, with no precise counterpart in the 'conventional' system, and which forms the basis of an extremely flexible method of succinctly indicating the precise logical functions of very many complex ICs. The dependency notation summarises the relationships between inputs and outputs of complex logic elements, and is in addition to the general qualifying symbols describing the overall element function. It thus defines a consistent framework for labelling the various inputs and outputs of the more complex logical elements.

The dependency notation is built around a number of distinct 'dependency types', indicated by a capital letter. These include, amongst others:

EN: Enable dependency,
G: AND dependency,

C: Control dependency,
S, R: Set and Reset dependency,
Z: Interconnection dependency (i.e., internal connections between elements),
M: Mode dependency (i.e., effects depend on the mode of operation),
A: Address dependency, and
N: Negate (XOR) dependency.

The general rules for the dependency notation are that each logic line *affecting* other logic lines is labelled with the appropriate letter chosen from the list of possible dependencies above, followed by a decimal integer or a Greek letter unique for the particular part of the logic element concerned and obviously usually chosen to describe the function of the element as clearly as possible. Furthermore, each logic line *affected by* the *affecting* logic lines is labelled with the same integer. Logical inversion between *affecting* and *affected* lines is indicated by placing a complementing bar over the label. The OR of two or more logic lines can be indicated by labelling them with the *same* letter and integer or Greek letter. If any line is *affected* by several *affecting* logic lines, then the order in which the effects are applied is indicated by the left-to-right ordering of the *affecting* labels.

As an example, an enable signal with EN dependency affects only the outputs of the element to which it is connected, even if the logic symbol used contains a number of elements, as shown in Figure A.9. When the internal enable signal is 1, the outputs of that element are enabled. When not enabled, the result depends upon the type of outputs employed. For a tri-state output, a disabling signal puts the outputs into the high impedance Z state externally, though the internal state is unaffected. Totem-pole outputs are taken to the logic 0 state, and open collector outputs are taken to the 'off' state.

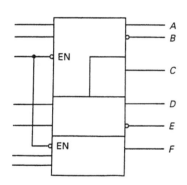

Figure A.9 *Control of outputs using Enable (EN) dependency. Only outputs A, B, C, and F are controlled by the Enable input*

It should be noted that, because of the flexibility of the system, any one device can often be described by several possible functional symbols; the symbol designer chooses the symbol that describes the function of the device most clearly, and there are some instances where the choice of the clearest symbol depends upon the particular use being made of the device.

A.4 Simple examples of G dependency in functional logic symbols

The rule for the G (AND) dependency is that when the *affecting* input is in logic state 1, the *affected* input behaves as it would without the dependency symbol; and when the *affecting* input is in the 0 state, the effect is as if the *affected* input were in the 0 state. This is the essence of an 'AND' gate, of course. A typical example of a functional logic symbol using the G dependency, for the 74251 8-to-1 line multiplexer, is shown in Figure A.10. Pin numbers on the IC package are given in parentheses outside the logic element outline. Polarity symbols indicate an active low Enable input $\bar{G}$ and

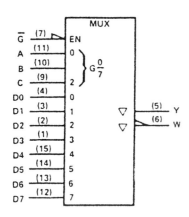

Figure A.10 *Functional logic symbol for the 74251 8-to-1 line multiplexer*

a complemented output W. Both the inverted and non-inverted outputs are tri-state. Select signals A, B, and C, having weights 2^0, 2^1, and 2^2 respectively, provide a range of binary input values from $(0)_{10}$ to $(7)_{10}$ inclusive. The internal qualifying symbol $G\frac{0}{7}$ indicates AND dependency between these select inputs and the data inputs D0 to D7 labelled with internal qualifying symbols 0 to 7. This dependency identifies the combination of select signals (C, B, A) required to select individual data lines so that, for example, the control signal combination $C\bar{B}\bar{A}$ is ANDed with data input D4. This is, of course, the correct function of a logic multiplexer; if select signals $(C, B, A) = (1, 0, 0)$ are applied, then data line D4 is selected.

A more complex example is the functional logic symbol for the 74153 dual 4-to-1 line multiplexer, shown in Figure A.11. The following points refer to this diagram.

1. The two main logic elements (specified by the general qualifying symbol MUX) are separated by a horizontal straight line, meaning that they are not interconnected.

2. The two logic elements are identical, and so to reduce clutter, only the first element is described in full on the symbol.

3. Each logic element has its own Enable signal, $1\bar{G}$ and $2\bar{G}$, controlling outputs 1Y and 2Y respectively. Polarity indicators are used at these inputs to indicate that these are active low inputs.

4. Control signals A and B, having weights 2^0 and 2^1 respectively and providing a range of binary input values from $(0)_{10}$ to $(3)_{10}$ inclusive, are supplied to a common control block. Since the control block is common to all logic elements, both MUX elements are controlled by the same address inputs.

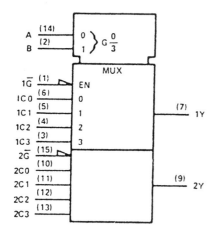

Figure A.11 *Functional logic symbol for the 74153 dual 4-to-1 line multiplexer*

5. G (AND) dependency is specified by the internal qualifying symbol $G\frac{0}{3}$ in the common control block, corresponding to the internal qualifying symbols 0, 1, 2 and 3 labelling the data inputs of the top MUX element. This dependency identifies the combination of control signals A and B required to select individual data lines. For example, if $(A, B) = (0, 0)$, then data lines 1C0 and 2C0 are selected; or if $(A, B) = (1, 1)$ then data lines 1C3 and 2C3 are selected.

The logic IC type 74139 can be used either as a dual 2-to-4 line decoder or as a 2-to-4 line demultiplexer. Because of its two somewhat different uses it has two alternative functional logic symbols using two different general qualifying symbols, as shown in Figure A.12. In principle, these two functional logic symbols are equivalent and

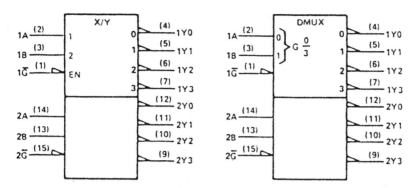

Figure A.12 *Equivalent functional logic symbols for the 74139 dual 2-to-4 line decoders/demultiplexers*

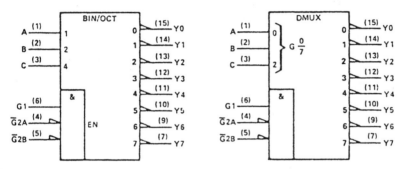

Figure A.13 *Equivalent functional logic symbols for the 74138 3-to-8 line decoder/demultiplexer*

interchangeable. In practice, a wise designer will choose to use the one closer to the actual function required of the IC in the circuit. Note that in both representations, all the output lines are active low. Pin (1) functions as an Enable input in the decoder symbol but, in the symbol for the same device used as a demultiplexer, pin (1) acts as the data input and so the designation EN is not needed.

Two possible functional logic symbols for the 74138 3-to-8 line decoder/demultiplexer are shown in Figure A.13. Note that instead of the general qualifying symbol for a decoder (X/Y), the more specific symbol BIN/OCT is used. Also, there is an 'embedded AND' gate within the main outline of both symbols. The symbol EN to the right of the gate embedded in the decoder symbol indicates that the implied horizontal connection, between the embedded gate and the main part of the logic element, takes the function of the Enable signal for the device. For the demultiplexer symbol, this connection acts as the data input to the main part of the element and so the designation EN is removed.

A.5 Control, Set, and Reset dependency

The rule for the C (Control) dependency is that when the *affecting* input is in logic state 1, the *affected* input behaves as it would without the dependency symbol; and when the *affecting* input is in the 0 state, the *affected* input is not permitted to control the element. An example of C dependency is shown by the controlled SR latch shown in

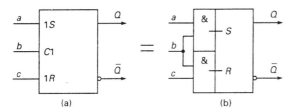

Figure A.14   *An example of the use of control dependency: the controlled SR latch*

Figure A.14(a). Control input b is the affecting input and a and c are the affected inputs. Normal operation of the SR latch only occurs when the control input is in the high state (logic level 1). The Control input is identified by the qualifying label C1 while the affected inputs are each given the qualifying label 1. Exactly the same functionality is indicated in a different manner, using embedded elements, in Figure A.14(b). In this case, internal signal $S = a \cdot b$ and internal signal $R = b \cdot c$, so that inputs a and c take the roles S and R respectively only when $b = 1$.

When $S = R = 1$ the output of a normal SR latch is unspecified, and in practice this is regarded as a forbidden input condition. However, S or R dependency can be used in a functional symbol when the corresponding component has a well-defined output for $S = R = 1$.

The rule for the S (Set) dependency is that when the *affecting* input is in logic state 1, the *affected* output behaves as it would with $S = 1$, $R = 0$; and when the *affecting* input is in the 0 state, there is no effect. Therefore, S dependency overrides whatever logic level is present on an R input. The S input is denoted by S1, indicating that it is now an affecting input. The two complementary outputs are labelled by 1 to indicate that they are both affected by input S1. For $S = R = 1$, the S input overrides the effect of R so that $Q = 1$ and $\bar{Q} = 0$. Since the latch is behaving as it would for the input conditions $S = 1$, $R = 0$, this is also known as 'Set overrides Reset'. A functional logic symbol for such a latch appears in Figure A.15(a).

The rule for the R (Reset) dependency is that when the *affecting* input is in logic state 1, the *affected* output behaves as it would with $S = 0$, $R = 1$; and when the *affecting* input is in the 0 state, there is no effect. Therefore, R dependency overrides whatever logic level is present on an S input. The R input is denoted by R1, indicating that it is now an affecting input. The two complementary outputs are labelled by 1 to indicate that they are both affected by input R1. For $S = R = 1$, the R input overrides the effect of S so that $Q = 0$ and $\bar{Q} = 1$. Since the latch is behaving as it would for the input conditions $S = 0$, $R = 1$, this is also known as 'Reset overrides Set'. A functional logic symbol for such a latch appears in Figure A.15(b).

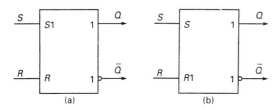

Figure A.15   *(a) Set (S) dependency (b) Reset (R) dependency*

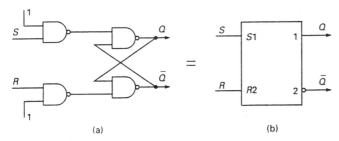

Figure A.16 *SR dependency of a NAND gate SR latch (a) Gate implementation of latch, and (b) its functional logic symbol*

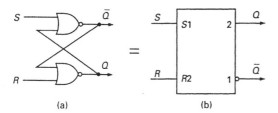

Figure A.17 *SR dependency of a NOR gate SR latch (a) Gate implementation of latch, and (b) its functional logic symbol*

In practice, a more common situation is that shown by the SR latch made using NAND gates, as shown in Figure A.16. For this circuit, $Q = \bar{Q} = 1$ when $S = R = 1$ and both Set and Reset dependency are present in the circuit. Both S and R are affecting inputs and the corresponding input pins are designated S1 and R2. Output Q is affected by input S1 and so is given label 1 to identify this dependency, while output $\bar{Q}$ is affected by input R2 and so is given label 2 to identify its dependency on R2. If inputs S1 and R2 are simultaneously held at *high* logic level, the notation specifies that output $Q = 1$ (overriding input R2) and simultaneously output $\bar{Q} = 1$ (overriding input S1), describing the circuit action correctly. However, if both S1 and R2 are simultaneously held at *low* logic level, then neither input is active and so either $Q = 0$, $\bar{Q} = 1$ or $Q = 1$, $\bar{Q} = 0$ are possible, corresponding to the previous state of the latch.

For the SR latch made from NOR gates (Figure A.17(a)), the outputs are $Q = \bar{Q} = 0$ when $S = R = 1$. Again, S and R are both affecting inputs and so, as before, the corresponding input pins are designated S1 and R2. Now, however, output $\bar{Q}$ is affected by input S1 and so is labelled 1, while output Q is affected by input R2 and so is labelled 2. If input S1 is in logic high state then output $\bar{Q}$ is 0 irrespective of input R2. If input R2 is in logic high state then output Q is 0 irrespective of input S1. This gives the functional symbol shown in Figure A.17(b).

A.6 Bistable logic elements and C dependency

The master/slave JK flip-flop is an example of a single-bit memory element that is 'pulse triggered' – that is, the data must be set up prior to the arrival of the first logic edge of the clock pulse and must remain stable at least until the clock has returned to

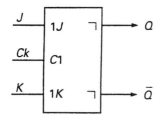

Figure A.18   *Functional logic symbol for master/slave JK flip-flop*

its original state. The final output signal appears when the clock signal returns to its original state and is said to be a 'postponed' output. This is indicated by the internal qualifying symbol '¬'. Figure A.18 shows the functional logic symbol for a master/slave JK flip-flop, having Control dependency provided by the clock C1; 1J and 1K are the affected inputs. (If there is a change in the logical value of 1J or 1K, or both, while the clock remains active, then the resulting output state is not described correctly by the functional logic symbol.)

The presence or absence of the qualifying symbol '¬' combined with the presence or absence of the symbol for an edge-triggered clock input gives the following four basic types of clocked flip-flops and latches that can be represented using the conventions of the functional logic symbols, as shown in Figure A.19.

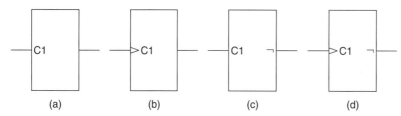

Figure A.19   *The distinguishing features of the four basic bistable elements (a) The transparent latch (b) The edge-triggered flip-flop (c) The pulse-triggered flip-flop (d) The data-lockout flip-flop*

a. *The transparent (controlled) latch.* Control dependency label C1 means that data input is only enabled when C1 = 1. For C1 = 0, the data inputs have no effect. Only changes of data input while C1 = 1 will result in a change of output state.

b. *The edge-triggered flip-flop.* Edge triggering is indicated by the internal qualifying symbol '>'. This means that the control signal C1 only enables data input on a rising edge (0 → 1) transition. As in the conventional symbol system, operation designed to take place on a falling edge (1 → 0) transition may be indicated by juxtaposing the internal qualifying symbol '>' by an inversion circle, an external qualifying symbol.

c. *The pulse-triggered flip-flop.* As in Figure A.18, this is identified by the presence of the output indicator '¬', and the *absence* of the edge-triggered input indicator '>'.

d. *The data-lockout flip-flop.* This is similar to a pulse-triggered device except that the control input C1 is considered dynamic, so that the functional logic symbol includes both the edge-triggered symbol '>' and the postponed output symbol '¬'. Shortly after C1 has made a 0 → 1 transition, the data inputs are disabled and data does not have to be maintained for the remainder of the clock pulse. However, the output is postponed until C1 returns to its original state. Type 74110 is an example of a flip-flop with data-lockout.

An example of a functional logic symbol for a single-bit memory element is shown in Figure A.20. The device illustrated is the 7470 positive edge-triggered JK flip-flop with

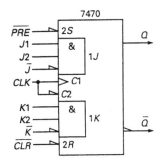

Figure A.20 *Functional logic symbol for the 7470 AND-gated positive edge-triggered JK flip-flop*

additional AND gating. The AND gates at the *J* and *K* inputs are incorporated within the logic symbol and edge-triggered operation is indicated by '>' at the clock input. Logic polarity indicators are used to denote active low inputs and an active low output. This device is unusual because, as well as the usual edge-triggered JK action, the complemented direct Clear and Preset inputs only operate when the Clock input is at logic low level. Therefore, the Clock and the inverted Clock are both regarded as control signals and are designated C1 and C2, respectively. The inputs 1J and 1K are affected by the dynamic Clock signal C1, where the control dependency is indicated by the numeral 1. The inverted Clock signal, control signal C2, affects the complemented Preset and Clear inputs, and the control dependency is therefore indicated by denoting these signals as 2S and 2R respectively. External polarity indicators show that these are actually complemented inputs.

A.7 Counters, Z and M dependency

The functional logic symbols for MSI counters are broadly similar to those described in section A.4 for multiplexers and comparators. For example, the 7468 is a dual 4-bit non-synchronous counter which may also be used for frequency division. The functional logic symbol for this MSI circuit, shown in Figure A.21(a), is divided into three separate elements with no logical interconnection. Each element is a separate counter and consequently has its own general qualifying symbol. The upper counter (CTR DIV 2) is a scale-of-2 counter; the centre block (CTR DIV 5) is a scale-of-5 counter; and the lower block (CTR DIV 10) is a scale-of-10 counter. All of these elements may also be used for frequency division by 2, 5, or 10 respectively, and the general qualifying symbols show that all the outputs have high current driving capability. In fact the detailed data for the 74LS68 show that its outputs each have approximately double the fan-out capability of normal LS outputs.

The bit grouping symbols indicate the weightings of the binary outputs, so that the potential binary range of the centre element is 0 to $(7)_{10}$, and for the lower element $(0)_{10}$ to $(15)_{10}$. However, the qualifying symbols CTR DIV 5 and CTR DIV 10 indicate that these counts are actually limited internally to $(5)_{10}$ and $(10)_{10}$ respectively.

The upper and centre counter elements have a common active low Clear input, while a separate active low Clear input is provided for the remaining counter. The symbol $CT = 0$ by each of these inputs indicates that, when active, these inputs clear the counter contents to zero. Separate edge-triggered clock inputs are provided for each counter element, and all three elements are capable only of counting up.

In practice, $1Q_A$ can be externally connected to 1CLKB, thus amalgamating the upper two counters into a single scale-of-10 counter. Together with the lowest element there are then two decade counters, and this is the reason that this IC is described as a *dual* counter. Frequency division by 100 can now be achieved by externally connecting $1Q_D$ to 2CLK.

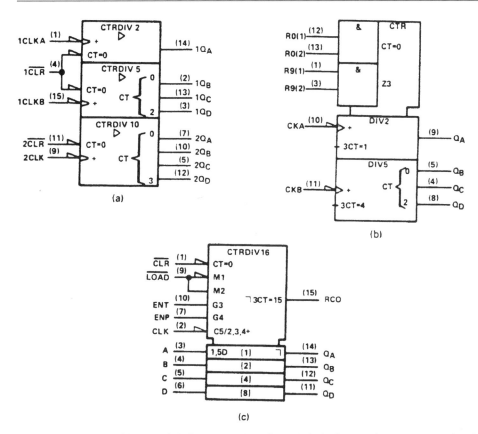

Figure A.21 *Functional logic symbols for counters (a) The 7468 dual 4-bit asynchronous counter (b) The 74290 decade counter (c) The 74161 synchronous 4-bit counter*

Another example is the 74290 non-synchronous decade counter, shown in Figure A.21(b). This MSI circuit consists of two counters, one a scale-of-2 and the other a scale-of-5, serviced by a common control block. Both counters have independent clock inputs and the potential range of the lower counter is $(0)_{10}$ to $(7)_{10}$. However, the qualifying symbol DIV 5 indicates that its count is actually limited to 5. If Q_A is connected externally to CKB, the operation of the two counters may be combined to give a scale-of-10 counter.

The two counter elements have a common Clear signal which clears both counts to zero when R0(1) AND R0(2) are high. Inputs R9(1) and R9(2) are used to set the combined count to $(9)_{10}$. These are both active high inputs to an AND gate embedded in the control block; this generates an output signal Z3 that controls both counters using the Interconnection (Z) dependency. This is specified by prefixing the CT symbols in both counter blocks with the same numeral. The rule for the Z dependency is that when the *affecting* input is in logic state 1, the *affected* input behaves as if logic state 1 has been imposed on it; and when the *affecting* input is in the 0 state, the *affected* input behaves as if logic state 0 has been imposed on it. For the upper counter, $3CT = 1$ indicates that Z3 active causes $Q_A = 1$; simultaneously, for the lower counter $3CT = 4$ indicates that Z3 active causes this counter's outputs to be set to $(Q_D, Q_C, Q_B) = (1, 0, 0)$. Therefore, when R9(1) and R9(2) are simultaneously active, the combined count is set to $(Q_D, Q_C, Q_B, Q_A) = (1, 0, 0, 1) \equiv (9)_{10}$.

Finally, the 74161 is a more complex MSI circuit having further control functions described by the Mode dependency, M. Its functional logic symbol is shown in Figure A.21(c) and the general qualifying symbol specifies that this device is a scale-of-16 counter. The rule for the M dependency is analogous to that for the C dependency: when the *affecting* input is in logic state 1, the corresponding mode is selected; and when the *affecting* input is in logic state 0, the corresponding mode is not selected. The basic counting function is indicated by the symbol '2, 3, 4+' adjacent to pin (2) (CLK). This denotes that the count will be incremented by unity when the clock makes a $0 \rightarrow 1$ transition provided inputs $\overline{\text{LOAD}}$, ENP, and ENT, marked with dependencies M2, G3, and G4 respectively, are high. The counter is directly cleared (CT = 0) by an active low input on pin (1). The counter block is divided into four similar sections, one for each bit of the counter, and the integers 1, 2, 4, and 8 indicate the weighting of each bit of the count. Surrounding these integers with brackets [] indicates that these integers are non-standard qualifying symbols, introduced purely for clarity.

The symbol 1,5D adjacent to pin (3) indicates that the counter is loaded with the logic values on inputs A, B, C, and D, when the $\overline{\text{LOAD}}$ input is low (Mode dependency M1) and pin (2) makes a $0 \rightarrow 1$ transition (Control dependency C5). Since pin (2) is associated with two separate functions (load and count), both of them are specified using the clock input line but separated by an oblique stroke (or forward slash, '/').

Each stage of the counter has its own individual postponed output, designated Q_A, Q_B, Q_C and Q_D. Finally, the ripple carry output RCO is described by qualifying label '¬3CT = 15', indicating that it is a postponed output dependent upon the enable signal ENT. RCO is active only when the counter contents CT = 15 and the enable signal at G3 is active.

A.8 Shift registers

A functional logic symbol for a basic 4-bit shift register with parallel outputs from each bit is shown in Figure A.22(a). The upper section of the diagram is the control block

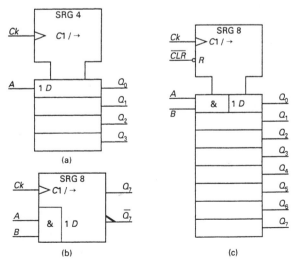

(a)

(b)

(c)

Figure A.22 *Functional logic symbols for shift registers (a) Four-bit SIPO (b) 7491 8-bit SISO (c) 74164 8-bit SIPO*

and the lower section contains the four flip-flops that make up the four stages of the shift register. Data input to the register is through line A. The qualifying symbol '$\rightarrow$' indicates that the data shifts one stage further away from the control block every time there is a rising logic edge at the clock input Ck. The enabling control of the flip-flops by the clock is indicated by Control dependency C1 and the label 1D in the lower section of the diagram.

A typical example of a serial-in, serial-out shift register is the 8-bit type 7491. Its functional logic symbol is shown in Figure A.22(b); again the flip-flops are controlled by the edge-triggered clock line Ck. Serial-out data is available in true or complemented form from the last stage of the register. Independent outputs are not available from the seven remaining flip-flops in this device. Serial data formed by the result A AND B enters the first stage of the shift register through an implied horizontal connection (labelled 1D to the right of the embedded AND gate). In practice, one of the inputs A and B can be used as the data input line while the other input is used as a line to enable or inhibit data input.

Lastly, the functional logic symbol for the 74164 parallel-out 8-bit shift register is shown in Figure A.22(c). Independent outputs are provided from each stage of the register and an additional active low clear line (R dependency) is provided for initialising the register contents to 0. Because the clear line resets *all* the flip-flops to 0, it is connected to the control block of the symbol. Note once again the implied horizontal connection out of the embedded AND gate in the top section of the register.

A.9   Programmable devices and A dependency

Programmable logic devices have the common feature (shared with certain other devices) that one of an array of elements (usually binary words) is selected by the use of a set of address or select inputs. This is indicated on a functional logic symbol by the use of the Address (A) dependency. The rule for the Address dependency is that the *affecting* inputs (the address lines) allow the element that is selected by the address to function fully and to react to the *affected* inputs, while the corresponding functions of all the other elements in the device are disallowed.

As an example, the functional logic symbol for the TMS27128 EPROM is shown in Figure A.23(a). All outputs are tri-state, and this EPROM has two control lines: an output enable $\bar{G}$ (an active low input) which gates data to the output lines, thus eliminating bus contention, and an active low chip enable input $\bar{E}$ which, besides selecting the chip, provides the additional facility of being able to put the EPROM in a standby mode. When $\bar{E} = 1$ (i.e., this signal is not *asserted*), the tri-state output buffers are placed in their high-impedance (Z) state (see Chapter 10) and simultaneously the EPROM power consumption is reduced to less than 10% of its value when fully operational. The address dependency indicates that valid addresses range from 0 to 16383, and the eight affected data outputs show that this is a byte-organised EPROM. Figure A.23(a) shows the function of the device as a ROM and so the active-low input used to program the EPROM (pin 27, denoted $\overline{PGM}$) is not shown.

A second example, the TMS47256 ROM, is shown in Figure A.23(b); there are two possible symbols as pin (20) has a dual function. It can either be operated as a chip enable/power down input ($\bar{E}$ or E) or, alternatively, as a secondary chip select pin ($\bar{S}_2$ or S_2). Both of these functions can be programmed as either active low or active

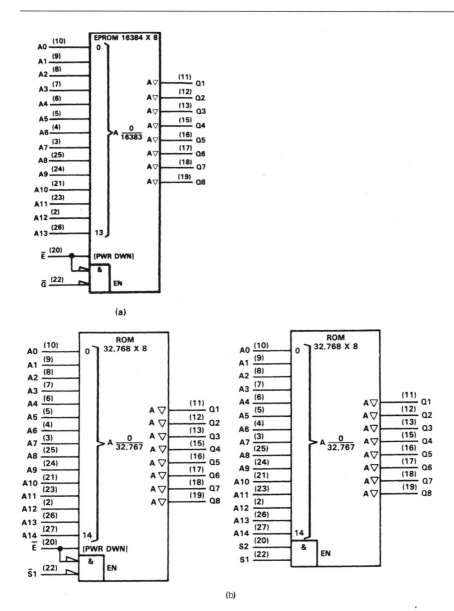

Figure A.23 *Logic symbols for (a) TMS27128 16 Kbyte EPROM, and (b) the two modes of operation of the TMS47256 32 Kbyte ROM*

high during mask fabrication. The chip select input (pin (22), labelled $\bar{S}_1$ or S_1) can also be programmed during mask fabrication to be either active high or active low. The address dependency indicates that valid addresses range from 0 to 32767, and the eight affected data outputs show that this is also a byte-organised ROM.

A.10 Arithmetic circuits and N dependency

A typical example of a 4-bit adder is type 74283, and its functional logic symbol is shown in Figure A.24. The preferred designations for the operands in a device

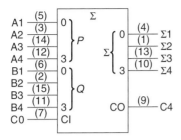

Figure A.24 *Functional logic symbol for the 74283 4-bit adder*

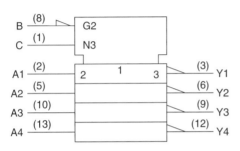

Figure A.25 *Functional logic symbol for the 7487 true/complement unit*

performing a mathematical operation are P and Q. Note that the binary weighting of the operand input lines and the sum output lines are indicated by the braces ('}'). In this diagram, the carry-in line is designated both by the acronym CI and as C0, i.e. the carry from the bit one place less significant than bit 1, and the carry-out line is designated both by the acronym CO and as C4.

IC type 7487 is a 'true/complement unit' intended for controlling one set of inputs to a 4-bit adder. The functional logic symbol for the 7487 is shown in Figure A.25. The functional logic symbol indicates that the device has a common block having an input C and an active low input B. The polarity indicator and the qualifying symbol G2 (AND dependency) at pin (8) denote that $\bar{B}$ is ANDed with the data inputs $A1$, $A2$, $A3$, and $A4$ (all labelled 2) inside each controlled element. The rule for the Negate or XOR (N) dependency is that when the *affecting* input is in logic state 1, the *affected* input is complemented; and when the *affecting* input is in the 0 state, the *affected* input is *not* complemented.

This is the essence of 'controlled inversion' provided by an XOR gate, of course (see Chapter 4). Input C has qualifying symbol N3, controlling the four outputs identified by the label 3 that are then complemented. Thus, the result from the topmost controlled element is

$$Y1 = \overline{C \oplus (A1 \cdot \bar{B})} \equiv C \oplus \overline{(A1 \cdot \bar{B})}.$$

If $C = 1$ then $Y1 = A1 \cdot \bar{B}$, whereas if $C = 0$ then $Y1 = \overline{(A1 \cdot \bar{B})}$. In this case, all the logic in the device is indicated by the dependency symbols G and N, so that each controlled element is given the general qualifying symbol 1, denoting a logic buffer with no further logic functions. Finally, as the controlled elements are all similar, it is actually only necessary to label the topmost controlled element in the symbol.

Answers to problems

Chapter 1

1.1   (a) $(173)_{10}$ (b) $(54.5)_{10}$ (c) $(1.156)_{10}$

1.2   (a) $(187)_{10}$ (b) $(529)_{10}$ (c) $(14.581)_{10}$

1.3   (a) $(325)_{10}$ (b) $(41665)_{10}$ (c) $(26.696)_{10}$

1.4   (a) $(1111010)_2$ (b) $(1100010)_2$ (c) $(110000.0111)_2$

1.5   (a) $(1012)_8$ (b) $(2137)_8$ (c) $(201.2)_8$

1.6   (a) $(479)_{16}$ (b) $(975)_{16}$ (c) $(C0.DC3)_{16}$

1.7   (a) base $6, 7, 8, 9, 10\ldots$ (b) $b = 4$ (c) $b = 5$ (d) $b = 7$

1.8   (a) 8   (b) 12   (c) 8

1.9   (a)
```
  101011
 +10111
 1000010
 //////
 111111
```
(b)
```
   1101
   1110
  +1001
 100100
  ////
1 111
```
(c)
```
  11101
 -10110
  00111
   //
  11
```
(d)
```
  1100.010
 -1000.111
  0011.011
  /////
  111 11
```

1.10   (a) 0,1110111 SM   (b) 1,1001101 SM   (c) 1,0000011 SM
           0,1110111 2'sC   1,0110011 2'sC   1,1111101 2'sC
           0,1110111 1'sC   1,0110010 1'sC   1,1111100 1'sC

1.11   (a) 0,0100101   (b) 0,1100101   (c) 1,1110100

1.12   (a) 10101 (2'sC)   (b) 4709 (10'sC)   (c) 3055 (8'sC)   (d) 543F (16'sC)
           10100 (1'sC)   4708 (9'sC)   3054 (7'sC)   543E (15'sC)

1.13   (a) 10110110   (b) 100100110   (c) 1000001101

1.14   (a) $-63 = 1,1000001$ (b) $+108 = 0,1101100$ (c) $+104 = 0,1101000$

1.15   (a) 1000 $R = 010$ (b) 1011 $R = 011$

1.16   (a)   0111   1001   (b)   1000   0111

               0001   <u>0000</u>   <u>0001</u>   0001   <u>0111</u>   <u>1001</u>

                          0111   1010   1111   0000

                            <u>1</u>   <u>0110</u>   1

               0001   1000   0000   <u>1</u>   <u>0110</u>   <u>0110</u>

                                    0001   0110   0110

     (c)   1001   1000

          9'sC of 43 = 56   <u>0101</u>   <u>0110</u>

                            1110   1110

                            0110   <u>0110</u>

                            <u>1</u>   0100

                        10101

                        └──────────→1

                        0101

1.17   (a) 0000   1000   0101   (b) 9'sC of 92 = 907

           <u>0000</u>   <u>0110</u>   <u>0111</u>   1001   0000   0111

           0000   1110   1100   <u>0000</u>   <u>0100</u>   <u>0011</u>

                0110   <u>0110</u>   1001   0100   1010

            <u>1</u>   <u>1</u>   0010   <u>1</u>   <u>0110</u>

           0001   0101   0101   0000

1.18

	7421	5211	2421	8	4−2−1
0	0000	0000	0000	0	0 0 0
1	0001	0001	0001	0	1 1 1
2	0010	0011	0010	0	1 1 0
3	0011	0110	0011	0	1 0 1
4	0100	0111	0100	0	1 0 0
5	0101	1000	1011	1	0 1 1
6	0110	1001	1100	1	0 1 0
7	1000	1100	1101	1	0 0 1
8	1001	1110	1110	1	0 0 0
9	1010	1111	1111	1	1 1 1

Chapter 2

2.1 $f_1 = B + C$ $f_2 = AB\bar{C} + D$ $f_3 = B(\bar{A} + \bar{C}) + \bar{A}C$

2.2 $f_1 = A\bar{C} + A\bar{B} + AD + B\bar{C}D$ $f_2 = \bar{A} + B + \bar{C} + D$

      $f_3 = B + AD$

2.4 $\bar{f}_1 = \bar{A}(\bar{B} + C)$ $\bar{f}_2 = \bar{A}\bar{C} + \bar{B}\bar{C} + \bar{A}\bar{B}$

      $\bar{f}_3 = \bar{A}B + B\bar{C}$ $\bar{f}_4 = AB\bar{D} + A\bar{B}D + B\bar{C}\bar{D} + \bar{B}\bar{C}D$

2.5 $\bar{f}_1 = \bar{A}(B + \bar{C})$ $\bar{f}_2 = (A + \bar{B}\bar{C})(\bar{B} + D + A\bar{C})$
$f_3 = (\bar{A} + \bar{B})[\bar{C} + A(\bar{D} + \bar{E})] + B[\bar{A} + \bar{C} + E(B + D)]$

2.6 (a) $f_1 = \bar{A}\bar{B}\bar{C}$ (b) $f_2 = \bar{B}\bar{C}$ (c) $f_3 = \bar{A} + \bar{C}$

2.8 f_1: 9 included, 7 not f_2: 13 included, 3 not

$$f_1 = m_5 + m_6 + m_7 + m_9 + m_{10} + m_{11} + m_{13} + m_{14} + m_{15}$$
$$f_2 = m_2 + m_3 + m_5 + m_6 + m_7 + m_8 + m_9 + m_{10} + m_{11} + m_{12} + m_{13} + m_{14} + m_{15}$$
$$f_1 = BD + BC + AD + AC$$
$$f_2 = A + C + BD$$

2.9 $f = \bar{A}\bar{B} + \bar{B}C$

2.10

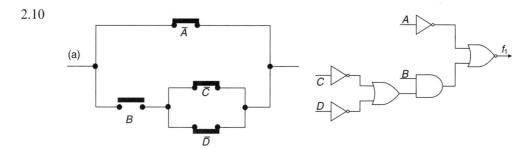

(a)

The solutions to (b) and (c) are left to the reader.

2.11 $A \oplus B \oplus C$.

2.12

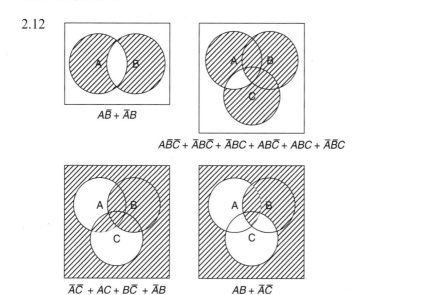

$A\bar{B} + \bar{A}B$

$A\bar{B}\bar{C} + \bar{A}B\bar{C} + \bar{A}BC + AB\bar{C} + ABC + \bar{A}\bar{B}C$

$\bar{A}\bar{C} + AC + B\bar{C} + \bar{A}B$

$AB + \bar{A}\bar{C}$

2.14 If D = passengers in doorway, P = passengers still moving, B = button pressed, then f = doors close = $\bar{D}\bar{P}B$

Chapter 3

3.1   (a) $P_1 + P_2 + P_3 + P_5 + P_7$
      (b) $P_1 + P_3 + P_5 + P_6 + P_7$
      (c) $P_2 + P_3 + P_4 + P_5 + P_6 + P_7 + P_{11} + P_{12} + P_{13} + P_{14} + P_{15}$

3.2   (a) $f_1 = \bar{B}C + B\bar{C}$ (b) $f_2 = \bar{A} + BC$ (c) $f_3 = AB + AC + BC$

3.3   (a) $f = S_2 S_3 S_4 S_5 S_6 S_7$ (b) $f = P_3 + P_4$ (c) $\bar{f} = S_3 S_5 S_6 S_7$

3.4   (a) $f = \bar{A}\bar{B} + A\bar{C} + BC$
      (b) $f = \bar{A}\bar{B} + \bar{B}\bar{C} + CD + A\bar{C}\bar{D}$
      (c) $f = AB + \bar{A}\bar{B}C + \bar{A}\bar{B}D + ACD$
      (d) $f = AC + \bar{C}\bar{D}\bar{E} + CDE + \bar{A}\bar{B}E + BCD + B\bar{C}\bar{D} + \bar{A}B\bar{C}\bar{E}$

3.5   (a) $f = AC + BC$ (b) $f = \bar{A}C + \bar{A}D + \bar{B}C$

$$(c)\ f = A\bar{C} + \bar{C}D + ABD + \begin{cases} \bar{A}\bar{B}C + A\bar{B}\bar{D}, & \text{or} \\ \bar{A}\bar{B}D + \bar{B}C\bar{D}, & \text{or} \\ \bar{A}\bar{B}C + \bar{B}C\bar{D} \end{cases}$$

      (d) $f = ABD + \bar{A}C\bar{E} + BC\bar{D} + \bar{A}\bar{B}C + \bar{B}DE + AB\bar{C}\bar{E}$

3.6   (a) $f = (\bar{A} + C)(A + \bar{B} + \bar{C})$
      (b) $f = \bar{B}(A + \bar{C})(\bar{A} + C + D)$
      (c) $f = (\bar{A} + \bar{B})(\bar{A} + \bar{D})(\bar{B} + D)(A + D + E)(B + \bar{D} + \bar{E})$
      (d) $f = (A + B)(C + D)(\bar{B} + D)$

3.7   (a) $F = \bar{A}CD$
      (b) $F = \bar{B}CD + B\bar{C}D + A\bar{B}\bar{C}\bar{D}$
      (c) $F = A\bar{B} + \bar{A}B\bar{C}$

3.8 $0 = \bar{A}\bar{B}, 1 = \bar{A}\bar{C}\bar{D}, 2 = \bar{A}\bar{C}D, 3 = \bar{A}C\bar{D}, 4 = BCD, 5 = \bar{B}\bar{C}\bar{D},$
      $6 = \bar{B}\bar{C}D, 7 = \bar{B}C\bar{D}, 8 = ACD, 9 = AB$

3.9   (a) $f = \bar{B}D + AC + \bar{A}\bar{B}\bar{C} + AB\bar{D} + BC\bar{D}$
      (b) $f = ACD + BC\bar{E} + \bar{B}CE + \bar{A}\bar{B}\bar{C}\bar{D} + \bar{A}BD\bar{E} + A\bar{B}D\bar{E} + B\bar{C}DE + AB\bar{C}E$
      (c) $f = (\bar{B} + C)(A + D)(B + \bar{C} + \bar{D})(\bar{A} + \bar{C} + \bar{D})$

3.10 (a) $f = \bar{A}D + \bar{B}C + \bar{A}B\bar{C} + B\bar{C}D + A\bar{B}\bar{D} + AC\bar{D}$
      (b) $f = AE + B\bar{E} + CE + DE + \bar{A}\bar{B}\bar{C} + BC\bar{D} + A\bar{B}C$
      (c) $f = (A + \bar{C} + D)(A + \bar{B} + D)(\bar{A} + \bar{B} + C)(\bar{B} + \bar{C} + D)(\bar{A} + \bar{C} + \bar{D})(\bar{A} + C + D)$

3.11

A \ BC	00	01	11	10
0	D	D	D	
1	$\bar{D}$	1	$\bar{D}$	D

3.12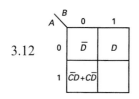

3.13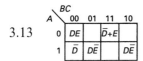

3.14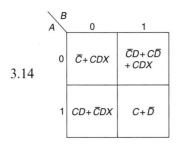

3.15   (a) $f_1 = \bar{A}\bar{B}\bar{D} + \bar{A}\bar{C}D + ABC$
       (b) $f_2 = \bar{C}\bar{D} + AB\bar{C} + \bar{B}CD$
       (c) $f_3 = A + B\bar{C} + \bar{B}C + \bar{C}\bar{D}$

Chapter 4

4.1

(a)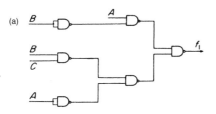

(b)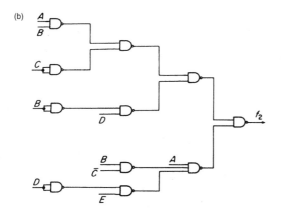

4.2

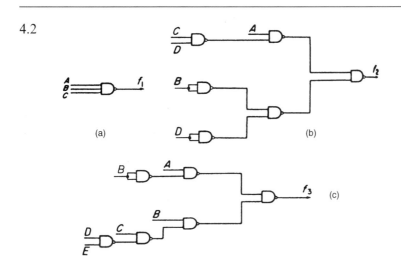

(a)

(b)

(c)

4.3   (a)

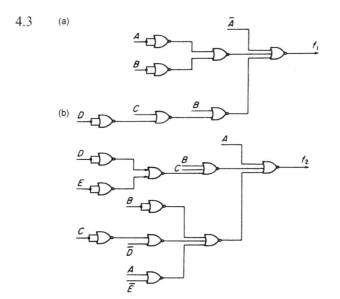

(b)

4.4   (a)

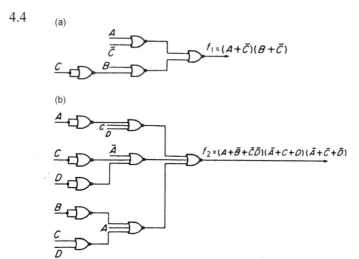

$f_1 = (A + \bar{C})(B + \bar{C})$

(b)

$f_2 = (A + \bar{B} + \bar{C}\bar{D})(\bar{A} + C + D)(\bar{A} + \bar{C} + \bar{D})$

4.5   (a) $f_1 = A + D(B\bar{C} + \bar{B}C)$

   (b) $f_2 = C(\bar{A} + B) + \bar{A}\bar{D}$ or $f_2 = \bar{A}(C + \bar{D}) + BC$

   (c) $f_3 = A\bar{B}D + \bar{C}(A + D)(\bar{B} + D)$ or $f_3 = A\bar{B}\bar{C} + D(A + \bar{C})(\bar{B} + \bar{C})$

4.6   (a) $f_1 = B\bar{C}(A + \bar{D}) + A(C + D)$

   (b) $f_2 = A(B + C) + D(\bar{A} + B + \bar{C})$

   (c) $f_3 = CD + \bar{A}BD + AB\bar{D}$

4.7 – 4.8   (a) $f_1 = ABC\bar{D}$

   (b) $f_2 = [\bar{A}\bar{B} + \bar{E}\bar{F} + (\bar{C} + \bar{D})]G\bar{E}\bar{F}$

   (c) $f_3 = (\bar{A} + \bar{B})(\bar{C} + DE) + G[F + C(\bar{D} + \bar{E})]$

4.9   (a) $f_1 = (\bar{A} + BC)(D + AC)(B + \bar{C})$

   (b) $f_2 = (\overline{\bar{A} + C} + \overline{\bar{B} + \bar{C}})(\bar{A} + \bar{C})$

   (c) $f_3 = B(\bar{A} + \bar{C}D) + A\bar{B}D$

4.10   (a) $ABC + A\bar{B}\bar{C}$

   (b) $ABC + A\bar{B}\bar{C}$

   (c) $A + \bar{B}C + B\bar{C}$

   (d) $A + \bar{B}C + B\bar{C}$

Chapter 5

5.1   (a) $D_0 = \bar{C},$ $D_1 = 1,$ $D_2 = C,$ $D_3 = C$

   (b) $D_0 = A,$ $D_1 = \bar{A},$ $D_2 = A,$ $D_3 = 1$

   (c) $D_0 = 1,$ $D_1 = 0,$ $D_2 = 1,$ $D_3 = 1$

5.2   (a) $D_0 = \bar{C} + D,$ $D_1 = \bar{C}D + C\bar{D},$ $D_2 = \bar{C} + D,$ $D_3 = \bar{C}$

   (b) $D_0 = A + \bar{D},$ $D_1 = A,$ $D_2 = 0,$ $D_3 = D$

   (c) $D_0 = \bar{A}\bar{B},$ $D_1 = 1,$ $D_2 = A,$ $D_3 = A + \bar{B}$

   (d) $D_0 = 0,$ $D_1 = \bar{B}\bar{C},$ $D_2 = 1,$ $D_3 = B + C$

5.3   (a) Level 1 4-to-1 MUX control signals D and E

   Level 2 four 4-to-1 MUXs;

   MUX1: $DE = 00.$ $D_0 = \bar{A},$ $D_1 = 1,$ $D_2 = \bar{A},$ $D_3 = \bar{A}$;

   MUX2: $DE = 01.$ $D_0 = \bar{A},$ $D_1 = A,$ $D_2 = 1,$ $D_3 = 1$;

   MUX3: $DE = 10.$ $D_0 = 1,$ $D_1 = 0,$ $D_2 = A, D_3 = 1$;

   MUX4: $DE = 11.$ $D_0 = 1,$ $D_1 = 0,$ $D_2 = \bar{A}, D_3 = A$

(b) Level 1 4-to-1 MUX control signals D and E
Level 2 four 4-to-1 MUXs;
MUX1: $DE = 00$. $D_0 = A, D_1 = 0, D_2 = 1, D_3 = 0$;
MUX2: $DE = 01$. $D_0 = A, D_1 = A, D_2 = 1, D_3 = A$;
MUX3: $DE = 10$. $D_0 = A, D_1 = 1, D_2 = 1, D_3 = 1$;
MUX4: $DE = 11$. $D_0 = A, D_1 = 0, D_2 = 1, D_3 = 1$

5.4 (a) Inputs to MUX1, $D_0 = \bar{A} + \bar{B}, D_1 = \bar{A}B, D_2 = 0, D_3 = \bar{A} + B$;
Inputs to MUX2, $D_0 = \bar{A}\bar{B}, D_1 = A + \bar{B}, D_2 = A\bar{B}, D_3 = A$;
Inputs to MUX3, $D_0 = A + B, D_1 = \bar{A}B, D_2 = \bar{A}B, D_3 = \bar{A} + B$;
Inputs to MUX4, $D_0 = \bar{A}\bar{B} + AB, D_1 = \bar{B}, D_2 = A\bar{B}, D_3 = AB$

(b) MUX5, E and F control signals; MUXs 1, 2, 3 and 4, B, C
and D control signals.
Inputs to MUX1, $D_0 = 1, D_1 = 0, D_2 = 0, D_3 = \bar{A}, D_4 = \bar{A}, D_5 = \bar{A}, D_6 = 0$,
$D_7 = 1$;
Inputs to MUX2, $D_0 = \bar{A}, D_1 = 1, D_2 = A, D_3 = A, D_4 = 0, D_5 = A, D_6 = 0$,
$D_7 = A$;
Inputs to MUX3, $D_0 = A, D_1 = 0, D_2 = 0, D_3 = \bar{A}, D_4 = 1, D_5 = \bar{A}, D_6 = \bar{A}$,
$D_7 = 1$;
Inputs to MUX4, $D_0 = \bar{A}, D_1 = 1, D_2 = A, D_3 = 0, D_4 = A, D_5 = 0, D_6 = 0$,
$D_7 = A$

5.5 Represent 8421 by $ABCD$ and 5421 by $PQRS$. MUX control signals A and B:

MUX1(P) $D_0 = 0$, $D_1 = \bar{C} + D$, $D_2 = \bar{C}$, $D_3 = 0$
MUX2(Q) $D_0 = 0$, $D_1 = \bar{C}\bar{D}$, $D_2 = \bar{C}D$, $D_3 = 0$
MUX3(R) $D_0 = C$, $D_1 = CD$, $D_2 = \bar{C}\bar{D}$, $D_3 = 0$
MUX4(S) $D_0 = D$, $D_1 = C\bar{D}$, $D_2 = \bar{C}\bar{D}$, $D_3 = 0$

5.6 The four NBCD digits are A, B, C and D, digit A the most significant. Segment
equations: $a = A + C + \overline{B \oplus D}$;
$b = \bar{B} + \overline{C \oplus D}$; $c = \bar{B}C\bar{D}$; $d = e + B\bar{C}D + \bar{B}C$; $e = \bar{D}(\bar{B} + C)$; $f = A + \bar{C}\bar{D} + B\bar{D} + B\bar{C}$; $g = A + C\bar{D} + (B \oplus C)$.

5.7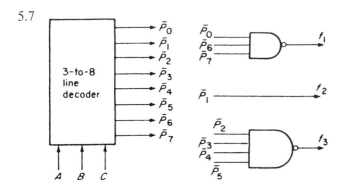

5.8 Requires one 4-to-10 line decoder generating the complement of the P terms, i.e. $\bar{P}_0$, $\bar{P}_1$ etc. in each case. Then

(a) $W = \overline{\bar{P}_5 \cdot \bar{P}_6 \cdot \bar{P}_7 \cdot \bar{P}_8 \cdot \bar{P}_9}$, $X = \overline{\bar{P}_4 \cdot \bar{P}_6 \cdot \bar{P}_7 \cdot \bar{P}_8 \cdot \bar{P}_9}$,

$Y = \overline{\bar{P}_2 \cdot \bar{P}_3 \cdot \bar{P}_5 \cdot \bar{P}_8 \cdot \bar{P}_9}$, $Z = \overline{\bar{P}_1 \cdot \bar{P}_3 \cdot \bar{P}_5 \cdot \bar{P}_7 \cdot \bar{P}_9}$,

where W is the left-most digit of the 2421 code.

(b) $W = \overline{\bar{P}_5 \cdot \bar{P}_6 \cdot \bar{P}_7 \cdot \bar{P}_8 \cdot \bar{P}_9}$, $X = \overline{\bar{P}_1 \cdot \bar{P}_2 \cdot \bar{P}_3 \cdot \bar{P}_4 \cdot \bar{P}_9}$,

$Y = \overline{\bar{P}_0 \cdot \bar{P}_3 \cdot \bar{P}_4 \cdot \bar{P}_7 \cdot \bar{P}_8}$, $Z = \overline{\bar{P}_0 \cdot \bar{P}_2 \cdot \bar{P}_4 \cdot \bar{P}_6 \cdot \bar{P}_8}$,

where W is the left-most digit of the XS3 code.

(c) $W = \overline{\bar{P}_5 \cdot \bar{P}_6 \cdot \bar{P}_7 \cdot \bar{P}_8 \cdot \bar{P}_9}$, $\bar{X} = \overline{\bar{P}_0 \cdot \bar{P}_9}$,

$\bar{Y} = \overline{\bar{P}_3 \cdot \bar{P}_4 \cdot \bar{P}_5 \cdot \bar{P}_6}$, $Z = \overline{\bar{P}_2 \cdot \bar{P}_3 \cdot \bar{P}_6 \cdot \bar{P}_7}$,

where W is the left-most digit of the XS3 Gray code.

5.9 Requires one 4-to-16 line decoder, then $f_1 = \overline{\bar{P}_2 \cdot \bar{P}_3 \cdot \bar{P}_{14} \cdot \bar{P}_{15}}$, $f_2 = \bar{P}_{12}$, $f_3 = \bar{P}_0 \cdot \bar{P}_2$ and $f_4 = \overline{\bar{P}_5 \cdot \bar{P}_6 \cdot \bar{P}_8 \cdot \bar{P}_{11} \cdot \bar{P}_{12} \cdot \bar{P}_{13} \cdot \bar{P}_{14} \cdot \bar{P}_{15}}$.

5.10

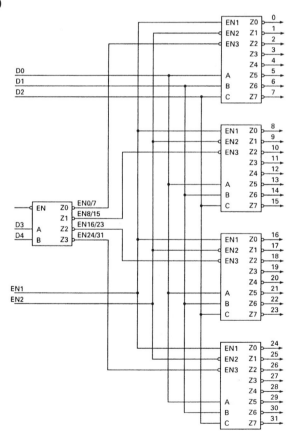

5.11 $f_1 = \overline{\bar{P}_0 \cdot \bar{P}_1 \cdot \bar{P}_3 \cdot \bar{P}_9 \cdot \bar{P}_{12} \cdot \bar{P}_{14}}$ $f_2 = \overline{\bar{P}_5 \cdot \bar{P}_9 \cdot \bar{P}_{10} \cdot \bar{P}_{12} \cdot \bar{P}_{13} \cdot \bar{P}_{15}}$

$f_3 = \overline{\bar{P}_0 \cdot \bar{P}_3 \cdot \bar{P}_8 \cdot \bar{P}_{11} \cdot \bar{P}_{12} \cdot \bar{P}_{15}}$ $f_4 = \overline{\bar{P}_1 \cdot \bar{P}_2 \cdot \bar{P}_7 \cdot \bar{P}_8 \cdot \bar{P}_{11} \cdot \bar{P}_{12} \cdot \bar{P}_{14}}$

5.12

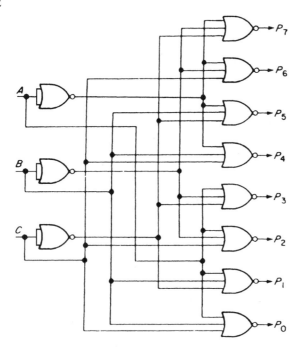

5.13   Solution obtained with a Gray code allocation of the addresses:
$$A = \bar{f}_7 + \bar{f}_6 + \bar{f}_5 + \bar{f}_4 \quad B = f_6 f_7 (\bar{f}_5 + \bar{f}_4 + \bar{f}_3 + \bar{f}_2)$$
$$C = f_7 [\bar{f}_6 + \bar{f}_5 + f_3 f_4 (\bar{f}_2 + \bar{f}_1)]$$

5.14 $C = \overline{f_4 f_5 f_6 f_7} \quad B = \overline{f_2 f_3 f_6 f_7} \quad A = \overline{f_1 f_3 f_5 f_7}$

5.15 $E_4 = \bar{A}_4 \bar{B}_4 + A_4 B_4$; similarly E_3, E_2, and E_1

Then $E = E_4 E_3 E_2 E_1$
$$A > B = A_4 \bar{B}_4 + E_4 A_3 \bar{B}_3 + E_4 E_3 A_2 \bar{B}_2 + E_4 E_3 E_2 A_1 \bar{B}_1$$
$$A < B = \overline{A > B} \cdot \bar{E}$$

5.16

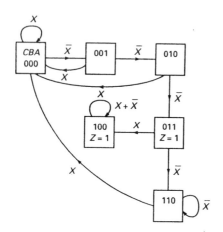

5.17

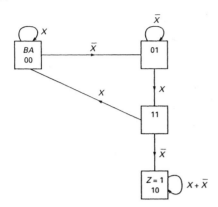

Chapter 6

6.1 (i) $a = 1, b = 0, c = 1$ (ii) $a = 0, b = 1, c = 1$
(iii) $a = 1, b = 0, c = 0$

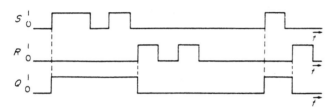

6.2

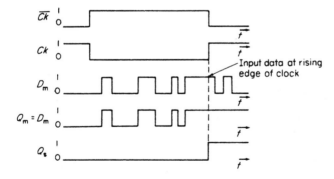

6.3 (i) Rising edge of clock pulse 1

(a) $1 \rightarrow 0$, (b) $0 \rightarrow 1$, (c) $1 \rightarrow 1$, (d) $0 \rightarrow 0$, (e) $1 \rightarrow 1$, (f) $1 \rightarrow 0$, (g) $0 \rightarrow 1$, (h) $1 \rightarrow 1$

Trailing edge of clock pulse 1

(a) $0 \rightarrow 1$, (b) $1 \rightarrow 1$, (c) $1 \rightarrow 0$, (d) $0 \rightarrow 1$, (e) $1 \rightarrow 1$, (f) $0 \rightarrow 0$, (g) $1 \rightarrow 1$, (h) $1 \rightarrow 0$

(ii), (iii) and (iv) repeat as in (i) above.

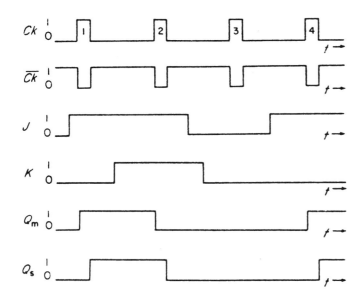

6.4

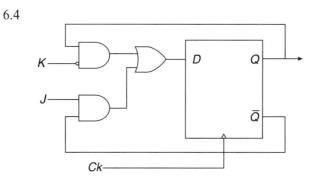

6.5 $Q^{t+\delta t} = (\bar{J}'\bar{Q} + \bar{K}Q)^t$

6.6   For both cases

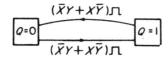

6.7

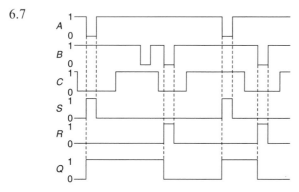

6.8

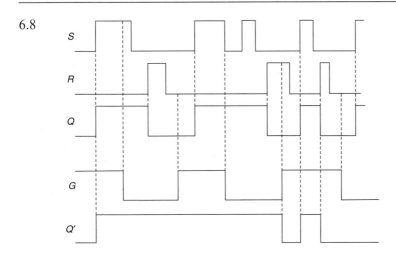

6.9 Each gate is a NAND gate, so the analysis is as shown in Figure 6.10.

Chapter 7

7.1 D is assumed to be the most significant bit of the counter.
$T_D = ABC + ABD, T_C = AB\bar{D}, T_B = A, T_A = 1, S_D = ABC,$
$R_D = ABD$ or $AB\bar{C}, S_C = AB\bar{C}\bar{D}, R_C = ABC, S_B = A\bar{B}, R_B = AB, S_A = \bar{A},$
$R_A = A, J_D = ABC, K_D = AB, J_C = AB\bar{D}, K_C = AB, J_B = A, K_B = A, J_A = 1,$
$K_A = 1, D_D = \bar{B}D + \bar{A}D + ABC, D_C = \bar{B}C + \bar{A}C + AB\bar{C}\bar{D}, D_B = A\bar{B} + \bar{A}B, D_A = \bar{A}.$

7.2 $J_C = \bar{A} + \bar{B}, K_C = \bar{A}, J_B = C, K_B = \bar{C}, J_A = B, K_A = \bar{B}.$ Lock-in state $CBA = 111$.

7.3 P is the most significant bit of the counter.
$J_P = R\bar{S}, K_P = \bar{Q}, J_Q = \bar{P}, K_Q = P R\bar{S}, J_R = \bar{P}S, K_R = PS, J_S = \bar{P}Q\bar{R} + PR,$
$K_S = P\bar{R} + \bar{P}R.$

7.4

Z^t	$Z^{t+\delta t}$	P	Q
0	0	1	X
0	1	0	X
1	0	X	1
1	1	X	0

C is assumed to be the most significant bit of the counter.
$P_C = \bar{A} + \bar{B}, Q_C = AB, P_B = \bar{A}, Q_B = A, P_A = 0, Q_A = 1.$

7.5

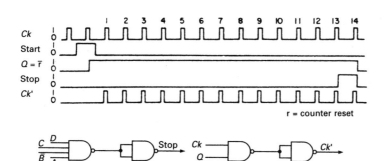

7.6   (a) $f_u = 5\,\text{MHz}$

(b) 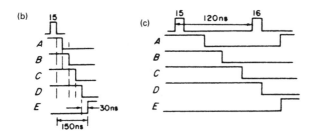   (c)

7.7

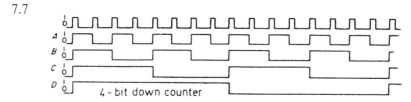

4-bit down counter

7.8

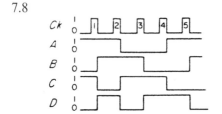

7.9   Chosen sequence: $S_0 - S_1 - S_2 - S_5 - S_{11} - S_7 - S_{15} - S_{14} - S_{13} - S_{10} - S_4 - S_8 - S_0$.

Feedback function assuming A is least significant flip-flop in register
$f = B\bar{D} + AC\bar{D} + \bar{A}BC + A\bar{C}D + \bar{A}\bar{C}D, 0 = \bar{A}\bar{B}\bar{C}\bar{D}, 1 = A\bar{C}\bar{D}, 2 = B\bar{C}\bar{D},$
$3 = A\bar{B}C\bar{D}, 4 = A\bar{C}D, 5 = AB\bar{D}, 6 = ABCD, 7 = \bar{A}BC, 8 = \bar{B}CD, 9 = \bar{A}B\bar{C}D,$
$10 = \bar{A}\bar{B}C, 11 = \bar{A}\bar{B}D$

7.10   Because of ambiguities in the shift register sequence developed from the given binary sequence it is necessary to design a modulo-11 *SR* counter. The modulo-11 sequence chosen from the De Bruijn diagram is $S_0 - S_1 - S_2 - S_5 - S_{11} - S_7 - S_{14} - S_{13} - S_{10} - S_4 - S_8 - S_0$. Feedback logic $f = \bar{A}\bar{C}\bar{D} + BCD + A\bar{C}D + A\bar{B}C\bar{D}$ where A is the least significant stage of the shift register. Output logic $g = D + A\bar{C}$

7.11   (a)

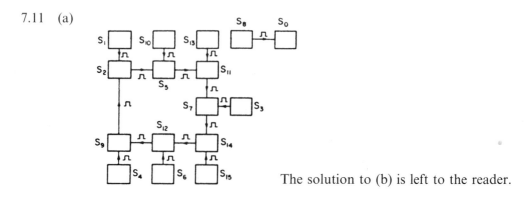

The solution to (b) is left to the reader.

7.12 Feedback logic $f = B \oplus C + \bar{A}\bar{B} + m\bar{B}$ (A is the least significant stage of the shift register)

7.13 $P = \bar{A}BC + AB\bar{C}, Q = \bar{A}\bar{C}, R = \bar{A}\bar{B} + \bar{B}\bar{C} + \bar{A}\bar{C} + ABC, S = A\bar{B}C + \bar{A}\bar{C}$

7.14

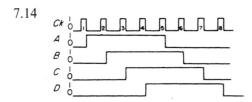

Reduced sequence requires $f = \bar{C}\bar{D}$.

Chapter 8

8.1

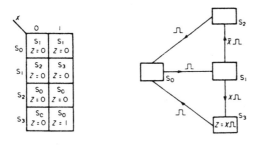

Examines 3-digit words to give an output $Z = 1$ if the last two bits of the word are 1's.

8.2

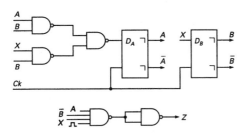

8.3 Circuit solution depends upon state assignment.

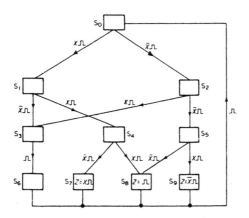

8.4 $J_A = m$
      $K_A = \bar{m}$
      $Z = AX$

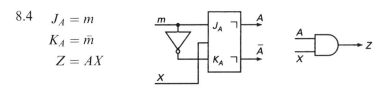

8.5

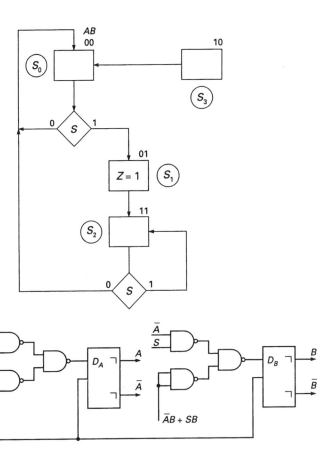

8.6

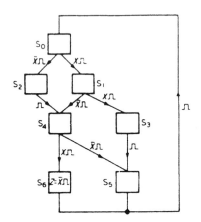

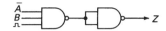

8.7

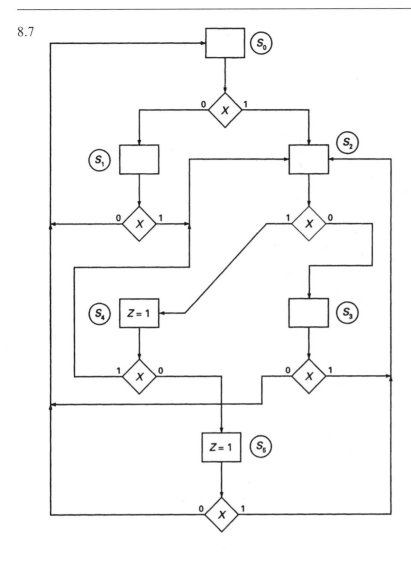

8.8

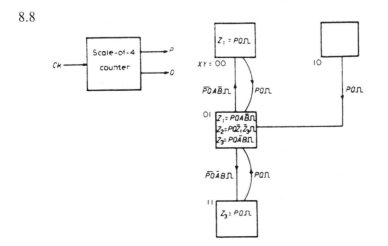

8.9

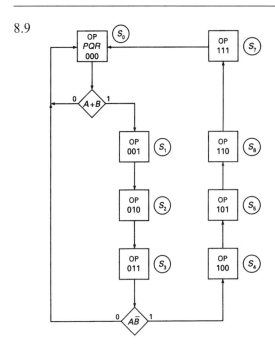

8.10   (a) $P_f = (S_4S_5)(S_0)(S_1)(S_2)(S_3)(S_6)$
       (b) $P_f = (S_3S_6S_{10}S_{11}S_{13})(S_2S_5S_{12})(S_4S_7S_{14})(S_0)(S_1)(S_8)(S_9)$

8.11

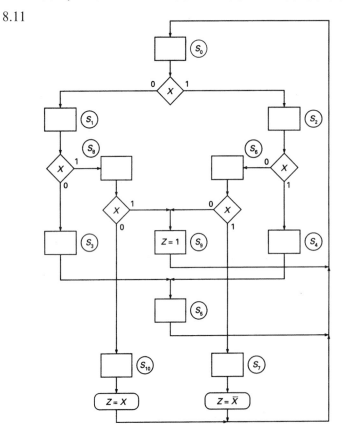

Chapter 9

9.1 The basic state diagram for the problem.

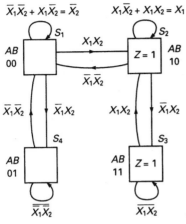

$$\overline{X_1}\overline{X_2} + X_1\overline{X_2} = \overline{X_2} \qquad X_1\overline{X_2} + X_1X_2 = X_1$$

A^tB^t	$X_1^t X_2^t$ 00	01	11	10
S_1 00	00	01	10	00
S_4 01	00	01	01	01
S_3 11	11	11	10	11
S_2 10	00	11	10	10

$$A^{t+\delta t}B^{t+\delta t}$$

$$A^{t+\delta t} = (AB + AX_2 + AX_1 + \overline{B}X_1X_2)^t$$

$$B^{t+\delta t} = (\overline{X_1}X_2 + \overline{A}BX_1 + AB\overline{X_2})^t$$

$$Z = A$$

State diagram

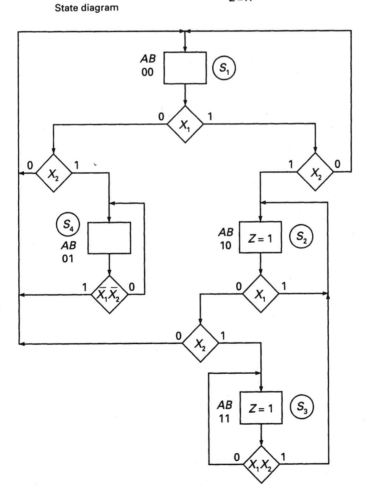

9.2

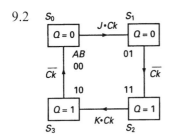

Turn-on set of $A = BC\overline{k}$
Turn-off set of $A = \overline{B}C\overline{k}$
$A^{t+\delta t} = [BC\overline{k} + (B + Ck)A]^t$
Turn-on set of $B = \overline{A}J \cdot Ck$
Turn-off set of $B = Ak \cdot Ck$
$B^{t+\delta t} = [\overline{A}J \cdot Ck + (\overline{A} + \overline{K} + \overline{Ck})B]^t$

9.3

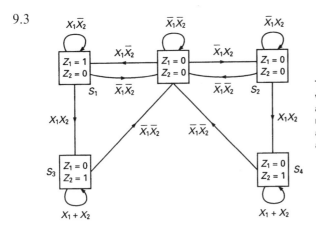

Three secondary
variables are required
and to obtain a
race-free assignment
a dummy state is
also required.

9.4

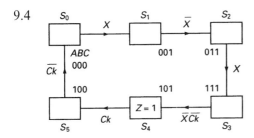

9.5

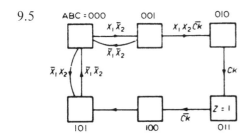

Basic state diagram.

9.6

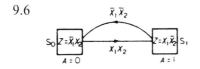

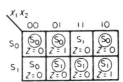

X_1X_2	00	01	11	10	11	01	00	10	00
State	S_0	S_0	S_1	S_1	S_1	S_1	S_0	S_0	S_0
Output	$Z=0$	$Z=1$	$Z=0$	$Z=1$	$Z=0$	$Z=0$	$Z=0$	$Z=0$	$Z=0$

9.7

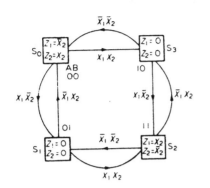
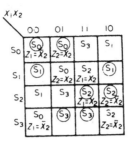

X_1X_2		00	01	11	10	00	01	11	01	11	10	00
State		S_0	S_0	S_3	S_2	S_1	S_0	S_3	S_3	S_3	S_2	S_1
Output Z_1Z_2		10	01	00	01	00	01	00	00	00	01	00

9.8　Races (1) $S_0 \rightarrow S_3$ on signal X_1X_2 – critical

(2) $S_1 \rightarrow S_2$ on signal X_1X_2 – critical

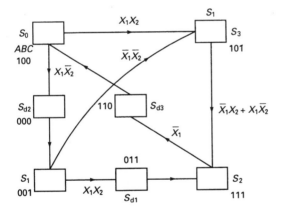

Turn-on condition for $A = \bar{B}C\bar{X}_1\bar{X}_2 + BC$

$$= C\bar{X}_1\bar{X}_2 + BC$$

Turn-off condition for $A = \bar{B}\bar{C}X_1X_2$

$A^{t+\delta t} = [C\bar{X}_1\bar{X}_2 + BC + (B + C + \bar{X}_1 + \bar{X}_2)A]^t$

$B^{t+\delta t}$ and $C^{t+\delta t}$ can be found in the same way.

9.9　(a) $f = \bar{C}\bar{D} + \bar{A}\bar{D} + B\bar{D} + \bar{A}B\bar{C} + ABC + ACD + A\bar{B}D + A\bar{B}C$

(b) $f = AB + B\bar{C} + B\bar{D} + ACD + \bar{B}CD$

9.10　(a) Static 1-hazards $A = B = C = 1$ and $A = 1, B = 1, C = 0$;

Static 0-hazards $B = C = 0, D = 1$.

(b) Static 1-hazards $A = 0, C = D = 1$ and $B = D = 0, C = 1$;

Static 0-hazards $A = B = C = 0$.

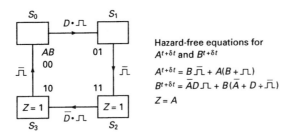

Hazard-free equations for
$A^{t+\delta t}$ and $B^{t+\delta t}$

$A^{t+\delta t} = B\,\bar{\sqcap} + A(B + \sqcap)$

$B^{t+\delta t} = \bar{A}D\,\sqcap + B(\bar{A} + D + \bar{\sqcap})$

$Z = A$

9.12   (a) Compatible pairs
$(S_4S_5)\ (S_2S_5)\ (S_1S_4)\ (S_2S_3)\ (S_1S_3)\ (S_0S_3)\ (S_1S_2)\ (S_0S_2)\ (S_0S_1)$

(b) Maximal compatibles
$(S_0S_1S_2S_3)\ (S_2S_5)\ (S_1S_4)\ (S_4S_5)$

Minimum state table can be formed from S_{01}, S_{23} and S_{45}.

Chapter 10

10.1   Assuming threshold voltage is $\approx 0.5V_{cc}$:

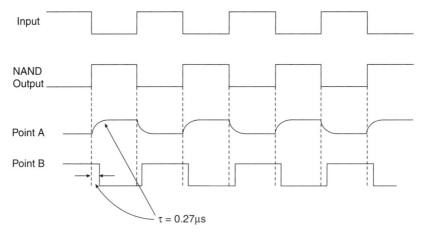

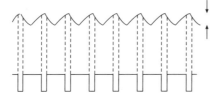

As f increases, waveform at point A becomes more like this. Amplitude decreases as it is governed by time available for RC charge/discharge cycle.

Waveform at B has mark/space ratio no longer 50/50 because threshold voltage is unlikely to be the average voltage at point A. Eventually point B becomes stuck at logic 0 or 1 level.

Finally, A = high, B = low.

10.2 ~ 400 Kbyte.

10.3 ~ 783 Mbyte.

10.4 $(2^{20} - 1) \cong 120$ dB.

10.5 n

10.6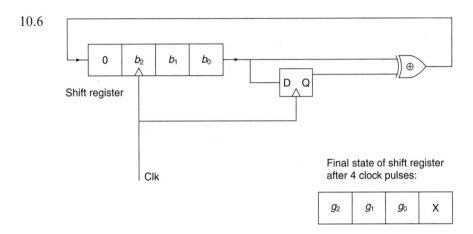

10.7 $D' = D + AC + BC$; $C' = C + B + A\bar{D} + \bar{A}D$; $B' = \bar{B}\bar{C} + \bar{A}\bar{C} + ABC$; $A' = B$

10.8   CLK = high except for low spikes, $\approx 0.4\mu s$ long, at each transition in X. Q = 0 for clockwise, Q = 1 for anticlockwise.

10.9 $\overline{(AB)} \cdot CD$.

10.10 Q3 on, Q4 off.

10.11 Connect LED in series with $R = 360\Omega$ between gate output and ground.

10.12 15 mA.

Chapter 11

11.1

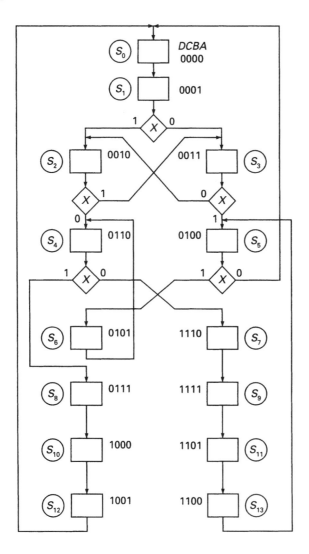

Present state					Output word				Present state					Output word			
X	D	C	B	A	D	C	B	A	X	D	C	B	A	D	C	B	A
0	0	0	0	0	0	0	0	1	0	0	1	1	1	1	0	0	0
1	0	0	0	0	0	0	0	1	1	0	1	1	1	1	0	0	0
0	0	0	0	1	0	0	1	1	0	1	0	0	0	1	0	0	1
1	0	0	0	1	0	0	1	0	1	1	0	0	0	1	0	0	1
0	0	0	1	0	0	1	1	0	0	1	0	0	1	0	0	0	0
1	0	0	1	0	0	0	1	1	1	1	0	0	1	0	0	0	0
0	0	0	1	1	0	0	1	0	0	1	0	1	0	0	1	0	0
1	0	0	1	1	0	1	0	0	1	1	0	1	0	0	1	0	0
0	0	1	0	0	0	0	0	0	0	1	1	0	0	1	1	0	0
1	0	1	0	0	0	1	0	1	1	1	1	0	0	1	1	0	0
0	0	1	0	1	0	1	1	0	0	1	1	0	1	1	1	1	1
1	0	1	0	1	0	1	1	0	1	1	1	0	1	1	1	1	1
0	0	1	1	0	1	1	1	0	0	1	1	1	1	1	1	0	1
1	0	1	1	0	0	1	1	1	1	1	1	1	1	1	1	0	1

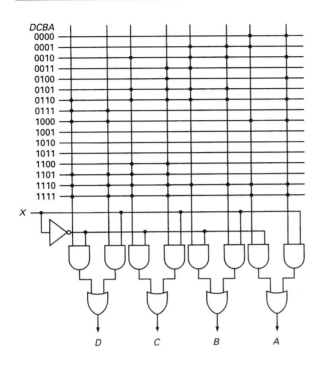

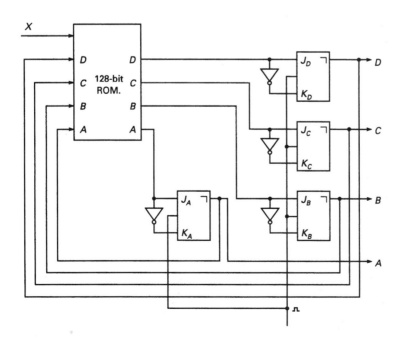

$D_D = BD + \overline{A}\overline{C}D + ACD + BC\overline{X} + ABC$, $D_C = CD + A\overline{B}C + \overline{A}B\overline{X} + \overline{A}CX + AB\overline{C}X$, $D_B = \overline{A}B + B\overline{C}\overline{X} + A\overline{B}\overline{D}$,
$D_A = BD + \overline{A}\overline{D}X + \overline{A}\overline{D}X + \overline{B}\overline{C}DX$

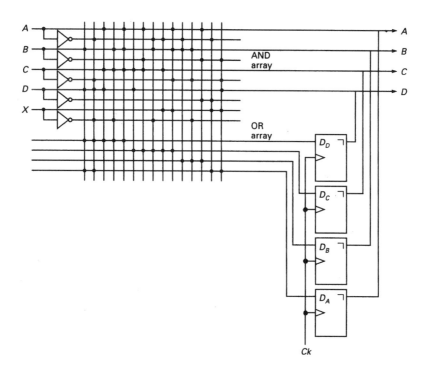

11.2

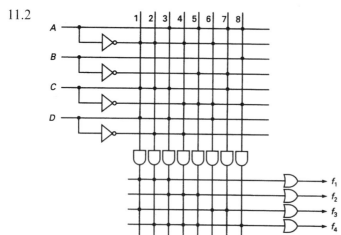

$f_1 = \overline{A}CD + \overline{A}C\overline{D} + ACD$
      1   2   3
$f_2 = \overline{A}\overline{C}\overline{D} + A\overline{B}\overline{C} + ACD$
      4   5   3
$f_3 = \overline{A}\overline{C}D + \overline{A}CD + A\overline{B}C$
      6   1   7
$f_4 = \overline{A}\overline{C}\overline{D} + \overline{A}C\overline{D} + AB\overline{C} + A\overline{B}\overline{C} + ACD$
      4   2   8   5   3

11.3

2-out-of-5 code					7-segment code						
E	D	C	B	A	a	b	c	d	e	f	g
0	0	0	1	1	1	1	1	1	1	1	0
0	0	1	0	1	0	1	1	0	0	0	0
0	0	1	1	0	1	1	0	1	1	0	1
0	1	0	0	1	1	1	1	1	0	0	1
0	1	0	1	0	0	1	1	0	0	1	1
0	1	1	0	0	1	0	1	1	0	1	1
1	0	0	0	1	0	0	1	1	1	1	1
1	0	0	1	0	1	1	1	0	0	0	0
1	0	1	0	0	1	1	1	1	1	1	1
1	1	0	0	0	1	1	1	0	0	1	1

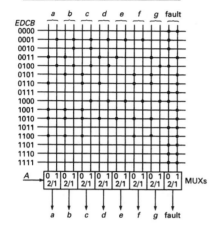

11.4

Dec.	A	B	C	D	E	F
0	0	0	0	0	0	0
3	0	0	0	0	1	1
4	0	0	0	1	0	0
7	0	0	0	1	1	1
11	0	0	1	0	1	1
16	0	1	0	0	0	0
18	0	1	0	0	1	0
19	0	1	0	0	1	1
20	0	1	0	1	0	0
31	0	1	1	1	1	1
36	1	0	0	1	0	0
41	1	0	1	0	0	1
43	1	0	1	0	1	1
50	1	1	0	0	1	0
51	1	1	0	0	1	1
52	1	1	0	1	0	0
55	1	1	0	1	1	1
57	1	1	1	0	0	1
63	1	1	1	1	1	1

C	D	E	F	X_1	X_2	X_3
0	0	0	0	0	0	0
0	0	1	1	0	0	1
1	0	1	1	0	1	0
0	1	0	0	0	1	1
1	1	1	1	1	0	0
1	0	0	1	1	0	1
0	0	1	0	1	1	0
0	1	1	1	1	1	1

A	B	X_1	X_2	X_3	f
0	0	0	0	0	1
0	0	0	0	1	1
0	0	0	1	0	1
0	0	0	1	1	1
0	0	1	1	1	1
0	1	0	0	0	1
0	1	0	0	1	1
0	1	0	1	1	1
0	1	1	0	0	1
0	1	1	1	0	1
1	0	0	1	0	1
1	0	0	1	1	1
1	0	1	0	1	1
1	1	0	0	1	1
1	1	0	1	1	1
1	1	1	0	1	1
1	1	1	1	1	1

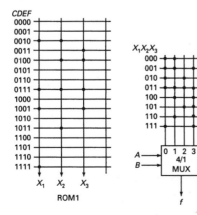

11.5

Present State			Next State						Flip–flop excitation signals											
			X=0			X=1			X=0		X=1		X=0		X=1		X=0		X=1	
A	B	C	A	B	C	A	B	C	J_A	K_A	J_A	K_A	J_B	K_B	J_B	K_B	J_C	K_C	J_C	K_C
0	0	0	0	0	0	0	0	1	0	X	0	X	0	X	0	X	0	X	1	X
0	0	1	0	1	1	0	0	0	0	X	0	X	1	X	0	X	X	0	X	1
0	1	1	0	1	0	0	0	0	0	X	0	X	X	0	X	1	X	1	X	1
0	1	0	0	0	0	1	1	0	0	X	1	X	X	1	X	0	0	X	0	X
1	1	0	0	0	0	1	1	1	X	1	X	0	X	1	X	0	0	X	1	X
1	1	1	1	0	1	0	0	0	X	0	X	1	X	1	X	1	X	0	X	1
1	0	1	1	0	0	0	0	0	X	0	X	1	0	X	0	X	X	1	X	1
1	0	0	0	0	0	0	0	0	X	1	X	1	0	X	0	X	0	X	0	X

From the table
$$J_A = XB\bar{C}$$
$$K_A = X\bar{B} + XC + \bar{X}C$$
$$J_B = \bar{X}\bar{A}C$$
$$K_B = AC + XC + \bar{X}\bar{C}$$
$$J_C = XAB + X\bar{A}\bar{B}$$
$$K_C = X + A\bar{B} + \bar{A}B$$

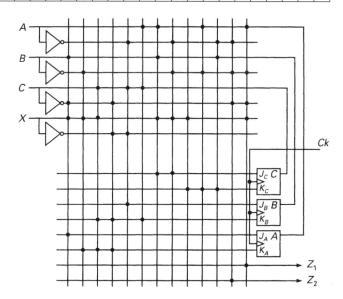

11.6

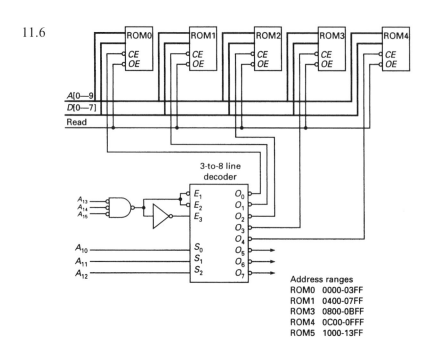

Address ranges
ROM0 0000–03FF
ROM1 0400–07FF
ROM3 0800–0BFF
ROM4 0C00–0FFF
ROM5 1000–13FF

Chapter 12

12.1 Binary number $= ABCD$, 2's complement form $= PQRS$,
$P = A \oplus (B + C + D), Q = B \oplus (C + D), R = C \oplus D, S = D$.

12.2 NBCD $= ABCD$, 10's complement form $= PQRS$,
$P = \bar{A}\bar{B}\bar{C} + \bar{A}\bar{B}\bar{D}, Q = B\bar{C} + B\bar{D} + \bar{B}CD, R = \bar{C}D + CD, S = D$.

12.3 (a) When a carry is not generated by the addition of two XS3 numbers, correction is add 1101.
(b) When a carry is generated by the addition, correction is add 0011.

12.4 NBCD $= ABCD$, XS3 $= PQRS$, $P = A + BD + BC$,
$Q = \bar{B}D + \bar{B}C + B\bar{C}\bar{D}, R = C \odot D, S = \bar{D}$

12.5

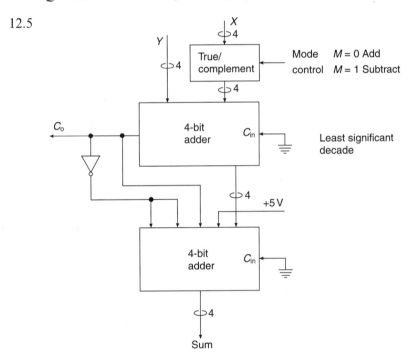

12.6 Algorithm for duodecimal addition
$0 < S \le (11)_{10}$ no correction required
$(11)_{10} < S \le (23)_{10}$ add 0100

12.7

d	2421	d	2421
0	0000	5	1011
1	0001	6	1100
2	0010	7	1101
3	0011	8	1110
4	0100	9	1111
	P		Q

Algorithm
(a) $0 < S \leq 4$ no correction required
(b) $4 < S \leq 8$ both digits from section P, sum falls in section Q, add 0110
(c) $4 < S \leq 13$ one digit from P and one from Q, no correction required
(d) $9 < S \leq 14$ both digits from Q, sum falls in P, add 1010
(e) $14 < S \leq 18$ both digits from Q, sum falls in Q, no correction required

12.8   One solution is
$$Y_i = \bar{S}_1 A_i + \bar{S}_0 S_1 \bar{A}_i$$
$$Z_i = S_0 \bar{B}_i + \bar{S}_0 \bar{S}_1 B_i$$

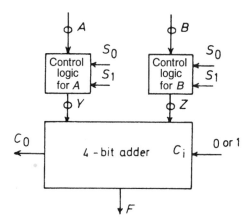

12.9

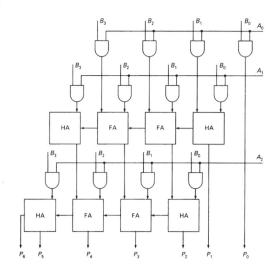

12.10

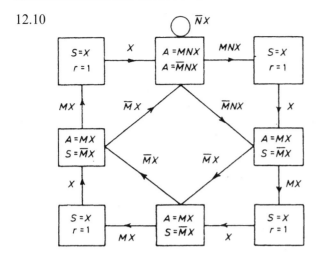

$\overline{N}X$

| $S=X$ $r=1$ | | X → | $A=MNX$ $A=\overline{M}NX$ | MNX → | $S=X$ $r=1$ |

MX ↑ $\overline{M}X$ $\overline{M}NX$ X ↓

| $A=MX$ $S=\overline{M}X$ | $\overline{M}X$ $\overline{M}X$ | $A=MX$ $S=\overline{M}X$ |

X ↑ MX ↓

| $S=X$ $r=1$ | ← MX | $A=MX$ $S=\overline{M}X$ | ← X | $S=X$ $r=1$ |

Chapter 13

13.1 T_0, T_1, T_2 and T_4 are not required. T_3 tests for p_1 and s_0. T_5 tests for q_1 and s_0. T_6 tests for r_1 and s_0. T_7 tests for p_0, q_0, r_0 and s_1.

13.2 T_6 tests for p_0, q_0, s_1 and u_0. T_2 tests for p_1, s_0 and u_1. T_5 tests for q_1, s_0 and u_1. T_1 or T_3 or T_5 test for r_0, t_1 and u_0. T_0 or T_2 or T_4 test for r_1, t_0 and u_1.

13.3 $T = \{25$ or 27 or $29\}$ faults detected $p_0 q_0 s_1 t_0$; $T = \{9$ or 11 or $13\}$ faults detected $p_1 s_0 t_1$; $T = \{17$ or 19 or $21\}$ faults detected $q_1 s_0 t_1$; $T = \{7$ or 15 or $23\}$ faults detected $r_0 u_0 v_1 t_0$; $T = \{3$ or 11 or $19\}$ faults detected $r_1 v_0 t_1$; $T = \{5$ or 13 or $21\}$ faults detected $u_1 v_0 t_1$; $T = \{1$ or 3 or 5 or 9 or 11 or 13 or 17 or 19 or $21\}$ faults detected $w_0 t_1$; $T = \{0$ or 2 or 4 or 8 or 10 or 12 or 16 or 18 or $20\}$ faults detected $w_1 t_0$.

13.4 $T_{p1} = \{2, 3, 8, 10, 11\}$

13.5 p s-a-0 $X_1 X_2 = 11$; q s-a-1 $X_1 X_2 = 11$; r s-a-0 $X_1 X_2 = 00$ or 11

13.6   (a) $T = \{2, 4, 6, 1$ or 3 or $5\}$
       (b) $T = \{6, 0$ or $1, 3$ or $7, 4$ or $5\}$
       (c) $T = \{1, 3, 5, 2$ or 4 or $6\}$

13.7 $T = \{6, 12, 9$ or $11, 0$ or 2 or $8\}$

13.8   (a) $T_0 = \{4$ or $6, 11, 13\}$ $T_1 = \{3, 9, 14\}$
       (b) $T_0 = \{2, 9, 12, 15\}$ $T_1 = \{2$ or $3, 4, 13, 6$ or $14\}$
       (c) $T_0 = \{0$ or 2 or $8, 5$ or $13, 11$ or 14 or $15\}$ $T_1 = \{1, 12, 7\}$
       (d) $T_0 = \{2, 4, 6, 8, 10, 12\}$ $T_1 = \{0, 3$ or 5 or 7 or 9 or 11 or $13\}$

13.9   (a) $\overline{X}_1 X_3 + X_1 \overline{X}_3$
       (b) $\overline{X}_1 X_3 + X_1 \overline{X}_3$

Bibliography

Almaini, A.E., *Electronic Logic Systems*, Prentice-Hall, 1986

Clare, C.R., *Designing Logic Systems Using State Machines*, McGraw-Hill, 1973

Crowe, J. and Hayes-Gill, B., *Introduction to Digital Electronics*, Arnold, 1998

Daniels, J.D., *Digital Design from Zero to One*, Wiley, 1996

Ercegovac, M.D. and Lang, T., *Digital Systems and Hardware/Firmware Algorithms*, Wiley, 1985

Fletcher, W.I., *An Engineering Approach to Digital Design*, Prentice-Hall, 1980

Floyd, T.L., *Digital Fundamentals*, 7th Ed., Prentice Hall, 2000

Garrod, S.A.R. and Borns, R.J., *Digital Logic: Analysis, Application & Design*, Saunders, 1991

Green, D., *Modern Logic Design*, Addison-Wesley, 1986

Hill, F.J. and Petersen, G.R., *Introduction to Switching Theory and Logical Design*, Wiley, 1974

Hill, F.J. and Petersen, G.R., *Digital Logic and Microprocessors*, Wiley, 1984

Hill, F.J. and Petersen, G.R., *Hardware Organisation and Design*, Wiley, 1987

Holdsworth, B., *Microprocessor Engineering*, Butterworth, 1987

Holdsworth, B. and Martin, G.R., *Digital Systems Reference Book*, Butterworth-Heinemann, 1991

Kohavi, Z., *Switching Theory and Finite Automata Theory*, McGraw-Hill, 1978

Lee, S.C., *Digital Circuits and Logic Design*, Prentice-Hall, 1976

Lee, S.C., *Modern Switching Theory and Digital Design*, Prentice-Hall, 1978

Lewin, D., *Design of Logic Systems*, Van Nostrand Reinhold, 1986

McCluskey, E.J., *Logic Design Principles*, Prentice-Hall, 1986

Malvino, A.P. and Leach, D.P., *Digital Principles and Applications*, 4th Ed., McGraw-Hill, 1986

Mann, F.A., *Using Functional Logic Symbols*, Texas Instruments, 1987

Mano, M., *Digital Principles and Applications*, McGraw-Hill, 1979

Mano, M., *Digital Design*, Prentice-Hall, 1984

Millman, J., *Microelectronics*, McGraw-Hill, 1979

Monk, R., *Digital Electronics: a Practical Approach*, Newnes, 1998

Nelson, V.P., Nagle, H.T., Carroll, B.D. and Irwin, J.D., *Digital Logic Circuit Analysis and Design*, Prentice-Hall, 1995

Peatman, J.B., *Design of Digital Systems*, McGraw-Hill, 1972

Peatman, J.B., *Digital Hardware Design*, McGraw-Hill, 1980

Roth, C.H., *Fundamentals of Logic Design*, 4th Ed., West, 1992

Taub, H., *Digital Circuits and Microprocessors*, McGraw-Hill, 1982

Tocci, R.J. and Widmer N.S., *Digital Systems, Principles and Applications*, 8th Ed., Prentice-Hall, 2001

Tokheim, R.L., *Digital Principles*, 3rd Ed., Schaum, 1994

Wakerly, J.F., *Digital Design and Practices*, Prentice-Hall, 1990

Warnes, L., *Analogue and Digital Electronics*, Macmillan, 1998
Wilkins, B.R., *Testing Digital Circuits*, Van Nostrand Reinhold, 1986
Wilkinson, B., *The Essence of Digital Design*, Prentice Hall, 1998
Woolvet, G.A., *Transducers in Digital Systems*, Peregrinus, 1977
Yarbrough, J.M., *Digital Logic Applications and Design*, West, 1997

Index